Exercises in Electricity and Magnetism

Teruo Matsushita

Exercises in Electricity and Magnetism

100 Examples and 400 Exercises

Teruo Matsushita
Department of Computer Science
and Electronics
Kyushu Institute of Technology
Iizuka, Japan

ISBN 978-3-031-67939-1   ISBN 978-3-031-67940-7 (eBook)
https://doi.org/10.1007/978-3-031-67940-7

© The Editor(s) (if applicable) and The Author(s), under exclusive license to Springer Nature Switzerland AG 2025

This work is subject to copyright. All rights are solely and exclusively licensed by the Publisher, whether the whole or part of the material is concerned, specifically the rights of reprinting, reuse of illustrations, recitation, broadcasting, reproduction on microfilms or in any other physical way, and transmission or information storage and retrieval, electronic adaptation, computer software, or by similar or dissimilar methodology now known or hereafter developed.
The use of general descriptive names, registered names, trademarks, service marks, etc. in this publication does not imply, even in the absence of a specific statement, that such names are exempt from the relevant protective laws and regulations and therefore free for general use.
The publisher, the authors and the editors are safe to assume that the advice and information in this book are believed to be true and accurate at the date of publication. Neither the publisher nor the authors or the editors give a warranty, expressed or implied, with respect to the material contained herein or for any errors or omissions that may have been made. The publisher remains neutral with regard to jurisdictional claims in published maps and institutional affiliations.

This Springer imprint is published by the registered company Springer Nature Switzerland AG
The registered company address is: Gewerbestrasse 11, 6330 Cham, Switzerland

If disposing of this product, please recycle the paper.

Preface

Electromagnetism is an important subject in today's physics. Abstract concepts are frequently used to describe it, however, and therefore, it is not easy for students to come to a complete understanding of electromagnetism, although various phenomena are concisely described with mathematics. For this reason, many textbooks have been published to assist students to understand electromagnetism better.

In *Exercises in Electricity and Magnetism* written by the author, the ***E-B*** analogy, which is now the common basis in Electromagnetism, is strengthened by the introduction of superconductors. That is, the electric property of $\boldsymbol{E} = 0$ in conductors corresponds to the magnetic property of $\boldsymbol{B} = 0$ in superconductors. Various analogies between electric properties in dielectric materials and magnetic properties in magnetic materials are also described in the textbook. In addition, many examples and exercises in the textbook are helpful for understanding the subject. The number of examples and exercises is still insufficient, however, to cover various aspects in electricity and magnetism.

This book includes a great variety of exercises that are not sufficiently picked up in the textbook. The equipotential surface in static electric phenomena and the equivector potential surface in static magnetic phenomena are treated for various cases. The image method, which is a useful method to determine the potential, is applied to various electric and magnetic phenomena. These will help the readers to deeply understand the ***E-B*** analogy. The inductance is usually determined from the magnetic energy for a system in which current flow is spread within some area. Comparison of this method with a method in which the mean magnetic flux is used will clarify the validity of the former method. It is also shown that the vector potential can be directly used to determine the induced electromotive force, similarly to the induction law and the motional law. This also shows that the vector potential is a useful quantity. It is shown that the Poynting vector is useful for understanding the energy flow into capacitors or transmission lines in a charging process or into resistors in a dissipation process. For example, a comparison between normal conducting and superconducting transmission lines shows a clear difference in the energy flow, although the final stored magnetic energy is the same. The exercises to determine the electromagnetic potential and the Poynting vector for various electromagnetic waves in wave guides will be useful for understanding the electromagnetism of electromagnetic wave. The number of exercises in each chapter is approximately the same, to attach importance to the right balance. Information on the examples and exercises is given in the tables of arrangement of problems attached in the bookend, in which the examples and exercises in the textbook are also shown.

To effectively use this book, the student is encouraged to supplement the exercises presented during a lecture on electricity and magnetism. The author hopes that this book helps to widen the field of education of electricity and magnetism.

Iizuka, Japan   Teruo Matsushita

Contents

1 Electrostatic Field 1
   1.1   Electric Charge in Vacuum 1
   1.2   Coulomb's Law 1
   1.3   Electric Field 4
   1.4   Gauss's Law 7
   1.5   Electric Potential 10
   1.6   Electric Dipole 12
   1.7   Exercises 15

2 Conductors 41
   2.1   Electric Properties of Conductors 41
   2.2   Special Solution Method for Electrostatic Field 43
   2.3   Electrostatic Induction 52
   2.4   Exercises 55

3 Conductor Systems in Vacuum 83
   3.1   Coefficients in Conductor System 83
   3.2   Capacitor 85
   3.3   Electrostatic Energy 88
   3.4   Electrostatic Force 92
   3.5   Exercises 94

4 Dielectric Materials 117
   4.1   Electric Polarization 117
   4.2   Electric Flux Density 119
   4.3   Boundary Conditions 120
   4.4   Electrostatic Energy in Dielectric Materials 130
   4.5   Exercises 132

5 Steady Current — 157
- 5.1 Current — 157
- 5.2 Ohm's Law — 158
- 5.3 Fundamental Equations for Steady Electric Current — 159
- 5.4 Resistance — 162
- 5.5 Electric Power — 169
- 5.6 Exercises — 171

6 Current and Magnetic Flux Density — 191
- 6.1 The Biot-Savart Law — 191
- 6.2 Force on Current — 194
- 6.3 Magnetic Flux Lines — 195
- 6.4 Ampere's Law — 196
- 6.5 Vector Potential — 198
- 6.6 Small Closed Current — 201
- 6.7 Equivector Potential Surface — 204
- 6.8 Magnetic Charge — 206
- 6.9 Exercises — 207

7 Superconductors — 237
- 7.1 Magnetic Properties of Superconductors — 237
- 7.2 Special Solution Method for Magnetic Flux Density — 241
- 7.3 The Meissner State — 249
- 7.4 Exercises — 252

8 Current Systems — 277
- 8.1 Inductance — 277
- 8.2 Coils — 278
- 8.3 Magnetic Energy — 280
- 8.4 Mean Magnetic Flux — 287
- 8.5 Exercises — 290

9 Magnetic Materials — 313
- 9.1 Magnetization — 313
- 9.2 Magnetic Field — 316
- 9.3 Boundary Conditions — 317
- 9.4 Magnetic Energy in Magnetic Materials — 329
- 9.5 Exercises — 330

10 Electromagnetic Induction — 355
- 10.1 Induction Law — 355
- 10.2 Potential — 363
- 10.3 Magnetic Energy — 365
- 10.4 Skin Effect — 367
- 10.5 Exercises — 370

11	**Displacement Current and Maxwell's Equations**	395
	11.1 Displacement Current	395
	11.2 Maxwell's Equations	398
	11.3 Electromagnetic Potential	400
	11.4 The Poynting Vector	401
	11.5 Exercises	408
12	**Electromagnetic Waves**	437
	12.1 Planar Electromagnetic Waves	437
	12.2 Reflection and Refraction of the Planar Electromagnetic Wave	439
	12.3 Energy of the Electromagnetic Wave	446
	12.4 Wave Guides	448
	12.5 Exercises	458

Appendix	487
Table of Arrangement of Problems	491
Index	503

11. Displacement Current and Maxwell's Equations
   11.1 Displacement Current
   11.2 Maxwell's Equations
   11.3 Electromagnetic Potential
   11.4 The Poynting Vector
   11.5 Ellipsoses

12. Electromagnetic Waves
   12.1 Planar Electromagnetic Waves
   12.2 Reflection and Refraction of the Plane Electromagnetic Wave
   12.3 Energy of the Electromagnetic Wave
   12.4 Wave Guides
   12.5 Exercises and Problems

Appendix

Table of Arrangement of Problems

Index

Chapter 1
Electrostatic Field

1.1 Electric Charge in Vacuum

There are two kinds of electric charge that cause electric phenomena: One is true electric charge, which can be transferred outside a substance, and the other is polarization charge, which cannot be transferred outside due to local binding around nuclei. The former appears on the surface of a conductor, and its flow gives a current, while the latter appears on the surface of a dielectric material.

The principle of conservation of charge holds, i.e., the total amount of electric charge is constant in a closed system. Even when positive and negative charges cancel each other, the algebraic sum of electric charge is unchanged. When the system is not closed and the total sum of electric charge changes, movement of electric charge, i.e., an electric current, occurs. A conservation relationship holds between the electric charge and the current, which will be described in Sect. 5.1.

1.2 Coulomb's Law

The electric force between electric charges is the Coulomb force. These are the following characteristic points relating to two electric charges in vacuum:

- The force between two electric charges of the same kind is repulsive, and that between electric charges of different kinds is attractive.
- The magnitude of the force is proportional to the product of the two electric charges.
- The magnitude of the force is inversely proportional to the square of the distance between the two electric charges.
- The direction of the force lies on the straight line connecting the two electric charges.

Fig. 1.1 The Coulomb force exerted on electric charge q by electric charge q'

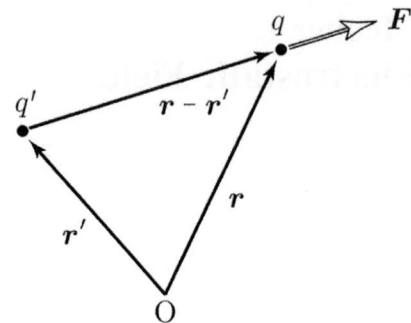

This is called Coulomb's law: The force that electric charge q staying at position r experiences from the electric charge q' at position r' is:

$$F = \frac{qq'(r-r')}{4\pi\epsilon_0|r-r'|^3}, \tag{1.1}$$

where $(r-r')/|r-r'|$ is a unit vector directed from q' to q. (See Fig. 1.1.) The force that q' experiences from q is obtained by replacing r by r' as $-F$. In the above, the constant ϵ_0 is the permittivity of vacuum,

$$\epsilon_0 = \frac{10^7}{4\pi c^2} = 8.8541878 \times 10^{-12} \, \text{C}^2/\text{Nm}^2, \tag{1.2}$$

where c is the light speed in vacuum.

When there are more than two electric charges, the total Coulomb force is given by the sum of each individual Coulomb force. For example, the force on electric charge q at position r exerted by electric charges q_i at $r_i (i = 1, 2, \ldots, n)$ is

$$F = \sum_{i=1}^{n} F_i = \frac{q}{4\pi\epsilon_0} \sum_{i=1}^{n} \frac{q_i(r-r_i)}{|r-r_i|^3}, \tag{1.3}$$

where F_i is the force exerted by the i-th electric charge. When the electric charge is distributed with density $\rho(r')$ within a region V in vacuum, we can regard the electric charge, $\rho dV'$, in an infinitesimal volume dV' as a point charge at position r'. Then, the Coulomb force that is exerted by the distributed electric charge on q outside the region V is:

$$F = \frac{q}{4\pi\epsilon_0} \int_V \frac{\rho(r')(r-r')}{|r-r'|^3} dV', \tag{1.4}$$

1.2 Coulomb's Law

where $\int dV'$ is a volume integral with respect to $\mathbf{r}'$.

Example 1.1 Electric charge is uniformly distributed with a linear density, λ, along a semicircle of radius a. Determine the Coulomb force on a point charge, Q, placed at the center of curvature of the semicircle.

Solution 1.1 An angle is defined as shown in Fig. 1.2. We treat an electric charge $\lambda a d\theta$ between θ and $\theta + d\theta$ as a point charge. The Coulomb force it exerts on charge Q at the center is

$$dF = \frac{Q\lambda a d\theta}{4\pi\epsilon_0 a^2} = \frac{Q\lambda d\theta}{4\pi\epsilon_0 a}.$$

From symmetry, the vertical components of the Coulomb forces exerted by infinitesimal arc elements on Q cancel out, and only the horizontal component remains. This component is $dF' = dF \cos\theta$. Hence, the total Coulomb force is

$$F = \frac{Q\lambda}{4\pi\epsilon_0 a} \int_{-\pi/2}^{\pi/2} \cos\theta d\theta = \frac{Q\lambda}{4\pi\epsilon_0 a}[\sin\theta]_{-\pi/2}^{\pi/2} = \frac{Q\lambda}{2\pi\epsilon_0 a}.$$

This force is directed to the right in the figure.

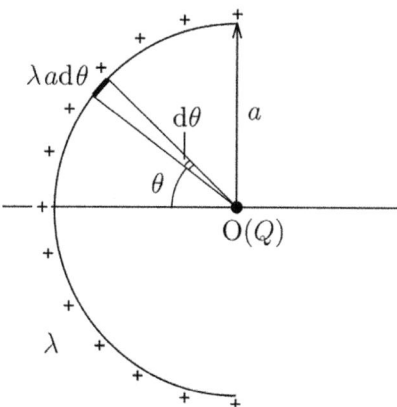

Fig. 1.2 Electric charge distributed uniformly with linear density λ on a semicircle and point charge Q placed at center O

◆

1.3 Electric Field

When electric charge Q is placed somewhere in space, the Coulomb force is exerted on another charge q. It can be considered that this force is caused by an electrical distortion in space produced by the electric charge Q. The electrical distortion caused by existing electric charges is called the electric field. The force on the electric charge q placed at r due to the electric charge Q at the origin is expressed as

$$F = qE. \tag{1.5}$$

Then, the vector E is called the electric field. The unit of the electric field strength is [N/C]. This also expressed as [V/m] using the unit [V] (volt) of electric potential, which will be defined later. Hence, the electric field strength is equal to the Coulomb force on a unit charge (1 C) that is placed at the given position. In the above case, we have

$$E(r) = \frac{Qr}{4\pi \epsilon_0 |r|^3}. \tag{1.6}$$

When electric charges q_i are placed at $r_i (i = 1, 2, \ldots, n)$, the electric field strength at r is

$$E(r) = \frac{1}{4\pi \epsilon_0} \sum_{i=1}^{n} \frac{q_i (r - r_i)}{|r - r_i|^3}. \tag{1.7}$$

When the electric charge is distributed with density $\rho(r')$ in area V, the electric field strength at r is

$$E(r) = \frac{1}{4\pi \epsilon_0} \int_V \frac{\rho(r')(r - r')}{|r - r'|^3} dV'. \tag{1.8}$$

Equations (1.7) and (1.8) for the electric field strength are also called Coulomb's law.

We can visualize the electric field using electric field lines. The electric field line is defined as follows: The direction of a tangential line at any point on the electric field line is the same as the direction of E, and its line density is defined as equal to the magnitude of E.

Example 1.2 Electric charge is uniformly distributed with a linear density, λ, on a straight segment of length $2a$ parallel to the y-axis, as shown in Fig. 1.3. Determine the electric field strength at a point, A, at distance b from the center of the segment to the direction of the x-axis.

1.3 Electric Field

Fig. 1.3 Electric charge distributed uniformly with linear density λ on a straight segment of length $2a$

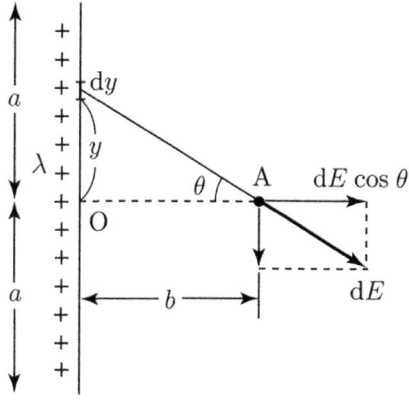

Solution 1.2 We define the y-axis along the length of the segment with the origin at its center. The electric field strength at A produced by the charge, λdy, in the region from y to $y + dy$ is

$$dE = \frac{\lambda dy}{4\pi\epsilon_0(y^2 + b^2)}.$$

The angle θ is defined as shown in the figure. From symmetry, the y-component of the electric field strength is cancelled and only the x-component, $dE\cos\theta$, remains. The relationship $y = b\tan\theta$ gives $dy = bd\theta/\cos^2\theta$ and $y^2 + b^2 = b^2/\cos^2\theta$. The electric field strength is given by

$$E = \frac{\lambda}{4\pi\epsilon_0 b} \int_{-\theta_a}^{\theta_a} \cos\theta\, d\theta$$

with $\theta_a = \tan^{-1}(a/b)$. After a simple calculation we have

$$E = \frac{\lambda}{4\pi\epsilon_0 b}[\sin\theta]_{-\theta_a}^{\theta_a} = \frac{\lambda a}{2\pi\epsilon_0 b(a^2 + b^2)^{1/2}}.$$

◆

Example 1.3 Electric charge is uniformly distributed with a linear density λ around a square of width a. Determine the electric field strength at point P at distance z above the center of the square (see Fig. 1.4).

Fig. 1.4 Square with uniformly distributed electric charge and point P above the center

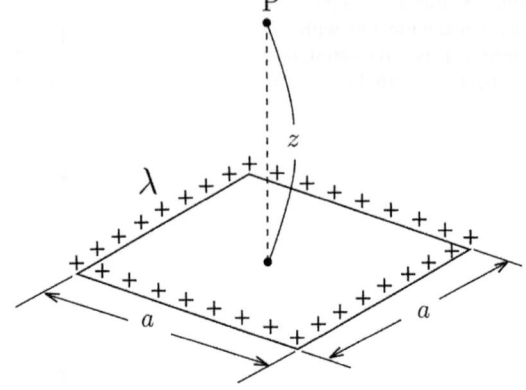

Solution 1.3 The distance from one side to point P is $r = \left[(a^2/4) + z^2\right]^{1/2}$ and the electric field strength due to the electric charge on one side is

$$E' = \frac{\lambda}{4\pi\epsilon_0} \int_{-a/2}^{a/2} \frac{dx}{r^2 + x^2} = \frac{\lambda a}{4\pi\epsilon_0 r \left[(a^2/4) + r^2\right]^{1/2}}.$$

From symmetry, only the vertical component of the electric field remains, and we have

$$E = \frac{4z}{r} E' = \frac{\lambda a}{4\pi\epsilon_0 \left[(a^2/4) + z^2\right]\left[(a^2/2) + z^2\right]^{1/2}}.$$

◆

Example 1.4 Suppose that electric charge is uniformly distributed on a circle of radius a with a linear density, λ, as illustrated in Fig. 1.5. Determine the electric field strength at a point P at distance z from the center of the circle.

Fig. 1.5 Circle with uniformly distributed electric charge and point P above the center

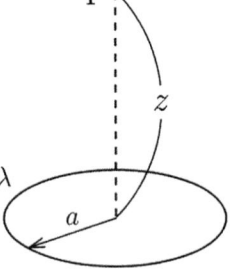

Solution 1.4 The azimuthal angle, φ, is defined from the center of the circle on the plane in which the circle exists. The electric field strength at P produced by the electric charge on a small segment of length $a d\varphi$ is

$$dE = \frac{a\lambda d\varphi}{4\pi \epsilon_0 (z^2 + a^2)}.$$

Only the vertical component, $\left[z/(z^2 + a^2)^{1/2} \right] dE$, remains from symmetry, and the electric field strength is

$$E = \int_0^{2\pi} \frac{az\lambda}{4\pi \epsilon_0 (z^2 + a^2)^{3/2}} d\varphi = \frac{az\lambda}{2\epsilon_0 (z^2 + a^2)^{3/2}}.$$

♦

1.4 Gauss's Law

Suppose a closed surface, S, includes a point charge, q, inside. In this case, the following relation holds:

$$\int_S \mathbf{E} \cdot d\mathbf{S} = \frac{q}{\epsilon_0}. \tag{1.9}$$

When electric charges $q_1, q_2, \ldots, q_m$ are inside S and $q_{m+1}, q_{m+2}, \ldots, q_{m+n}$ are outside S, Eq. (1.9) can be extended to be

$$\int_S \mathbf{E} \cdot d\mathbf{S} = \frac{1}{\epsilon_0} \sum_{i=1}^m q_i. \tag{1.10}$$

When the electric charge is distributed with density $\rho(r)$, we have

$$\int_S \mathbf{E} \cdot d\mathbf{S} = \frac{1}{\epsilon_0} \int_V \rho(r) dV, \tag{1.11}$$

where V is the volume surrounded by S. Equations (1.9)–(1.11) are called Gauss's law. Using Gauss's theorem on the left-hand side of Eq. (1.11), we have

$$\nabla \cdot \mathbf{E} = \frac{\rho}{\epsilon_0}. \tag{1.12}$$

This is called Gauss's divergence law.

While Coulomb's law describes the local electric field produced by electric charge, Gauss's law gives the global relationship between electric field and electric charge. Gauss's law can sometimes be used to determine the electric field strength.

Example 1.5 Electric charge is uniformly distributed with a density, ρ, inside an infinitely long cylinder of radius a. Determine the electric field strength inside and outside the cylinder.

Solution 1.5 We apply Gauss's law, Eq. (1.11), to an imaginary finite cylindrical closed surface, S, of radius R and length l with a common axis with the infinite cylinder (see Fig. 1.6). The electric field, E, is directed radially from the central axis and perpendicular to it. Hence, E is perpendicular to the elementary surface vector, dS, on the top and bottom surfaces, and there is no contribution to the surface integral of the electric field strength from these surfaces. On the other hand, E is parallel to dS on the side surface and the strength, E, is constant and depends only on the distance from the axis. Hence, the surface integral in Eq. (1.11) gives

$$\int_S E \cdot dS = 2\pi R l E.$$

The total charge inside S is $\pi R^2 l \rho$ for $R < a$ and $\pi a^2 l \rho$ for $R > a$. Hence, the electric field strength is

$$E = \frac{\rho R}{2\epsilon_0}; \quad 0 \leq R < a,$$

$$= \frac{\rho a^2}{2\epsilon_0 R}; \quad R > a.$$

Fig. 1.6 Cylinder with distributed electric charge and cylindrical closed surface S (case for $R > a$)

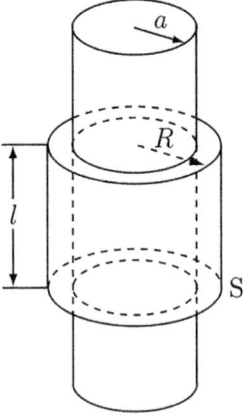

1.4 Gauss's Law

Example 1.6 Electric charge is uniformly distributed with a density, σ, on a wide flat plane. Determine the electric field strength at a point, A, at distance h from the plane.

Solution 1.6 From symmetry, we can assume that the electric field, $\boldsymbol{E}$, is directed normally to the plane with its strength dependent only on the distance from the plane. Assume a closed cylindrical surface, S, of radius a and length $2h$, as shown in Fig. 1.7: Its side surface is normal to and the top and bottom surfaces are parallel to the plane, and A is on the top surface. We apply Gauss's law to this cylindrical surface. The electric field, $\boldsymbol{E}$, is parallel to the side surface, and there are no field lines passing through this surface. The numbers of field lines that go out of the top and bottom surfaces are the same from symmetry. Thus, we have

$$\int_S \boldsymbol{E} \cdot \mathrm{d}\boldsymbol{S} = 2ES$$

with $S = \pi a^2$ denoting the area of the top or bottom surface. Since the total electric charge included inside S is σS, the electric field strength is

$$E = \frac{\sigma}{2\epsilon_0}. \tag{1.13}$$

This result shows that the electric field strength does not change with the distance from the plane.

Fig. 1.7 Cylindrical closed surface S with point A on top surface

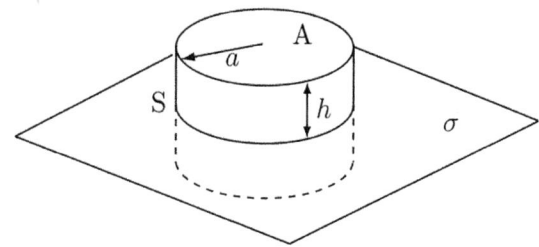

1.5 Electric Potential

The electric potential, ϕ, which causes the electric field, is defined as

$$\phi(\mathbf{r}) = -\int_{\mathbf{r}_0}^{\mathbf{r}} \mathbf{E} \cdot d\mathbf{s}, \tag{1.14}$$

where $\mathbf{r}_0$ is a reference point satisfying $\phi(\mathbf{r}_0) = 0$ and is usually taken at infinity. The electric potential that causes the electric field of (1.6) is given by

$$\phi(\mathbf{r}) = \frac{Q}{4\pi \epsilon_0 |\mathbf{r}|}. \tag{1.15}$$

For the electric field given by Eq. (1.7), the electric potential is

$$\phi(\mathbf{r}) = \frac{1}{4\pi \epsilon_0} \sum_{i=1}^{n} \frac{q_i}{|\mathbf{r} - \mathbf{r}_i|}. \tag{1.16}$$

The electric potential for a continuously distributed electric charge with density $\rho(\mathbf{r})$ is

$$\phi(\mathbf{r}) = \frac{1}{4\pi \epsilon_0} \int_V \frac{\rho(\mathbf{r}')}{|\mathbf{r} - \mathbf{r}'|} dV'. \tag{1.17}$$

A virtual surface composed of points with the same electric potential is an equipotential surface. The electric field is normal to the equipotential surface.

From Eq. (1.14), the electric field is given by

$$\mathbf{E} = -\nabla \phi. \tag{1.18}$$

So, the electric field obeys

$$\nabla \times \mathbf{E} = 0. \tag{1.19}$$

The electric field is an irrotational field.

From Eqs. (1.12) and (1.18), we have

$$\Delta \phi = -\frac{\rho}{\epsilon_0}. \tag{1.20}$$

This is called Poisson's equation. When there is no electric charge, Eq. (1.20) is reduced to Laplace's equation:

$$\Delta \phi = 0. \tag{1.21}$$

1.5 Electric Potential

Example 1.7 Electric charge is uniformly distributed with a density, ρ, inside a sphere of radius a. Determine the electric potential inside and outside the sphere.

Solution 1.7 We apply Gauss's law to a virtual sphere, S, of radius r concentric with the charged sphere (see Fig. 1.8). The electric field, **E**, is parallel to the elementary surface vector, d**S**, with a constant strength, E, on the surface of S. Thus, we have

$$\int_S \mathbf{E} \cdot d\mathbf{S} = 4\pi r^2 E.$$

Since the total electric charge inside S is $(4\pi/3)r^3\rho$ and $(4\pi/3)a^3\rho$ for $r < a$ and $r > a$, respectively, we obtain the electric field strength as.

$$E = \frac{\rho R}{3\epsilon_0}; \quad 0 \leq r < a$$

$$= \frac{\rho a^2}{2\epsilon_0 R}; \quad r > a.$$

Substituting these results into Eq. (1.14) gives

$$\phi = -\int_\infty^r \frac{\rho a^3}{3\epsilon_0 r^2} dr = \frac{\rho a^3}{3\epsilon_0 r}; \quad r > a,$$

$$= -\int_a^r \frac{\rho r}{3\epsilon_0} dr + \frac{\rho a^2}{3\epsilon_0} = \frac{\rho}{2\epsilon_0}\left(a^2 - \frac{r^2}{3}\right); \quad 0 \leq r < a.$$

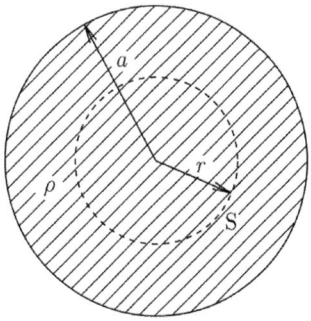

Fig. 1.8 Virtual sphere S with radius r smaller than a

◆

Example 1.8 Electric charge is uniformly distributed with a linear density, λ, around a circle of radius a, as shown in Fig. 1.9. Determine the electric potential at the center, O.

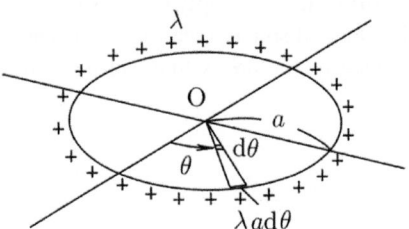

Fig. 1.9 Electric charge distributed uniformly around a circle of radius a

Solution 1.8 Here, we directly calculate the electric potential with Eq. (1.17). The electric charge in an infinitesimal region between θ and $\theta + d\theta$, $\lambda a d\theta$, is regarded as a point charge, $\rho dV'$, with θ denoting the azimuthal angle. Thus, we have

$$\phi = \frac{1}{4\pi\epsilon_0} \int_0^{2\pi} \lambda d\theta = \frac{\lambda}{2\epsilon_0}.$$

◆

1.6 Electric Dipole

Most materials are electrically neutral with equal amounts of positive and negative electric charges. When an electric field is applied to such materials, positive and negative charges are displaced in and against the direction of the electric field, respectively, resulting in local deviations from neutrality. The fundamental element for such electric phenomena is a pair of positive and negative point charges that are separated by a small distance. This is called an electric dipole.

Suppose that electric charges q and $-q$ are displaced by $d/2$ in the positive and negative directions from the origin along the z-axis, as shown in Fig. 1.10 We use spherical coordinates with the origin at the center of the dipole and the zenithal angle θ is measured from the z-axis. The electric potential due to the pair of charges is given by

$$\phi(r) = \frac{\boldsymbol{p}\cdot\boldsymbol{r}}{4\pi\epsilon_0 r^3} = \frac{p\cos\theta}{4\pi\epsilon_0 r^2}, \qquad (1.22)$$

1.6 Electric Dipole

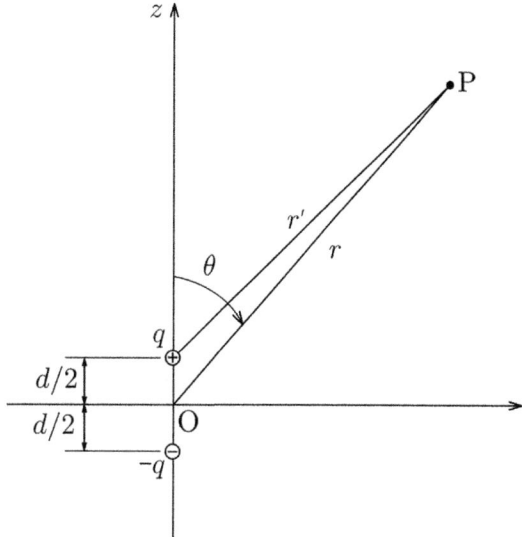

Fig. 1.10 Pair of positive and negative electric charges near the origin and point P sufficiently far from the charges

where $\boldsymbol{p} = qd\boldsymbol{i}_z$ and $p = qd$, where p is called the electric dipole moment. The resultant electric field is

$$E_r = -\frac{\partial \phi}{\partial r} = \frac{p \cos \theta}{2\pi \epsilon_0 r^3},$$
$$E_\theta = -\frac{1}{r} \cdot \frac{\partial \phi}{\partial \theta} = \frac{p \sin \theta}{4\pi \epsilon_0 r^3}, \quad (1.23)$$
$$E_\varphi = -\frac{1}{r \sin \theta} \cdot \frac{\partial \phi}{\partial \varphi} = 0.$$

Figure 1.11 shows the electric potential and electric field produced by the electric dipole. The surface on which the following relationship holds is the equipotential surface:

Fig. 1.11 Equipotential surfaces (*dotted lines*) and electric field lines (*solid lines*) produced by an electric dipole on a plane including positive and negative electric charges

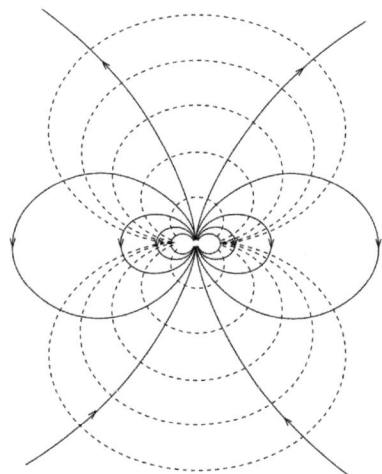

$$\frac{\cos\theta}{r^2} = \text{const.} \tag{1.24}$$

Assume that electric charge is uniformly distributed with a linear density λ and $-\lambda$ on the lines at $x = d/2$ and $x = -d/2$ parallel to the z-axis, as shown in Fig. 1.12. Such a pair of electric charges is called an electric dipole line. We use cylindrical coordinates with the central line along the z-axis and the azimuthal angle θ is measured from the x-axis. The electric potential is given by

$$\phi(R, \varphi) = \frac{\hat{p}\cos\varphi}{2\pi\epsilon_0 R}, \tag{1.25}$$

where $\hat{p} = \lambda d$ is the moment of an electric dipole line. The resultant electric field is

$$\begin{aligned} E_R &= -\frac{\partial\phi}{\partial R} = \frac{\hat{p}\cos\varphi}{2\pi\epsilon_0 R^2}, \\ E_\varphi &= -\frac{1}{R}\cdot\frac{\partial\phi}{\partial\varphi} = \frac{\hat{p}\sin\varphi}{2\pi\epsilon_0 R^2}, \\ E_z &= -\frac{\partial\phi}{\partial z} = 0. \end{aligned} \tag{1.26}$$

The surface on which the following relationship holds is the equipotential surface:

$$\frac{\cos\varphi}{R} = \text{const.} \tag{1.27}$$

Fig. 1.12 Electric dipole line

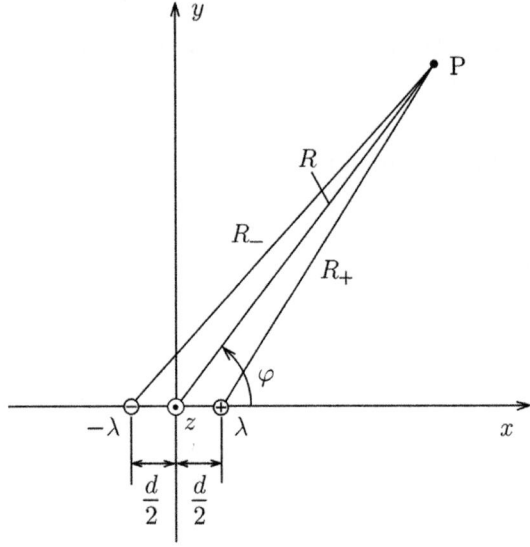

1.7 Exercises

Fig. 1.13 Equipotential surfaces (*dotted lines*) and electric field lines (*solid lines*) in a plane normal to the line charges in Fig. 1.12

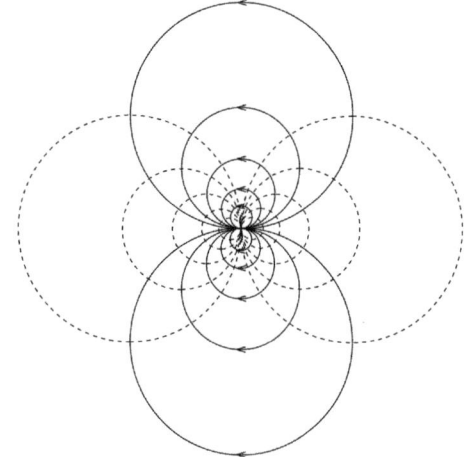

Figure 1.13 shows intersection lines between the equipotential surfaces and the sheet and electric field lines.

1.7 Exercises

1.1. Two linear electric charges of densities λ and $-\lambda$ are placed with distance $2a$ apart, as shown in Fig. E1.1. Determine the force on linear charge of density λ' in a unit length that is equidistant from each linear charge.

Fig. E1.1 Two liner electric charges $\pm\lambda$ and a linear electric charge λ' that is equidistance from both

Fig. E1.2 Linear electric charge and square electric charge

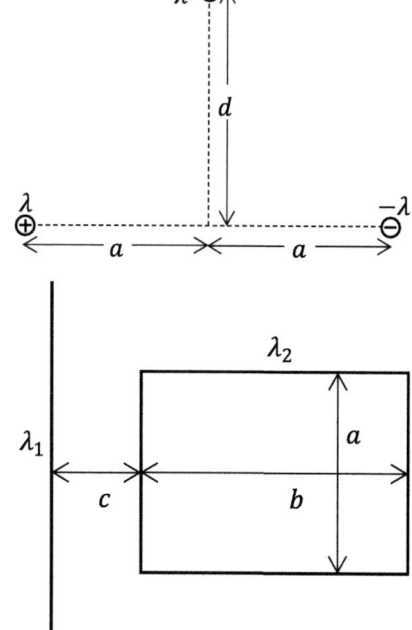

1.2. Determine the Coulomb force between an electric charge of linear density λ_1 on a straight line and that of linear density λ_2 on a square, as shown in Fig. E1.2. These are placed on a common plane.

1.3. Determine the Coulomb force between an electric charge of linear density λ_1 on a straight line and that of linear density λ_2 on a circle of radius a. These are placed on a common plane and the distance between the line and the center of the circle is d, as shown in Fig. E1.3.

1.4. Determine the Coulomb force between an electric charge of linear density λ_1 on a straight line and that of linear density λ_2 on a closed half-circle of radius a. These are placed on a common plane and the distance between the line and the center of the half-circle is d, as shown in Fig. E1.4.

1.5. Electric charge Q and unknown electric charges Q_x and Q_y are placed at each vertex of the equilateral triangle with side length a, as shown in Fig. E1.5. The electric

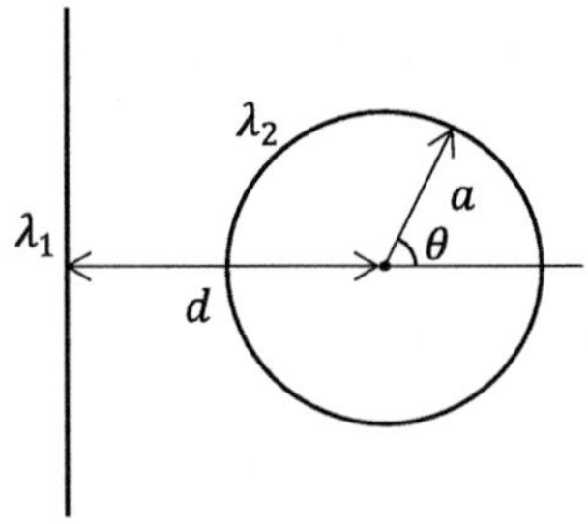

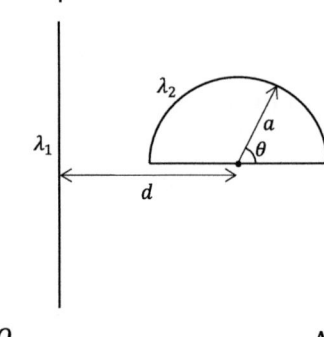

Fig. E1.3 Linear electric charge and circular electric charge

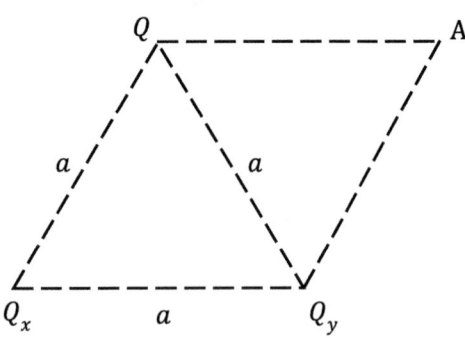

Fig. E1.4 Linear electric charge and electric charge on a closed half-circle

Fig. E1.5 Electric charge Q and unknown electric charges Q_x and Q_y

1.7 Exercises

field strength at point A, which is symmetric with the position of the charge Q_x, is zero. Determine the values of Q_x and Q_y and the electric potential at point A.

1.6. Electric charge is uniformly distributed with linear density λ on a closed triangle, as shown in Fig. E1.6. Determine the electric field at point A.

1.7. Electric charge is uniformly distributed with a linear density λ on an equilateral triangle of side length a, as shown in Fig. E1.7. Determine the electric field strength at point P at distance z above the center O of the triangle.

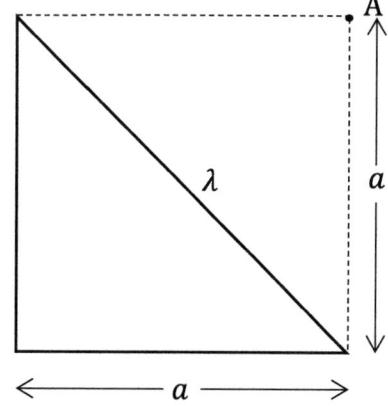

Fig. E1.6 Triangle with uniformly distributed electric charge and observation point A

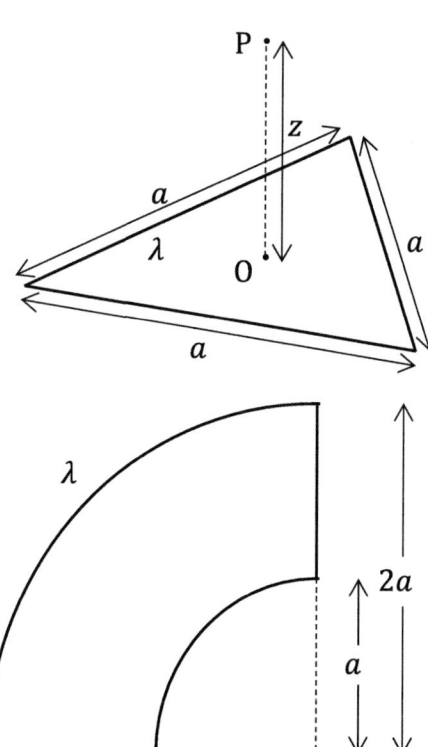

Fig. E1.7 Triangular linear charge and observation point above the center

Fig. E1.8 Closed linear electric charge and observation point O

1.8. Suppose that electric charge is uniformly distributed with linear density λ on a closed circuit composed of two quarter circles and two straight lines on a common plane as shown in Fig. E1.8. Determine the electric field at point O.

1.9. Solve the problem of Example 1.6 with Coulomb's law.

1.10. Suppose that electric charge is uniformly distributed with density ρ_0 inside a wide slab of thickness $2a$ parallel to the y-z plane, as shown in Fig. E1.9. Determine the electric field in each region using Gauss's law.

1.11. Electric point charges Q and $-Q$ are placed at $(a, 0, 0)$ and $(-a, 0, 0)$, respectively, in the space. Determine the electric potential and electric field in the space.

1.12. Determine the electric potential produced by a circular line charge in Example 1.8 using the electric field.

1.13. Electric charge is uniformly distributed with a linear density λ on an equilateral triangle of side length a, as shown in Fig. E1.10. Determine the electric field strength and electric potential at the center O of the triangle.

Fig. E1.9 Substance in which electric charge is uniformly distributed

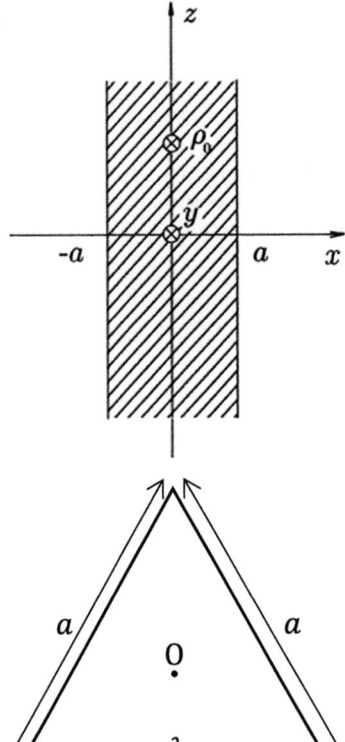

Fig. E1.10 Equilateral triangle with linear electric charge and its center O

1.7 Exercises

Fig. E1.11 Parallel lines with positive and negative linear electric charges

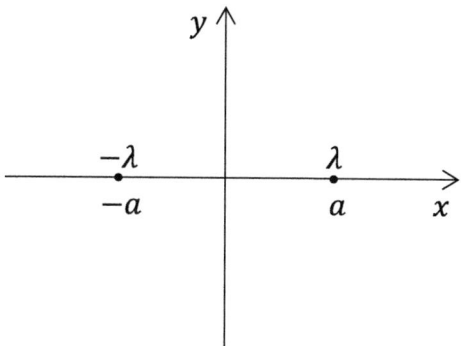

1.14. Determine the electric potential at point P at distance z above the center O of the triangle with linear electric charge of density λ treated in Exercise 1.7.

1.15. Suppose that electric charge is uniformly distributed with a linear density λ on a regular square of side length a, as discussed in Example 1.3. Determine the electric potential at point P at a distance z above the center of the square (see Fig. 1.4).

1.16. Linear electric charges of densities λ and $-\lambda$ are placed at $(a, 0)$ and $(-a, 0)$ in the x-y plane, respectively, as shown in Fig. E1.11. Determine the electric potential and electric field in the space.

1.17. Three linear electric charges are placed parallel at each vertex of an equilateral triangle with side length a as shown in Fig. E1.12. Determine the electric potential and electric field at the center of the equilateral triangle.

1.18. Assume that linear electric charges of densities $\pm\lambda$ are placed in parallel, as shown in Fig. E1.13. Determine the electric potential and electric field at the center O.

Fig. E1.12 Linear electric charges placed at each vertex of the equilateral triangle and the center

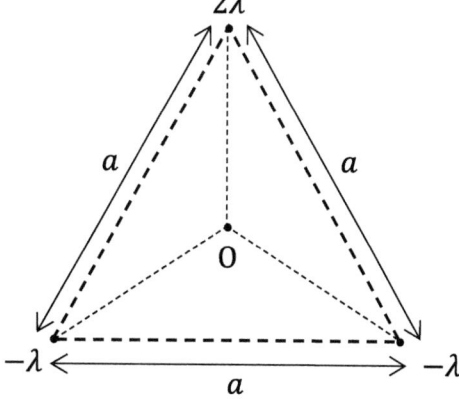

Fig. E1.13 Linear electric charges placed at each vertex of the square and the center O

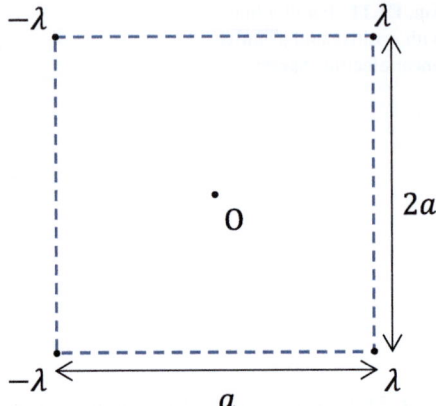

1.19. Electric charge is uniformly distributed with density σ on a flat plane except in the region surrounded by a circle of radius a, as shown in Fig. E1.14. Determine the electric field strength at point P at distance z above the center of the circle.

1.20. Suppose that electric charge is uniformly distributed with a surface density σ on a half spherical surface of radius a, as shown in Fig. E1.15. Determine the electric field and electric potential at point P at distance z above the center of the sphere.

1.21. Determine the electric field and electric potential at center O ($z = 0$) in Exercise 1.20.

1.22. The electric potential described with spherical coordinates is given by

$$\phi = \frac{\gamma}{2\epsilon_0}\left(a^2 - \frac{r^2}{3}\right); \quad 0 \leq r < a,$$
$$= \frac{\gamma a^3}{3\epsilon_0 r}; \quad r > a.$$

Determine the distribution of electric charge.

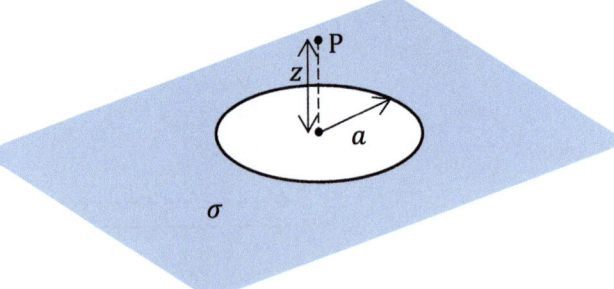

Fig. E1.14 Uniformly distributed electric charge with density σ on a flat plane except for the vacant region and point P

Fig. E1.15 Electric charge on a half spherical surface and observation point P

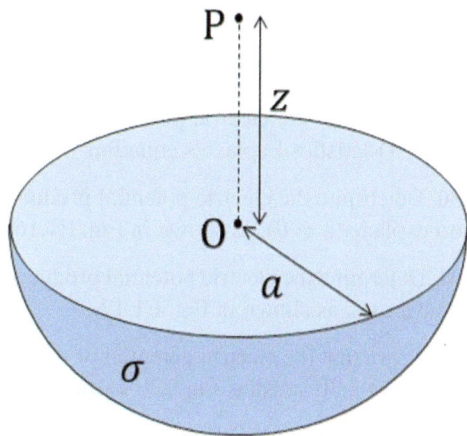

1.23. The electric potential described with cylindrical coordinates is given by

$$\phi = \frac{\gamma}{4\epsilon_0}\left[(a^2 - R^2) + 2a^2 \log \frac{R_0}{a}\right]; \quad 0 \leq R < a,$$

$$= \frac{a^2\gamma}{2\epsilon_0} \log \frac{R_0}{R}; \qquad R > a.$$

Determine the electric charge distribution.

1.24. Suppose that electric charge is distributed around the origin as

$$\rho(r) = a - br^2; \quad 0 \leq r < (a/b)^{1/2},$$
$$= 0; \qquad r > (a/b)^{1/2}.$$

Determine the electric field and electric potential.

1.25. The electric potential described with spherical coordinates is given by

$$\phi(r) = \alpha\left(-r^2 + \frac{1}{2a}r^3\right); \quad 0 \leq r < a,$$

$$= -\frac{\alpha a^3}{2r}; \qquad r > a.$$

Determine the electric charge distribution.

1.26. Electric charge is distributed around the origin with the density given by

$$\rho(r) = a - br; \quad 0 \leq r < r_0,$$
$$= 0; \qquad r > r_0.$$

Determine the electric potential using Poisson's equation.

1.27. Determine the electric field lines of a two-dimensional electric dipole line.

1.28. Determine the electric field lines of a three-dimensional electric dipole.

1.29. Prove that the electric potential produced by an electric dipole line given by Eq. (1.25) satisfies Laplace's equation.

1.30. Determine the electric potential produced by an electric quadrupole placed on the x-y plane ($z = 0$), as shown in Fig. E1.16.

1.31. Determine the electric potential produced by a linear electric quadrupole placed on the z-axis, as shown in Fig. E1.17.

1.32. Prove that the electric potential of the transverse electric quadrupole discussed in Exercise 1.30 satisfies Laplace's equation.

1.33. Prove that the electric potential of the linear electric quadrupole discussed in Exercise 1.31 satisfies Laplace's equation.

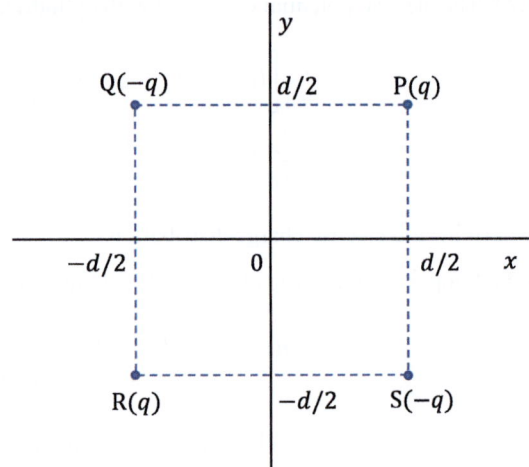

Fig. E1.16 Arrangement of electric charges on each vertex of a square on the x-y plane

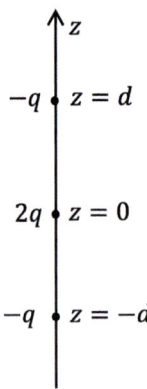

Fig. E1.17 Arrangement of electric charges on a line along the z-axis

1.7 Exercises

Answers to Exercises

1.1. From the electric field obtained using Gauss's law, λ' receives a repulsive force of magnitude $\lambda\lambda'/\left[2\pi\epsilon_0(a^2+d^2)^{1/2}\right]$ in a unit length from the electric charge of the same sign and an attractive force of the same magnitude from the electric charge of the opposite sign. As a result, the combined force is $a\lambda\lambda'/\left[\pi\epsilon_0(a^2+d^2)\right]$ in a unit length and is directed to the right, as shown in Fig. B1.1.

1.2. The Coulomb force between the long line charge and an electric charge of a part of length dy on the left side of the square is $\mathrm{d}F = \lambda_1\lambda_2\mathrm{d}y/2\pi\epsilon_0 c$. Hence, the force on this side is

$$F_1 = \frac{\lambda_1\lambda_2 a}{2\pi\epsilon_0 c}.$$

The force on the right side is similarly obtained to be

$$F_2 = \frac{\lambda_1\lambda_2 a}{2\pi\epsilon_0(b+c)}.$$

The force on an electric charge of the region from x to $x + \mathrm{d}x$ on the horizontal side from the long line charge is $\mathrm{d}F = \lambda_1\lambda_2\mathrm{d}x/2\pi\epsilon_0 x$. So, the force on one horizontal side is

$$F_3 = \frac{\lambda_1\lambda_2}{2\pi\epsilon_0}\int_c^{b+c}\frac{\mathrm{d}x}{x} = \frac{\lambda_1\lambda_2}{2\pi\epsilon_0}\log\frac{b+c}{c}.$$

So, the total force is

$$F = F_1 + F_2 + 2F_3 = \frac{\lambda_1\lambda_2}{2\pi\epsilon_0}\left(\frac{a}{c} + \frac{a}{b+c} + 2\log\frac{b+c}{c}\right),$$

and directed to the right.

Fig. B1.1 Coulomb's forces from each electric charge and the combined force

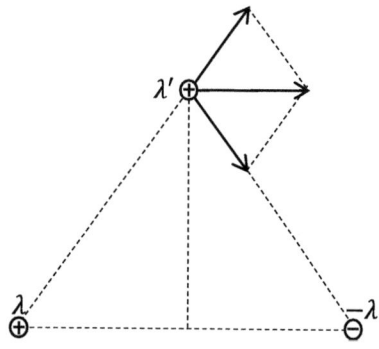

1.3. The angle θ is defined as shown in Fig. E1.3. The Coulomb force between the electric charge in the region from θ to $\theta + d\theta$ of the circle and the straight linear charge is

$$dF = \frac{\lambda_1 \lambda_2 a d\theta}{2\pi \epsilon_0 (d + a\cos\theta)}.$$

So, the total Coulomb force is

$$F = \frac{\lambda_1 \lambda_2 a}{\pi \epsilon_0} \int_0^\pi \frac{d\theta}{d + a\cos\theta} = \frac{a \lambda_1 \lambda_2}{\sqrt{d^2 - a^2} \epsilon_0}$$

and is repulsive, where we used Eq. (23) in the Appendix.

1.4. The angle θ is defined as shown in Fig. E1.4. The Coulomb force that the electric charge in the region from θ to $\theta + d\theta$ of the half-circle receives from the straight linear charge is

$$dF_1 = \frac{\lambda_1 \lambda_2 a d\theta}{2\pi \epsilon_0 (d + a\cos\theta)}.$$

So, the total Coulomb force on the half-circle is

$$F_1 = \frac{\lambda_1 \lambda_2 a}{2\pi \epsilon_0} \int_0^\pi \frac{d\theta}{d + a\cos\theta} = \frac{a \lambda_1 \lambda_2}{2\sqrt{d^2 - a^2} \epsilon_0},$$

where we used Eq. (23) in the Appendix. The force on the straight section is

$$F_2 = \frac{\lambda_1 \lambda_2}{2\pi \epsilon_0} \int_{d-a}^{d+a} \frac{dR}{R} = \frac{\lambda_1 \lambda_2}{2\pi \epsilon_0} \log \frac{d+a}{d-a}.$$

Thus, the total force is

$$F = F_1 + F_2 = \frac{\lambda_1 \lambda_2}{2\epsilon_0} \left(\frac{a}{\sqrt{d^2 - a^2}} + \frac{1}{\pi} \log \frac{d+a}{d-a} \right)$$

and is repulsive.

1.5. Since the electric field at point a produced by Q and Q_y must lie on the line connecting Q_x and point a, we have $Q_y = Q$. Then, the electric field strength produced by two Q's is

$$E = \frac{\sqrt{3} Q}{4\pi \epsilon_0 a^2}$$

and the electric field strength at point A produced by Q_x is

$$E_x = \frac{Q_x}{4\pi\epsilon_0 \left(\sqrt{3}a\right)^2} = \frac{Q_x}{12\pi\epsilon_0 a^2}.$$

From the requirement $E + E_x = 0$, we have $Q_x = -3\sqrt{3}Q$. The electric potential at A is

$$\phi = \frac{Q}{4\pi\epsilon_0 a} \times 2 + \frac{Q_x}{4\sqrt{3}\pi\epsilon_0 a} = -\frac{Q}{4\pi\epsilon_0 a}.$$

1.6. First, we treat the contribution from the left side. We denote the distance from the upper vertex by y. Then, the horizontal and vertical components of the electric field produced by the electric charge in the region from y to $y + dy$ is

$$dE_{1x} = \frac{\lambda a\, dy}{4\pi\epsilon_0 (y^2 + a^2)^{3/2}}, \quad dE_{1y} = \frac{\lambda y\, dy}{4\pi\epsilon_0 (y^2 + a^2)^{3/2}}.$$

Integrating them in the region from $y = 0$ to $y = a$, we have

$$E_{1x} = \frac{\lambda}{4\sqrt{2}\pi\epsilon_0 a}, \quad E_{1y} = \frac{(\sqrt{2}-1)\lambda}{4\sqrt{2}\pi\epsilon_0 a}.$$

The contribution from the electric charge on the bottom side is similarly calculated as

$$E_{2x} = \frac{(\sqrt{2}-1)\lambda}{4\sqrt{2}\pi\epsilon_0 a}, \quad E_{2y} = \frac{\lambda}{4\sqrt{2}\pi\epsilon_0 a}.$$

Then, the contribution from the longer side is calculated. This is the electric field at distance $a/\sqrt{2}$ produced by the electric charge of the length $\sqrt{2}a$. From symmetry, the electric field is directed normal to this side. Using the same method as in Example 1.2, we have

$$E_{3x} = E_{3y} = \frac{\lambda}{2\sqrt{2}\pi\epsilon_0 a}.$$

Thus, the electric field is determined to be

$$E_x = E_{1x} + E_{2x} + E_{3x} = \frac{(1+\sqrt{2})\lambda}{4\pi\epsilon_0 a}, \quad E_y = E_{1y} + E_{2y} + E_{3y} = \frac{(1+\sqrt{2})\lambda}{4\pi\epsilon_0 a}.$$

1.7. The distance between one side and point P is $(z^2 + a^2/12)^{1/2}$. Using the result of Example 1.2, the electric field produced by the electric charge of one side is

$$E' = \frac{\lambda a}{4\pi\epsilon_0 (z^2 + a^2/12)^{1/2} (z^2 + a^2/3)^{1/2}}.$$

The vertical component remains from symmetry, and we have

$$E = \frac{3zE'}{(z^2 + a^2/12)^{1/2}} = \frac{3\lambda az}{4\pi\epsilon_0 (z^2 + a^2/12)(z^2 + a^2/3)^{1/2}}.$$

1.8. First, we calculate the contribution from the electric charge on the quarter circle of radius a. The horizontal component of the electric field at point O is given by

$$E_{ax} = \frac{1}{4\pi\epsilon_0} \int_0^{\pi/2} \frac{\lambda a}{a^2} \cos\theta d\theta = \frac{\lambda}{4\pi\epsilon_0 a}.$$

The vertical component is similarly calculated as

$$E_{ay} = -\frac{1}{4\pi\epsilon_0} \int_0^{\pi/2} \frac{\lambda a}{a^2} \sin\theta d\theta = -\frac{\lambda}{4\pi\epsilon_0 a}.$$

The horizontal and vertical components of the electric field produced by the quarter circle of radius $2a$ are obtained as

$$E_{2ax} = \frac{\lambda}{8\pi\epsilon_0 a}, \quad E_{2ay} = -\frac{\lambda}{8\pi\epsilon_0 a}.$$

The horizontal line produces only the horizontal component:

$$E_{hx} = \frac{1}{4\pi\epsilon_0} \int_a^{2a} \frac{\lambda dx}{x^2} = \frac{\lambda}{8\pi\epsilon_0 a},$$

and the vertical line produces only the vertical component:

$$E_{vy} = -\frac{1}{4\pi\epsilon_0} \int_a^{2a} \frac{\lambda dy}{y^2} = -\frac{\lambda}{8\pi\epsilon_0 a}.$$

Summing each electric field, the electric field is determined to be

$$E_x = E_{ax} + E_{2ax} + E_{hx} = \frac{\lambda}{2\pi\epsilon_0 a}, \quad E_y = E_{ay} + E_{2ay} + E_{vy} = -\frac{\lambda}{2\pi\epsilon_0 a}.$$

1.7 Exercises

1.9. Suppose a cluster of straight line charges. The x-y axes are defined on the plane with the origin (0, 0) at the foot of a vertical line from the observation point. The electric charge in the region from x to $x + dx$ is regarded as a line charge with density $\sigma\,dx$. We can repeat the calculation in Example 1.2, and then, the electric field strength by this line charge at the observation point is obtained to be $dE' = \sigma\,dx / \left[2\pi\epsilon_0(x^2 + h^2)^{1/2}\right]$. From the symmetry, only the vertical component given by $dE = dE'h/(x^2 + h^2)^{1/2}$ remains. (See Fig. B1.2.) Thus, we have.

$$E = \int_{-\infty}^{\infty} \frac{\sigma h\,dx}{2\pi\epsilon_0(x^2 + h^2)}.$$

If we put $x = h\tan\theta$, θ is the angle shown in Fig. B1.2. From the relationships $x^2 + h^2 = h^2/\cos^2\theta$ and $dx = h\,d\theta/\cos^2\theta$, we have

$$E = \frac{\sigma}{2\pi\epsilon_0} \int_{-\pi/2}^{\pi/2} d\theta = \frac{\sigma}{2\epsilon_0},$$

which agrees with the result in Example 1.6.

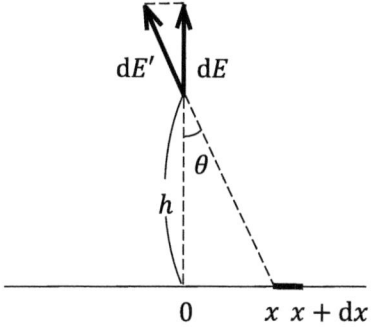

Fig. B1.2 The electric field produced by linear electric charge in the region from x to $x + dx$ and its vertical component

1.10. We assume closed parallelepiped S a plane of which stays on the central plane, $x = 0$, as shown in Fig. B1.3. Then, the electric field must be zero on this plane from symmetry with respect to the x-axis. That is, the same result must be obtained when the right and left sides are reversed. It can be concluded that the electric field has only the x component (E_x) from symmetry with respect to the y- and z-axes. Hence, the electric field is parallel to the surface on the four surfaces parallel to the x-axis and the surface integral of the electric field on these surfaces is zero. The surface integral on the remaining surface only has a nonzero value. The position of and the electric field on this surface is denoted by x and $E_x(x)$, respectively. Then, the surface integral of the electric field is $AE_x(x)$, where A is the area of the surface parallel to the

Fig. B1.3 Closed parallelepiped S for $x > a$ on which Gauss's law is applied

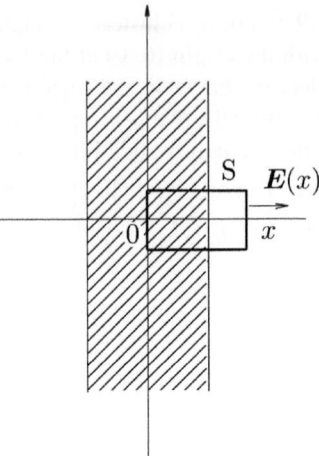

y-z plane. Now we estimate the total electric charge inside S. A simple calculation leads to $Ax\rho_0$ for $0 \le x \le a$ and $Aa\rho_0$ for $x > a$. Thus, we have

$$E_x = -\frac{a\rho_0}{\epsilon_0}; \quad x < -a,$$

$$= \frac{x\rho_0}{\epsilon_0}; \quad -a \le x \le a,$$

$$= \frac{a\rho_0}{\epsilon_0}; \quad x > a,$$

where we have used the symmetric condition with respect to $x = 0$.

1.11. The electric potential is given by

$$\phi(x, y, z) = \frac{Q}{4\pi \epsilon_0} \left\{ \frac{1}{\left[(x-a)^2 + y^2 + z^2\right]^{1/2}} - \frac{1}{\left[(x+a)^2 + y^2 + z^2\right]^{1/2}} \right\}.$$

Then, the electric field is determined:

$$E_x = -\frac{\partial \phi}{\partial x} = \frac{Q}{4\pi \epsilon_0} \left\{ \frac{x-a}{\left[(x-a)^2 + y^2 + z^2\right]^{1/2}} - \frac{x+a}{\left[(x+a)^2 + y^2 + z^2\right]^{1/2}} \right\},$$

$$E_y = -\frac{\partial \phi}{\partial y} = \frac{Q}{4\pi \epsilon_0} \left\{ \frac{y}{\left[(x-a)^2 + y^2 + z^2\right]^{1/2}} - \frac{y}{\left[(x+a)^2 + y^2 + z^2\right]^{1/2}} \right\},$$

$$E_z = -\frac{\partial \phi}{\partial z} = \frac{Q}{4\pi \epsilon_0} \left\{ \frac{z}{\left[(x-a)^2 + y^2 + z^2\right]^{1/2}} - \frac{z}{\left[(x+a)^2 + y^2 + z^2\right]^{1/2}} \right\}.$$

1.7 Exercises

1.12. Suppose a path normal to the circle, as shown in Fig. B1.4. We assume a point P on the path and denote the distance between point P and the origin by z. The electric field strength at point P is determined to be (see Example 1.4)

$$E = \frac{az\lambda}{2\epsilon_0 (z^2 + a^2)^{3/2}}.$$

Thus, the electric potential on the origin is determined:

$$\phi = -\int_{\infty}^{0} \frac{az\lambda}{2\epsilon_0 (z^2 + a^2)^{3/2}} dz = -\frac{a\lambda}{2\epsilon_0} \left[\frac{1}{(z^2 + a^2)^{1/2}} \right]_0^{\infty} = \frac{\lambda}{2\epsilon_0}.$$

This is the same result directly obtained using Eq. (1.17).

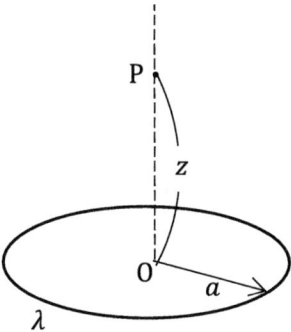

Fig. B1.4 Circle with uniformly distributed electric charge and point P above the center

1.13. It is clear that the electric field at the center is zero from symmetry. In fact, the result obtained in Exercise 1.7 shows that E reduces to 0 at $z = 0$.

We denote the distance of a point on one side from its center by x. Then, the distance from this point to the center O is $(x^2 + a^2/12)^{1/2}$. So, the electric potential at the center O produced by the electric charge from x to $x + dx$ is

$$d\phi = \frac{\lambda dx}{4\pi \epsilon_0 (x^2 + a^2/12)^{1/2}}.$$

Hence, the contribution from one side to the electric potential is

$$\phi' = \frac{\lambda}{4\pi \epsilon_0} \int_{-a/2}^{a/2} \frac{dx}{(x^2 + a^2/12)^{1/2}} = \frac{\lambda}{4\pi \epsilon_0} \left[\log\left(x + \sqrt{x^2 + a^2/12}\right) \right]_{-a/2}^{a/2}$$

$$= \frac{\log(2 + \sqrt{3})\lambda}{2\pi \epsilon_0},$$

where Eq. (22) in the Appendix is used. The electric potential at the center is determined to be

$$\phi = 3\phi' = \frac{3\log(2+\sqrt{3})\lambda}{2\pi\epsilon_0}.$$

1.14. The distance between one side and point P is $(z^2 + a^2/12)^{1/2}$. Then, the distance between a point of distance x from the center of this side and point P is $(x^2 + z^2 + a^2/12)^{1/2}$. Hence, the contribution from this side to the electric potential is

$$\phi' = \frac{\lambda}{4\pi\epsilon_0} \int_{-a/2}^{a/2} \frac{dx}{(x^2 + z^2 + a^2/12)^{1/2}} = \frac{\lambda}{4\pi\epsilon_0} \left[\log\left(x + \sqrt{x^2 + z^2 + a^2/12}\right)\right]_{-a/2}^{a/2}$$

$$= \frac{\lambda}{4\pi\epsilon_0} \log\left[1 + \frac{a}{\sqrt{z^2 + (a^2/3)} - a/2}\right],$$

where we used Eq. (22) in the Appendix. Then, the electric potential at the center is

$$\phi = 3\phi' = \frac{3\lambda}{4\pi\epsilon_0} \log\left[1 + \frac{a}{\sqrt{z^2 + (a^2/3)} - a/2}\right].$$

Using Eq. (1.18), we can derive the electric field obtained in Exercise 1.7.

1.15. We denote the distance of a point on one side from its center by x. Then, the distance from this point to point P is $(x^2 + z^2 + a^2/4)^{1/2}$. So, the electric potential at the center produced by the electric charge from x to $x + dx$ is

$$d\phi = \frac{\lambda dx}{4\pi\epsilon_0 (x^2 + z^2 + a^2/4)^{1/2}}.$$

Hence, the contribution from one side to the electric potential is

$$\phi' = \frac{\lambda}{4\pi\epsilon_0} \int_{-a/2}^{a/2} \frac{dx}{(x^2 + z^2 + a^2/4)^{1/2}} = \frac{\lambda}{4\pi\epsilon_0} \left[\log\left(x + \sqrt{x^2 + z^2 + \frac{a^2}{4}}\right)\right]_{-\frac{a}{2}}^{\frac{a}{2}}$$

$$= \frac{\lambda}{4\pi\epsilon_0} \log\left[1 + \frac{a}{\sqrt{z^2 + (a^2/2)} - a/2}\right],$$

where we used Eq. (22) in the Appendix. The electric potential at the center is determined to be

1.7 Exercises

$$\phi = 4\phi' = \frac{\lambda}{\pi\epsilon_0} \log\left[1 + \frac{a}{\sqrt{z^2 + (a^2/2)} - a/2}\right].$$

Using Eq. (1.18), we can derive the electric field obtained in Example 1.3.

1.16. In this case, the electric potential is given by

$$\phi(x, y) = \frac{\lambda}{4\pi\epsilon_0} \log \frac{(x+a)^2 + y^2}{(x-a)^2 + y^2},$$

since the reference point can be chosen at infinity. The electric field is

$$E_x = -\frac{\partial \phi}{\partial x} = \frac{\lambda}{4\pi\epsilon_0}\left[\frac{x-a}{(x-a)^2 + y^2} - \frac{x+a}{(x+a)^2 + y^2}\right],$$

$$E_y = -\frac{\partial \phi}{\partial y} = \frac{\lambda}{4\pi\epsilon_0}\left[\frac{y}{(x-a)^2 + y^2} - \frac{y}{(x+a)^2 + y^2}\right].$$

1.17. All the linear charges are placed at the same distance, $a/\sqrt{3}$, from the center, and the total electric charge is zero. So, the electric potential at the center is zero. The electric field due to the linear charge 2λ is directed downward and its strength is $\sqrt{3}\lambda/\pi\epsilon_0 a$. The horizontal component of the electric field of the two linear charges $-\lambda$ is cancelled, and only the vertical component remains. Each vertical component is $\sqrt{3}\lambda/4\pi\epsilon_0 a$ and is directed downward. The total electric field is directed downward and its strength is

$$\frac{\sqrt{3}\lambda}{\pi\epsilon_0 a} + \frac{\sqrt{3}\lambda}{4\pi\epsilon_0 a} \times 2 = \frac{3\sqrt{3}\lambda}{2\pi\epsilon_0 a}.$$

1.18. All the linear charges are placed at the same distance, $\sqrt{2}a$, from the center, and the total electric charge is zero. So, the electric potential at the center is zero. The vertical component of the electric field due to the linear charges λ is cancelled and only the horizontal component remains. Its value for each is $\lambda/4\pi\epsilon_0 a$ and the electric field is directed to the left. The same result is obtained for the two linear charges $-\lambda$. Thus, the total electric field is directed to the left and its strength is

$$\frac{\lambda}{4\pi\epsilon_0 a} \times 4 = \frac{\lambda}{\pi\epsilon_0 a}.$$

1.19. The electric field produced by a circular electric charge in the region of radius r to $r + dr$ from the center is directed above and its strength is

$$dE = \frac{\sigma z r dr}{2\epsilon_0 \left(r^2 + z^2\right)^{3/2}}.$$

Hence, the total electric field strength is

$$E = \int dE = \frac{\sigma z}{2\epsilon_0} \int_a^\infty \frac{r \, dr}{(r^2 + z^2)^{3/2}} = \frac{\sigma z}{2\epsilon_0 (a^2 + z^2)^{1/2}}.$$

1.20. We define angle θ as shown in Fig. B1.5. The electric charge of a narrow ring in the region from θ to $\theta + d\theta$ is $2\pi a^2 \sigma \sin\theta \, d\theta$, and the distance between this ring and point P is

$$l = (z^2 + a^2 + 2az \cos\theta)^{1/2}.$$

So, the electric potential at point P is

$$\phi = \frac{a^2 \sigma}{2\epsilon_0} \int_0^{\pi/2} \frac{\sin\theta \, d\theta}{(z^2 + a^2 + 2az \cos\theta)^{1/2}} = -\frac{a\sigma}{2\epsilon_0 z} \left[(z^2 + a^2 + 2az \cos\theta)^{1/2} \right]_0^{\pi/2}$$

$$= \frac{a\sigma}{2\epsilon_0 z} \left[z + a - (z^2 + a^2)^{1/2} \right].$$

The electric field is vertical upward and the strength is given by

$$E = -\frac{\partial \phi}{\partial z} = \frac{a\sigma}{2\epsilon_0} \left[\frac{1}{(z^2 + a^2)^{1/2}} - \frac{(z^2 + a^2)^{1/2} - a}{z^2} \right].$$

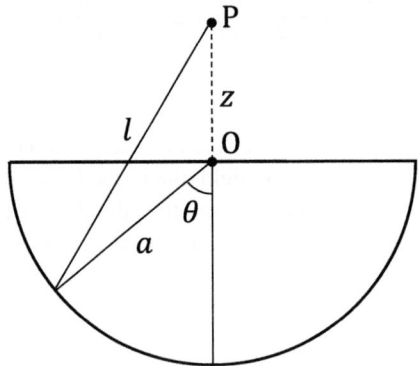

Fig. B1.5 Definition of angle θ

1.21. Since all electric charge, $Q = 2\pi a^2 \sigma$, is placed at the distance a from the center, the electric potential is simply given by

1.7 Exercises

$$\phi = \frac{Q}{4\pi\epsilon_0 a} = \frac{a\sigma}{2\epsilon_0}.$$

We calculate the electric field produced by electric charge of a narrow ring in the region from θ to $\theta+d\theta$. We define azimuthal angle φ around the vertical axis extended downward from the center. The electric field in a small segment from φ to $\varphi + d\varphi$ on this ring is

$$dE = \frac{a^2\sigma \sin\theta\, d\theta\, d\varphi}{4\pi\epsilon_0 a^2} = \frac{\sigma \sin\theta\, d\theta\, d\varphi}{4\pi\epsilon_0}.$$

Only the vertical component, $dE \cos\theta$, remains from symmetry. Integrating this with respect to φ, we have the electric field produced by the electric charge of the narrow ring. So, the total electric field is

$$E = \frac{\sigma}{2\epsilon_0} \int_0^{\pi/2} \sin\theta \cos\theta\, d\theta = \frac{\sigma}{4\epsilon_0}.$$

These results can also be obtained from the results in the limit $z \to 0$ in Exercise 1.20.

1.22. Using Eq. (1.20), the electric charge distribution is determined to be

$$\rho(r) = -\epsilon_0 \Delta\phi = -\epsilon_0 \frac{1}{r^2} \cdot \frac{\partial}{\partial r}\left(r^2 \frac{\partial \phi}{\partial r}\right).$$

Substituting the given ϕ yields the distribution:

$$\rho(r) = \gamma; \quad 0 \le r < a,$$
$$= 0; \quad r > a.$$

1.23. Using Eq. (1.20), the electric charge distribution is determined to be

$$\rho(R) = -\epsilon_0 \Delta\phi = -\frac{\epsilon_0}{R} \cdot \frac{\partial}{\partial R}\left(R \frac{\partial \phi}{\partial R}\right).$$

Substituting the given ϕ yields the distribution:

$$\rho(R) = \gamma; \quad 0 \le R < a,$$
$$= 0; \quad R > a.$$

1.24. The electric field has a radial component, E_r. Equation (1.12) leads to

$$\frac{1}{r^2} \cdot \frac{\partial}{\partial r}(r^2 E_r) = \frac{1}{\epsilon_0}(a - br^2)$$

for $0 \leq r < (a/b)^{1/2}$. After a simple calculation we have

$$E_r = \frac{1}{\epsilon_0}\left(\frac{a}{3}r - \frac{b}{5}r^3 + \frac{c}{r^2}\right),$$

where c is a constant. Since the electric field must be finite at $r = 0$, we have $c = 0$. The total electric charge in the region $0 \leq r < (a/b)^{1/2}$ is

$$Q = \int_0^{(a/b)^{1/2}} \rho(r) 4\pi r^2 dr = \frac{8\pi a}{15}\left(\frac{a}{b}\right)^{3/2}.$$

Thus, the electric field is determined to be

$$E_r = \frac{1}{\epsilon_0}\left(\frac{a}{3}r - \frac{b}{5}r^3\right); \quad 0 \leq r < \left(\frac{a}{b}\right)^{1/2},$$

$$= \frac{2a(a/b)^{3/2}}{15\epsilon_0 r^2}; \quad r > \left(\frac{a}{b}\right)^{1/2}.$$

Then, the electric potential is given by

$$\phi = \frac{2a(a/b)^{3/2}}{15\epsilon_0 r}; \quad r > \left(\frac{a}{b}\right)^{1/2},$$

$$= \frac{1}{\epsilon_0}\left(-\frac{a}{6}r^2 + \frac{b}{20}r^4 + \frac{a^2}{4b}\right); \quad 0 \leq r < \left(\frac{a}{b}\right)^{1/2}.$$

1.25. Using Eq. (1.18), the electric field is determined to be

$$E_r = \alpha\left(2r - \frac{3r^2}{2a}\right); \quad 0 \leq r < a,$$

$$= -\frac{\alpha a^3}{2r^2}; \quad r > a.$$

Then, the electric charge distribution is obtained to be

$$\rho(r) = 6\alpha\epsilon_0\left(1 - \frac{r}{a}\right); \quad 0 \leq r < a,$$

$$= 0; \quad r > a.$$

1.26. Since the electric charge changes only along the radial direction, we can assume that the electric potential depends only on r. So, Poisson's equation is given by

$$\frac{1}{r^2} \cdot \frac{\partial}{\partial r}\left(r^2 \frac{\partial \phi}{\partial r}\right) = -\epsilon_0 \rho(r).$$

1.7 Exercises

In the region $0 \leq r < r_0$, we have

$$\phi = -\epsilon_0 \left(\frac{a}{6} r^2 - \frac{b}{12} r^3 + \frac{c}{r} \right),$$

where c is a constant. Since the electric potential must be finite at $r = 0$, we have $c = 0$.

In the region $r > r_0$, Poisson's equation gives

$$\phi = \frac{d}{r},$$

with d denoting a constant. From the continuity of the electric potential at $r > r_0$, we can determine d.

The electric potential is obtained to be

$$\phi = -\epsilon_0 \left(\frac{a}{6} r^2 - \frac{b}{12} r^3 \right); \quad 0 \leq r < r_0,$$

$$= -\epsilon_0 \left(\frac{a}{6} r_0^3 - \frac{b}{12} r_0^4 \right) \frac{1}{r}; \quad r > r_0.$$

1.27. The equipotential surface is given by Eq. (1.27):

$$\phi'(R, \varphi) = \frac{\cos \varphi}{R} = \text{const.}$$

Here, we assume electric field lines of the form:

$$\psi(R, \varphi) = \frac{\sin \varphi}{R^n} = \text{const.}$$

The exponent n is determined from the orthogonal condition:

$$\nabla \phi' \cdot \nabla \psi = 0.$$

Using

$$\nabla \phi' = \frac{\partial \phi'}{\partial R} i_R + \frac{1}{R} \cdot \frac{\partial \phi'}{\partial \varphi} i_\varphi = -\frac{\cos \varphi}{R^2} i_R - \frac{\sin \varphi}{R^2} i_\varphi,$$

$$\nabla \psi = \frac{\partial \psi}{\partial R} i_R + \frac{1}{R} \cdot \frac{\partial \psi}{\partial \varphi} i_\varphi = -\frac{n \sin \varphi}{R^{n+1}} i_R + \frac{\cos \varphi}{R^{n+1}} i_\varphi,$$

the above condition leads to

$$\frac{(n-1) \sin \varphi \cos \varphi}{R^{n+3}} = 0.$$

Thus, we have $n = 1$ and the electric field lines are given by

$$\psi(R, \varphi) = \frac{\sin \varphi}{R} = \text{const.} \tag{B1.1}$$

In the following, we show this really describes the electric field lines. The electric field lines in Cartesian coordinates are described by

$$\frac{E_y}{E_x} = \frac{dy}{dx}.$$

The electric field components are obtained from Eq. (1.26) as

$$E_x = E_R \cos \varphi - E_\varphi \sin \varphi = \frac{\hat{p} \cos 2\varphi}{2\pi \epsilon_0 R^2}, \quad E_y = E_R \sin \varphi + E_\varphi \cos \varphi = \frac{\hat{p} \sin 2\varphi}{2\pi \epsilon_0 R^2}.$$

Hence, the above equation leads to

$$\frac{dy}{dx} = \tan 2\varphi,$$

which is written as

$$\frac{dy}{dx} = \frac{2xy}{x^2 - y^2}, \tag{B1.2}$$

where we used $\tan \varphi = y/x$. On the other hand, with the relation $R^2 = x^2 + y^2$, Eq. (B1.1) is written as

$$\frac{y}{x^2 + y^2} = c,$$

where c is a constant. This relation is rewritten as

$$x^2 + \left(y - \frac{1}{2c}\right)^2 = \frac{1}{4c^2}.$$

This exactly shows the electric field lines shown by the solid lines in Fig. 1.13. This leads to

$$\frac{dy}{dx} = -\frac{x}{y - 1/2c}.$$

It is easy to show that this is identical with Eq. (B1.2). Hence, Eq. (B1.1) describes the electric field lines.

1.28. The equipotential surface is given by Eq. (1.24):

$$\phi'(r, \theta) = \frac{\cos \theta}{r^2} = \text{const.}$$

1.7 Exercises

Here, we assume the electric field lines of the form:

$$\psi(r,\theta) = \frac{\sin\theta}{r^n} = \text{const.}$$

The exponent n is determined from the orthogonal condition:

$$\nabla\phi' \cdot \nabla\psi = 0.$$

Using

$$\nabla\phi' = \frac{\partial\phi'}{\partial r}\mathbf{i}_r + \frac{1}{r}\cdot\frac{\partial\phi'}{\partial\theta}\mathbf{i}_\theta = -\frac{2\cos\theta}{r^3}\mathbf{i}_r - \frac{\sin\theta}{r^3}\mathbf{i}_\theta,$$

$$\nabla\psi = \frac{\partial\psi}{\partial r}\mathbf{i}_r + \frac{1}{r}\cdot\frac{\partial\psi}{\partial\theta}\mathbf{i}_\theta = -\frac{n\sin\theta}{r^{n+1}}\mathbf{i}_r + \frac{\cos\theta}{r^{n+1}}\mathbf{i}_\theta,$$

the above condition leads to

$$\frac{(2n-1)\sin\theta\cos\theta}{r^{n+4}} = 0.$$

Thus, we have $n = 1/2$ and the electric field lines are given by

$$\psi(r,\theta) = \frac{\sin\theta}{r^{1/2}} = \text{const.}$$

On the plane $y = 0$, this leads to

$$(x^2 + z^2)^{3/2} = cx^2,$$

with c denoting a constant. The electric field lines pass through the origin and are symmetric with respect to $z = 0$, as shown by solid lines in Fig. 1.11.

1.29. Using Eq. (15) in the Appendix, we have

$$\Delta\phi = \frac{1}{R}\cdot\frac{\partial}{\partial R}\left(R\frac{\partial\phi}{\partial R}\right) + \frac{1}{R^2}\cdot\frac{\partial^2\phi}{\partial\varphi^2}$$

$$= \frac{\hat{p}}{2\pi\epsilon_0}\cdot\frac{\cos\varphi}{R^3} - \frac{\hat{p}}{2\pi\epsilon_0}\cdot\frac{\cos\varphi}{R^3} = 0.$$

Thus, Laplace's equation holds.

1.30. We represent the position of the observation point a as (r, θ, φ) using spherical coordinates. In Cartesian coordinates the position is rewritten as

$$(r\sin\theta\cos\varphi, r\sin\theta\sin\varphi, r\cos\theta)$$

We represent for example the distance of point A from point P as r_P. Then, the electric potential at A is given by

$$\phi = \frac{q}{4\pi\epsilon_0}\left(\frac{1}{r_P} - \frac{1}{r_Q} + \frac{1}{r_R} - \frac{1}{r_S}\right).$$

The distance r_P is calculated as

$$r_P = \left[\left(r\sin\theta\cos\varphi - \frac{d}{2}\right)^2 + \left(r\sin\theta\sin\varphi - \frac{d}{2}\right)^2 + r^2\cos^2\theta\right]^{1/2}$$

$$= \left[r^2 - dr\sin\theta(\sin\varphi + \cos\varphi) + \frac{d^2}{2}\right]^{1/2}.$$

If we use approximation of $(1+\delta)^{-1/2} \simeq 1 - \delta/2 + 3\delta^2/8$ for small quantity δ, we have

$$\frac{1}{r_P} \simeq \frac{1}{r} + \frac{d}{2r^2}\sin\theta(\sin\varphi + \cos\varphi) - \frac{d^2}{4r^3} + \frac{3d^2}{8r^3}\sin^2\theta(\sin\varphi + \cos\varphi)^2$$

under the condition $r \gg d$. Similar calculation is done for $1/r_Q$, etc., and finally we have

$$\phi = \frac{3d^2 q}{8\pi\epsilon_0 r^3}\sin^2\theta\sin 2\varphi.$$

1.31. We use spherical coordinates with the origin at the middle electric charge $2q$. The position of observation point A is denoted by (r, θ, φ). The distances from the point charges at $z = d$ and $z = -d$ to point A are denoted by r_+ and r_-, respectively, the electric potential is given by

$$\phi = \frac{q}{4\pi\epsilon_0}\left(\frac{2}{r} - \frac{1}{r_+} - \frac{1}{r_-}\right)$$

with

$$\frac{1}{r_\pm} = [r^2\sin^2\theta + (r\cos\theta \mp d)^2]^{-1/2} = [r^2 \mp 2rd\cos\theta + d^2]^{-1/2}$$

$$\simeq \frac{1}{r} \pm \frac{d}{r^2}\cos\theta - \frac{d^2}{2r^3} + \frac{3d^2}{2r^3}\cos^2\theta.$$

Thus, the electric potential is given by

$$\phi = \frac{d^2 q}{4\pi\epsilon_0 r^3}(1 - 3\cos^2\theta).$$

1.7 Exercises

1.32. The electric potential of the electric quadrupole is given by

$$\phi = \frac{3d^2q}{8\pi \epsilon_0 r^3} \sin^2 \theta \sin 2\varphi.$$

Each component of the second derivative is:

$$\left(\nabla^2 \phi\right)_r = \frac{1}{r^2} \cdot \frac{\partial}{\partial r}\left(r^2 \frac{\partial \phi}{\partial r}\right) = \frac{9d^2q}{4\pi \epsilon_0 r^5} \sin^2 \theta \sin 2\varphi,$$

$$\left(\nabla^2 \phi\right)_\theta = \frac{1}{r^2 \sin \theta} \cdot \frac{\partial}{\partial \theta}\left(\sin \theta \frac{\partial \phi}{\partial \theta}\right) = \frac{3d^2q}{4\pi \epsilon_0 r^5}\left(2 - 3\sin^2 \theta\right) \sin 2\varphi,$$

$$\left(\nabla^2 \phi\right)_\varphi = \frac{1}{r^2 \sin^2 \theta} \cdot \frac{\partial^2}{\partial \varphi^2} = -\frac{3d^2q}{2\pi r^3} \sin 2\varphi.$$

Thus, we have

$$\nabla^2 \phi = \frac{3d^2q}{4\pi \epsilon_0 r^5}\left(3\sin^2 \theta + 2 - 3\sin^2 \theta - 2\right) \sin 2\varphi = 0,$$

and Laplace's equation is satisfied.

1.33. The electric potential of the electric quadrupole is given by

$$\phi = \frac{d^2q}{4\pi \epsilon_0 r^3}\left(1 - 3\cos^2 \theta\right).$$

Each component of the second derivative is:

$$\left(\nabla^2 \phi\right)_r = \frac{1}{r^2} \cdot \frac{\partial}{\partial r}\left(r^2 \frac{\partial \phi}{\partial r}\right) = \frac{3d^2q}{2\pi \epsilon_0 r^5}\left(1 - 3\cos^2 \theta\right),$$

$$\left(\nabla^2 \phi\right)_\theta = \frac{1}{r^2 \sin \theta} \cdot \frac{\partial}{\partial \theta}\left(\sin \theta \frac{\partial \phi}{\partial \theta}\right) = \frac{3d^2q}{2\pi \epsilon_0 r^5}\left(3\cos^2 \theta - 1\right).$$

Thus, we have

$$\nabla^2 \phi = 0,$$

and Laplace's equation is satisfied.

Chapter 2
Conductors

2.1 Electric Properties of Conductors

A conductor is a material in which the electric field is zero in a steady state:

$$\mathbf{E} = 0. \tag{2.1}$$

From Eq. (1.12) we have

$$\rho = 0 \tag{2.2}$$

in the conductor. So, electric charge stays only on the surface of the conductor. Equation (2.1) shows that the conductor is equipotential:

$$\phi = \text{const.} \tag{2.3}$$

Hence, the electric field is normal to the conductor surface, and there is a relationship between the electric field strength and the surface density σ of the electric charge:

$$E = \frac{\sigma}{\epsilon_0}. \tag{2.4}$$

Example 2.1 Suppose a pair of concentric spherical conductors, as shown in Fig. 2.1. Determine the electric field strength and electric potential when the electric charges Q_1 and Q_2 are given to the inner and outer conductors, respectively.

Fig. 2.1 Isolated concentric spherical conductors

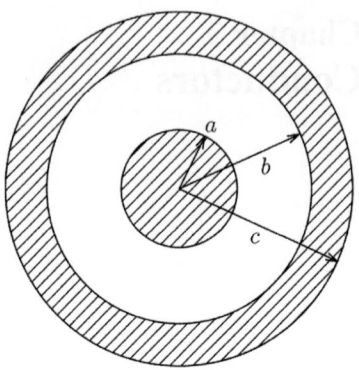

Solution 2.1 Electric charge Q_1 is distributed uniformly on the surface of the inner sphere ($r = a$) and the electric charge $-Q_1$ induced by the electrostatic induction is distributed uniformly on the inner surface of the outer sphere ($r = b$). Thus, the electric charge $Q_1 + Q_2$ appears on the outer surface of the outer sphere ($r = c$), following the principle of conservation of electric charge. The electric field is directed radially, and its strength is

$$\begin{aligned}
E &= 0; & 0 \leq r < a, \\
&= \frac{Q_1}{4\pi\epsilon_0 r^2}; & a < r < b, \\
&= 0; & b < r < c, \\
&= \frac{Q_1 + Q_2}{4\pi\epsilon_0 r^2}; & r < c.
\end{aligned}$$

Then, we obtain the electric potential as

$$\begin{aligned}
\phi &= \frac{Q_1 + Q_2}{4\pi\epsilon_0 r}; & r > c, \\
&= \frac{Q_1 + Q_2}{4\pi\epsilon_0 c}; & b < r < c, \\
&= \frac{Q_1}{4\pi\epsilon_0}\left(\frac{1}{r} - \frac{1}{b} + \frac{1}{c}\right) + \frac{Q_2}{4\pi\epsilon_0 c}; & a < r < b, \\
&= \frac{Q_1}{4\pi\epsilon_0}\left(\frac{1}{a} - \frac{1}{b} + \frac{1}{c}\right) + \frac{Q_2}{4\pi\epsilon_0 c}; & 0 \leq r < a.
\end{aligned}$$

◆

2.2 Special Solution Method for Electrostatic Field

Suppose that an electric point charge, q, is put at a position at distance a from a flat infinite conductor surface, as shown in Fig. 2.2a. An electric charge with the opposite sign appears on the conductor surface because of the electrostatic induction and exerts an attractive force on q. The x-y plane is defined as the conductor surface, with the origin, O, at the foot of a perpendicular line from the electric charge. The electric potential is constant on the conductor surface ($z = 0$). Such an electric potential can be realized in the following way: The conductor is virtually removed, and then an electric charge, $-q$, is put at the point $(0, 0, -a)$, which is symmetric to the location of q with respect to the conductor surface, as shown in Fig. 2.2b. We now check the validity of this speculation.

The electric potential that the two electric charges produce outside the conductor ($z > 0$) is

$$\phi(x, y, z) = \frac{1}{4\pi\epsilon_0} \left\{ \frac{q}{\left[x^2 + y^2 + (z-a)^2\right]^{1/2}} - \frac{q}{\left[x^2 + y^2 + (z+a)^2\right]^{1/2}} \right\}. \quad (2.5)$$

It is easily found that this satisfies the condition, $\phi = 0$, on the conductor surface ($z = 0$). Since this satisfies Laplace's equation outside the conductor and the boundary condition of Eq. (2.3) on the conductor surface, this is the solution. This shows that the above intuitive method is useful. In the conductor ($z < 0$), the electric potential is not given by Eq. (2.5) but by $\phi = 0$. This solution method is called the method of images, and the virtual electric charge is called an image charge.

From Eq. (2.5) we obtain the electric field strength outside the conductor as

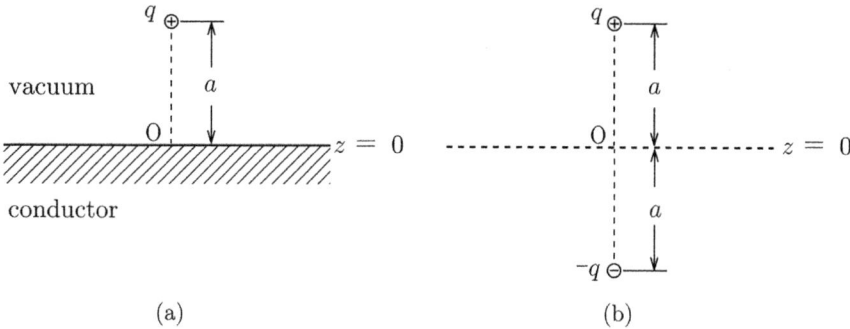

Fig. 2.2 **a** Point charge, q, at distance a from a wide flat conductor surface and **b** image charge, $-q$, put at the symmetric point of the given charge with respect to the conductor surface

$$E_x = -\frac{q}{4\pi\epsilon_0}\left\{\frac{x}{[x^2+y^2+(z-a)^2]^{1/2}} - \frac{x}{[x^2+y^2+(z+a)^2]^{1/2}}\right\},$$

$$E_y = -\frac{q}{4\pi\epsilon_0}\left\{\frac{y}{[x^2+y^2+(z-a)^2]^{1/2}} - \frac{y}{[x^2+y^2+(z+a)^2]^{1/2}}\right\},$$

$$E_z = -\frac{q}{4\pi\epsilon_0}\left\{\frac{z-a}{[x^2+y^2+(z-a)^2]^{1/2}} - \frac{z+a}{[x^2+y^2+(z+a)^2]^{1/2}}\right\}. \quad (2.6)$$

On the conductor surface this reduces to

$$E_x(x,y,0) = E_y(x,y,0) = 0, \quad E_z(x,y,0) = \frac{qa}{2\pi\epsilon_0(x^2+y^2+a^2)^{3/2}}. \quad (2.7)$$

Figure 2.3 shows the electric field lines. Then, from Eq. (2.4), we obtain the density of electric charge induced on the conductor surface as.

$$\sigma = -\frac{qa}{2\pi(x^2+y^2+a^2)^{3/2}}. \quad (2.8)$$

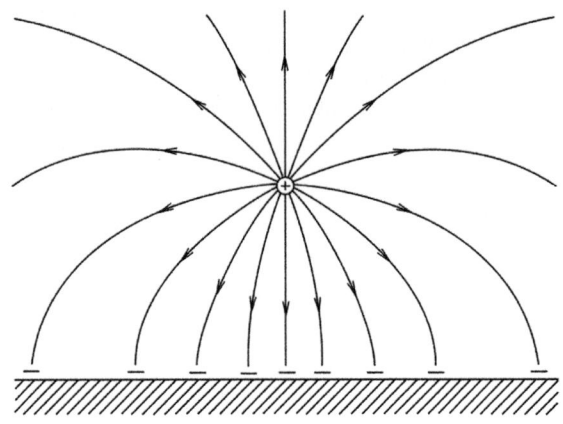

Fig. 2.3 Electric field lines between a point charge and electric charges induced on the conductor surface

Example 2.2 A point charge Q is placed at a point at distances a and b from the two flat conductor surfaces that are perpendicular to each other, as shown in Fig. 2.4. Determine the electric potential and electric field strength in the vacuum.

2.2 Special Solution Method for Electrostatic Field

Fig. 2.4 Two perpendicular flat conductor surfaces and a point charge

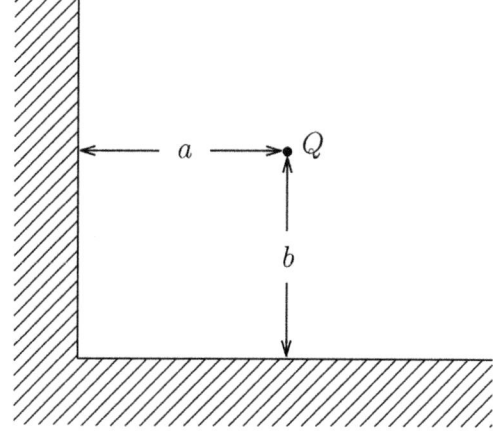

Solution 2.2 We denote two conductor surfaces that are perpendicular to each other by the *x-y* and *y-z* planes, as shown in Fig. 2.5. Assume the given electric charge is located on the plane $y = 0$. We virtually remove the conductor and place three electric charges, $-Q$, $-Q$, and Q, at $(a, 0, -b)$, $(-a, 0, b)$, and $(-a, 0, -b)$, respectively. Then, the electric potential in the vacuum region ($x > 0, z > 0$) is

$$\phi = \frac{Q}{4\pi\epsilon_0}\left\{\frac{1}{\left[(x-a)^2+y^2+(z-b)^2\right]^{1/2}} - \frac{1}{\left[(x-a)^2+y^2+(z+b)^2\right]^{1/2}} \right. \\ \left. - \frac{1}{\left[(x+a)^2+y^2+(z-b)^2\right]^{1/2}} + \frac{1}{\left[(x+a)^2+y^2+(z+b)^2\right]^{1/2}}\right\}.$$

This satisfies $\phi = 0$ on the surfaces $x = 0$ and $z = 0$, and hence, this gives the correct electric potential. We determine the electric charge density on the *x-y* and *y-z* planes to be

Fig. 2.5 True electric charge Q and three image charges

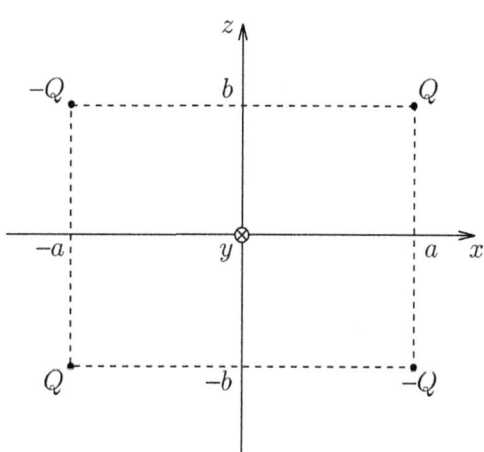

$$\sigma(x, y, 0) = -\epsilon_0 \left(\frac{\partial \phi}{\partial z}\right)_{z=0}$$

$$= -\frac{Qb}{2\pi} \left\{ \frac{1}{\left[(x-a)^2 + y^2 + b^2\right]^{3/2}} - \frac{1}{\left[(x-a)^2 + y^2 + b^2\right]^{3/2}} \right\},$$

$$\sigma(0, y, z) = -\epsilon_0 \left(\frac{\partial \phi}{\partial x}\right)_{x=0}$$

$$= -\frac{Qa}{2\pi} \left\{ \frac{1}{\left[y^2 + (z-b)^2 + a^2\right]^{3/2}} - \frac{1}{\left[y^2 + (z+b)^2 + a^2\right]^{3/2}} \right\}.$$

◆

Suppose that a point charge, q, is placed at point A at distance d from the center, O, of a grounded spherical conductor of radius a ($d > a$), as illustrated in Fig. 2.6a. Now, we determine the electric potential outside the spherical conductor. Assume that the conductor is removed and that an image charge, Q, is placed at point B at a distance h from the center, as shown in Fig. 2.6b. The quantities Q and h are unknown and need to be determined. Then, the electric potential on the conductor surface is

$$\phi = \frac{1}{4\pi \epsilon_0} \left[\frac{q}{(a^2 + d^2 - 2ad \cos\theta)^{1/2}} + \frac{Q}{(a^2 + h^2 - 2ah \cos\theta)^{1/2}} \right]$$

$$= \frac{1}{4\pi \epsilon_0} \left\{ \frac{q/(a^2 + d^2)^{1/2}}{[1 - 2ad \cos\theta/(a^2 + d^2)]^{1/2}} + \frac{Q/(a^2 + h^2)^{1/2}}{[1 - 2ah \cos\theta/(a^2 + h^2)]^{1/2}} \right\}, \quad (2.9)$$

where the angle θ is defined in Fig. 2.6b. Hence, $\phi = 0$ is realized at any point on the conductor surface ($r = a$), and the boundary condition is satisfied, if the following conditions are fulfilled:

$$\frac{q}{(a^2 + d^2)^{1/2}} + \frac{Q}{(a^2 + h^2)^{1/2}} = 0, \quad (2.10)$$

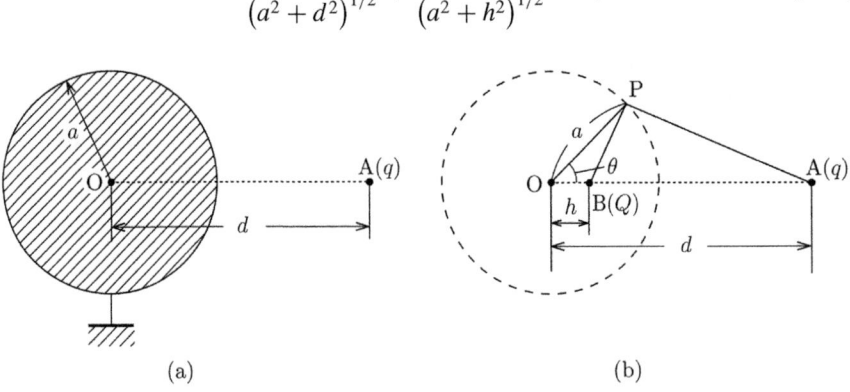

(a) (b)

Fig. 2.6 **a** Grounded spherical conductor and point charge at point A and **b** image charge at point B after removal of the conductor

2.2 Special Solution Method for Electrostatic Field

and

$$\frac{2ad}{a^2 + d^2} = \frac{2ah}{a^2 + h^2}, \tag{2.11}$$

which reduce to

$$h = \frac{a^2}{d}, \quad Q = -\frac{aq}{d}. \tag{2.12}$$

Thus, the electric potential at point (r, θ) outside the conductor is given by

$$\phi(r, \theta) = \frac{q}{4\pi\epsilon_0} \left\{ \frac{1}{(r^2 + d^2 - 2rd\cos\theta)^{1/2}} - \frac{a}{d\left[r^2 + (a^2/d)^2 - 2(a^2 r/d)\cos\theta\right]^{1/2}} \right\}. \tag{2.13}$$

Figure 2.7 shows the electric field calculated using the electric potential. The density of electric charge on the conductor surface is obtained as

$$\sigma(\theta) = \epsilon_0 E_r(r = a) = -\frac{q(d^2 - a^2)}{4\pi (a^2 + d^2 - 2ad\cos\theta)^{3/2}}. \tag{2.14}$$

Fig. 2.7 Electric field lines between the point electric charge and grounded spherical conductor

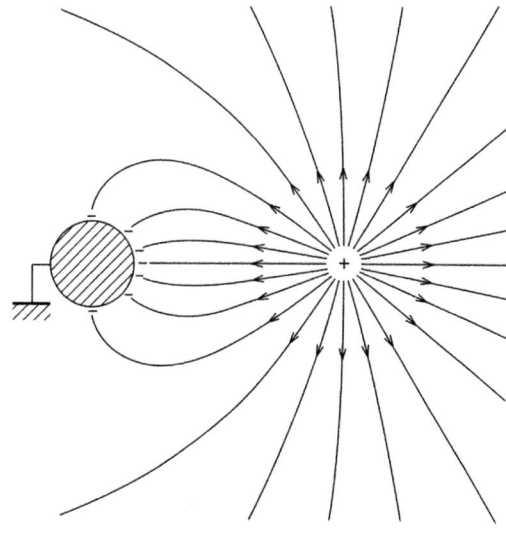

Example 2.3 Suppose that the conductor is not grounded in the problem shown in Fig. 2.6a. Determine the electric potential outside the conductor.

Solution 2.3 In this case, the total electris charge on the conductor surface is zero. This problem is solved using the method of superposition. That is, this situation is obtained by a superposition of the electric charge $-aq/d$, which is distributed according to Eq. (2.14), and the charge aq/d, which is uniformly distributed on the surface. The distributed charge $-aq/d$ and point charge q give the zero electric potential of the conductor, and the distributed charge aq/d makes the conductor equipotential, $q/(4\pi\epsilon_0 d)$. Hence, this situation satisfies the conductor condition. If the electric potential given by Eq. (2.13) is denoted by $\phi_1(r, \theta)$, the electric potential outside the conductor is

$$\phi(r, \theta) = \phi_1(r, \theta) + \frac{qa}{4\pi\epsilon_0 dr}.$$

◆

Example 2.4 An electric charge Q is placed at a point at a distance h from the center, O, of a hollow spherical conductor, as shown in Fig. 2.8. Determine the electric potential in the hollow interior and the electric charge density on the inner surface of the conductor.

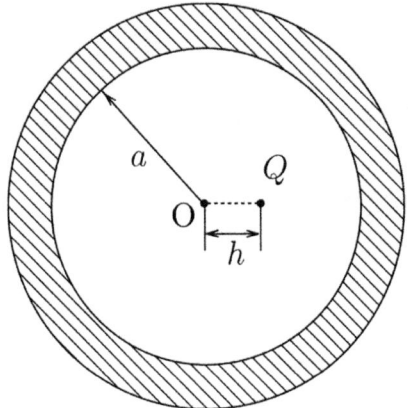

Fig. 2.8 Hollow spherical conductor and an electric charge Q at a position inside the conductor

Solution 2.4 We remove the conductor and place an image electric charge, q, on a line extending from the center to the electric charge Q (see Fig. 2.9). We denote the distance of this point from the center by d. The electric potential at point P on the inner surface of the conductor is

2.2 Special Solution Method for Electrostatic Field

Fig. 2.9 Image electric charge q after removal of the conductor

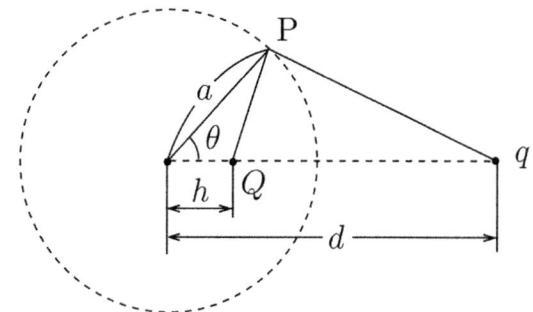

$$\phi(a, \theta) = \frac{1}{4\pi\epsilon_0} \left[\frac{q}{(a^2 + d^2 - 2ad\cos\theta)^{1/2}} - \frac{Q}{(a^2 + h^2 - 2ah\cos\theta)^{1/2}} \right].$$

The conditions that satisfy $\phi(a, \theta) = 0$ are

$$d = \frac{a^2}{h}, \quad q = -\frac{dQ}{a} = -\frac{aQ}{h}.$$

Thus, the electric potential in the hollow space is

$$\phi(r, \theta) = \frac{Q}{4\pi\epsilon_0} \left\{ -\frac{a}{h\left[r^2 + (a^2/h)^2 - 2(a^2 r/h)\cos\theta\right]^{1/2}} - \frac{1}{(r^2 + h^2 - 2rh\cos\theta)^{1/2}} \right\}.$$

The electric charge density on the inner surface is

$$\sigma(\theta) = -\epsilon_0 E_r(r = a) = \epsilon_0 \left(\frac{\partial \phi}{\partial r} \right)_{r=a} = -\frac{Q(a^2 - h^2)}{4\pi a(a^2 + h^2 - 2ah\cos\theta)^{3/2}}.$$

◆

***

Example 2.5 A long electric line charge of uniform density λ is placed at distance d from the central axis of a grounded parallel long cylindrical conductor of radius $a(< d)$. Determine the electric charge induced on the conductor surface.

Solution 2.5 We virtually remove the conductor and place an electric line charge of density λ' at the line located at distance h from the center O, as shown in Fig. 2.10. The electric potential at point P on the surface of the cylindrical conductor is

$$\phi = \frac{\lambda}{2\pi\epsilon_0} \log \frac{R_0}{(a^2 + d^2 - 2ad\cos\varphi)^{1/2}} + \frac{\lambda'}{2\pi\epsilon_0} \log \frac{R'_0}{(a^2 + d^2 - 2ad\cos\varphi)^{1/2}},$$

where φ is the angle defined in Fig. 2.10, and R_0 and R'_0 are the distances from O to reference points of the electric potential. So that the electric potential does not depend on φ, the conditions $\lambda' = -\lambda$ and $h = a^2/d$ must be fulfilled. In addition, $R'_0 = (a/d)R_0$ is required so that the electric potential of the conductor is zero. Thus, the electric potential outside the conductor is

$$\phi(R, \varphi) = \frac{\lambda}{2\pi\epsilon_0} \log \frac{d\left[R^2 + (a^2/d)^2 - 2(a^2 R/d)\cos\varphi\right]^{1/2}}{a(R^2 + d^2 - 2Rd\cos\varphi)^{1/2}}.$$

The electric charge density on the conductor surface is

$$\sigma = -\epsilon_0 \left(\frac{\partial \phi}{\partial R}\right)_{R=a} = -\frac{\lambda(d^2 - a^2)}{2\pi a(a^2 + d^2 - 2ad\cos\varphi)}.$$

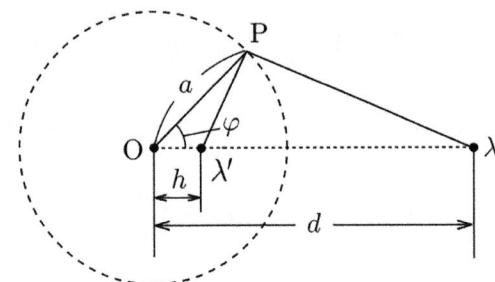

Fig. 2.10 Image of electric line charge

♦

Example 2.6 A long cylindrical conductor of radius a is placed at distance $l(> a)$ from an infinite flat conductor surface, as shown in Fig. 2.11, and an electric charge of linear density λ is given to the cylindrical conductor. Determine the density of electric charge on the surfaces of the two conductors.

2.2 Special Solution Method for Electrostatic Field

Fig. 2.11 Cylindrical conductor parallel to an infinite flat conductor surface

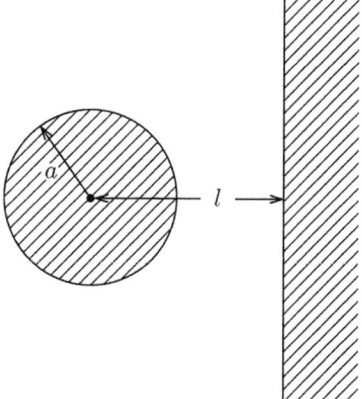

Solution 2.6 We assume that the electric field produced by the electric charge on the cylindrical conductor surface is the same as that produced by a line charge of density λ placed at a distance h from the center of the cylinder after virtually removing the cylinder (see Fig. 2.12). If we place an image line charge of density $-\lambda$ in the infinite conductor at distance $l - h$ from its surface after virtually removing the infinite conductor, the infinite conductor surface is equipotential. Hence, if the distance $2l - h$ between the image charge $-\lambda$ and the cylinder center corresponds to d in Example 2.5, the cylindrical conductor surface is also equipotential, and all the required conditions are satisfied. The result in Example 2.6 gives $h = a^2/d$. From the above conditions we have

$$d = l + \sqrt{l^2 - a^2}, \quad h = l - \sqrt{l^2 - a^2}.$$

Substituting these into the result in Example 2.5 yields the electric potential outside the conductors:

Fig. 2.12 Image line charges placed in two conductors

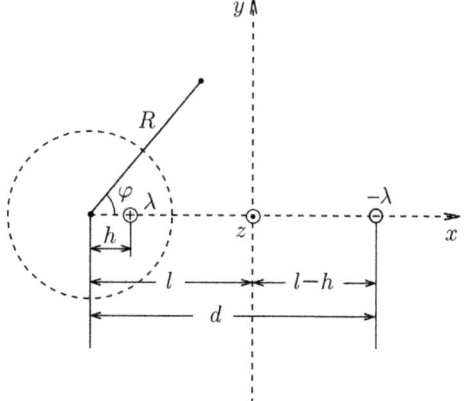

$$\phi(R, \varphi) = \frac{\lambda}{4\pi\epsilon_0} \log \frac{R^2 + \left(l + \sqrt{l^2 - a^2}\right)^2 - 2R\left(l + \sqrt{l^2 - a^2}\right)\cos\varphi}{R^2 + \left(l - \sqrt{l^2 - a^2}\right)^2 - 2R\left(l - \sqrt{l^2 - a^2}\right)\cos\varphi}.$$

We find that the electric potential on the surface, $\phi(a, \varphi)$, is constant. The electric charge density on the cylindrical surface is

$$\sigma = -\epsilon_0 \left(\frac{\partial \phi}{\partial R}\right)_{R=a} = \frac{\lambda\sqrt{l^2 - a^2}}{2\pi a(l - a\cos\varphi)}.$$

Next, we define Cartesian coordinates with the y-z plane ($x = 0$) on the infinite conductor surface and the central axis of the cylindrical conductor at $y = 0$. From the relationships $R\cos\varphi = x + l$ and $R\sin\varphi = y$, the electric potential is also expressed as

$$\phi(x, y) = \frac{\lambda}{4\pi\epsilon_0} \log \frac{\left(x - \sqrt{l^2 - a^2}\right)^2 + y^2}{\left(x + \sqrt{l^2 - a^2}\right)^2 + y^2}.$$

Thus, we can easily confirm that $\phi(x = 0) = 0$ is satisfied. The electric charge density on the infinite conductor surface is

$$\sigma = \epsilon_0 \left(\frac{\partial \phi}{\partial x}\right)_{x=0} = -\frac{\lambda\sqrt{l^2 - a^2}}{\pi\left(y^2 + l^2 - a^2\right)}.$$

It should be noted that the sign is opposite, since the normal vector on the conductor surface is directed along the negative x-axis.

◆

2.3 Electrostatic Induction

Suppose that a spherical conductor of radius a is placed in a uniform electric field of strength E_0 (see Fig. 2.13). An electric charge appears on the conductor surface and cancels out the electric field in the conductor. This phenomenon is the electrostatic induction. Here, we determine the surface electric charge density and the electric field around the conductor. We use spherical coordinates with the origin at the center of the conductor and the zenith along the direction of the applied electric field.

2.3 Electrostatic Induction

Fig. 2.13 Spherical conductor placed in a uniform electric field

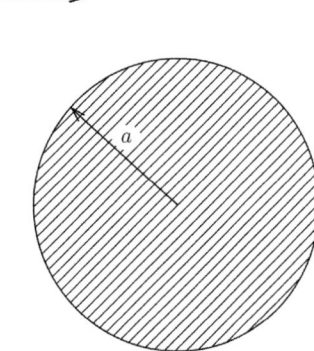

The effect of the surface electric charge can be realized by an electric dipole placed at the center of the conductor after virtually removing the spherical conductor. Then, the electric potential outside the conductor is composed of the electric potential, ϕ_f, due to the applied electric field, E_0, and the electric potential, ϕ_d, due to the electric dipole with a moment p placed at the center. These are given by

$$\phi_f = -E_0 r \cos\theta, \tag{2.15}$$

$$\phi_d = \frac{p \cos\theta}{4\pi \epsilon_0 r^2}, \tag{2.16}$$

where θ is the zenithal angle measured from the direction of the applied electric field. The electric potential is given by

$$\phi = \phi_f + \phi_d = \left(-E_0 r + \frac{p}{4\pi \epsilon_0 r^2}\right) \cos\theta. \tag{2.17}$$

From the requirement that $\phi(r = a) = 0$, we have

$$p = 4\pi \epsilon_0 a^3 E_0. \tag{2.18}$$

Thus, the electric potential outside the conductor is determine to be

$$\phi = -E_0 \left(r - \frac{a^3}{r^2}\right) \cos\theta. \tag{2.19}$$

The electric field strength outside the conductor is given by

$$E_r = -\frac{\partial \phi}{\partial r} = E_0\left(1 + \frac{2a^3}{r^3}\right)\cos\theta,$$

$$E_\theta = -\frac{1}{r}\cdot\frac{\partial \phi}{\partial \theta} = -E_0\left(1 - \frac{a^3}{r^3}\right)\sin\theta,$$

$$E_\varphi = -\frac{1}{r\sin\theta}\cdot\frac{\partial \phi}{\partial \varphi} = 0. \tag{2.20}$$

Then, the surface electric charge density is determined to be

$$\sigma = \epsilon_0 E_r(r = a) = 3\epsilon_0 E_0 \cos\theta. \tag{2.21}$$

Example 2.7 Derive Eqs. (2.18) and (2.21) for a conducting sphere in a uniform electric field using Eq. (2.1) with the boundary condition that the electric field is perpendicular to the conductor surface and its value is given by Eq. (2.4).

Solution 2.7 The radial and zenithal components of the applied electric field outside the spherical conductor are $E_0 \cos\theta$ and $-E_0 \sin\theta$, respectively. The radial and zenithal components due to the electric dipole moment p at a point at distance r from the origin are $p\cos\theta/(2\pi\epsilon_0 r^3)$ and $p\sin\theta/(4\pi\epsilon_0 r^3)$, respectively. The condition that the zenithal component of the electric field just outside the surface is zero is written as

$$-E_0 \sin\theta + \frac{p\sin\theta}{4\pi\epsilon_0 a^3} = 0,$$

which gives

$$p = 4\pi\epsilon_0 a^3 E_0.$$

The normal component of the electric field just outside the surface is equal to the surface electric charge density σ divided by ϵ_0. Thus, we have

$$\sigma = \epsilon_0\left(E_0 \cos\theta + \frac{p\cos\theta}{2\pi\epsilon_0 a^3}\right) = 3\epsilon_0 E_0 \cos\theta.$$

These results agree with Eqs. (2.18) and (2.21). ◆

Example 2.8 A long cylindrical conductor of radius a is placed in a uniform normal electric field of strength E_0. Determine the electric potential and electric field outside the conductor and the density of the electric charge on the conductor surface.

Solution 2.8 We use cylindrical coordinates and define the z-axis at the central axis of the conductor, and measure the azimuthal angle φ from the direction of the applied electric field. The electric potential outside the conductor can be determined by putting the electric dipole line on the central axis after removing the conductor similarly to what we did in the above analysis. We denote by $\hat{p}$ the moment of the electric dipole line in a unit length along the z-axis. Then, the electric potential outside the conductor is given by

$$\phi(R, \varphi) = \left(-E_0 R + \frac{\hat{p}}{2\pi \epsilon_0 R}\right) \cos \varphi.$$

The first and second terms are the electric potential due to the applied electric field and the electric dipole line, respectively. Hence, the requirement that $\phi = 0$ at $R = a$ leads to $\hat{p} = 2\pi \epsilon_0 a^2 E_0$. Then, the electric potential outside the conductor is determined to be

$$\phi(R, \varphi) = -E_0 \left(R - \frac{a^2}{R}\right) \cos \varphi.$$

Thus, we obtain the electric field strength as

$$E_R = E_0 \left(1 + \frac{a^2}{R^2}\right) \cos \varphi, \quad E_\varphi = -E_0 \left(1 - \frac{a^2}{R^2}\right) \sin \varphi, \quad E_z = 0.$$

The surface electric charge density is

$$\sigma = \epsilon_0 E_R(R = a) = 2\epsilon_0 E_0 \cos \varphi.$$

◆

2.4 Exercises

2.1. Electric charges Q and Q_x are given to the left and right slab conductors shown in Fig. E2.1. The electric field in the region $x > b$ is $Q/2\epsilon_0 S$ in the direction of the x-axis, where S is the area of each surface. Determine Q_x, the electric charge on each surface, and the electric field in each space.

Fig. E2.1 Two parallel slab conductors

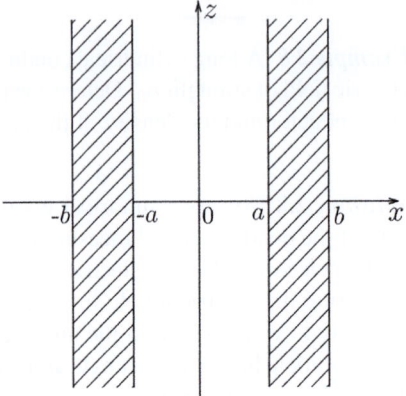

2.2. Electric charges $2Q$, $-Q$, and Q are given to the left, middle, and right slab conductors shown in Fig. E2.2 Determine the electric charge on each surface, and the electric field in each space. The area of each surface is S.

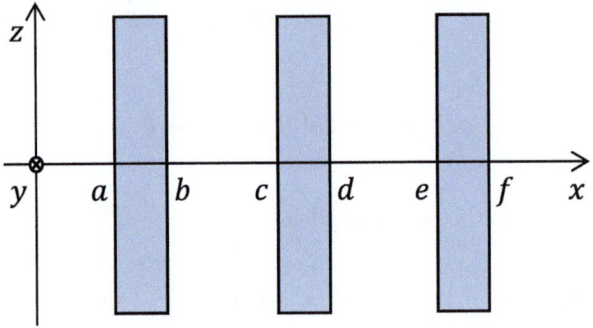

Fig. E.2 Three slab conductors

2.3. Electric charges $2Q$ and $-Q$ are given to the left and right slab conductors shown in Fig. E2.2 in Exercise 2.2. When we give some electric charge to the middle slab conductor, the electric field in the space $d < x < e$ is zero. Determine the electric charge given to the middle conductor.

2.4. Prove that Eq. (2.4) is fulfilled on the conductor surfaces in Example 2.1.

2.5. We give electric charges q_1 and q_3 to the inner and outer conductors 1 and 3, respectively, in the arrangement of concentric spherical conductors shown in Fig. E2.3. Determine the electric charge of the middle conductor 2 that causes the electric potential of the middle conductor to be zero.

2.4 Exercises

Fig. E2.3 Cross section of concentric spherical conductors

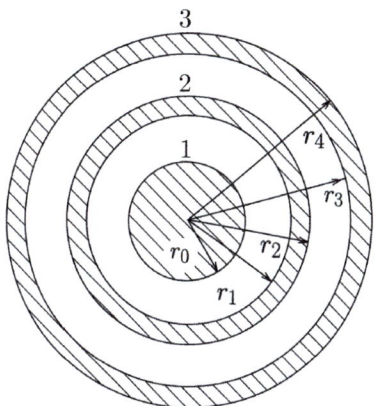

2.6. We suppose a pair of spherical concentric conductors 1 and 2 and a point conductor 3 outside the spherical conductors, as shown in Fig. E2.4. We give electric charge Q to the inner conductor 1. Determine the electric charge to give to conductor 3 that causes the electric potential of conductor 1 to be zero.

Fig. E2.4 Spherical concentric conductors 1 and 2 and a point conductor 3

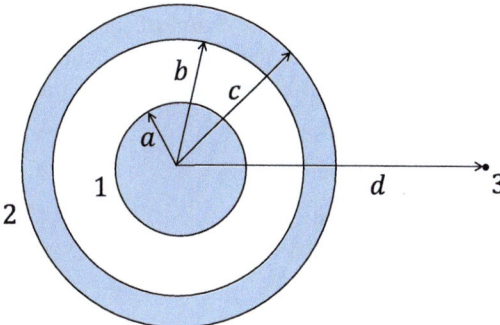

2.7. We suppose a pair of spherical concentric conductors 1 and 2 and a point conductor 3 outside the spherical conductors, as shown in Fig. E2.4. We give electric charges Q and q to conductors 1 and 3, respectively. Determine the electric charge to give to conductor 2 that causes the electric potential of conductor 2 to be zero.

2.8. The electric charge on a conductor surface was determined using the image method in Sect. 2.2. Prove that the interior of the conductor is completely shielded by the induced electric charge given by Eq. (2.8).

2.9. The image method is employed in solving the problem for two conductor surfaces at 90 degrees to each other, as shown in Example 2.2. Discuss the applicability of this method for other angles between the two surfaces.

2.10. We apply electric charge Q to a spherical conductor facing an infinitely wide conductor surface, as shown in Fig. E2.5. Is it possible to determine the electric potential and the charge distribution on the conductor surfaces?

Fig. E2.5 Spherical conductor and infinitely wide conductor surface

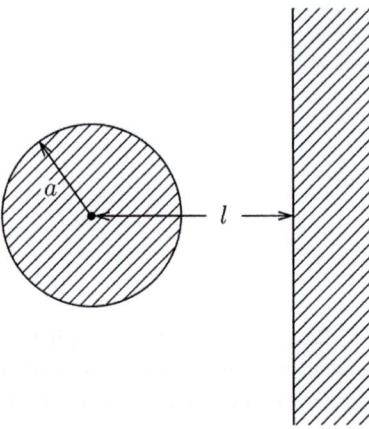

2.11. A straight electric line charge of density λ is placed at distances a and b from two flat conductor surfaces that are perpendicular to each other, as shown in Fig. E2.6. Determine the electric potential in vacuum and the electric charge density on the conductor surfaces.

Fig. E2.6 Two flat conductor surfaces and straight electric line charge λ

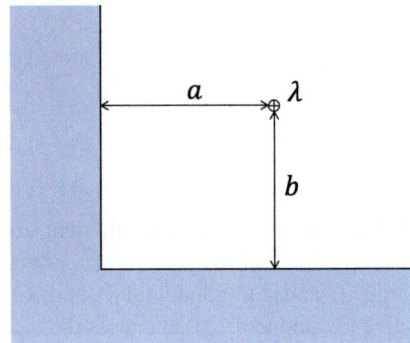

2.12. Suppose three flat conductor surfaces, which are the x-y, y-z, and z-x planes, as shown in Fig. E2.7. An electric charge Q is placed on a point $(x, y, z) = (a, b, c)$. Show the suitable image charges and prove the validity of these image charges.

2.13. Determine the force on the given electric charge exerted by the electric charge induced on the conductor surfaces in Example 2.2.

2.4 Exercises

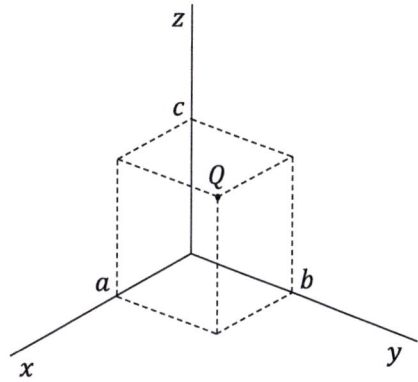

Fig. E2.7 Three conductor surfaces perpendicular to each other and a point charge

2.14. An electric line charge of density λ is placed at a distance h from the central axis O of a long hollow conductor with an internal radius a, as shown in Fig. E2.8. Determine the electric potential in the hollow interior and the electric charge density on the inner surface of the conductor.

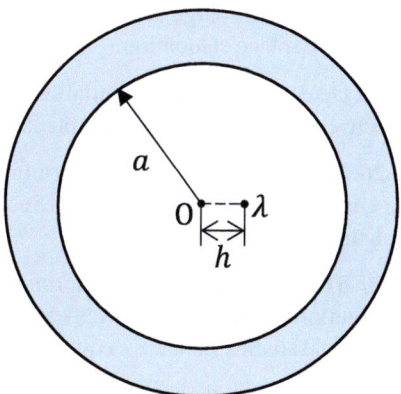

Fig. E2.8 Long conductor with a cylindrical hollow interior and an electric line charge placed at distance h from the central axis

2.15. Suppose that the long cylindrical conductor treated in Example 2.5 is not grounded. Determine the electric potential outside the conductor.

2.16. The electric potential is discussed for a long conducting hollow cylinder and a straight line charge inside the hollow interior in Exercise 2.14. Discuss the equipotential surface.

2.17. The electric condition is treated in Example 2.6 for a long cylindrical conductor with an electric charge and a flat conductor surface. Determine the equipotential surface in the space between the cylindrical conductor and the conductor surface.

2.18. Discuss the equipotential surface for the grounded cylindrical conductor and the electric line charge of density λ discussed in Example 2.5.

2.19. The electric potential in a vacuum region is given by Eq. (2.5), when a point charge is put above an infinite conductor surface. Prove that the electric potential satisfies Laplace's equation.

2.20. The electric potential in a vacuum region is given by Eq. (2.13), when a point charge is put near a grounded spherical conductor. Prove that the electric potential satisfies Laplace's equation.

2.21. The electric condition around a cylindrical conductor near a linear electric charge is discussed in Example 2.5. Prove that the electric potential satisfies Laplace's equation.

2.22. The electric condition due to an electric charge inside the spherical hollow interior of a conductor is treated in Example 2.4. Prove that the electric potential satisfies Laplace's equation.

2.23. The electric condition due to a linear electric charge inside the cylindrical hollow conductor is discussed in Exercise 2.14. Prove that the electric potential satisfies Laplace's equation.

2.24. The electric condition is treated in Example 2.6 for a long cylindrical conductor with an electric charge and a flat conductor surface. Prove that the electric potential satisfies Laplace's equation.

2.25. Determine the equipotential surface for the cylindrical conductor in a transverse electric field discussed in Example 2.8.

2.26. The electric potential is determined inside and outside of a spherical conductor in a uniform applied electric field in Sect. 2.3. Prove that the obtained electric potential satisfies Laplace's equation.

2.27. The electric potential is determined inside and outside the cylindrical conductor in a uniform transverse electric field in Example 2.8. Prove that the obtained electric potential satisfies Laplace's equation.

2.28. Determine the induced electric charge density on the inner surface of a conductor with a spherical hollow interior of radius a, when a dipole moment p is placed at the center of the hollow space.

2.29. In Exercise 2.28, the electric condition is investigated when an electric dipole moment p is placed at the center of the spherical hollow interior of a conductor. Prove that the electric potential satisfies Laplace's equation in the hollow interior.

2.30. Determine the induced electric charge density on the inner surface of a long conductor with a cylindrical hollow interior of radius a, when an electric dipole line of moment $\hat{p}$ in a unit length is placed at the central axis of the hollow interior.

2.31. In Exercise 2.30, the electric condition is investigated in the cylindrical hollow interior of a long conductor, when an electric dipole line of moment $\hat{p}$ in a unit length is placed at the central axis of the hollow interior. Prove that the electric potential satisfies Laplace's equation in the hollow space.

2.4 Exercises

2.32. The electric condition is treated in Exercise 2.30 when an electric dipole line of moment $\hat{p}$ in a unit length is placed at the central axis of the cylindrical hollow interior of a long conductor. Determine the equipotential surface in the hollow space.

2.33. The electric condition around a grounded spherical conductor near a point charge is discussed in Sect. 2.2. Solve the same problem using the boundary condition that the electric field is normal to the conductor surface.

2.34. The electric condition around a cylindrical conductor near a parallel straight line-charge is discussed in Example 2.5. Solve the same problem using the boundary condition that the electric field is normal to the conductor surface.

2.35. Electrostatic induction for a cylindrical conductor in a normal electric field is discussed in Example 2.8. Solve the same problem using the boundary condition that the electric field is normal to the conductor surface.

Answers to Exercises

2.1. From the electric field strength in $x > b$, the electric charge on the surface at $x = b$ is $Q/2$. Then, the electric charge at $x = a$ is $Q_x - Q/2$. So that the electric field is zero in the right conductor, the total electric charge in the region $x \leq a$ must be equal to $Q/2$, i.e., $Q_x + Q/2 = Q/2$, and we have $Q_x = 0$. So, the electric charge at $x = a$ is $-Q/2$. Since the total electric charge in the region $-a \leq x \leq a$ must be zero, the electric charge on the surface at $x = -a$ is $Q/2$, and the electric charge on the surface at $x = -b$ is $Q/2$.

The obtained electric field in each space is

$$E = -\frac{Q}{2\epsilon_0 S}; \quad x < -b,$$

$$= \frac{Q}{\epsilon_0 S}; \quad -a < x < a,$$

$$= \frac{Q}{2\epsilon_0 S}; \quad x > b.$$

2.2. We denote the electric charges on the surfaces at $x = a$, $x = b$, $x = c$, $x = d$, $x = e$, and $x = f$ by Q_a, Q_b, Q_c, Q_d, Q_e, and Q_f, respectively. Then, the condition that the electric field strength is zero in the left conductor is $Q_a = 2Q - Q_a$, which leads to $Q_a = Q$ and we have $Q_b = Q$. The condition for the middle conductor is $Q_c + 2Q = -Q_c$, which leads to $Q_c = -Q$ and we have $Q_d = 0$. The condition for the right conductor is $Q_e + Q = -Q_e + Q$, which leads to $Q_e = 0$ and we have $Q_f = Q$.

The electric field along the x-axis in each space is determined to be

$$E = -\frac{Q}{\epsilon_0 S}; \quad x < a,$$

$$= \frac{Q}{\epsilon_0 S}; \quad b < x < c,$$

$$= 0; \quad d < x < e,$$

$$= \frac{Q}{\epsilon_0 S}; \quad x > f.$$

2.3. We denote the electric charge given to the middle conductor by Q_x. From the condition that $E = 0$ in the space $d < x < e$, the electric charges on the surfaces at $x = d$ and $x = e$ are zero. So, the electric charges at $x = a$ and $x = f$ are $2Q + Q_x$ and $-Q$, respectively. From the total condition that the electric fields in the regions $x < a$ and $x > f$ are of the same magnitude with opposite signs to each other, the electric charges at $x = a$ and $x = f$ must be the same. So, we have $Q_x = -3Q$.

2.4. On the surface at $r = a$, the density of electric charge is $\sigma_a = Q_1/(4\pi a^2)$, and the electric field has strength $E(a) = Q_1/(4\pi \epsilon_0 a^2)$ and is directed outward. Thus, the relationship of Eq. (2.4), $E(a) = \sigma_a/\epsilon_0$, holds. On the surface at $r = c$, the density of electric charge and electric field strength are $\sigma_c = (Q_1 + Q_2)/(4\pi c^2)$ and $E(c) = (Q_1 + Q_2)/(4\pi \epsilon_0 c^2)$, respectively. So, the same relationship holds. On the surface at $r = b$, the density of electric charge is $\sigma_b = -Q_1/(4\pi b^2)$ and the electric field strength directed inward is $E(b) = Q_1/(4\pi \epsilon_0 b^2)$. Thus, Eq. (2.4) holds taking account of the direction of the electric field and the sign of the electric charge density.

2.5. The electric charge of the middle conductor is denoted by q_x. Then, the distribution of the electric charge is as follows:

$$\begin{aligned}
q_1; & \quad r = r_0, \\
-q_1; & \quad r = r_1, \\
q_1 + q_x; & \quad r = r_2, \\
-q_1 - q_x; & \quad r = r_3, \\
q_1 + q_x + q_3; & \quad r = r_4.
\end{aligned}$$

So, the electric field for $r > r_2$ is given by

$$E = \frac{q_1 + q_x + q_3}{4\pi \epsilon_0 r^2}; \quad r > r_4,$$

$$= \frac{q_1 + q_x}{4\pi \epsilon_0 r^2}; \quad r_2 < r < r_3,$$

and the electric potential of the middle conductor is given by

2.4 Exercises

$$\phi(r_2) = \frac{q_1 + q_x + q_3}{4\pi \epsilon_0 r_4} + \frac{q_1 + q_x}{4\pi \epsilon_0}\left(\frac{1}{r_2} - \frac{1}{r_3}\right).$$

So, from the condition $\phi(r_2) = 0$, we have

$$q_x = -q_1 - \frac{r_2 r_3}{r_2 r_3 + (r_3 - r_2) r_4} q_3.$$

2.6. The electric potential of conductor 1 produced by charge Q is

$$\phi_1' = \frac{Q}{4\pi \epsilon_0}\left(\frac{1}{a} - \frac{1}{b} + \frac{1}{c}\right).$$

The electric charge of conductor 3 is denoted by q. The electric potential of conductor 2 produced by this charge is given by $q/4\pi\epsilon_0 d$, as shown in Example 2.3. So, the same electric potential is produced for conductor 1:

$$\phi_1'' = \frac{q}{4\pi \epsilon_0 d},$$

Then, the electric potential of conductor 1 is

$$\phi_1 = \phi_1' + \phi_1'' = \frac{1}{4\pi \epsilon_0}\left[Q\left(\frac{1}{a} - \frac{1}{b} + \frac{1}{c}\right) + \frac{q}{d}\right].$$

From the condition $\phi_1 = 0$, the electric charge of conductor 3 is determined to be

$$q = -\left(\frac{1}{a} - \frac{1}{b} + \frac{1}{c}\right) dQ.$$

2.7. The electric charge of conductor 2 is denoted by Q_x. The contribution to the electric potential of conductor 2 from this charge and Q of conductor 1 is

$$\phi_2' = \frac{Q + Q_x}{4\pi \epsilon_0 c}.$$

As shown in Example 2.3 the electric potential of conductor 2 produced by charge q is given by

$$\phi_2'' = \frac{q}{4\pi \epsilon_0 d}.$$

Then, the electric potential of conductor 2 is

$$\phi_2 = \phi_2' + \phi_2'' = \frac{1}{4\pi\epsilon_0}\left(\frac{Q+Q_x}{c} + \frac{q}{d}\right).$$

From the condition $\phi_2 = 0$, the electric charge of conductor 2 is determined to be

$$Q_x = -Q - \frac{c}{d}q.$$

2.8. The electric charge induced on the conductor surface produces also the electric field inside the conductor. This electric field is symmetric with that in vacuum with respect to the surface. Hence, the electric field in the conductor produced by the induced electric charge is equal to that produced by the electric charge $-q$ placed at the position of q. Namely, the total electric field strength is equal to that produced by q and $-q$ at the same place, i.e., the electric field strength when no electric charge is given. Thus, the zero electric field strength in the conductor can be proved.

2.9. This method is useful only for special angles. This is the case in which the boundary conditions of the two surfaces are satisfied. If we say a conclusion, it is useful only when the angle between the two surfaces is π/n with n being an integer. That is, $n = 1$ in the problem in Sect. 2.2 and $n = 2$ in Example 2.2.

The case of $n = 3$ is shown in Fig. B2.1. The couple of the given charge $Q_1(Q)$ and the image charge $Q_2(-Q)$, that of $Q_3(Q)$ and $Q_6(-Q)$, and that of $Q_4(-Q)$ and $Q_5(Q)$ satisfy the boundary condition on the surface S_1. The couple of $Q_1(Q)$ and $Q_6(-Q)$, that of $Q_2(-Q)$ and $Q_5(Q)$, and that of $Q_3(Q)$ and $Q_4(-Q)$ satisfy the boundary condition on the surface S_2.

In the case of $n = 3/2$, it may seem to be possible. But there are only three points on which we can put image charges, and the image method cannot be applied.

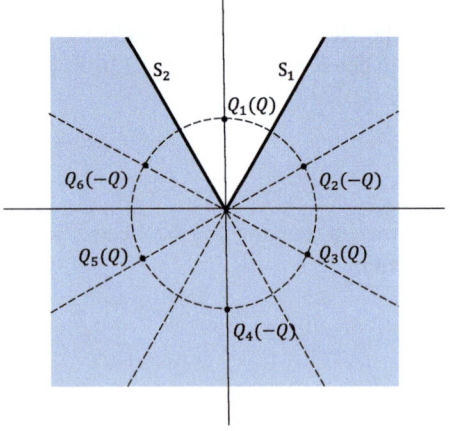

Fig. B2.1 Given charge $Q_1(Q)$ and other five image charges for the case of $n = 3$

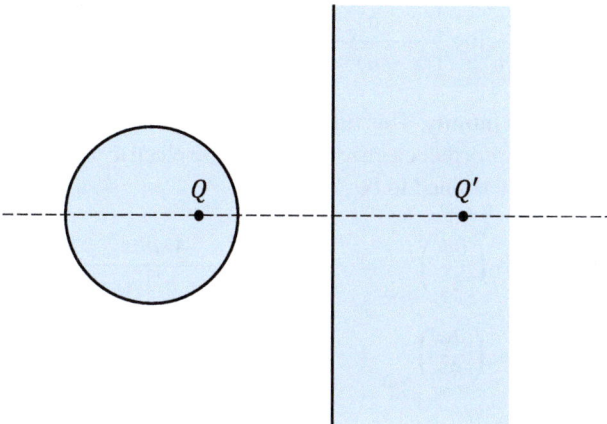

Fig. B2.2 Image point charges Q and Q'

2.10. It is expected to assume image point charges Q and Q' inside the spherical conductor and the wide superconductor, respectively, as shown in Fig. B2.2. These electric charges must be determined so that the boundary conditions of constant electric potentials are satisfied on the surfaces of the spherical conductor and wide superconductor. However, Q must be smaller than Q' so as to satisfy the boundary condition on the surface of the spherical conductor as shown in Fig. 2.7, while Q must be equal to Q' so as to satisfy the boundary condition on the wide superconductor surface as shown in Fig. 2.3. These are contradictory. So, this problem cannot be solved simply using the image method.

2.11. We denote two conductor surfaces that are perpendicular to each other by the x-y and y-z planes. Assume that the given linear electric charge is located at (a, b) on the x-z plane. We virtually remove the conductor and place three image line charges, $-\lambda, -\lambda$, and λ at $(a, -b), (-a, b)$, and $(-a, -b)$, respectively, as shown in Fig. B2.3. Then, the electric potential in the vacuum region ($x > 0, z > 0$) is

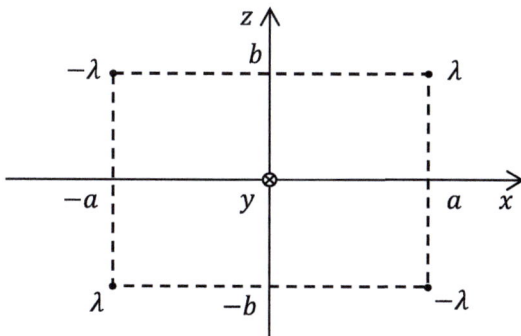

Fig. B2.3 Linear electric charge λ and three image charges

$$\phi(x,z) = \frac{\lambda}{4\pi\epsilon_0} \log \frac{[(x-a)^2 + (z+b)^2][(x+a)^2 + (z-b)^2]}{[(x-a)^2 + (z-b)^2][(x+a)^2 + (z+b)^2]},$$

which satisfies $\phi = 0$ at infinity. This fulfills $\phi = 0$ on the surfaces $x = 0$ and $z = 0$, and hence, this gives the correct electric potential. The electric charge density on the x-y and y-z planes is determined to be

$$\sigma(x, y, 0) = -\epsilon_0 \left(\frac{\partial \phi}{\partial z}\right)_{z=0} = -\frac{4\lambda abx}{\pi[(x-a)^2 + b^2][(x+a)^2 + b^2]},$$

$$\sigma(0, y, z) = -\epsilon_0 \left(\frac{\partial \phi}{\partial x}\right)_{x=0} = -\frac{4\lambda abz}{\pi[(z-b)^2 + a^2][(z+b)^2 + a^2]}.$$

2.12. First, we put three image charges $-Q$ at $(-a, b, c)$, $(a, -b, c)$, $(a, b, -c)$ so that the boundary condition of the normal electric field is satisfied on the y-z, z-x, and x-y planes, respectively. But this is not sufficient, since the boundary condition is not satisfied, and we need additional image charges. Next, we put additionally three image charges Q at $(-a, -b, c)$, $(a, -b, -c)$, and $(-a, b, -c)$, and another image charge $-Q$ at $(-a, -b, -c)$. As to the boundary condition on the x-y plane ($z = 0$), the couple of Q at (a, b, c) and $-Q$ at $(a, b, -c)$, that of $-Q$ at $(-a, b, c)$ and Q at $(-a, b, -c)$, that of $-Q$ at $(a, -b, c)$ and Q at $(a, -b, -c)$, and that of Q at $(-a, -b, c)$ and $-Q$ at $(-a, -b, -c)$ satisfy the condition. The boundary conditions on other planes are similarly obtained. So, the image method is useful.

2.13. We define Cartesian coordinates as in Example 2.2. The force on the given charge due to the image charge $-Q$ at $(x, z) = (-a, b)$ is

$$F_{-a,b} = \frac{Q^2}{16\pi\epsilon_0 a^2}$$

and is directed to the left (the negative x-axis). The force on the given charge due to the image charge $-Q$ at $(x, z) = (a, -b)$ is

$$F_{a,-b} = \frac{Q^2}{16\pi\epsilon_0 b^2}$$

and is directed to downward (the negative z-axis). The force on the given charge due to the image charge Q at $(x, z) = (-a, -b)$ is

$$F_{-a,-b} = \frac{Q^2}{16\pi\epsilon_0(a^2 + b^2)}$$

and is directed to the positive x- and z-axes along the direction connecting the given and image charges. So, the x- and z-axis components of the total force are

$$F_x = -\frac{Q^2}{16\pi\epsilon_0}\left[\frac{1}{a^2} - \frac{a}{(a^2+b^2)^{3/2}}\right], \quad F_z = -\frac{Q^2}{16\pi\epsilon_0}\left[\frac{1}{b^2} - \frac{b}{(a^2+b^2)^{3/2}}\right].$$

2.14. We remove the conductor and place an image electric line charge of density λ' on a line extending from the center to the electric line charge λ, as shown in Fig. B2.4. We denote the distance of this line charge from the central axis by d. The electric potential at point P on the inner surface of the conductor is

$$\phi(a,\varphi) = \frac{\lambda}{2\pi\epsilon_0}\log\frac{R_0}{(a^2+h^2-2ah\cos\varphi)^{1/2}}$$
$$+ \frac{\lambda'}{2\pi\epsilon_0}\log\frac{R'_0}{(a^2+d^2-2ad\cos\varphi)^{1/2}},$$

where φ is the angle defined in Fig. B2.4, and R_0 and R'_0 are distances from O to the reference points of the electric potential. So that the electric potential does not depend on φ, the conditions $\lambda' = -\lambda$ and $d = a^2/h$ must be fulfilled. In addition, $R'_0 = (a/h)R_0$ is required so that the electric potential of the conductor is zero. Thus, the electric potential in the hollow space is

$$\phi(R,\varphi) = \frac{\lambda}{2\pi\epsilon_0}\log\frac{\left[R^2 + (a^2/h)^2 - (2a^2R/h)\cos\varphi\right]^{1/2}}{(R^2+h^2-2hR\cos\varphi)^{1/2}}.$$

The electric charge density on the inner surface is

$$\sigma = \epsilon_0\left(\frac{\partial\phi}{\partial R}\right)_{R=a} = -\frac{\lambda(a^2-h^2)}{2\pi a(a^2+h^2-2ah\cos\varphi)}.$$

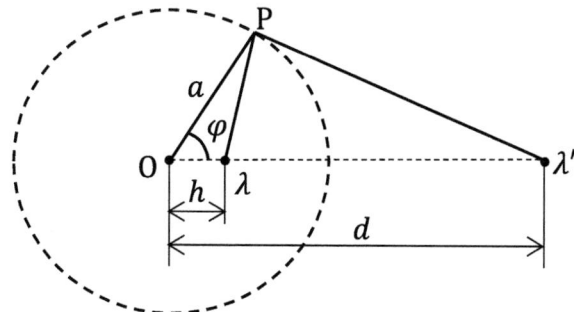

Fig. B2.4 Image line charge λ' outside the hollow space

Using Eq. (23) in the Appendix, we obtain the total electric charge in a unit length as

$$\int_0^{2\pi} \sigma a\, d\varphi = -\frac{\lambda(a^2 - h^2)}{\pi} \int_0^{\pi} \frac{d\varphi}{a^2 + h^2 - 2ah \cos\varphi} = -\lambda.$$

2.15. The total electric charge on the conductor surface must be zero under the condition that it shields the interior of the conductor. This situation can be realized by superposing two charge distributions: One is the distribution determined in Example 2.5, which shields the electric field produced by the electric line charge, and the other is a uniform distribution that does not produce electric field inside the conductor. The latter electric charge must be λ from the above condition. So, the electric potential outside the cylindrical conductor is

$$\phi(R, \varphi) = \phi'(R, \varphi) + \frac{\lambda}{2\pi \epsilon_0} \log \frac{R_0}{R},$$

where $\phi'(R, \varphi)$ is the electric potential obtained in Example 2.5 and R_0 is the distance to the reference point with zero electric potential.

2.16. The electric potential in the hollow space is given by

$$\phi(R, \varphi) = \frac{\lambda}{2\pi \epsilon_0} \log \frac{\left[R^2 + (a^2/h)^2 - 2(a^2 R/h) \cos \varphi\right]^{1/2}}{(R^2 + h^2 - 2Rh \cos \varphi)^{1/2}},$$

where R and φ are the distance and azimuthal angle from the central axis of the conductor. So, the equipotential surface is given by

$$\frac{R^2 + (a^2/h)^2 - 2(a^2 R/h) \cos \varphi}{R^2 + h^2 - 2Rh \cos \varphi} = K,$$

where K is a constant. If we use Cartesian coordinates, we have $R^2 = x^2 + y^2$ and $R \cos \varphi = x$. So, the above equation is rewritten as

$$\left[x - \frac{1}{K-1}\left(Kh - \frac{a^2}{h}\right)\right]^2 + y^2$$
$$= \frac{1}{K-1}\left[\left(\frac{a^2}{h}\right)^2 - Kh^2\right] + \frac{1}{(K-1)^2}\left(Kh - \frac{a^2}{h}\right)^2.$$

Thus, the equipotential surfaces are cylindrical surfaces parallel to the cylindrical conductor. Especially, the conductor surface ($x^2 + y^2 = a^2$) is an equipotential surface, which is obtained for $K = a^2/h^2$. The maximum value of K is infinity and corresponds to the equipotential surface just around the given line charge. So, the above solution holds for $K \geq a^2/h^2$.

2.17. The electric potential is given by

$$\phi(x, y) = \frac{\lambda}{4\pi\epsilon_0} \log \frac{\left(x - \sqrt{l^2 - a^2}\right)^2 + y^2}{\left(x + \sqrt{l^2 - a^2}\right)^2 + y^2}$$

for $x < 0$, where λ is the electric charge given to the cylindrical conductor of a unit length. So, the equipotential surface is given by

$$\frac{\left(x - \sqrt{l^2 - a^2}\right)^2 + y^2}{\left(x + \sqrt{l^2 - a^2}\right)^2 + y^2} = K,$$

where K is a positive constant. This leads to

$$\left(x + \frac{K+1}{K-1}\sqrt{l^2 - a^2}\right)^2 + y^2 = \frac{4K(l^2 - a^2)}{(K-1)^2}. \tag{B2.1}$$

Thus, the equipotential surfaces are cylindrical surfaces parallel to the cylindrical conductor. The surface of the cylindrical conductor, which is an equipotential surface expressed as

$$(x + l)^2 + y^2 = a^2,$$

is obtained for $K = \left[\left(l + \sqrt{l^2 - a^2}\right)/a\right]^2$. In the limit $K \to 1$, neglecting small terms such as x^2 and y^2, Eq. (B2.1) leads to

$$2\alpha x = 0,$$

where $\alpha = (K+1)\sqrt{l^2 - a^2}/(K-1)$. Thus, the wide conductor surface ($x = 0$) is obtained. The range of K is $1 < K \leq \left[\left(l + \sqrt{l^2 - a^2}\right)/a\right]^2$.

2.18. The electric potential outside the conductor is given by

$$\phi(R, \varphi) = \frac{\lambda}{2\pi\epsilon_0} \log \frac{d\left[R^2 + (a^2/d)^2 - 2(a^2R/d)\cos\varphi\right]^{1/2}}{a(R^2 + d^2 - 2Rd\cos\varphi)^{1/2}},$$

where R and φ are the distance and azimuthal angle from the central axis of the conductor. So, the equipotential surface is given by

$$\frac{R^2 + (a^2/d)^2 - 2(a^2R/d)\cos\varphi}{R^2 + d^2 - 2Rd\cos\varphi} = \frac{1}{K},$$

where K is a constant. If we use Cartesian coordinates, we have $R^2 = x^2 + y^2$ and $R\cos\varphi = x$. So, the above equation is rewritten as

$$\left[x + \frac{1}{K-1}\left(d - \frac{Ka^2}{d}\right)\right]^2 + y^2 = \frac{1}{(K-1)^2}\left(d - \frac{Ka^2}{d}\right)^2 + \frac{1}{K-1}\left[d^2 - \left(\frac{Ka^2}{d}\right)^2\right].$$

Thus, equipotential surfaces are cylindrical surfaces parallel to the cylindrical conductor. Especially, the conductor surface ($x^2 + y^2 = a^2$) is an equipotential surface, which is obtained for $K = d^2/a^2$. The minimum value of K is 0 and corresponds to the equipotential surface just around the given line charge. So, the above solution holds for $0 < K \leq d^2/a^2$.

2.19. For simplicity, we denote as $K = 4\pi\epsilon_0/q$. Simple calculation leads to

$$K\frac{\partial^2\phi}{\partial x^2} = \frac{2x^2 - y^2 - (z-a)^2}{[x^2 + y^2 + (z-a)^2]^{5/2}} - \frac{2x^2 - y^2 - (z+a)^2}{[x^2 + y^2 + (z+a)^2]^{5/2}},$$

$$K\frac{\partial^2\phi}{\partial y^2} = \frac{-x^2 + 2y^2 - (z-a)^2}{[x^2 + y^2 + (z-a)^2]^{5/2}} - \frac{-x^2 + 2y^2 - (z+a)^2}{[x^2 + y^2 + (z+a)^2]^{5/2}},$$

$$K\frac{\partial^2\phi}{\partial z^2} = \frac{-x^2 - y^2 + 2(z-a)^2}{[x^2 + y^2 + (z-a)^2]^{5/2}} - \frac{-x^2 - y^2 + 2(z+a)^2}{[x^2 + y^2 + (z+a)^2]^{5/2}}.$$

Thus, we have

$$K\left(\frac{\partial^2\phi}{\partial x^2} + \frac{\partial^2\phi}{\partial y^2} + \frac{\partial^2\phi}{\partial z^2}\right) = 0$$

and Laplace's equation is satisfied.

2.20. The electric potential is given by

$$\phi(r,\theta) = \frac{q}{4\pi\epsilon_0}(\phi_1 - \phi_2);$$

$$\phi_1 = \frac{1}{(r^2 + d^2 - 2dr\cos\theta)^{1/2}},$$

$$\phi_2 = \frac{a}{d\left[r^2 + (a^2/d)^2 - 2(a^2r/d)\cos\theta\right]^{1/2}}.$$

2.4 Exercises

Since ϕ_1 and ϕ_2 are functions of the same form, it is enough to show that ϕ_1 satisfies Laplace's equation. $\Delta\phi_1$ is given by

$$\Delta\phi_1 = \frac{1}{r^2} \cdot \frac{\partial}{\partial r}\left(r^2 \frac{\partial \phi_1}{\partial r}\right) + \frac{1}{r^2 \sin\theta} \cdot \frac{\partial}{\partial \theta}\left(\sin\theta \frac{\partial \phi_1}{\partial \theta}\right).$$

Substituting ϕ_1, we have

$$\frac{1}{r^2} \cdot \frac{\partial}{\partial r}\left(r^2 \frac{\partial \phi_1}{\partial r}\right) = \frac{d\left[2r^2 \cos\theta - d^2 r(3+\cos\theta) + 2d^2 \cos\theta\right]}{r\left(r^2 + d^2 - 2dr\cos\theta\right)^{5/2}},$$

$$\frac{1}{r^2 \sin\theta} \cdot \frac{\partial}{\partial \theta}\left(\sin\theta \frac{\partial \phi_1}{\partial \theta}\right) = -\frac{d\left[2r^2 \cos\theta - d^2 r(3+\cos\theta) + 2d^2 \cos\theta\right]}{r\left(r^2 + d^2 - 2dr\cos\theta\right)^{5/2}}.$$

So, the equation, $\Delta\phi_1 = 0$, holds. Hence, it is concluded that the electric potential satisfies Laplace's equation.

2.21. The electric potential is given by

$$\phi(R, \varphi) = \frac{\lambda}{4\pi\epsilon_0}(\phi_1 - \phi_2) + \text{const.};$$

$$\phi_1(R, \varphi) = \log\left[R^2 + \left(\frac{a^2}{d}\right)^2 - 2\left(\frac{a^2 R}{d}\right)\cos\varphi\right],$$

$$\phi_2(R, \varphi) = \log(R^2 + d^2 - 2Rd\cos\varphi).$$

Since ϕ_1 and ϕ_2 are functions of the same form, it is enough to show that ϕ_2 satisfies Laplace's equation. $\Delta\phi_2$ is given by

$$\Delta\phi_2 = \frac{1}{R} \cdot \frac{\partial}{\partial R}\left(R \frac{\partial \phi_2}{\partial R}\right) + \frac{1}{R^2} \cdot \frac{\partial^2 \phi_2}{\partial \varphi^2}.$$

Substituting ϕ_2, we have

$$\frac{1}{R} \cdot \frac{\partial}{\partial R}\left(R \frac{\partial \phi_2}{\partial R}\right) = -\frac{2(R^2 d \cos\varphi - 2Rd^2 + d^3 \cos\varphi)}{R(R^2 + d^2 - 2Rd\cos\varphi)^2},$$

$$\frac{1}{R^2} \cdot \frac{\partial^2 \phi_2}{\partial \varphi^2} = \frac{2(R^2 d \cos\varphi - 2Rd^2 + d^3 \cos\varphi)}{R(R^2 + d^2 - 2Rd\cos\varphi)^2}.$$

So, the equation, $\Delta\phi_2 = 0$, holds. Hence, it is concluded that the electric potential satisfies Laplace's equation.

2.22. The electric potential in the hollow space is given by

$$\phi(r,\theta) = -\frac{1}{4\pi\epsilon_0}(\phi_1 + \phi_2);$$

$$\phi_1 = \frac{a}{h\left[r^2 + (a^2/h)^2 - 2(a^2r/h)\cos\theta\right]^{1/2}},$$

$$\phi_2 = \frac{1}{(r^2 + h^2 - 2hr\cos\theta)^{1/2}}.$$

Since ϕ_1 and ϕ_2 are functions of the same form, it is enough to show that ϕ_2 satisfies Laplace's equation. $\Delta\phi_2$ is given by

$$\Delta\phi_2 = \frac{1}{r^2} \cdot \frac{\partial}{\partial r}\left(r^2 \frac{\partial \phi_1}{\partial r}\right) + \frac{1}{r^2 \sin\theta} \cdot \frac{\partial}{\partial \theta}\left(\sin\theta \frac{\partial \phi_1}{\partial \theta}\right).$$

Substituting ϕ_2, we have

$$\frac{1}{r^2} \cdot \frac{\partial}{\partial r}\left(r^2 \frac{\partial \phi_2}{\partial r}\right) = \frac{h\left[2r^2\cos\theta - h^2r(3+\cos\theta) + 2h^2\cos\theta\right]}{r(r^2 + h^2 - 2hr\cos\theta)^{5/2}},$$

$$\frac{1}{r^2 \sin\theta} \cdot \frac{\partial}{\partial \theta}\left(\sin\theta \frac{\partial \phi_1}{\partial \theta}\right) = -\frac{h\left[2r^2\cos\theta - h^2r(3+\cos\theta) + 2h^2\cos\theta\right]}{r(r^2 + h^2 - 2hr\cos\theta)^{5/2}}.$$

So, the equation, $\Delta\phi_2 = 0$, holds. Hence, it is concluded that the electric potential satisfies Laplace's equation.

2.23. The electric potential in the hollow space is given by

$$\phi(R,\varphi) = \frac{\lambda}{4\pi\epsilon_0}(\phi_1 - \phi_2);$$

$$\phi_1(R,\varphi) = \log\left[R^2 + \left(\frac{a^2}{h}\right)^2 - 2\left(\frac{a^2R}{h}\right)\cos\varphi\right],$$

$$\phi_2(R,\varphi) = \log(R^2 + h^2 - 2Rh\cos\varphi).$$

Since ϕ_1 and ϕ_2 are functions of the same form, it is enough to show that ϕ_2 satisfies Laplace's equation. $\Delta\phi_2$ is given by

$$\Delta\phi_2 = \frac{1}{R} \cdot \frac{\partial}{\partial R}\left(R \frac{\partial \phi_2}{\partial R}\right) + \frac{1}{R^2} \cdot \frac{\partial^2 \phi_2}{\partial \varphi^2}.$$

2.4 Exercises

Substituting ϕ_2, we have

$$\frac{1}{R} \cdot \frac{\partial}{\partial R}\left(R \frac{\partial \phi_2}{\partial R}\right) = -\frac{2(R^2 h \cos\varphi - 2Rh^2 + h^3 \cos\varphi)}{R(R^2 + h^2 - 2Rh \cos\varphi)^2},$$

$$\frac{1}{R^2} \cdot \frac{\partial^2 \phi_2}{\partial \varphi^2} = \frac{2(R^2 h \cos\varphi - 2Rh^2 + h^3 \cos\varphi)}{R(R^2 + h^2 - 2Rh \cos\varphi)^2}.$$

So, the equation, $\Delta\phi_2 = 0$, holds. Hence, it is concluded that the electric potential satisfies Laplace's equation.

2.24. The electric potential is given by

$$\phi(x, y) = \frac{\lambda}{4\pi\epsilon_0} \log \frac{\left(x - \sqrt{l^2 - a^2}\right)^2 + y^2}{\left(x + \sqrt{l^2 - a^2}\right)^2 + y^2}$$

for $x < 0$, where λ is the electric charge given to the cylindrical conductor of a unit length. So, simple calculation leads to

$$\frac{\partial^2 \phi}{\partial x^2} = \frac{\lambda}{2\pi\epsilon_0} \left\{ \frac{-\left(x - \sqrt{l^2 - a^2}\right)^2 + y^2}{\left[\left(x - \sqrt{l^2 - a^2}\right)^2 + y^2\right]^2} - \frac{-\left(x + \sqrt{l^2 - a^2}\right)^2 + y^2}{\left[\left(x + \sqrt{l^2 - a^2}\right)^2 + y^2\right]^2} \right\},$$

$$\frac{\partial^2 \phi}{\partial y^2} = \frac{\lambda}{2\pi\epsilon_0} \left\{ \frac{\left(x - \sqrt{l^2 - a^2}\right)^2 - y^2}{\left[\left(x - \sqrt{l^2 - a^2}\right)^2 + y^2\right]^2} - \frac{\left(x + \sqrt{l^2 - a^2}\right)^2 - y^2}{\left[\left(x + \sqrt{l^2 - a^2}\right)^2 + y^2\right]^2} \right\}.$$

Thus, we have

$$\frac{\partial^2 \phi}{\partial x^2} + \frac{\partial^2 \phi}{\partial y^2} = 0,$$

and Laplace's equation is satisfied.

2.25. We define the x-axis in the direction of the applied electric field. The equipotential surface is given by

$$E_0\left(R - \frac{a^2}{R}\right) \cos\varphi = K,$$

where K is a constant. Using the relationships $R^2 = x^2 + y^2$ and $R\cos\varphi = x$, the above relation leads to

$$\left(1 - \frac{a^2}{x^2 + y^2}\right)x = \frac{K}{E_0}.$$

So, we have

$$y = \pm a\left(1 + \frac{K}{E_0 x - K} - \frac{x^2}{a^2}\right)^{1/2}$$

in the x-y plane. The surface of the conductor $(x^2 + y^2 = a^2)$ is an equipotential surface obtained for $K = 0$.

2.26. The electric potential inside the spherical conductor is $\phi = 0$ and satisfies Laplace's equation. The electric potential outside the conductor is given by Eq. (2.19). Using Eq. (19) in the Appendix, we have

$$\Delta\phi = \frac{1}{r^2} \cdot \frac{\partial}{\partial r}\left(r^2 \frac{\partial\phi}{\partial r}\right) + \frac{1}{r^2 \sin\theta} \cdot \frac{\partial}{\partial\theta}\left(\sin\theta \frac{\partial\phi}{\partial\theta}\right)$$

$$= -2E_0\left(\frac{1}{r} - \frac{a^3}{r^4}\right)\cos\theta + 2E_0\left(\frac{1}{r} - \frac{a^3}{r^4}\right)\cos\theta = 0.$$

Thus, Laplace's equation holds inside and outside the spherical conductor.

2.27. The electric potential inside the cylindrical conductor is $\phi = 0$ and satisfies Laplace's equation. The electric potential outside the conductor is given by

$$\phi(R, \varphi) = -E_0\left(R - \frac{a^2}{R}\right)\cos\varphi.$$

Using Eq. (2.15) in the Appendix, we have

$$\Delta\phi = \frac{1}{R} \cdot \frac{\partial}{\partial R}\left(R \frac{\partial\phi}{\partial R}\right) + \frac{1}{R^2} \cdot \frac{\partial^2\phi}{\partial\varphi^2}$$

$$= -E_0\left(\frac{1}{R} - \frac{a^2}{R^3}\right)\cos\varphi + E_0\left(\frac{1}{R} - \frac{a^2}{R^3}\right)\cos\varphi = 0.$$

Thus, Laplace's equation holds inside and outside the cylindrical conductor.

2.28. The dipole moment is assumed to be directed to the zenith ($\theta = 0$). The radial electric field must be cancelled in the conductor region ($r > a$). So, a shielding electric charge must be induced to produce an electric field of the same magnitude with the opposite sign. This means that the electric charge is equivalent to the dipole moment $-p$ placed at the center for $r > a$. So, the electric charge density on the surface is given by

2.4 Exercises

$$\sigma = \frac{3p}{4\pi a^3} \cos\theta. \tag{B2.2}$$

On the other hand, the electric charge on the surface produces a uniform electric field in the hollow space. If we denote the uniform electric field density by E_0, Eq. (2.18) gives

$$E_0 = \frac{p}{4\pi\epsilon_0 a^3}.$$

The radial and zenithal components are

$$E_{0r} = \frac{p}{4\pi\epsilon_0 a^3} \cos\theta, \quad E_{0\theta} = -\frac{p}{4\pi\epsilon_0 a^3} \sin\theta.$$

The electric field produced in the hollow space by the electric moment p is

$$E_{mr} = \frac{p}{2\pi\epsilon_0 r^3} \cos\theta, \quad E_{m\theta} = \frac{p}{4\pi\epsilon_0 r^3} \sin\theta.$$

Thus, the electric field on the surface ($r = a$) is

$$E_r = \frac{3p}{4\pi\epsilon_0 a^3} \cos\theta, \quad E_\theta = 0.$$

So, the relationship holds:

$$\sigma = -\epsilon_0 E_r(r = a),$$

and the boundary condition is satisfied. Note that the normal vector of the inner surface is opposite to the radial direction. This means that the induced electric charge density on the surface is correctly given by Eq. (B2.2).

2.29. The electric potential produced by the electric dipole moment is

$$\phi_d = \frac{p}{4\pi\epsilon_0 r^2} \cos\theta,$$

and the electric potential produced by the induced electric charge that produces a uniform electric field E_0 is

$$\phi_f = -E_0 r \cos\theta = -\frac{pr}{4\pi\epsilon_0 a^3} \cos\theta.$$

The electric potential in the hollow space ($0 \le r < a$) is given by

$$\phi = \phi_d + \phi_f = \frac{p}{4\pi\epsilon_0} \left(\frac{1}{r^2} - \frac{r}{a^3}\right) \cos\theta.$$

So, $\Delta\phi$ is determined to be

$$\Delta\phi = \frac{1}{r^2} \cdot \frac{\partial}{\partial r}\left(r^2 \frac{\partial \phi}{\partial r}\right) + \frac{1}{r^2 \sin\theta} \cdot \frac{\partial}{\partial \theta}\left(\sin\theta \frac{\partial \phi}{\partial \theta}\right)$$

$$= \frac{p}{2\pi\epsilon_0}\left[\frac{\cos\theta}{r^4}\left(1 - \frac{r^3}{a^3}\right) - \frac{\cos\theta}{r^4}\left(1 - \frac{r^3}{a^3}\right)\right] = 0.$$

Thus, Laplace's equation is satisfied.

2.30. The moment of the electric dipole line is assumed to be directed to the azimuth ($\varphi = 0$). The radial electric field must be cancelled in the conductor region ($R > a$). So, the shielding electric charge must be induced to produce the electric field of the same magnitude with the opposite sign. It means that the electric charge is equivalent to the linear dipole moment $-\hat{p}$ placed at the central axis for $R > a$. So, the electric charge density on the inner surface is given by

$$\sigma = \frac{\hat{p}}{\pi a^2} \cos\varphi. \tag{B2.3}$$

On the other hand, the electric charge on the surface produces a uniform electric field in the hollow space. If we denote the uniform electric field density by E_0, from the result in Example 2.7, it is given by

$$E_0 = \frac{\hat{p}}{2\pi\epsilon_0 a^2}.$$

The radial and azimuthal components are

$$E_{0R} = \frac{\hat{p}}{2\pi\epsilon_0 a^2}\cos\varphi, \quad E_{0\varphi} = -\frac{\hat{p}}{2\pi\epsilon_0 a^2}\sin\varphi.$$

From Eq. (1.26), the electric field in the hollow space produced by the electric moment $\hat{p}$ is

$$E_{mR} = \frac{\hat{p}}{2\pi\epsilon_0 R^2}\cos\varphi, \quad E_{m\varphi} = \frac{\hat{p}}{2\pi\epsilon_0 R^2}\sin\varphi.$$

Thus, the electric field on the inner surface ($R = a$) is

$$E_R = \frac{\hat{p}}{\pi\epsilon_0 a^2}\cos\varphi, \quad E_\varphi = 0.$$

So, the relationship holds:

$$\sigma = -\epsilon_0 E_R(R = a),$$

2.4 Exercises

and the boundary condition is satisfied. Note that the normal vector of the inner surface is opposite to the radial direction. This means that the induced electric charge density on the surface is correctly given by Eq. (B2.3).

2.31. The electric potential produced by the electric dipole line of moment $\hat{p}$ is

$$\phi_d = \frac{\hat{p}}{2\pi \epsilon_0 R} \cos \varphi,$$

and the electric potential produced by the induced electric charge that produces a uniform electric field E_0 is

$$\phi_f = -E_0 R \cos \varphi = -\frac{\hat{p} R}{2\pi \epsilon_0 a^2} \cos \varphi.$$

The electric potential in the space hollow ($0 \leq r < a$) is given by

$$\phi = \phi_d + \phi_f = \frac{\hat{p}}{2\pi \epsilon_0} \left(\frac{1}{R} - \frac{R}{a^2} \right) \cos \varphi.$$

So, $\Delta \phi$ is determined to be

$$\Delta \phi = \frac{1}{R} \cdot \frac{\partial}{\partial R} \left(R \frac{\partial \phi}{\partial R} \right) + \frac{1}{R^2} \cdot \frac{\partial^2 \phi}{\partial \varphi}$$

$$= \frac{\hat{p}}{2\pi \epsilon_0} \left[\frac{\cos \varphi}{R^3} \left(1 - \frac{R^2}{a^2} \right) - \frac{\cos \varphi}{R^3} \left(1 - \frac{R^2}{a^2} \right) \right] = 0.$$

Thus, Laplace's equation is satisfied.

2.32. The electric potential due to the dipole line moment $\hat{p}$ is

$$\phi_m = \frac{\hat{p}}{2\pi \epsilon_0 R} \cos \varphi$$

and that due to the surface electric charge that produces uniform electric field E_0 is

$$\phi_f = -E_0 R \cos \varphi = -\frac{\hat{p} R}{2\pi \epsilon_0 a^2} \cos \varphi.$$

Hence, the electric potential is given by

$$\phi = \phi_m + \phi_f = \frac{\hat{p}}{2\pi \epsilon_0} \left(\frac{1}{R} - \frac{R}{a^2} \right) \cos \varphi.$$

Hence, the equipotential surface is expressed as

$$\left(\frac{1}{R} - \frac{R}{a^2} \right) \cos \varphi = K,$$

with K denoting a constant. Noting $R^2 = x^2 + y^2$ and $R \cos\varphi = x$, the above equation is written as

$$\left(\frac{1}{x^2+y^2} - \frac{1}{a^2}\right)x = K. \tag{B2.4}$$

So, we have

$$y = \pm\left(a^2 - \frac{Ka^4}{x+Ka^2} - x^2\right)^{1/2}.$$

The surface of the conductor $(x^2 + y^2 = a^2)$ is an equipotential surface obtained for $K = 0$. When $|K|$ is very large, the second term on the left-hand side in Eq. (B2.4) can be neglected, and the equipotential surface approaches

$$\left(x - \frac{1}{2K}\right)^2 + y^2 = \frac{1}{4K^2}.$$

2.33. The electric potential due to the given electric charge q and the image charge Q inside the conductor is given by

$$\phi = \frac{q}{4\pi\epsilon_0\left(r^2 + d^2 - 2rd\cos\theta\right)^{1/2}} + \frac{Q}{4\pi\epsilon_0\left(r^2 + h^2 - 2rh\cos\theta\right)^{1/2}}.$$

The parallel component of the electric field on the conductor surface ($r = a$) is

$$E_\theta(r = a) = -\frac{1}{a}\cdot\left(\frac{\partial\phi}{\partial\theta}\right)_{r=a}$$

$$= \frac{1}{4\pi\epsilon_0}\sin\theta\left[\frac{qd}{(a^2+d^2-2ad\cos\theta)^{3/2}} + \frac{Qh}{(a^2+h^2-2ah\cos\theta)^{3/2}}\right]$$

$$= \frac{\sin\theta}{4(2a)^{3/2}\pi\epsilon_0}\left[\frac{q/d^{1/2}}{[(a^2+d^2)/2ad - \cos\theta]^{3/2}} + \frac{Q/h^{1/2}}{[(a^2+h^2)/2ah - \cos\theta]^{3/2}}\right].$$

The following conditions must be satisfied so that the above electric field is identically zero:

$$Q = -\left(\frac{h}{d}\right)^{1/2}q, \quad \frac{a^2+h^2}{2ah} = \frac{a^2+d^2}{2ad}.$$

The latter condition is reduced to $h = a^2/d$. Then, the former condition leads to $Q = -(a/d)q$. Thus, the obtained electric potential agrees with that in Eq. (2.13), and the correct result is obtained using the boundary condition.

2.4 Exercises

2.34. The electric potential due to the given line charge λ and the image line charge λ' inside the conductor is given by

$$\phi = \frac{\lambda}{2\pi\epsilon_0} \log \frac{R_0}{\left(R^2 + d^2 - 2Rd \cos\varphi\right)^{1/2}} + \frac{\lambda'}{2\pi\epsilon_0} \log \frac{R_0'}{\left(R^2 + h^2 - 2Rh \cos\varphi\right)^{1/2}},$$

where R_0 and R_0' are constants denoting distances to a reference point of the electric potential. The azimuthal component of the electric field on the conductor surface ($R = a$) is

$$E_\varphi(R = a) = \frac{1}{a} \cdot \left(\frac{\partial \phi}{\partial \varphi}\right)_{R=a}$$

$$= \frac{1}{2\pi\epsilon_0} \sin\varphi \left(\frac{\lambda d}{a^2 + d^2 - 2ad \cos\varphi} + \frac{\lambda' h}{a^2 + h^2 - 2ah \cos\varphi}\right)$$

$$= \frac{1}{4\pi\epsilon_0 a} \sin\varphi \left[\frac{\lambda}{(a^2 + d^2)/2ad - \cos\varphi} + \frac{\lambda'}{(a^2 + h^2)/2ah - \cos\varphi}\right].$$

The following conditions must be satisfied so that the above electric field is identically zero:

$$\lambda' = -\lambda, \quad \frac{a^2 + h^2}{2ah} = \frac{a^2 + d^2}{2ad}.$$

The latter condition is reduced to $h = a^2/d$. Thus, the obtained electric potential agrees with that in Example 2.5, and the correct result is obtained using the boundary condition.

2.35. Using the electric potential in vacuum ($R > a$),

$$\phi(R, \varphi) = \left(-E_0 R + \frac{\hat{p}}{2\pi\epsilon_0 R}\right) \cos\varphi,$$

the azimuthal component of the electric field in this region is given by

$$E_\varphi = -\frac{1}{R} \cdot \frac{\partial \phi}{\partial \varphi} = \left(-E_0 + \frac{\hat{p}}{2\pi\epsilon_0 R^2}\right) \sin\varphi.$$

From the requirement that this component is zero on the conductor surface ($R = a$), we have

$$\hat{p} = 2\pi\epsilon_0 a^2 E_0,$$

which agrees with the result obtained in Example 2.7. Thus, the same result is obtained as in Example 2.7 using the boundary condition on the electric field.

Coffee Break: Reason for the Reversed Mirror Images

The image method is known to be useful for solving not only the electric field around a conductor or dielectric material but also for solving the magnetic field around a superconductor or magnetic material. The mirror image, which is the origin of this method, can be clearly described by optics. The reason why the image is reversed with respect to right and left has not yet been clarified, however. For example, when a man stands with a flower in his right hand in front of a mirror, his image appears to have a flower in his left hand. The reason for such a reverse is subject to question.

The mirror image is a virtual one. This can be explained using coordinates. Suppose that a mirror is placed on the x-y plane at $z = 0$. The position of a point is represented as (x, y, z). Assume that the observer stays on the side of $z > 0$ (see Fig. C2.1). A light that starts from a light source at point A (a, b, c) and hits the origin $(0, 0, 0)$ on the mirror passes through point B $(-a, -b, c)$. This light seems to start from the virtual image at C $(a, b, -c)$ (see Fig. C2.2). In this case, there is no change between the upper and lower sides and between the right and left sides, and the z-axis only changes because of the reflection on the plane at $z = 0$. This explains the fact that the real flower and the virtual flower in the mirror stay on the right side from the observer.

Then, the question is why the observer feels that his mirror image has a flower in his left hand. Note that he can only see himself with the flower from the mirror image. He feels that this is different from the reality by comparing this mirror image with something. What is it? This is a supposition of himself when someone sees him, or a photograph taken by other persons or by himself. Here, we discuss the case of a photograph taken by himself. He moves to a position at which a photograph is taken, turns round to the direction of his standing point, sets a camera to take a photograph, goes back to the original standing point, and then takes a photograph. The key point

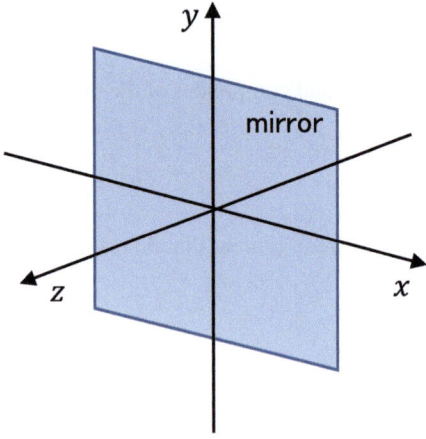

Fig. C2.1 A mirror placed on the x-y plane ($z = 0$)

2.4 Exercises

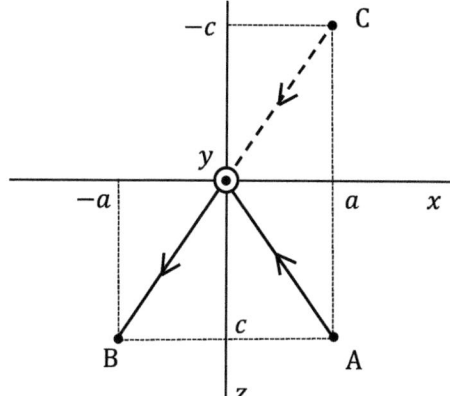

Fig. C2.2 A light that starts from a light source at A and is reflected on the origin on the mirror ($z = 0$)

is that he needs to turn round, and the right and left sides are reversed by this action. That is, if the right side is the east side at the original standing point, the left side changes to the east side by the turning round.

If he takes a photograph of himself with the flower in his left hand, the photograph is similar to the mirror image of himself with the flower in his right hand. In conclusion, the reason why one feels that the right and left sides change comes from comparison of the mirror image which he really sees with a supposed image that other persons will have.

Chapter 3
Conductor Systems in Vacuum

3.1 Coefficients in Conductor System

For a single conductor, the electric potential ϕ is generally proportional to the electric charge Q according to

$$\phi = p_c Q, \qquad (3.1)$$

where p_c is a proportional constant called the coefficient of electric potential and is uniquely determined when the shape of the conductor is given. Its unit is [V/C].

When we rewrite Eq. (3.1) as

$$Q = C\phi, \qquad (3.2)$$

focusing on the electric charge, $C = 1/p_c$ is called the capacity or capacitance, which is equal to the electric charge stored by a unit electric potential, and its unit is [F] (farad).

Suppose a conductor system composed of n conductors and assume that conductor i has electric charge Q_i and electric potential ϕ_i ($i = 1, 2, \ldots, n$). The electric potential of conductor i is expressed as

$$\phi_i = \sum_{j=1}^{n} p_{ij} Q_j. \qquad (3.3)$$

The p_{ij}'s are the coefficients of electric potential determined by the geometrical arrangement of the conductors and fulfill the following conditions:

$$p_{ii} > 0, \quad p_{ij} = p_{ji} \geq 0 (i \neq j), \quad p_{ii} \geq p_{ij}. \qquad (3.4)$$

The electric charge is given by the inverse of Eq. (3.3):

$$Q_i = \sum_{j=1}^{n} C_{ij}\phi_j. \tag{3.5}$$

The C_{ij}'s in Eq. (3.5) are capacity coefficients or capacitance coefficients and have the same unit as the capacitance. These coefficients fulfill the following conditions:

$$C_{ii} > 0, \quad C_{ij} = C_{ji} \leq 0 (i \neq j). \tag{3.6}$$

The matrix of coefficients of electric potential, $\hat{P} = \{p_{ij}\}$, and the matrix of capacitance coefficients, $\hat{C} = \{C_{ij}\}$, are inverses of each other and satisfy

$$\sum_{k=1}^{n} p_{ik} C_{kj} = \sum_{k=1}^{n} C_{ik} p_{kj} = \delta_{ij}, \tag{3.7}$$

where δ_{ij} is the Kronecker delta. The equalities $p_{ij} = p_{ji}$ and $C_{ij} = C_{ji}$ in Eqs. (3.4) and (3.6) are called the reciprocity theorem.

Example 3.1 The electric condition around a grounded spherical conductor near a point charge was discussed in Sect. 2.2 (see Fig. 2.6). Determine the electric charge induced on the spherical conductor using the coefficients of electric potential.

Solution 3.1 The spherical conductor is named conductor 1, and a very small imaginary conductor placed at the position of the point charge is called conductor 2. Under the conditions $Q_1 = 1$ and $Q_2 = 0$, we have

$$\phi_1 = p_{11} = \frac{1}{4\pi\epsilon_0 a}, \quad \phi_2 = p_{21} = \frac{1}{4\pi\epsilon_0 d}.$$

Under the conditions $Q_1 = Q$ and $Q_2 = q$, the electric potential of conductor 1 is

$$\phi_1 = p_{11} Q + p_{21} q.$$

Since conductor 1 is grounded, $\phi_1 = 0$. This and the reciprocity theorem $p_{12} = p_{21}$ give

$$Q = -\frac{p_{21}}{p_{11}} q = -\frac{a}{d} q,$$

which agrees with Eq. (2.12). ◆

3.2 Capacitor

The component used to store electric charge is called a condenser or a capacitor. Larger capacitance is usually desirable for this purpose. Common capacitors consist of two parallel sheet conductors separated by a small distance, as schematically shown in Fig. 3.1. Each sheet conductor is connected to an outer circuit with a lead line for transporting electric charge. Such a capacitor is called a parallel-plate capacitor. The sheet conductor that stores the electric charge is called an electrode. Suppose that the surface area of each electrode is S and that the distance between the two electrodes is d, which is very small compared with the electrode size: $d \ll \sqrt{S}$. When an electric charge, Q, is given to one electrode and $-Q$ is given to the other, the electric charges are distributed uniformly on the inner surfaces of the electrodes (facing each other) but do not appear on the outer surfaces. Hence, the electric field is almost entirely concentrated in the space between the electrodes and perpendicular to them except at the edges.

The surface density of positive electric charge is

$$\sigma = \frac{Q}{S}, \tag{3.8}$$

and the electric field strength in the space between the two electrodes is

$$E = \frac{\sigma}{\epsilon_0} = \frac{Q}{\epsilon_0 S}. \tag{3.9}$$

The electric potential difference between the two electrodes is

$$V = Ed = \frac{Qd}{\epsilon_0 S}. \tag{3.10}$$

Hence, the electric charge that can be stored in the capacitor by a unit electric potential difference, i.e., the capacitance, is given by

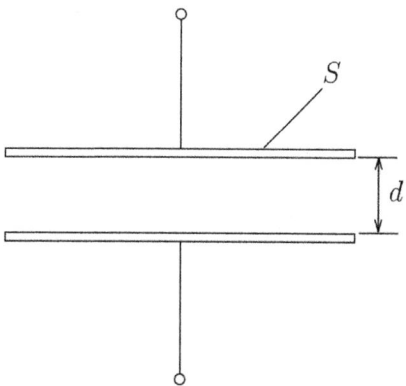

Fig. 3.1 Parallel-plate capacitor

$$C = \frac{Q}{V} = \frac{\epsilon_0 S}{d}. \qquad (3.11)$$

We connect capacitors in series, apply an electric charge, Q, to the top capacitor, and then ground the lowest capacitor. Then, electric charges $\pm Q$ appear in the electrodes of each capacitor. When the capacitance of each capacitor is $C_i (i = 1, 2, \ldots, n)$, the electric potential difference across each capacitor is $V_i = Q/C_i$. Hence, the total electric potential difference through the series of capacitor is

$$V = \sum_{i=1}^{n} V_i = Q \sum_{i=1}^{n} \frac{1}{C_i}. \qquad (3.12)$$

If the capacitance of capacitors connected in series is denoted by C, we have

$$\frac{1}{C} = \sum_{i=1}^{n} \frac{1}{C_i}. \qquad (3.13)$$

Next, we connect capacitors in parallel. When we apply a voltage, V, between the terminals, the electric charges that appear in the capacitor of capacitance C_i are $\pm Q_i = \pm C_i V$. Hence, the total amount of positive electric charge is

$$Q = \sum_{i=1}^{n} Q_i = V \sum_{i=1}^{n} C_i. \qquad (3.14)$$

If the capacitance of capacitors connected in parallel is denoted by C, we have

$$C = \sum_{i=1}^{n} C_i. \qquad (3.15)$$

Example 3.2 Determine the capacitance of the concentric spherical capacitor in Fig. 3.2.

Fig. 3.2 Concentric spherical capacitor

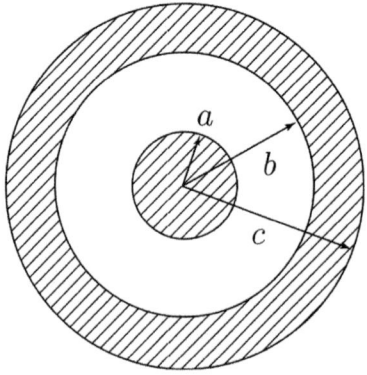

3.2 Capacitor

Solution 3.2 An electric charge Q is given to the inner conductor and the outer conductor is grounded. Then, an electric charge $-Q$ appears on the inner surface of the outer conductor. Applying Gauss's law, we determine the electric field strength to be

$$E = \frac{Q}{4\pi\epsilon_0 r^2}; \quad a < r < b,$$

$$= 0; \quad 0 \le r < a, \quad r > b,$$

where r is the distance from the center. The electric potential difference between the electrodes is

$$V = \int_a^b \frac{Q}{4\pi\epsilon_0 r^2} dr = \frac{Q}{4\pi\epsilon_0}\left(\frac{1}{a} - \frac{1}{b}\right).$$

Hence, the capacitance is given by

$$C = \frac{4\pi\epsilon_0 ab}{b-a}.$$

◆

Example 3.3 Two long and thin cylindrical conductors of radius a are placed parallel to each other with distance d, as shown in Fig. 3.3. Determine the capacitance in a unit length of these conductors. Assume that a is sufficiently smaller than d.

Fig. 3.3 Two long and thin cylindrical conductors parallel to each other

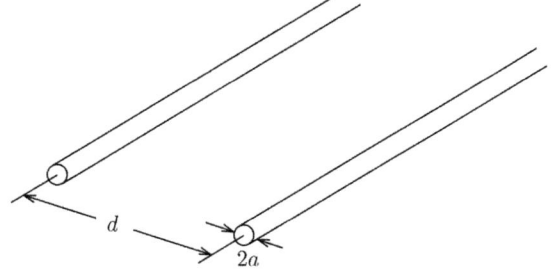

Solution 3.3 Suppose that electric charges $\pm\lambda$ are given to each conductor in a unit length. We define the x-axis normal to these conductors in such a way that it passes through the centers of these conductors. We denote the positions of the centers of the conductors with negative and positive electric charges by $x = 0$ and $x = d$, respectively. Since the diameter of these conductors are much smaller than the interval d, we can approximate the electric charge as being uniformly distributed on each surface. Hence, the electric field at position x is

$$E = -\frac{\lambda}{2\pi\epsilon_0}\left(\frac{1}{x} + \frac{1}{d-x}\right)$$

under the definition of positive electric field directed along the positive x-axis. The electric potential difference between the two conductors is

$$V = -\int_a^{d-a} E\,dx = \frac{\lambda}{\pi\epsilon_0}\log\left(\frac{d}{a} - 1\right).$$

The capacitance in a unit length is

$$C' = \frac{\lambda}{V} = \frac{\pi\epsilon_0}{\log[(d/a) - 1]}.$$

♦

3.3 Electrostatic Energy

Suppose that there is a conductor system with electric charges. The electric charges produce an electric field in the space outside the conductors. Hence, we can regard the conductor system as having some kind of energy. This is called the electrostatic energy or electric energy. It is reasonable to assume that there is no electrostatic energy in this system when there is no electric charge. Thus, the electrostatic energy is equivalent to the mechanical work necessary to bring electric charges in from infinity until we attain the desired distribution of electric charge.

Now, we determine the electrostatic energy for an isolated conductor of capacitance C and electric charge Q. Suppose that the conductor has an electric charge, q, in an intermediate state while electric charge is being brought in from infinity. The electric potential of the conductor is

$$\phi(q) = \frac{q}{C}.$$

3.3 Electrostatic Energy

A small amount of electric charge, dq, is additionally carried from infinity to the conductor. If this amount is sufficiently small, the transfer of this charge does not change the electric potential $\phi(q)$. Hence, the mechanical work needed for carrying this charge is given by

$$dW = \phi(q)dq = \frac{q}{C}dq.$$

Thus, the total work needed to carry all electric charge, Q, is

$$W = \int_0^Q \frac{q}{C}dq = \frac{1}{2C}Q^2. \qquad (3.16)$$

This gives the electrostatic energy, U_e. In terms of the electric potential, $\phi = Q/C$, this can also be written as

$$U_e = \frac{1}{2C}Q^2 = \frac{1}{2}Q\phi = \frac{1}{2}C\phi^2. \qquad (3.17)$$

Assume that conductor i has an electric charge Q_i and an electric potential $\phi_i (i = 1, 2, \ldots, n)$. The electrostatic energy of this system is given by

$$U_e = \frac{1}{2}\sum_{i=1}^{n} Q_i \phi_i = \frac{1}{2}\sum_{i=1}^{n}\sum_{j=1}^{n} p_{ij} Q_i Q_j. \qquad (3.18)$$

We can extend this result to the case where electric charge is continuously distributed with the density $\rho(r)$ and the electric potential is given by $\phi(r)$ in region V, so the electrostatic energy is

$$U_e = \frac{1}{2}\int_V \rho(r)\phi(r)dV. \qquad (3.19)$$

In the case where a capacitor with a capacitance C has electric charges $\pm Q$ under an electric potential difference V, substituting $Q_1 = Q$, $Q_2 = -Q$, and $\phi_1 = \phi_2 + V$ into Eq. (3.18) for $n = 2$, we have

$$U_e = \frac{1}{2C}Q^2 = \frac{1}{2}QV = \frac{1}{2}CV^2. \qquad (3.20)$$

Assume that a voltage V is given to the parallel-plate capacitor shown in Fig. 3.1. The electrostatic energy is given by

$$U_e = \frac{1}{2}\epsilon_0 E^2 Sd, \qquad (3.21)$$

where E is the electric field strength given by Eq. (3.9). Since Sd is the volume of the region in which the electric field is concentrated with a constant strength E, we can regard electrostatic energy of density

$$u_e = \frac{1}{2}\epsilon_0 E^2 \tag{3.22}$$

as filling this region. This is called the electrostatic energy density or electric energy density.

Even for a case where the electric field strength is not uniform in space, the electrostatic energy in volume V is given by

$$U_e = \int_V \frac{1}{2}\epsilon_0 E^2 dV. \tag{3.23}$$

Example 3.4 Electric charges Q_1 and Q_2 are given to the inner and outer spherical conductors shown in Fig. 3.2, respectively. Determine the electrostatic energy using the coefficients of electric potential.

Solution 3.4 We denote the inner and outer conductors as conductors 1 and 2, respectively. When we give a unit charge only to conductor 1 ($Q_1 = 1, Q_2 = 0$), the electric potential of each conductor is

$$\phi_1 = p_{11} = \frac{1}{4\pi\epsilon_0}\left(\frac{1}{a} - \frac{1}{b} + \frac{1}{c}\right),$$

$$\phi_2 = p_{21} = p_{12} = \frac{1}{4\pi\epsilon_0 c}.$$

When, we give a unit charge only to conductor 2 ($Q_1 = 0, Q_2 = 1$), the electric potential of conductor 2 is

$$\phi_2 = p_{22} = \frac{1}{4\pi\epsilon_0 c}.$$

The electrostatic energy when Q_1 and Q_2 are given to the inner and outer spherical conductors is

$$U_e = \frac{1}{2}p_{11}Q_1^2 + p_{12}Q_1 Q_2 + \frac{1}{2}p_{22}Q_2^2$$

$$= \frac{1}{8\pi\epsilon_0}\left[Q_1^2\left(\frac{1}{a} - \frac{1}{b}\right) + \frac{(Q_1 + Q_2)^2}{c}\right].$$

◆

3.3 Electrostatic Energy

Example 3.5 We denote the inner, middle, and outer spherical conductors in Fig. E2.3 in Exercise 2.5 as conductors 1–3. (a) Determine the coefficients of electric potential, and (b) determine the electrostatic energy using the coefficients of electric potential, when we give electric charges Q_1, Q_2, and Q_3 to conductors 1, 2, and 3, respectively.

Solution 3.5 (a) The coefficients of electric potential are

$$p_{11} = \frac{1}{4\pi\epsilon_0}\left(\frac{1}{r_0} - \frac{1}{r_1} + \frac{1}{r_2} - \frac{1}{r_3} + \frac{1}{r_4}\right),$$

$$p_{21} = p_{12} = p_{22} = \frac{1}{4\pi\epsilon_0}\left(\frac{1}{r_2} - \frac{1}{r_3} + \frac{1}{r_4}\right),$$

$$p_{31} = p_{32} = p_{13} = p_{23} = p_{33} = \frac{1}{4\pi\epsilon_0 r_4}.$$

(b) From Eq. (3.18) we calculate the electrostatic energy as

$$U_e = \frac{1}{2}\left(p_{11}Q_1^2 + p_{22}Q_2^2 + p_{33}Q_3^2\right) + p_{12}Q_1Q_2 + p_{23}Q_2Q_3 + p_{31}Q_3Q_1$$

$$= \frac{(Q_1 + Q_2 + Q_3)^2}{8\pi\epsilon_0 r_4} + \frac{(Q_1 + Q_2)^2}{8\pi\epsilon_0}\left(\frac{1}{r_2} - \frac{1}{r_3}\right) + \frac{Q_1^2}{8\pi\epsilon_0}\left(\frac{1}{r_0} - \frac{1}{r_1}\right).$$

◆

Example 3.6 An electric charge λ in a unit length is uniformly distributed on the inner conductor of the coaxial conductors in Fig. 3.4, and the outer conductor is grounded. Determine the electrostatic energy and capacitance in a unit length of the coaxial conductors.

Fig. 3.4 Cylindrical coaxial conductors

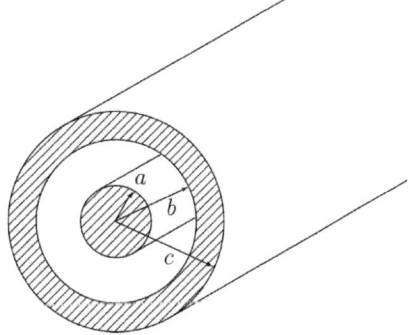

Solution 3.6 The electric charge distributed in a unit length of the concentric conductors is λ on the surface of the inner conductor ($R = a$) and $-\lambda$ on the inner surface of the outer conductor ($R = b$). As a result, the electric field is $E = \lambda/(2\pi\epsilon_0 R)$ in the region $a < R < b$ and is zero in other regions. Hence, the electrostatic energy density in this region is $u_e = \epsilon_0 E^2/2 = \lambda^2/(8\pi^2\epsilon_0 R^2)$ and the electrostatic energy in the conductors of a unit length is

$$U'_e = \int_a^b \frac{\lambda^2}{8\pi^2\epsilon_0 R^2} 2\pi R dR = \frac{\lambda^2}{4\pi\epsilon_0} \log \frac{b}{a}.$$

The electric potential of the outer conductor is zero and that of the inner conductor is

$$\phi = \frac{\lambda}{2\pi\epsilon_0} \log \frac{b}{a}.$$

Hence, we obtain the same electrostatic energy from $U'_e = \lambda\phi/2$ corresponding to Eq. (3.17).

Using this result and $U'_e = \lambda^2/(2C')$ corresponding to Eq. (3.20), the capacitance in a unit length is

$$C' = \frac{2\pi\epsilon_0}{\log(b/a)}.$$

◆

3.4 Electrostatic Force

The electrostatic force between charged conductors is the sum of the Coulomb force on individual electric charges distributed on the conductor surface. We can also determine this force using the principle of virtual displacement and the electrostatic energy.

Suppose that part of an isolated conductor system is forced to move a small distance $\Delta \mathbf{s}$ by an electrostatic force $\mathbf{F}$. The work done by the system, $\mathbf{F} \cdot \Delta \mathbf{s}$, is the energy that the system loses. Hence, if we use ΔU_e to denote the variation in electrostatic energy caused by the movement, the work is equal to $-\Delta U_e$. Thus, we have

$$\mathbf{F} \cdot \Delta \mathbf{s} + \Delta U_e = 0. \tag{3.24}$$

3.4 Electrostatic Force

In the limit $|\Delta s| = \Delta s \to 0$ for displacement along the direction of the electrostatic force, this gives

$$F = -\frac{\partial U_e}{\partial s}. \qquad (3.25)$$

Example 3.7 Suppose that electric charges $\pm Q$ are given to the electrodes of the parallel-plate capacitor shown in Fig. 3.1. Determine the electrostatic force between the electrodes.

Solution 3.7 Assume that the distance between the electrodes is changed to x. In this case, the electric charge Q and the electric field strength $E = Q/\epsilon_0 S$ are unchanged, and it is reasonable to describe the electrostatic energy of Eq. (3.21) as

$$U_e = \frac{Q^2 x}{2\epsilon_0 S}$$

in terms of Q. It should be noted that the voltage V changes with the distance, and it is not suitable to describe U_e in terms of V. Hence, we determine the electrostatic force on the electrode as

$$F = -\left(\frac{\partial U_e}{\partial x}\right)_{x=d} = -\frac{Q^2}{2\epsilon_0 S}.$$

Since this force is negative for expansion (increasing x), it is attractive. ◆

Example 3.8 Electric charges $\pm Q$ are given to each electrode of the capacitor shown in Fig. 3.5. Determine the mechanical work needed to change the distance between the electrodes from d to l. Determine also the change in the electrostatic energy during this change.

Fig. 3.5 Parallel-plate capacitor. The area of each electrode is S

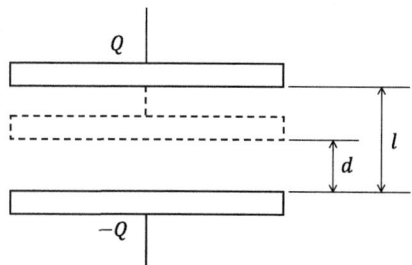

Solution 3.8 The electric field strength that the electric charge on one electrode gives to the other is $E = Q/2\epsilon_0 S$ and is independent of the distance between the electrodes x. The electrostatic force that one electrode exerts on the other is $F = QE = Q^2/2\epsilon_0 S$. Thus, the mechanical work needed to change the distance from d to l is

$$W = \int_d^l F \mathrm{d}x = \frac{Q^2}{2\epsilon_0 S}(l-d).$$

The electrostatic energy density in the space between the two electrodes is $u_\mathrm{e} = \epsilon_0 E^2/2$. Since the volume of this space changes by $S(l-d)$, the change in the electrostatic energy is

$$\Delta U_\mathrm{e} = u_\mathrm{e} S(l-d) = \frac{Q^2}{2\epsilon_0 S}(l-d),$$

which is equal to the mechanical work done on the system.

♦

3.5 Exercises

3.1. We give electric charges q_1 and q_3 to the inner (conductor 1) and outer (conductor 3) conductors, respectively, in the arrangement of concentric spherical conductors shown in Fig. E2.3 in Exercise 2.5. Determine the electric charge of the middle conductor (conductor 2) that causes the electric potential of the middle conductor to be zero with the coefficients of electric potential.

3.2. We give electric charges q_2 and q_3 to the middle (conductor 2) and outer (conductor 3) conductors, respectively, in the arrangement of concentric spherical conductors in Fig. E2.3 in Exercise 2.5. Determine the electric charge of the inner conductor (conductor 1) that causes the electric potential of the inner conductor to be zero with the coefficients of electric potential.

3.3. Solve the problem in Exercise 2.6 with the coefficients of electric potential.

3.4. Solve the problem in Exercise 2.7 with the coefficients of electric potential.

3.5. We treat a pair of concentric spherical conductors, as shown in Fig. 3.2 in Example 3.2, and the inner and outer conductors are denoted by conductors 1 and 2, respectively. Determine the capacity coefficients.

3.6. Determine the capacitance in a unit length of the parallel cylindrical conductors shown in Fig. E3.1.

3.5 Exercises

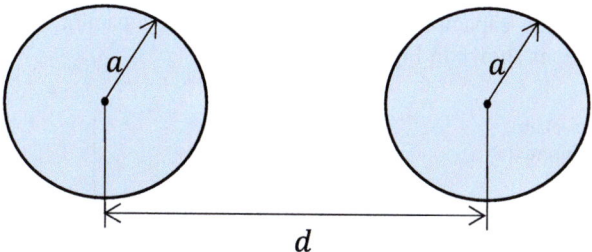

Fig. E3.1 Parallel cylindrical conductors

3.7. Determine the capacitance in a unit length between a flat conductor surface and a cylindrical conductor parallel to it, as treated in Example 2.6. (See Fig. 2.11.)

3.8. Suppose three spherical concentric conductors, as shown in Fig. E3.2. The outer conductor 3 and inner conductor 1 are grounded. Determine the capacitance between the middle conductor 2 and the ground. The thickness of each conductor is negligible.

Fig. E3.2 Spherical conductors with the outer and inner conductors grounded

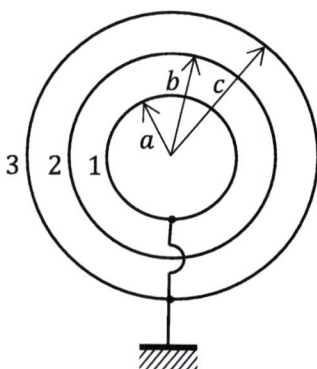

3.9. Solve the problem of Exercise 3.8 with the electric potential.

3.10. The upper electrode is slightly tilted in a capacitor, as shown in Fig. E3.3. The distance between the electrodes changes as $d(x) = d_0 + kx$. Determine the capacitance of this capacitor. The dimension of the capacitor normal to the sheet is b.

Fig. E3.3 Capacitor with a tilted electrode

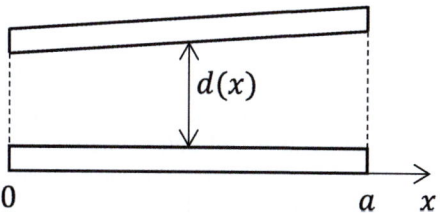

3.11. Determine the capacitance of the parallel-plate capacitor with an inserted conducting plate, as shown in Fig. E3.4.

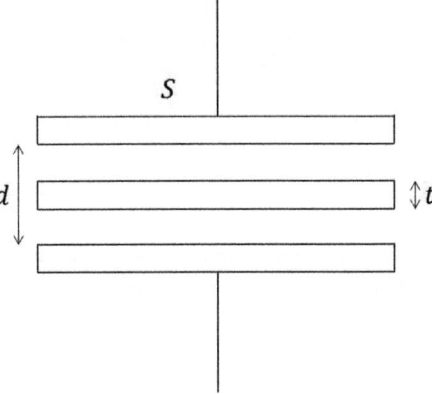

Fig. E3.4 Parallel-plate capacitor with an inserted conducting plate

3.12. The inserted conducting plate is slightly tilted in a capacitor, as shown in Fig. E3.5. The distances between the electrodes change as $d(x) = d_0 - kx$ and $d'(x) = d_0 + kx$. Determine the capacitance of this capacitor. The dimension of the capacitor normal to the sheet is b.

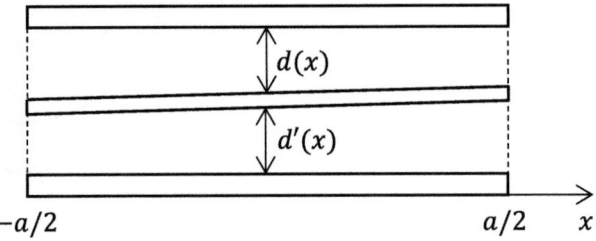

Fig. E3.5 Parallel-plate capacitor with a tilted conducting plate

3.13. Prove Eq. (3.18) using the coefficients of electric potential by mathematical induction.

3.14. Derive Eq. (3.18) for the electrostatic energy of a conductor system from general Eq. (3.19).

3.15. Express the electrostatic energy using the capacity coefficients.

3.16. The electrostatic energy is calculated when electric charges Q_1 and Q_2 are given to the inner and outer spherical conductors in Example 3.4. Determine the electrostatic energy using the capacity coefficients discussed in Exercise 3.15. The capacity coefficients of this system are determined in Exercise 3.5.

3.5 Exercises

3.17. Determine the electrostatic energy in a unit length when we give electric charges $\pm\lambda$ to the parallel long, thin cylindrical conductors that are shown in Fig. 3.3 in Example 3.3. The radius a of the conductor is much smaller than the distance d.

3.18. Electric charges Q_1 and Q_2 are given to the inner and outer spherical concentric conductors, respectively, that are shown in Fig. 3.2 in Example 3.2. Determine the electrostatic energy using the electric potential.

3.19. The electrostatic energy of spherical concentric conductors with electric charges is determined with the coefficients of electric potential in Example 3.5. Solve the same problem with the electric energy density.

3.20. Solve the same problem of Exercise 3.19 with the electric potential.

3.21. Determine the electrostatic energy stored in the capacitor treated in Exercise 3.10 as an assembly of small capacitors, when voltage V is applied.

3.22. Determine the electrostatic energy stored in the capacitor treated in Exercise 3.10 using the electric energy density, when electric charges $\pm Q$ are given.

3.23. Solve the problem of in Exercise 3.21 using the electric potential.

3.24. Suppose that electric charges λ_1 and λ_2 in a unit length are given to the inner and outer conductors of the coaxial conductors shown in Fig. 3.4. Determine the electrostatic energy in a unit length in the space up to the reference point of the electric potential, $R_0(>c)$.

3.25. Suppose that electric charges $\pm\lambda$ in a unit length are given to the electrodes of the coaxial capacitor treated in Example 3.6. Determine the electrostatic energy in a unit length of this capacitor with the coefficients of electric potential.

3.26. Electric charges Q'_1, Q'_2, and Q'_3 are given to the inner (1), middle (2), and outer (3) cylindrical coaxial conductors, respectively, that are shown in Fig. E3.6. Determine the electrostatic energy in the region up to the reference point of the electric potential, $R_\infty(\gg R_5)$, using Eq. (3.23).

Fig. E3.6 Cylindrical coaxial conductors

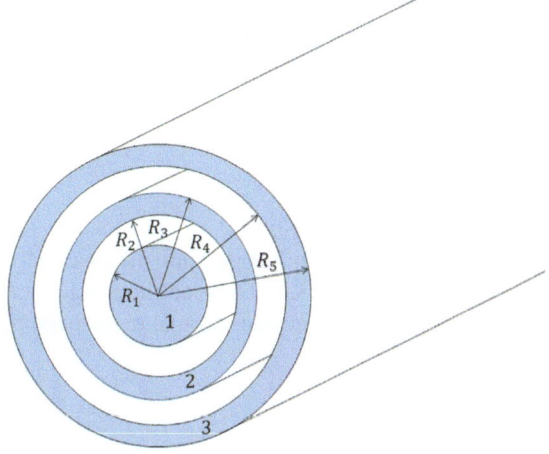

3.27. Solve the same problem as in Exercise 3.26 using the electric potential.

3.28. Solve the same problem as in Exercise 3.26 using the coefficients of the electric potential.

3.29. Electric charges $\pm Q$ are given to cylindrical inner and outer concentric conductors. When the inner conductor is displaced, as shown in Fig. E3.7, determine the force on the inner conductor.

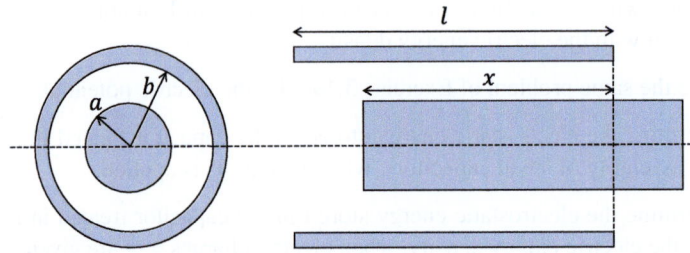

Fig. E3.7 Cylindrical concentric conductors with displaced inner conductor

3.30. Solve the same problem as in Example 3.8 in the condition where the capacitor is connected to a power source.

3.31. Mass m is hanging from the lower electrode of a parallel-plate capacitor connected to a power source with voltage V, as shown in Fig. E3.8. Determine the distance x in the balanced condition. The area of the electrode is S, and neglect the weight of the lower electrode. The gravitational acceleration is denoted by g.

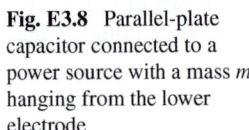

Fig. E3.8 Parallel-plate capacitor connected to a power source with a mass m hanging from the lower electrode

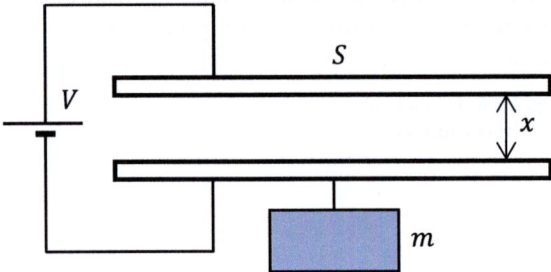

3.32. A conducting plate is inserted into the space between the electrodes of a capacitor, as shown in Fig. E3.9. The electric charges $\pm Q$ are given to the electrodes. The dimension of the electrodes normal to the page is b. Determine the force on the electrodes.

3.5 Exercises

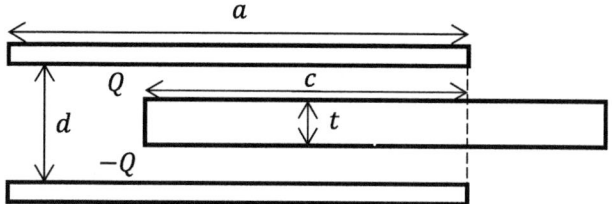

Fig. E3.9 Parallel-plate capacitor with conducting plate inserted into the space between the electrodes

3.33. Solve the same problem as in Exercise 3.32 for the case in which the capacitor is connected to an electric power source with output voltage V.

Answers to Exercises

3.1. We denote the electric charge on conductor 2 by q_2. Using the coefficients of electric potential determined in Example 3.5, the electric potential of conductor 2 is given by

$$\phi_2 = p_{21}q_1 + p_{22}q_2 + p_{23}q_3.$$

Thus, the condition $\phi_2 = 0$ leads to

$$q_2 = -\frac{1}{p_{22}}(p_{21}q_1 + p_{23}q_3) = -q_1 - \frac{r_2 r_3}{r_2 r_3 + (r_3 - r_2)r_4} q_3.$$

3.2. We denote the electric charge on conductor 1 by q_1. Using the coefficients of electric potential determined in Example 3.5, the electric potential of conductor 1 is given by

$$\phi_1 = p_{11}q_1 + p_{12}q_2 + p_{13}q_3.$$

Thus, the condition $\phi_1 = 0$ leads to

$$q_1 = -\frac{1}{p_{11}}(p_{12}q_2 + p_{13}q_3)$$

$$= -\left(\frac{1}{r_0} - \frac{1}{r_1} + \frac{1}{r_2} - \frac{1}{r_3} + \frac{1}{r_4}\right)^{-1} \left[\left(\frac{1}{r_2} - \frac{1}{r_3} + \frac{1}{r_4}\right)q_2 + \frac{q_3}{r_4}\right].$$

3.3. Under the given conditions of $Q_1 = Q$, $Q_2 = 0$, and $Q_3 = q$, the electric potential of conductor 1 is

$$\phi_1 = p_{11}Q + p_{13}q,$$

where the coefficients of electric potential are

$$p_{11} = \frac{1}{4\pi\epsilon_0}\left(\frac{1}{a} - \frac{1}{b} + \frac{1}{c}\right), \quad p_{13} = \frac{1}{4\pi\epsilon_0 d}.$$

The condition $\phi_1 = 0$ leads to

$$q = -\left(\frac{1}{a} - \frac{1}{b} + \frac{1}{c}\right)dQ.$$

3.4. Under the given conditions of $Q_1 = Q$, $Q_2 = Q_x$, and $Q_3 = q$, the Electric potential of conductor 1 is

$$\phi_2 = p_{21}Q + p_{22}Q_x + p_{23}q,$$

where the coefficients of electric potential are

$$p_{21} = p_{22} = \frac{1}{4\pi\epsilon_0 c}, \quad p_{23} = \frac{1}{4\pi\epsilon_0 d}.$$

The condition $\phi_2 = 0$ leads to

$$Q_x = -Q - \frac{c}{d}q.$$

3.5. We give electric charges Q_1 and Q_2 to conductors 1 and 2, respectively. Then, the electric potentials of respective conductors are:

$$\phi_1 = \frac{Q_1}{4\pi\epsilon_0}\left(\frac{1}{a} - \frac{1}{b} + \frac{1}{c}\right) + \frac{Q_2}{4\pi\epsilon_0 c},$$

$$\phi_2 = \frac{Q_1 + Q_2}{4\pi\epsilon_0 c}.$$

Hence, we have

$$Q_1 = \frac{4\pi\epsilon_0 ab}{b-a}(\phi_1 - \phi_2)$$

and

$$Q_2 = -\frac{4\pi\epsilon_0 ab}{b-a}\phi_1 + \left(\frac{4\pi\epsilon_0 ab}{b-a} + 4\pi\epsilon_0 c\right)\phi_2.$$

Thus, the matrix of capacitance coefficients is given by

3.5 Exercises

$$\begin{pmatrix} C_{11} & C_{12} \\ C_{21} & C_{22} \end{pmatrix} = \begin{pmatrix} \dfrac{4\pi \epsilon_0 ab}{b-a} & -\dfrac{4\pi \epsilon_0 ab}{b-a} \\ -\dfrac{4\pi \epsilon_0 ab}{b-a} & \dfrac{4\pi \epsilon_0 ab}{b-a} + 4\pi \epsilon_0 c \end{pmatrix}.$$

On the other hand, the matrix of coefficients of electric potential is

$$\begin{pmatrix} p_{11} & p_{12} \\ p_{21} & p_{22} \end{pmatrix} = \begin{pmatrix} \dfrac{1}{4\pi \epsilon_0}\left(\dfrac{1}{a}-\dfrac{1}{b}+\dfrac{1}{c}\right) & \dfrac{1}{4\pi \epsilon_0 c} \\ \dfrac{1}{4\pi \epsilon_0 c} & \dfrac{1}{4\pi \epsilon_0 c} \end{pmatrix}.$$

Thus, we have

$$\begin{pmatrix} p_{11} & p_{12} \\ p_{21} & p_{22} \end{pmatrix}\begin{pmatrix} C_{11} & C_{12} \\ C_{21} & C_{22} \end{pmatrix} = \begin{pmatrix} 1 & 0 \\ 0 & 1 \end{pmatrix}$$

and Eq. (3.7) is satisfied.

3.6. We apply electric charges $\pm\lambda$ in a unit length to each conductor in a unit length. The image linear charges are assumed as in the condition shown in Fig. B3.1. We define cylindrical coordinates with the z-axis on the central axis of the left conductor. Using the result of Example 2.5, the electric potential in the space outside the conductors is

$$\phi(R, \varphi) = \frac{\lambda}{4\pi \epsilon_0} \log \frac{R^2 + (d-h)^2 - 2R(d-h)\cos\varphi}{R^2 + h^2 - 2Rh\cos\varphi},$$

where R and φ are the distance and azimuthal angle from the central axis, and h is given by

$$h = \frac{1}{2}\left(d - \sqrt{d^2 - 4a^2}\right).$$

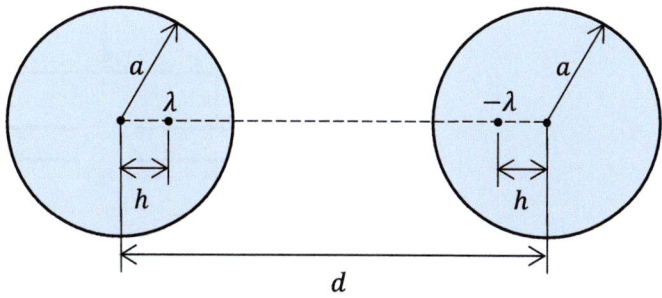

Fig. B3.1 Image linear charges in each conductor

The electric potential on the left conductor surface is

$$\phi(a, 0) = \frac{\lambda}{2\pi \epsilon_0} \log \frac{d - a - h}{a - h} = \frac{\lambda}{2\pi \epsilon_0} \log \frac{d + \sqrt{d^2 - 4a^2}}{2a}.$$

From symmetry, the electric potential difference between two conductors is

$$V = 2\phi(a, 0) = \frac{\lambda}{\pi \epsilon_0} \log \frac{d + \sqrt{d^2 - 4a^2}}{2a}.$$

Hence, the capacitance in a unit length is

$$C' = \frac{V}{\lambda} = \frac{1}{\pi \epsilon_0} \log \frac{d + \sqrt{d^2 - 4a^2}}{2a}.$$

3.7. As discussed in Example 2.6, the electric potential of the flat conductor is zero and that of the cylindrical conductor is

$$\phi = \frac{\lambda}{2\pi \epsilon_0} \log \frac{l + \sqrt{l^2 - a^2}}{a}.$$

So, ϕ gives the potential difference between the conductor surface and the cylindrical conductor. Thus, the capacitance in a unit length is determined to be

$$C' = \frac{\phi}{\lambda} = \frac{1}{2\pi \epsilon_0} \log \frac{l + \sqrt{l^2 - a^2}}{a}.$$

3.8. These conductors are regarded as two capacitances connected in parallel, as shown in Fig. B3.2. The capacitance between conductors 1 and 2 is

$$C_{12} = \frac{4\pi \epsilon_0 ab}{b - a}$$

Fig. B3.2 Equivalent connection

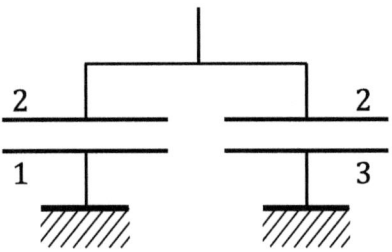

3.5 Exercises

and that between conductors 2 and 3 is

$$C_{23} = \frac{4\pi\epsilon_0 bc}{c-b}.$$

So, the capacitance is determined to be

$$C = C_{12} + C_{23} = \frac{4\pi\epsilon_0 ab}{b-a} + \frac{4\pi\epsilon_0 bc}{c-b} = \frac{4\pi\epsilon_0 b^2(c-a)}{(b-a)(c-b)}.$$

3.9. The electric charges that appear in conductors 1 and 3 are denoted by Q_1 and Q_3, when electric charge Q is given to conductor 2. The electric potential of each conductor is

$$\phi_1 = \frac{1}{4\pi\epsilon_0}\left(\frac{Q_1}{a} + \frac{Q}{b} + \frac{Q_3}{c}\right),$$

$$\phi_2 = \frac{1}{4\pi\epsilon_0}\left(\frac{Q_1}{b} + \frac{Q}{b} + \frac{Q_3}{c}\right),$$

$$\phi_3 = \frac{1}{4\pi\epsilon_0}\left(\frac{Q_1}{c} + \frac{Q}{c} + \frac{Q_3}{c}\right).$$

From the conditions of $\phi_1 = 0$ and $\phi_3 = 0$, electric charges Q_1 and Q_3 are obtained as

$$Q_1 = -\frac{a(c-b)}{b(c-a)}Q, \quad Q_3 = -\frac{c(b-a)}{b(c-b)}Q.$$

Then, the electric potential of conductor 2 is given by

$$\phi_2 = \frac{(b-a)(c-b)Q}{4\pi\epsilon_0 b^2(c-a)}.$$

The capacitance is determined to be

$$C = \frac{Q}{\phi_2} = \frac{4\pi\epsilon_0 b^2(c-a)}{(b-a)(c-b)}.$$

3.10. This capacitor can be regarded as an assembly of thin capacitors with slightly different electrode distances, as shown in Fig. B3.3. The capacitance of each thin capacitor is

$$dC = \frac{\epsilon_0 b dx}{d(x)}.$$

Fig. B3.3 Assembly of thin capacitors with slightly different electrode distances

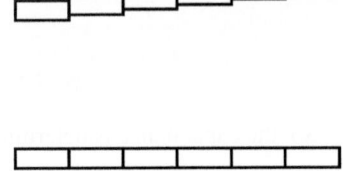

The aimed capacitor is composed by connecting these capacitors in parallel, and hence, the capacitance is given by

$$C = \int_0^a \frac{\epsilon_0 b dx}{d_0 + kx} = \frac{\epsilon_0 b}{k} \log\left(1 + \frac{ka}{d_0}\right).$$

This approaches the value of a usual parallel-plate capacitor, $\epsilon_0 ab/d_0$, in the limit $k \to 0$.

3.11. Suppose that the electric charges on the electrodes are $\pm Q$. Then, the electric field in vacuum between the electrodes is $E = Q/\epsilon_0 S$, while that in the conducting plate is zero. Thus, the voltage between the electrodes is $V = (d - t)E = (d - t)Q/\epsilon_0 S$. The capacitance is determined to be

$$C = \frac{Q}{V} = \frac{\epsilon_0 S}{d - t}.$$

3.12. We determine the capacitance of the upper part. The voltage between the upper plate and the conducting plate is denoted by V. Then, the electric field is $E(x) = V/d(x)$ and the electric charge density is $\sigma(x) = \epsilon_0 E(x) = \epsilon_0 V/d(x)$. The total electric charge is

$$Q = b\int_{-a/2}^{a/2} \sigma(x)dx = b\epsilon_0 V \int_{-a/2}^{a/2} \frac{dx}{d_0 - kx} = \frac{b\epsilon_0 V}{k}\log\frac{d_0 + (ka/2)}{d_0 - (ka/2)}.$$

Thus, the capacitance of the upper part is

$$C' = \frac{b\epsilon_0}{k} \log \frac{d_0 + (ka/2)}{d_0 - (ka/2)}.$$

The capacitance of the lower part is the same. So, the total capacitance is

$$C = \frac{C'}{2} = \frac{b\epsilon_0}{2k} \log \frac{d_0 + (ka/2)}{d_0 - (ka/2)}.$$

3.5 Exercises

3.13. It is clear that the relationship holds for $n = 1$. Then, we assume that the relationship holds for a system with n conductors. In this case, we assume that there are sufficient number of conductors and charges are given to n conductors. This is because the coefficients of electric potentials will change when the number of conductors is really changed. We assume that the i-th conductor has electric charge Q_i. Under this condition, the electric potential of the i-th conductor and the total electrostatic energy of n conductors are respectively given by

$$\phi_i = \sum_{j=1}^{n} p_{ij} Q_j$$

and

$$U_e(n) = \frac{1}{2} \sum_{i=1}^{n} \sum_{j=1}^{n} p_{ij} Q_i Q_j.$$

Then, we add electric charge Q_{n+1} to the $(n+1)$-th conductor. The electric potentials of other conductors are given by

$$\phi_i = \sum_{j=1}^{n+1} p_{ij} Q_j \quad (i = 1, \ldots, n),$$

and the increment of the electrostatic energy of these conductors is

$$\Delta U_e' = \frac{1}{2} \sum_{i=1}^{n} \sum_{j=1}^{n} p_{ij} Q_i Q_{n+1}.$$

The electrostatic energy of the $(n+1)$-th conductor is

$$\Delta U_e'' = \frac{1}{2} p_{n+1\,n+1} Q_{n+1}^2.$$

Then, the total electrostatic energy is given by

$$U_e(n+1) = U_e(n) + \Delta U_e' + \Delta U_e'' = \sum_{i=1}^{n+1} \sum_{j=1}^{n+1} p_{ij} Q_i Q_j = \frac{1}{2} \sum_{i=1}^{n+1} Q_i \phi_i.$$

Thus, the form of Eq. (3.18) holds also for $n+1$ conductors. So, Eq. (3.18) is proved to be valid generally.

3.14. The electrostatic energy of the conductor system is written as

$$U_e = \frac{1}{2} \sum_i \int_{V_i} \rho(r)\phi(r)dV,$$

where V_i represents the region of the i-th conductor. Since the electric potential of each conductor is a constant, the electric potential of the i-th conductor is denoted by ϕ_i. So, the electrostatic energy is written as

$$U_e = \frac{1}{2} \sum_i \phi_i \int_{V_i} \rho(r)dV = \frac{1}{2} \sum_i \phi_i Q_i,$$

where

$$Q_i = \int_{V_i} \rho(r)dV$$

is the electric charge of the i-th conductor.

3.15. Substitution of

$$Q_i = \sum_{j=1}^n C_{ij}\phi_j$$

for the electric charge into the expression of the electrostatic energy,

$$U_e = \frac{1}{2} \sum_{i=1}^n Q_i\phi_i,$$

leads to

$$U_e = \frac{1}{2} \sum_{i=1}^n \sum_{j=1}^n C_{ij}\phi_i\phi_j.$$

3.16. The electric potentials of the inner and outer conductors are denoted by ϕ_1 and ϕ_2. Then, with the capacity coefficients determined in Exercise 3.5, the electrostatic energy is given by

$$U_e = \frac{1}{2}C_{11}\phi_1^2 + C_{12}\phi_1\phi_2 + \frac{1}{2}C_{22}\phi_2^2$$

$$= \frac{2\pi\epsilon_0 ab}{b-a}(\phi_1 - \phi_2)^2 + 2\pi\epsilon_0 c\phi_2^2.$$

3.5 Exercises

Using the relationships obtained in Exercise 3.5,

$$\phi_1 - \phi_2 = \frac{(b-a)Q_1}{4\pi\epsilon_0 ab}, \quad \phi_2 = \frac{Q_1 + Q_2}{4\pi\epsilon_0 c},$$

the electrostatic energy is written as

$$U_e = \frac{1}{8\pi\epsilon_0}\left[\left(\frac{1}{a} - \frac{1}{b}\right)Q_1^2 + \frac{1}{c}(Q_1 + Q_2)^2\right].$$

This agrees with the result obtained in Example 3.4.

3.17. It can be approximated that the electric charge is uniformly distributed on each conductor surface. The electric potential of the conductor with electric charge λ is

$$\phi_+ = -\frac{\lambda}{2\pi\epsilon_0}\log a + \frac{\lambda}{2\pi\epsilon_0}\log(d-a) = \frac{\lambda}{2\pi\epsilon_0}\log\frac{d-a}{a}.$$

The electric potential of the conductor with electric charge $-\lambda$ is similarly obtained as

$$\phi_- = \frac{\lambda}{2\pi\epsilon_0}\log a - \frac{\lambda}{2\pi\epsilon_0}\log(d-a) = -\frac{\lambda}{2\pi\epsilon_0}\log\frac{d-a}{a}.$$

So, the electrostatic energy in a unit length is determined to be

$$U_e' = \frac{1}{2}(\phi_+\lambda - \phi_-\lambda) = \frac{\lambda^2}{2\pi\epsilon_0}\log\frac{d-a}{a}.$$

3.18. The electric field is given by

$$E(R) = \frac{Q_1 + Q_2}{4\pi\epsilon_0}; \quad r > c,$$

$$= \frac{Q_1}{4\pi\epsilon_0}; \quad a < r < b,$$

$$= 0; \quad \text{otherwise.}$$

So, the electric potentials of the inner and outer conductors are

$$\phi_1 = \frac{Q_1}{4\pi\epsilon_0}\left(\frac{1}{a} - \frac{1}{b}\right) + \frac{Q_1 + Q_2}{4\pi\epsilon_0 c},$$

$$\phi_2 = \frac{Q_1 + Q_2}{4\pi\epsilon_0 c}.$$

The electrostatic energy is determined to be

$$U_e = \frac{1}{2}(\phi_1 Q_1 + \phi_2 Q_2) = \frac{1}{8\pi\epsilon_0}\left[Q_1^2\left(\frac{1}{a} - \frac{1}{b}\right) + \frac{(Q_1 + Q_2)^2}{c}\right].$$

3.19. The electric field is

$$E(r) = \frac{Q_1}{4\pi\epsilon_0 r^2}; \qquad r_0 < r < r_1,$$

$$= \frac{Q_1 + Q_2}{4\pi\epsilon_0 r^2}; \qquad r_2 < r < r_3,$$

$$= \frac{Q_1 + Q_2 + Q_3}{4\pi\epsilon_0 r^2}; \qquad r > r_4.$$

Hence, the electrostatic energy density is

$$u_e = \frac{Q_1^2}{32\pi^2\epsilon_0 r^4}; \qquad r_0 < r < r_1,$$

$$= \frac{(Q_1 + Q_2)^2}{32\pi^2\epsilon_0 r^4}; \qquad r_2 < r < r_3,$$

$$= \frac{(Q_1 + Q_2 + Q_3)^2}{32\pi^2\epsilon_0 r^4}; \qquad r > r_4.$$

The electric energy is determined to be

$$U_e = \frac{Q_1^2}{32\pi^2\epsilon_0}\int_{r_0}^{r_1}\frac{4\pi r^2}{r^4}dr + \frac{(Q_1 + Q_2)^2}{32\pi^2\epsilon_0}\int_{r_2}^{r_3}\frac{4\pi r^2}{r^4}dr$$

$$+ \frac{(Q_1 + Q_2 + Q_3)^2}{32\pi^2\epsilon_0}\int_{r_4}^{\infty}\frac{4\pi r^2}{r^4}dr$$

$$= \frac{Q_1^2}{8\pi\epsilon_0}\left(\frac{1}{r_0} - \frac{1}{r_1}\right) + \frac{(Q_1 + Q_2)^2}{8\pi\epsilon_0}\left(\frac{1}{r_2} - \frac{1}{r_3}\right) + \frac{(Q_1 + Q_2 + Q_3)^2}{8\pi\epsilon_0 r_4}.$$

3.20. The electric potentials of conductors 1, 2, and 3 are

$$\phi_3 = \frac{Q_1 + Q_2 + Q_3}{4\pi\epsilon_0 r_4},$$

$$\phi_2 = \frac{Q_1 + Q_2 + Q_3}{4\pi\epsilon_0 r_4} + \frac{Q_1 + Q_2}{4\pi\epsilon_0}\left(\frac{1}{r_2} - \frac{1}{r_3}\right),$$

3.5 Exercises

$$\phi_1 = \frac{Q_1 + Q_2 + Q_3}{4\pi\epsilon_0 r_4} + \frac{Q_1 + Q_2}{4\pi\epsilon_0}\left(\frac{1}{r_2} - \frac{1}{r_3}\right) + \frac{Q_1}{4\pi\epsilon_0}\left(\frac{1}{r_0} - \frac{1}{r_1}\right).$$

Thus, the electrostatic energy is determined to be

$$U_e = \frac{1}{2}\sum_{i=1}^{3}\phi_i Q_i$$

$$= \frac{Q_1^2}{8\pi\epsilon_0}\left(\frac{1}{r_0} - \frac{1}{r_1}\right) + \frac{(Q_1+Q_2)^2}{8\pi\epsilon_0}\left(\frac{1}{r_2} - \frac{1}{r_3}\right) + \frac{(Q_1+Q_2+Q_3)^2}{8\pi\epsilon_0 r_4}.$$

3.21. The capacity of a piece of capacitor at position x is

$$dC = \frac{\epsilon_0 b dx}{(d_0 + kx)}.$$

So, the electrostatic energy stored in this capacitor is

$$dU_e = \frac{1}{2}dCV^2 = \frac{\epsilon_0 bV^2 dx}{2(d_0 + kx)}.$$

The total electrostatic energy is determined to be

$$U_e = \int dU_e = \int_0^a \frac{\epsilon_0 bV^2 dx}{2(d_0 + kx)} = \frac{\epsilon_0 bV^2}{2k}\log\left(1 + \frac{ka}{d_0}\right),$$

which is rewritten as $U_e = CV^2/2$ in terms of the capacitance obtained in Exercise 3.10.

3.22. The voltage between the electrodes and the positive electric charge density on the electrode surface are denoted by V and $\sigma(x)$, respectively. Then, the voltage between the electrodes is $E(x) = \sigma(x)/\epsilon_0$, and we have

$$\sigma(x) = \frac{\epsilon_0 V}{d_0 + kx}.$$

Hence, the total positive electric charge is

$$Q = b\int_0^a \sigma(x)dx = \frac{\epsilon_0 bV}{k}\log\left(1 + \frac{ka}{d_0}\right).$$

Thus, the electric field is given by

$$E(x) = \frac{kQ}{\epsilon_0 b\log[1 + (ka/d_0)](d_0 + kx)},$$

and the electrostatic energy is determined to be

$$U_e = \frac{\epsilon_0 b}{2} \int_0^a E^2(x) d(x) dx = \frac{k^2 Q^2}{2\epsilon_0 b \{\log[1 + (ka/d_0)]\}^2} \int_0^a \frac{dx}{d_0 + kx}$$

$$= \frac{kQ^2}{2\epsilon_0 b \log[1 + (ka/d_0)]}.$$

3.23. The electric potential of the electrode with a negative electric charge is denoted by ϕ_1. Then, the potential of another electrode with a positive electric charge is $\phi_2 = \phi_1 + V$. The density of the positive electric charge on the electrode surface, σ, is given by $\epsilon_0 V/d(x)$. Hence, the total positive charge is

$$Q = b \int_0^a \frac{\epsilon_0 V}{d_0 + kx} dx = \frac{\epsilon_0 b V}{k} \log\left(1 + \frac{ka}{d_0}\right).$$

The electrostatic energy of the capacitor is determined to be

$$U_e = -\frac{Q}{2} \phi_1 + \frac{Q}{2}(\phi_1 + V) = \frac{QV}{2} = \frac{\epsilon_0 b V^2}{2k} \log\left(1 + \frac{ka}{d_0}\right).$$

3.24. The electric field strength is

$$E = \frac{\lambda_1}{2\pi \epsilon_0 R}; \quad a < R < b,$$

$$= \frac{\lambda_1 + \lambda_2}{2\pi \epsilon_0 R}; \quad R > c,$$

$$= 0; \quad \text{otherwise.}$$

Hence, the electrostatic energy in a unit length is determined to be

$$U'_e = \frac{1}{8\pi^2 \epsilon_0} \left[\int_a^b \frac{\lambda_1^2}{R^2} 2\pi R dR + \int_c^{R_0} \frac{(\lambda_1 + \lambda_2)^2}{R^2} 2\pi R dR \right]$$

$$= \frac{1}{4\pi \epsilon_0} \left[\lambda_1^2 \log \frac{b}{a} + (\lambda_1 + \lambda_2)^2 \log \frac{R_0}{c} \right].$$

3.5 Exercises

The electric potential of each conductor is

$$\phi_{\text{out}} = \frac{\lambda_1 + \lambda_2}{2\pi\epsilon_0} \log \frac{R_0}{c},$$

$$\phi_{\text{in}} = \frac{\lambda_1 + \lambda_2}{2\pi\epsilon_0} \log \frac{R_0}{c} + \frac{\lambda_1}{2\pi\epsilon_0} \log \frac{b}{a}.$$

Then, the electrostatic energy is also obtained as

$$U_e' = \frac{1}{2}\phi_{\text{in}}\lambda_1 + \frac{1}{2}\phi_{\text{out}}\lambda_2 = \frac{1}{4\pi\epsilon_0}\left[\lambda_1^2 \log \frac{b}{a} + (\lambda_1 + \lambda_2)^2 \log \frac{R_0}{c}\right].$$

3.25. We assume the reference point of the electric potential at $R = R_0$. We denote the inner and outer conductors as conductor 1 and conductor 2. Then, the coefficients of the electric potential are

$$\hat{p}_{11} = \frac{1}{2\pi\epsilon_0} \log \frac{bR_0}{ac}, \quad \hat{p}_{21} = \hat{p}_{12} = \hat{p}_{22} = \frac{1}{2\pi\epsilon_0} \log \frac{R_0}{c}.$$

Hence, the electrostatic energy in a unit length in the given condition is

$$U_e' = \frac{1}{2}\hat{p}_{11}\lambda^2 - \hat{p}_{21}\lambda^2 + \frac{1}{2}\hat{p}_{22}\lambda^2 = \frac{\lambda^2}{4\pi\epsilon_0} \log \frac{b}{a}.$$

3.26. The electric field is

$$E = \frac{Q_1'}{2\pi\epsilon_0 R}; \quad R_1 < R < R_2,$$

$$= \frac{Q_1' + Q_2'}{2\pi\epsilon_0 R}; \quad R_3 < R < R_4,$$

$$= \frac{Q_1' + Q_2' + Q_3'}{2\pi\epsilon_0 R}; \quad R > R_5,$$

$$= 0; \quad \text{otherwise}.$$

Hence, the electrostatic energy in a unit length of the coaxial conductor is given by

$$U_e' = \frac{\epsilon_0}{2} \int_0^{R_\infty} E^2 2\pi R dR$$

$$= \frac{Q_1'^2}{4\pi\epsilon_0} \log \frac{R_2}{R_1} + \frac{(Q_1' + Q_2')^2}{4\pi\epsilon_0} \log \frac{R_4}{R_3} + \frac{(Q_1' + Q_2' + Q_3')^2}{4\pi\epsilon_0} \log \frac{R_\infty}{R_5}.$$

3.27. The electric potential of each conductor is

$$\phi_3 = \frac{Q'_1 + Q'_2 + Q'_3}{2\pi \epsilon_0} \log \frac{R_\infty}{R_5},$$

$$\phi_2 = \frac{Q'_1 + Q'_2}{2\pi \epsilon_0} \log \frac{R_4}{R_3} + \frac{Q'_1 + Q'_2 + Q'_3}{2\pi \epsilon_0} \log \frac{R_\infty}{R_5},$$

$$\phi_1 = \frac{Q'_1}{2\pi \epsilon_0} \log \frac{R_2}{R_1} + \frac{Q'_1 + Q'_2}{2\pi \epsilon_0} \log \frac{R_4}{R_3} + \frac{Q'_1 + Q'_2 + Q'_3}{2\pi \epsilon_0} \log \frac{R_\infty}{R_5}.$$

Hence, the electrostatic energy in a unit length of the coaxial conductors is determined to be

$$U'_e = \frac{1}{2}(Q'_1\phi_1 + Q'_2\phi_2 + Q'_3\phi_3)$$

$$= \frac{Q'^2_1}{4\pi \epsilon_0} \log \frac{R_2}{R_1} + \frac{(Q'_1 + Q'_2)^2}{4\pi \epsilon_0} \log \frac{R_4}{R_3} + \frac{(Q'_1 + Q'_2 + Q'_3)^2}{4\pi \epsilon_0} \log \frac{R_\infty}{R_5}.$$

3.28. The coefficients of the electric potential are

$$\hat{p}_{11} = \frac{1}{2\pi \epsilon_0} \log \frac{R_2 R_4 R_\infty}{R_1 R_3 R_5},$$

$$\hat{p}_{21} = \hat{p}_{12} = \hat{p}_{22} = \frac{1}{2\pi \epsilon_0} \log \frac{R_4 R_\infty}{R_3 R_5},$$

$$\hat{p}_{31} = \hat{p}_{13} = \hat{p}_{32} = \hat{p}_{23} = \hat{p}_{33} = \frac{1}{2\pi \epsilon_0} \log \frac{R_\infty}{R_5}.$$

Hence, the electrostatic energy in a unit length of the coaxial conductors is determined to be

$$U'_e = \frac{1}{2}\hat{p}_{11}Q'^2_1 + \frac{1}{2}\hat{p}_{22}Q'^2_2 + \frac{1}{2}\hat{p}_{33}Q'^2_3 + \hat{p}_{12}Q'_1 Q'_2 + \hat{p}_{23}Q'_2 Q'_3 + \hat{p}_{31}Q'_3 Q'_1$$

$$= \frac{Q'^2_1}{4\pi \epsilon_0} \log \frac{R_2}{R_1} + \frac{(Q'_1 + Q'_2)^2}{4\pi \epsilon_0} \log \frac{R_4}{R_3} + \frac{(Q'_1 + Q'_2 + Q'_3)^2}{4\pi \epsilon_0} \log \frac{R_\infty}{R_5}.$$

3.29. Electric charges are concentrated within the overlapped region of length x in both conductors, and the charge density on the surface of the inner conductor is $\sigma = Q/2\pi a x$, and the electric field in vacuum in this region is

$$E(R) = \frac{a\sigma}{\epsilon_0 R} = \frac{Q}{2\pi \epsilon_0 R x}; \quad a < R < b.$$

The electrostatic energy is

$$U_e = \int_a^b \frac{1}{2}\epsilon_0 E^2(R) 2\pi R dR x = \frac{Q^2}{4\pi \epsilon_0 x} \log \frac{b}{a}.$$

Hence, the force on the inner conductor is

$$F = -\frac{\partial U_e}{\partial x} = \frac{Q^2}{4\pi\epsilon_0 x^2} \log\frac{b}{a}.$$

The force is positive for increasing x, indicating an attractive force.

3.30. Suppose that the capacitor is connected to a power source with the voltage V and that the electric charges of the electrodes are $\pm Q$ when the distance between the electrodes is d. The change in the distance between the electrodes is denoted by x. When x is changed, V is kept constant, and Q changes. Thus, the electric charges move between the capacitor and the power source. If the electric charge is denoted by $Q'(x)$, we have $Q'(x) = (d/x)Q = \epsilon_0 SV/x$. First, we shall determine the mechanical work needed to increase the distance between the electrodes. Since the total electric field strength is $E = V/x$, one electrode experiences the electric field strength $E/2$ from the other electrode and the corresponding Coulomb force is $EQ'(x)/2 = \epsilon_0 SV^2/2x^2$. Hence, the needed work is

$$W = \frac{\epsilon_0 SV^2}{2}\int_d^l \frac{dx}{x^2} = \frac{\epsilon_0 SV^2}{2}\left(\frac{1}{d} - \frac{1}{l}\right).$$

Second, we shall determine the change in the electrostatic energy between the electrodes. The energy density changes from $\epsilon_0 V^2/2d^2$ to $\epsilon_0 V^2/2l^2$, and the volume of the space changes from Sd to Sl. Hence, the increase in the electrostatic energy is

$$\Delta U_e = -\frac{\epsilon_0 SV^2}{2}\left(\frac{1}{d} - \frac{1}{l}\right).$$

Thus, whereas we do mechanical work, the electrostatic energy decreases. Thus, the total lost energy is

$$W - \Delta U_e = \epsilon_0 SV^2\left(\frac{1}{d} - \frac{1}{l}\right).$$

On the other hand, the electric charge goes back to the power source, which is $\Delta Q = Q - Q'(l) = \epsilon_0 SV(d^{-1} - l^{-1})$. The corresponding energy that moves to the power source is

$$\Delta U_b = \Delta QV = \epsilon_0 SV^2\left(\frac{1}{d} - \frac{1}{l}\right).$$

This is equal to the lost energy estimated above. That is, the energy due to the mechanical work and a part of the electrostatic energy move to the power source.

3.31. Since the electric field between the electrodes is $E = V/x$, the electrostatic energy of the capacitor is given by

$$U_e = \frac{1}{2}\epsilon_0 E^2 S x = \frac{V^2 S}{2x}.$$

So, the increase in the electrostatic energy when the distance increases by Δx is

$$\Delta U_e = -\frac{\epsilon_0 V^2 S}{2x^2} \Delta x.$$

This change introduces the change in the electric charges on the electrodes. The change in the electric field is $\Delta E = -(V/x^2)\Delta x$ and the change in the charge density is $\Delta \sigma = -(\epsilon_0 V/x^2)\Delta x$. Hence, the change in the electric charge is $\Delta Q = -(\epsilon_0 V S/x^2)\Delta x$. Thus, the energy to the power source, $-V\Delta Q$, must be considered. That is, the sum of the mechanical work to the outside $F\Delta x$ and the energy to the power source $-V\Delta Q$ must be equal to the decrease in the energy of the capacitor: $F\Delta x - V\Delta Q = -\Delta U_e$. Thus, we have

$$F = -\frac{\Delta U_e}{\Delta x} + V\frac{\Delta Q}{\Delta x} = -\frac{\epsilon_0 V^2 S}{2x^2}.$$

This force is negative for increasing x, showing an attractive force between the electrodes and is balanced with the gravitational force, mg. Thus, we have

$$x = \left(\frac{\epsilon_0 S}{2mg}\right)^{1/2} V.$$

3.32. The positive electric charge densities in the area with and without the conducting plate are denoted by σ' and σ, respectively. Then, the corresponding electric field is σ'/ϵ_0 and σ/ϵ_0, respectively. From the requirement that the voltage is the same between these areas, we have $\sigma' = d\sigma/(d-t)$. The positive electric charge is given by $Q = b[(a-c)\sigma + c\sigma']$. So, we have

$$\sigma = \frac{(d-t)Q}{b[a(d-t)+ct]}, \quad \sigma' = \frac{dQ}{b[a(d-t)+ct]}.$$

The electrostatic energy is

$$U_e = \frac{1}{2\epsilon_0}\left[bd(a-c)\sigma^2 + b(d-t)c\sigma'^2\right]$$

$$= \frac{Qd(d-t)}{2\epsilon_0 b[a(d-t)+ct]}.$$

The force on the electrode is determined to be

3.5 Exercises

$$F = -\frac{\partial U_e}{\partial d} = -\frac{Q[ad^2 - 2(a-c)td + (a-c)t^2]}{2\epsilon_0 b[a(d-t) + ct]^2}.$$

It is obvious that the force is negative for increasing d, showing an attractive force.

3.33. The electric fields in the area with and without the conducting plate are denoted by E' and E, respectively. These are given by

$$E' = \frac{V}{d-t}, \quad E = \frac{V}{d}.$$

The electrostatic energy is

$$U_e = \frac{\epsilon_0}{2}\left[\left(\frac{V}{d}\right)^2 bd(a-c) + \left(\frac{V}{d-t}\right)^2 b(d-t)c\right]$$

$$= \frac{\epsilon_0 V^2 b}{2}\left(\frac{a-c}{d} + \frac{c}{d-t}\right).$$

Hence, when the distance d increases by Δd, the electrostatic energy changes by

$$\Delta U_e = -\frac{\epsilon_0 V^2 b}{2}\left[\frac{a-c}{d^2} + \frac{c}{(d-t)^2}\right]\Delta d.$$

In this case, the electric charge densities in the area with and without the conducting plate are $\epsilon_0 V/(d-t)$ and $\epsilon_0 V/d$, respectively. Hence, the total positive electric charge is

$$Q = \epsilon_0 Vb\left(\frac{a-c}{d} + \frac{c}{d-t}\right).$$

Hence, during the change in the distance the total electric charge changes by

$$\Delta Q = -\epsilon_0 Vb\left[\frac{a-c}{d^2} + \frac{c}{(d-t)^2}\right]\Delta d.$$

That is, the electric charge $-\Delta Q$ and the corresponding energy $-V\Delta Q$ go back to the power source. Thus, the increase in the energy of the capacitor, ΔU_e, must be balanced with the sum of the mechanical work $F\Delta x$ and the energy to the power source $-V\Delta Q$. So, we have $\Delta U_e + F\Delta x - V\Delta Q = 0$. The force is determined to be

$$F = -\frac{\Delta U_e}{\Delta x} + V\frac{\Delta Q}{\Delta x} = -\frac{\epsilon_0 V^2 b}{2}\left[\frac{a-c}{d^2} + \frac{c}{(d-t)^2}\right].$$

Chapter 4
Dielectric Materials

4.1 Electric Polarization

When an electric field is applied across a dielectric material, the nuclei and electrons bonded in it are slightly displaced opposite to each other, and innumerable electric dipoles appear inside the dielectric material. As a result, the electrically neutral state is maintained inside the material but electric charges appear on the surfaces. This phenomenon is called electric polarization or dielectric polarization, and the electric charge that appears on the surface is called the polarization charge. The above example caused by relative displacements of electrons and nuclei under an applied electric field is called electronic polarization.

In ionic crystals composed of cations and anions, the relative displacement of positive and negative ions in the electric field also brings about the type of electric polarization called ionic polarization. There is another case in which the molecule itself has an electric dipole moment, and its inclination along the direction of the applied electric field causes electric polarization. This is called orientation polarization. In many cases the magnitude of the electric polarization is proportional to the electric field strength.

We use $\boldsymbol{P}$ to represent the electric dipole moment appearing in a unit volume of the dielectric material in an electric field $\boldsymbol{E}$. This quantity is also called the electric polarization. Its unit is [C/m^2]. Usually, $\boldsymbol{P}$ is proportional to $\boldsymbol{E}$:

$$\boldsymbol{P} = \epsilon_0 \chi_e \boldsymbol{E}, \tag{4.1}$$

where χ_e is a dimensionless constant of proportionality called the electric susceptibility.

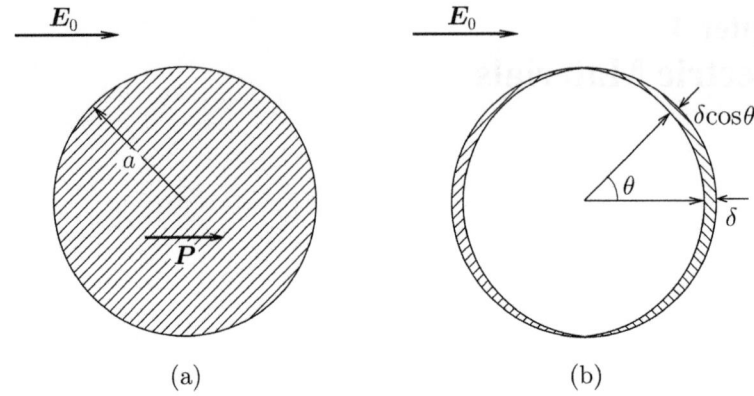

Fig. 4.1 **a** Electric polarization induced in a dielectric sphere and **b** polarization charge that appears on the surface

Suppose a small rectangular parallelepiped with two surfaces perpendicular to the direction of the electric polarization with strength P. If the surface densities of polarization charge that appears on these surfaces are $\pm\sigma_p$, we have

$$\sigma_p = P. \tag{4.2}$$

When the electric polarization is uniform in space, the polarization charge appears only on the surface, as shown above. When the electric polarization is not uniform, however, the polarization charge also appears inside the dielectric material. If the density of electric polarization is denoted by ρ_p, we have.

$$\nabla \cdot \boldsymbol{P} = -\rho_p. \tag{4.3}$$

Suppose that a dielectric sphere of diameter a is placed in a uniform electric field of strength, $\boldsymbol{E}_0$, as shown in Fig. 4.1a. Here, we determine the surface density of the polarization charge that appears on the surface of the dielectric sphere. The magnitude of the electric polarization, $\boldsymbol{P}$, is denoted by P.

We define spherical coordinates with the origin at the center of the sphere and the axis along the direction of the applied electric field . We denote by θ the zenithal angle measured from this axis, as shown in Fig. 4.1b. We also denote the positive and negative electric charge densities and the relative displacement of these charges by $\pm\rho_p$ and δ, respectively. Thus, the surface density of the polarization charge is given by $\sigma_p(\theta) = \rho_p \delta \cos\theta$. Since the magnitude of the electric polarization is equal to the amount of positive polarization charge that crosses the surface of a unit area, it is given by $P = \rho_p \delta$. Thus, we have

$$\sigma_p(\theta) = P\cos\theta. \tag{4.4}$$

4.2 Electric Flux Density

Suppose that true electric charge and polarization charge coexist with densities ρ and ρ_p, respectively, in region V. Both of them contribute to the electric field, and Gauss's law is written as

$$\int_S \boldsymbol{E} \cdot \mathrm{d}\boldsymbol{S} = \frac{1}{\epsilon_0} \int_V (\rho + \rho_p) \mathrm{d}V, \tag{4.5}$$

where S is the surface of region V. Substituting Eq. (4.3) for ρ_p gives

$$\int_S (\epsilon_0 \boldsymbol{E} + \boldsymbol{P}) \cdot \mathrm{d}\boldsymbol{S} = \int_V \rho \, \mathrm{d}V. \tag{4.6}$$

Here, we define

$$\boldsymbol{D} = \epsilon_0 \boldsymbol{E} + \boldsymbol{P} \tag{4.7}$$

and call $\boldsymbol{D}$ the electric flux density or electric displacement. Its unit is the same as for $\boldsymbol{P}$ and is [C/m^2]. Over a relatively wide range of electric field strength, $\boldsymbol{P}$ is proportional to $\boldsymbol{E}$, and $\boldsymbol{D}$ is also proportional to $\boldsymbol{E}$. Thus, we can write $\boldsymbol{D}$ as

$$\boldsymbol{D} = \epsilon_0 (1 + \chi_e) \boldsymbol{E} = \epsilon \boldsymbol{E}, \tag{4.8}$$

where ϵ is a constant inherent to each material called the dielectric constant. Its unit is the same as for ϵ_0.

Using the electric flux density, Eq. (4.6) gives

$$\int_S \boldsymbol{D} \cdot \mathrm{d}\boldsymbol{S} = \int_V \rho \, \mathrm{d}V. \tag{4.9}$$

This is an extension of Eq. (1.11) and is generally called Gauss's law. It should be noted that there is no constant in Eq. (4.9) and that only true electric charge is involved. Equation (1.11) describes only phenomena in vacuum or conductors, and is Gauss's law in a narrow sense. The surface integral of the electric flux density on the left side of Eq. (4.9) is called the electric flux. From Eq. (4.9), we have

$$\nabla \cdot \boldsymbol{D} = \rho. \tag{4.10}$$

This is the general form of Gauss's divergence law.

Substituting Eqs. (1.18) and (4.8) into Eq. (4.10) leads to Poisson's equation:

$$\Delta\phi = -\frac{\rho}{\epsilon}. \tag{4.11}$$

4.3 Boundary Conditions

Here, we describe the boundary conditions to be fulfilled for the electric field and electric flux density at an interface between two different dielectric materials with dielectric constants ϵ_1 and ϵ_2. Assume that a true electric charge with a surface density, σ, exists on the interface.

The boundary condition on the electric flux density is given by

$$\boldsymbol{n} \cdot (\boldsymbol{D}_1 - \boldsymbol{D}_2) = \sigma, \tag{4.12}$$

where $\boldsymbol{n}$ is the normal unit vector on the boundary directed from dielectric material 1 to 2. The boundary condition on the electric field is given by

$$\boldsymbol{n} \times (\boldsymbol{E}_1 - \boldsymbol{E}_2) = 0. \tag{4.13}$$

Using the electric potential, the boundary conditions (4.12) and (4.13) are, respectively, written as

$$\boldsymbol{n} \cdot (\epsilon_1 \nabla \phi_1 - \epsilon_2 \nabla \phi_2) = \sigma, \tag{4.14}$$

and

$$\phi_1 = \phi_2. \tag{4.15}$$

Here, we discuss refraction of electric field lines at a boundary using the boundary conditions. Suppose an interface between two dielectric materials with dielectric constants ϵ_1 and ϵ_2. Assume that an electric field of strength E_1 is applied to dielectric material 1 in the direction of an angle θ_1 measured from the normal direction to the interface, as shown in Fig. 4.2. The strength and angle of the electric field in dielectric material 2 are denoted by E_2 and θ_2. Since true electric charge does not usually exist on the interface, the normal component of the electric flux density is continuous on it:

$$\epsilon_1 E_1 \cos\theta_1 = \epsilon_2 E_2 \cos\theta_2. \tag{4.16}$$

The continuity of the parallel component of the electric field is written as

$$E_1 \sin\theta_1 = E_2 \sin\theta_2. \tag{4.17}$$

4.3 Boundary Conditions

Fig. 4.2 Refraction of electric field lines at an interface

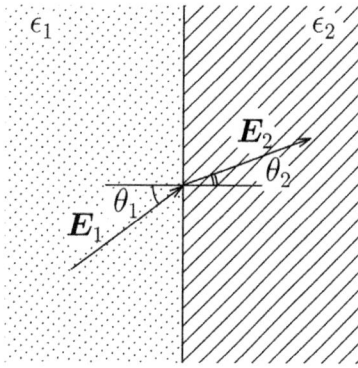

These equations give

$$\frac{\tan \theta_1}{\tan \theta_2} = \frac{\epsilon_1}{\epsilon_2}. \tag{4.18}$$

This is the law of refraction. We obtain E_2 and θ_2 as

$$E_2 = E_1 \left[\sin^2 \theta_1 + \left(\frac{\epsilon_1}{\epsilon_2}\right)^2 \cos^2 \theta_1 \right]^{1/2}, \tag{4.19}$$

$$\theta_2 = \tan^{-1}\left(\frac{\epsilon_2}{\epsilon_1} \tan \theta_1\right). \tag{4.20}$$

Example 4.1 The space between the electrodes in a concentric spherical capacitor is occupied by two dielectric materials with dielectric constants ϵ_1 and ϵ_2, as shown in Fig. 4.3. Determine the capacitance of the capacitor.

Fig. 4.3 Concentric spherical capacitor with two dielectric materials

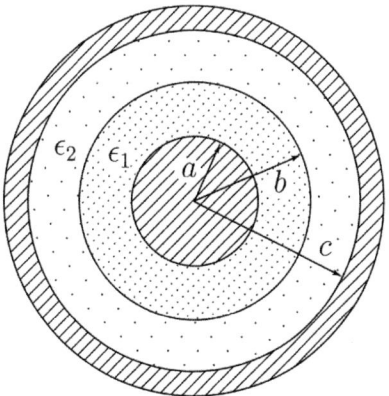

Solution 4.1 Assume that electric charges Q and $-Q$ appear on the inner and outer electrodes, respectively, when we apply potential difference V between the two electrodes. The electric flux density is directed radially between the two electrodes and its values are $D_1 = D_2 = Q/(4\pi r^2)$ in each region of the different dielectric materials. Hence, the electric fields in each region are $E_1 = Q/(4\pi \epsilon_1 r^2)$ and $E_3 = Q/(4\pi \epsilon_2 r^2)$. The electric potential difference between the two electrodes is

$$V = \int_a^b E_1 dr + \int_b^c E_2 dr = \frac{Q}{4\pi \epsilon_1}\left(\frac{1}{a} - \frac{1}{b}\right) + \frac{Q}{4\pi \epsilon_2}\left(\frac{1}{b} - \frac{1}{c}\right).$$

We obtain the capacitance as

$$C = \frac{Q}{V} = \frac{4\pi \epsilon_1 \epsilon_2 abc}{\epsilon_1 a(c-b) + \epsilon_2 c(b-a)}.$$

◆

Example 4.2 The space between the electrodes in a concentric spherical capacitor is occupied by two dielectric materials with dielectric constants, ϵ_1 and ϵ_2, as shown in Fig. 4.4. Determine the capacitance of the capacitor.

Fig. 4.4 Concentric spherical capacitor with two dielectric materials

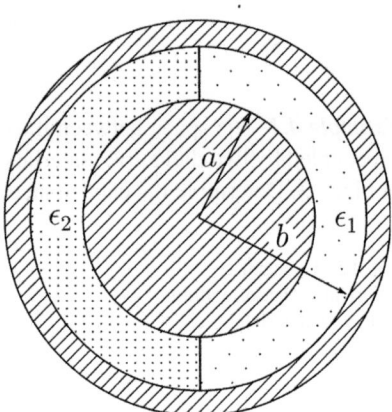

Solution 4.2 Assume that electric charges of density σ_1 and σ_2 appear on the inner electrode surface regions ($r = a$) facing the dielectric materials with dielectric

4.3 Boundary Conditions

constants ϵ_1 and ϵ_2, respectively, when we apply a potential difference V between the two electrodes. The electric flux density is directed radially between the two electrodes, and its values in dielectric materials 1 and 2 are $D_1 = a^2\sigma_1/r^2$ and $D_2 = a^2\sigma_2/r^2$. The electric fields in the respective regions are $E_1 = a^2\sigma_1/(\epsilon_1 r^2)$ and $E_2 = a^2\sigma_2/(\epsilon_2 r^2)$. Since the result of integration of these electric fields between the two electrodes is V, we have

$$\frac{a^2\sigma_1}{\epsilon_1}\left(\frac{1}{a}-\frac{1}{b}\right) = \frac{a^2\sigma_2}{\epsilon_2}\left(\frac{1}{a}-\frac{1}{b}\right) = V.$$

Thus, we determine the surface charge densities to be

$$\sigma_1 = \frac{b\epsilon_1 V}{a(b-a)}, \quad \sigma_2 = \frac{b\epsilon_2 V}{a(b-a)}.$$

This yields the total electric charge on the internal electrode,

$$Q = 2\pi a^2(\sigma_1 + \sigma_2) = \frac{2\pi ab(\epsilon_1 + \epsilon_2)V}{b-a}.$$

We obtain the capacitance as

$$C = \frac{Q}{V} = \frac{2\pi ab(\epsilon_1 + \epsilon_2)}{b-a}.$$

◆

Example 4.3 A dielectric sphere of radius a is placed in a uniform electric field of strength E_0, as shown in Fig. 4.1a. Determine the electric field strength, electric flux density, and electric polarization inside and outside the dielectric sphere, and the polarization charge density on the surface.

Solution 4.3 Spherical coordinates are defined. We can assume that a uniform electric polarization occurs in the dielectric material. The electric field outside the sphere is given by the sum of the applied electric field and the contribution of the electric dipole with a moment p along the applied electric field placed at the origin after virtual removal of the sphere. The electric field strength inside the sphere is expected to be uniform because of the uniform electric polarization. From Eq. (1.23), the radial and zenithal components of the electric field outside the sphere produced by the electric dipole are

$$E_r = \frac{p\cos\theta}{2\pi\epsilon_0 r^3}, \quad E_\theta = \frac{p\sin\theta}{4\pi\epsilon_0 r^3}.$$

We denote the internal electric field strength by E. The continuities of the parallel component of the electric field and the normal component of the electric flux density at the surface ($r = a$) are given by

$$-E_0 \sin\theta + \frac{p \sin\theta}{4\pi\epsilon_0 a^3} = -E \sin\theta, \quad \epsilon_0 \left(E_0 \cos\theta + \frac{p \cos\theta}{2\pi\epsilon_0 a^3} \right) = \epsilon E \cos\theta.$$

From these equations we have

$$p = \frac{\epsilon - \epsilon_0}{\epsilon + 2\epsilon_0} 4\pi\epsilon_0 a^3 E_0, \quad E = \frac{3\epsilon_0}{\epsilon + 2\epsilon_0} E_0.$$

We can see that E is smaller than E_0 because of $\epsilon > \epsilon_0$. This means that the dielectric material is shielded by the polarization charge. Using these results, the electric field is

$$E_r = \frac{D_r}{\epsilon_0} = \left(1 + \frac{\epsilon - \epsilon_0}{\epsilon + 2\epsilon_0} \cdot \frac{2a^3}{r^3} \right) E_0 \cos\theta,$$

$$E_\theta = \frac{D_\theta}{\epsilon_0} = -\left(1 - \frac{\epsilon - \epsilon_0}{\epsilon + 2\epsilon_0} \cdot \frac{a^3}{r^3} \right) E_0 \sin\theta$$

outside the sphere ($r > a$) and

$$E_r = \frac{D_r}{\epsilon} = \frac{3\epsilon_0}{\epsilon + 2\epsilon_0} E_0 \cos\theta, \quad E_\theta = \frac{D_\theta}{\epsilon} = -\frac{3\epsilon_0}{\epsilon + 2\epsilon_0} E_0 \sin\theta$$

inside the sphere ($r < a$). We obtain the electric polarization as

$$P = (\epsilon - \epsilon_0) E = \frac{3\epsilon_0 (\epsilon - \epsilon_0)}{\epsilon + 2\epsilon_0} E_0.$$

Here, we apply Eq. (4.5) to a small shell that contains a part of the surface (see Fig. 4.5). Since there is no true electric charge, the difference in the normal component of the electric field is equal to the surface polarization charge density divided by ϵ_0.

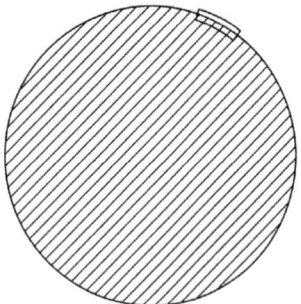

Fig. 4.5 Small shell containing a part of the surface of a dielectric sphere

4.3 Boundary Conditions

Thus, we have

$$\sigma_p(\theta) = \frac{3\epsilon_0(\epsilon - \epsilon_0)}{\epsilon + 2\epsilon_0} E_0 \cos\theta = P \cos\theta.$$

This agrees with Eq. (4.4).

♦

Example 4.4 A point charge, q, is placed at a position of distance a from a wide flat surface of a dielectric material with dielectric constant ϵ, as shown in Fig. 4.6. Determine the electric potential in the vacuum and in the dielectric material.

Fig. 4.6 Point charge q placed at distance a from the surface of a dielectric material

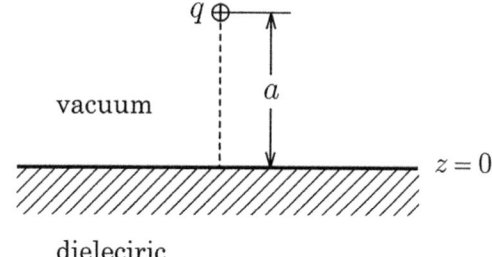

Solution 4.4 We draw the z-axis from the point charge in the direction normal to the dielectric material's surface, which is defined to be $z = 0$. We define the x-y plane on the surface with the origin at the foot of a perpendicular line from the point charge. This problem can also be solved using the method of images shown in Sect. 2.2.

We assume that the electric potential in the vacuum region ($z > 0$) is given by the sum of a contribution from q and that from a virtual point charge, q', placed at the symmetric position with respect to the surface with virtual removal of the dielectric material, as shown in Fig. 4.7a. Thus, the electric potential at point (x, y, z) is

$$\phi_v = \frac{1}{4\pi\epsilon_0}\left\{\frac{q}{\left[x^2 + y^2 + (z-a)^2\right]^{1/2}} + \frac{q'}{\left[x^2 + y^2 + (z+a)^2\right]^{1/2}}\right\}.$$

We assume that the electric potential in the dielectric material ($z < 0$) is equal to that produced by a point charge, q'', placed at the original position with virtual occupation of the vacuum region by the same dielectric material, as shown in Fig. 4.7b. Thus, the electric potential at point (x, y, z) is

$$\phi_d = \frac{1}{4\pi\epsilon} \cdot \frac{q''}{\left[x^2 + y^2 + (z-a)^2\right]^{1/2}}.$$

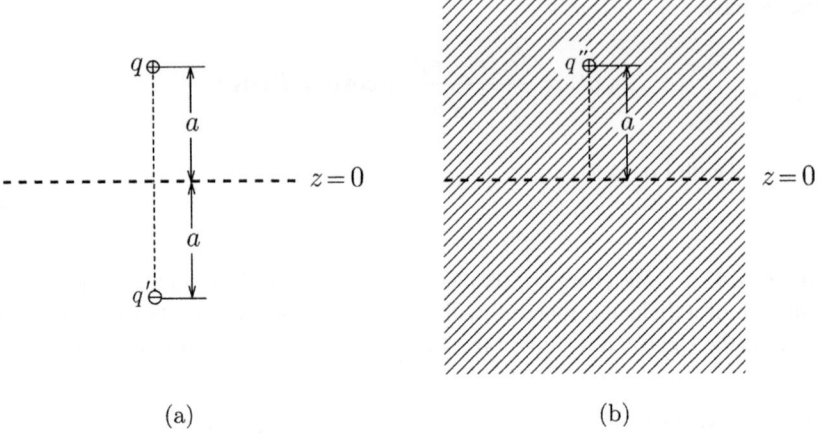

Fig. 4.7 Solution using the method of images; assumed conditions for **a** vacuum and **b** dielectric material

From Eq. (4.15), the continuity condition for the parallel component of the electric field is expressed as $\phi_v(z=0) = \phi_d(z=0)$, which gives

$$\frac{q+q'}{\epsilon_0} = \frac{q''}{\epsilon}.$$

The continuity condition for the normal component of the electric flux density is given by Eq. (4.14) with $\sigma = 0$: $\epsilon_0(\partial \phi_v/\partial z)_{z=0} = \epsilon(\partial \phi_d/\partial z)_{z=0}$, which gives

$$q - q' = q''.$$

From these equations we have

$$q' = -\frac{\epsilon - \epsilon_0}{\epsilon + \epsilon_0} q, \quad q'' = \frac{2\epsilon_0}{\epsilon + \epsilon_0} q.$$

Thus, the electric potential is

$$\phi = \frac{q}{4\pi \epsilon_0 (\epsilon + \epsilon_0)} \left\{ \frac{\epsilon + \epsilon_0}{[x^2 + y^2 + (z-a)^2]^{1/2}} - \frac{\epsilon - \epsilon_0}{[x^2 + y^2 + (z+a)^2]^{1/2}} \right\}; z > 0,$$

$$= \frac{q}{2\pi (\epsilon + \epsilon_0)[x^2 + y^2 + (z-a)^2]^{1/2}}; \quad z < 0.$$

◆

4.3 Boundary Conditions

Example 4.5 A long dielectric cylinder with radius a and dielectric constant ϵ is placed in a uniform normal electric field with strength E_0. Determine the electric field strength, electric flux density, electric polarization, and surface polarization charge density.

Solution 4.5 We define cylindrical coordinates with the z-axis at the central axis of the dielectric cylinder and the azimuthal angle measured from the direction of the applied electric field. We assume that the electric field outside the dielectric cylinder ($R > a$) produced by the polarized charge is given by the linear electric dipole of moment $\hat{p}$ in a unit length placed at the central axis after virtually removing the dielectric cylinder. The direction of the dipole moment is the same as that of the applied electric field. We assume that the electric field inside the dielectric cylinder ($R < a$) has a uniform strength E and is directed parallel to the applied electric field ($\varphi = 0$). The continuity conditions for the parallel (azimuthal) component of the electric field and the normal (radial) component of the electric flux density give

$$\hat{p} = \frac{\epsilon - \epsilon_0}{\epsilon + \epsilon_0} \cdot 2\pi \epsilon_0 a^2 E_0, \quad E = \frac{2\epsilon_0 E_0}{\epsilon + \epsilon_0}.$$

The electric field is

$$E_R = \frac{D_R}{\epsilon_0} = \left(1 + \frac{\epsilon - \epsilon_0}{\epsilon + \epsilon_0} \cdot \frac{a^2}{R^2}\right) E_0 \cos\varphi, \quad E_\varphi = \frac{D_\varphi}{\epsilon_0} = -\left(1 - \frac{\epsilon - \epsilon_0}{\epsilon + \epsilon_0} \cdot \frac{a^2}{R^2}\right) E_0 \sin\varphi$$

outside the dielectric cylinder ($R > a$) and

$$E_R = \frac{D_R}{\epsilon} = \frac{2\epsilon_0}{\epsilon + \epsilon_0} E_0 \cos\varphi, \quad E_\varphi = \frac{D_\varphi}{\epsilon} = -\frac{2\epsilon_0}{\epsilon + \epsilon_0} E_0 \sin\varphi$$

inside the dielectric cylinder ($0 \leq R < a$). The electric polarization inside the dielectric cylinder is

$$P = (\epsilon - \epsilon_0)E = \frac{2\epsilon_0(\epsilon - \epsilon_0)}{\epsilon + \epsilon_0} E_0.$$

Here, we apply Eq. (4.5) to a small shell that includes the surface of the dielectric cylinder, as shown in Fig. 4.5. Since there is no true electric charge on the surface, the surface polarization charge density is given by the difference in the normal component of the electric field on the surface multiplied by ϵ_0:

$$\sigma_p(\varphi) = \frac{2\epsilon_0(\epsilon - \epsilon_0)}{\epsilon + \epsilon_0} E_0 \cos\varphi = P \cos\varphi.$$

◆

Example 4.6 Solve the problem of Example 4.3 using the electric potential.

Solution 4.6 It is assumed that the electric field is applied along the x-axis. Hence, its electric potential is given by

$$\phi_f = -E_0 x = -E_0 r \cos\theta.$$

We put an electric dipole moment p directed along the applied electric field on the origin after the dielectric material is virtually removed. Its electric potential is given by

$$\phi_p = \frac{p\cos\theta}{4\pi\epsilon_0 r^2}.$$

Thus, the electric potential outside the dielectric sphere is

$$\phi_1 = \phi_f + \phi_p = \left(-E_0 r + \frac{p}{4\pi\epsilon_0 r^2}\right)\cos\theta.$$

The electric field inside the dielectric sphere is uniform along the x-axis and its value is denoted by E. Thus, the electric potential inside is

$$\phi_2 = -E r \cos\theta.$$

Equations (4.14) and (4.15) lead to

$$\epsilon_0\left(E_0 + \frac{p}{2\pi\epsilon_0 a^3}\right) = \epsilon E, \quad E_0 - \frac{p}{4\pi\epsilon_0 a^3} = E.$$

Thus, we have

$$p = \frac{\epsilon - \epsilon_0}{\epsilon + \epsilon_0} 4\pi\epsilon_0 a^3 E_0, \quad E = \frac{3\epsilon_0}{\epsilon + \epsilon_0} E_0.$$

◆

Example 4.7 A uniform line charge with density λ is placed at a distance a from the wide flat surface of a dielectric material with dielectric constant ϵ. Determine the electric potential in vacuum and the dielectric material.

Solution 4.7 We define the x-y plane ($z = 0$) on the dielectric material surface and the position of the line current as $x = 0$. To determine the electric potential in the

4.3 Boundary Conditions

vacuum region ($z > 0$), we assume that all the space is vacuum and that the electric potential is produced by both the line charge of density λ and a virtual line charge of density λ' located at the symmetric position with respect to the dielectric material surface:

$$\phi_v(x,z) = \frac{1}{2\pi\epsilon_0}\left\{\lambda \log \frac{R_0}{\left[x^2 + (z-a)^2\right]^{1/2}} + \lambda' \log \frac{R_0}{\left[x^2 + (z+a)^2\right]^{1/2}}\right\}.$$

In the above, R_0 is the distance of the reference point from the line at $x = 0$ on the surface. To determine the electric potential inside the dielectric material ($z < 0$), we assume that all the space is occupied by the dielectric material and that the electric potential is given by a line charge of density λ'' placed at the original position. Hence, the electric potential at (x, z) inside the dielectric material is

$$\phi_d(x,z) = \frac{\lambda''}{2\pi\epsilon} \log \frac{R_0}{\left[x^2 + (z-a)^2\right]^{1/2}}.$$

The continuity condition of the parallel component of the electric field, Eq. (4.15), gives $\phi_v(z=0) = \phi_d(z=0)$. This yields

$$\frac{\lambda + \lambda'}{\epsilon_0} = \frac{\lambda''}{\epsilon}.$$

Since there is no true electric charge on the surface, the normal component of the electric flux density is continuous. Then, $\epsilon_0(\partial\phi_v/\partial z)_{z=0} = \epsilon(\partial\phi_d/\partial z)_{z=0}$, given by Eq. (4.14), yields

$$\lambda - \lambda' = \lambda''.$$

From these conditions we obtain the linear electric charge densities as

$$\lambda' = -\frac{\epsilon - \epsilon_0}{\epsilon + \epsilon_0}\lambda, \quad \lambda'' = \frac{2\epsilon_0}{\epsilon + \epsilon_0}\lambda.$$

The electric potential is

$$\phi = \frac{\lambda}{2\pi\epsilon_0}\left\{\log \frac{R_0}{\left[x^2 + (z-a)^2\right]^{1/2}} - \frac{\epsilon - \epsilon_0}{\epsilon + \epsilon_0} \log \frac{R_0}{\left[x^2 + (z+a)^2\right]^{1/2}}\right\}; \quad z > 0,$$

$$= \frac{\lambda}{\pi(\epsilon + \epsilon_0)} \log \frac{R_0}{\left[x^2 + (z-a)^2\right]^{1/2}}; \quad z < 0.$$

◆

4.4 Electrostatic Energy in Dielectric Materials

The electrostatic energy in a system made of dielectric materials is formally the same as that in a conductor system due to the change from Eqs. (1.12) to (4.10). That is, the electrostatic energy density in dielectric materials is

$$u_e = \frac{1}{2}\epsilon E^2 = \frac{1}{2}\mathbf{E}\cdot\mathbf{D} = \frac{1}{2\epsilon}D^2 \qquad (4.21)$$

and the electrostatic energy is given by its volume integral,

$$U_e = \int_V \frac{1}{2}\epsilon E^2 dV = \int_V \frac{1}{2}\mathbf{E}\cdot\mathbf{D}\,dV = \int_V \frac{1}{2\epsilon}D^2 dV. \qquad (4.22)$$

***

Example 4.8 The space between the electrodes in a long cylindrical capacitor is occupied by two dielectric materials with dielectric constants ϵ_1 and ϵ_2, as shown in Fig. 4.8. Determine the electrostatic energy in a unit length of the capacitor, when electric charges $\pm Q'$ are given to the respective electrodes in a unit length.

Fig. 4.8 Cylindrical capacitor with two dielectric materials

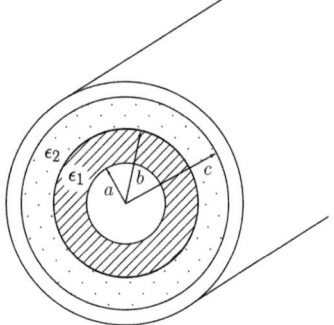

***

Solution 4.8 The electric field strength in the region between the electrodes is

$$E(R) = \frac{Q'}{2\pi\epsilon_1 R}, \quad a < R < b,$$

$$= \frac{Q'}{2\pi\epsilon_2 R}; \quad b < R < c.$$

Thus, the electric potential difference between the electrodes is

4.4 Electrostatic Energy in Dielectric Materials

$$V = \int_a^c E(R)dR = \frac{Q'}{2\pi}\left(\frac{1}{\epsilon_1}\log\frac{b}{a} + \frac{1}{\epsilon_2}\log\frac{c}{b}\right),$$

and the electrostatic energy in a unit length is given by

$$U'_e = \frac{1}{2}Q'V = \frac{Q'^2}{4\pi}\left(\frac{1}{\epsilon_1}\log\frac{b}{a} + \frac{1}{\epsilon_2}\log\frac{c}{b}\right).$$

The same result is obtained using Eq. (4.22) as

$$U'_e = \int_a^b \frac{Q'^2}{8\pi^2\epsilon_1 R^2}2\pi RdR + \int_b^c \frac{Q'^2}{8\pi^2\epsilon_2 R^2}2\pi RdR = \frac{Q'^2}{4\pi}\left(\frac{1}{\epsilon_1}\log\frac{b}{a} + \frac{1}{\epsilon_2}\log\frac{c}{b}\right).$$

◆

***

Example 4.9 A dielectric plate with thickness t and dielectric constant ϵ is inserted into the gap of a wide parallel-plate capacitor to a distance x from the edge, as shown in Fig. 4.9. The surface area and distance of the electrodes are S and d, and the dimensions of the dielectric plate and electrode in the direction normal to the sheet are the same. When electric charges $\pm Q$ are given to the two electrodes, determine the force on the dielectric plate.

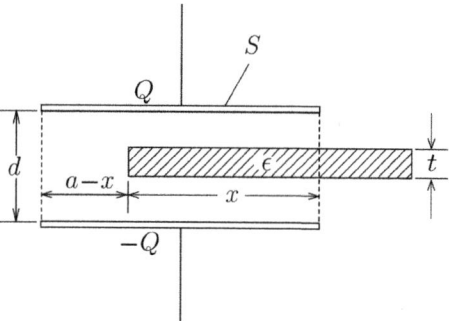

Fig. 4.9 Parallel-plate capacitor with inserted dielectric plate

***

Solution 4.9 The part in which the dielectric plate is not inserted can be regarded as one capacitor, and from Eq. (3.11) we obtain its capacitance as

$$C_1 = \frac{\epsilon_0 S}{d}\left(1 - \frac{x}{a}\right).$$

The capacitance of the remaining part, in which the dielectric plate is inserted, is similarly given by

$$C_2 = \frac{\epsilon_0 \epsilon S x}{a[\epsilon d - (\epsilon - \epsilon_0)t]}.$$

Then, we obtain the total capacitance as

$$C = C_1 + C_2 = \frac{\epsilon_0 S}{d}\left(1 - \frac{x}{a}\right) + \frac{\epsilon_0 \epsilon S x}{a[\epsilon d - (\epsilon - \epsilon_0)t]}.$$

The variation rate of the electrostatic energy determines the force on the dielectric plate:

$$\begin{aligned} F &= -\frac{d}{dx}\left(\frac{Q^2}{2C}\right) = \frac{Q^2}{2C^2} \cdot \frac{dC}{dx} \\ &= \frac{Q^2 a t d}{2\epsilon_0 S} \cdot \frac{(\epsilon - \epsilon_0)[\epsilon(d - t) + \epsilon_0 t]}{\{a[\epsilon(d - t) + \epsilon_0 t] + (\epsilon - \epsilon_0)t x\}^2}. \end{aligned}$$

This force is positive for increasing x because $\epsilon > \epsilon_0$, showing that it is attractive.

◆

4.5 Exercises

4.1. A dielectric cylinder of diameter a is placed in a uniform transverse electric field of strength E_0. Determine the surface density of polarization charge that appears on the surface of the dielectric material. The magnitude of electric polarization, P, is P.

4.2. The space between the electrodes in a long cylindrical capacitor is occupied by two dielectric materials with dielectric constants ϵ_1 and ϵ_2, as shown in Fig. E4.1. Determine the capacitance in a unit length.

Fig. E4.1 Cylindrical capacitor with two dielectric materials

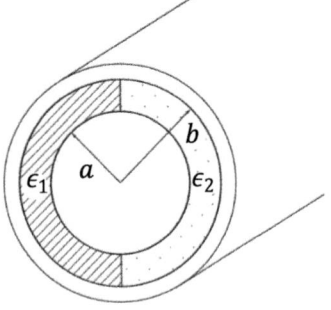

4.5 Exercises

4.3. Determine the self-inductance of the cylindrical capacitor with two kinds of dielectric material shown in Fig. 4.8 in Example 4.8.

4.4. The capacitor shown in Fig. 4.3 (Example 4.1) can be regarded as a capacitor composed of two capacitors connected in series. Determine the capacitance with Eq. (3.13).

4.5. The capacitor shown in Fig. 4.4 (Example 4.2) can be regarded as a capacitor composed of two capacitors connected in parallel. Determine the capacitance with Eq. (3.15).

4.6. In the parallel-plate capacitor shown in Fig. E4.2, the dielectric constant varies along the width as $\epsilon'(x) = \epsilon(1 + kx)$. Determine the capacitance of the capacitor. The dimension normal to the sheet is l.

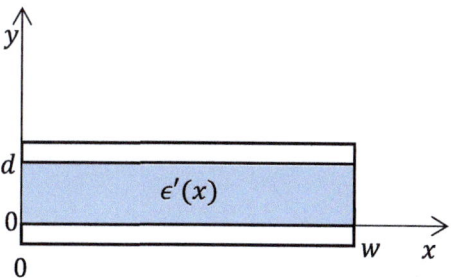

Fig. E4.2 Parallel-plate capacitor with inhomogeneous dielectric material

4.7. In the parallel-plate capacitor shown in Fig. E4.3, the dielectric constant varies with its thickness as $\epsilon'(y) = \epsilon(1 + ky)$. Determine the capacitance of the capacitor. The dimension normal to the sheet is l.

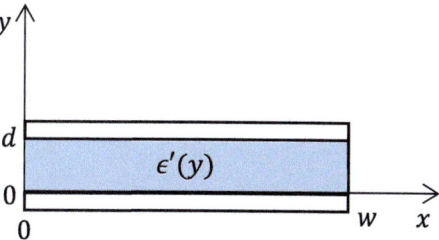

Fig. E4.3 Parallel-plate capacitor with inhomogeneous dielectric material

4.8. In the coaxial cylindrical capacitor shown in Fig. E4.4, the dielectric constant varies radially as $\epsilon'(R) = \epsilon(1 + kR)$. Determine the capacitance in a unit length.

4.9. In the concentric spherical capacitor shown in Fig. E4.5, the dielectric constant varies radially as $\epsilon'(r) = \epsilon(1 + kr)$. Determine the capacitance.

Fig. E4.4 Cylindrical capacitor with dielectric material with a dielectric constant that varies radially

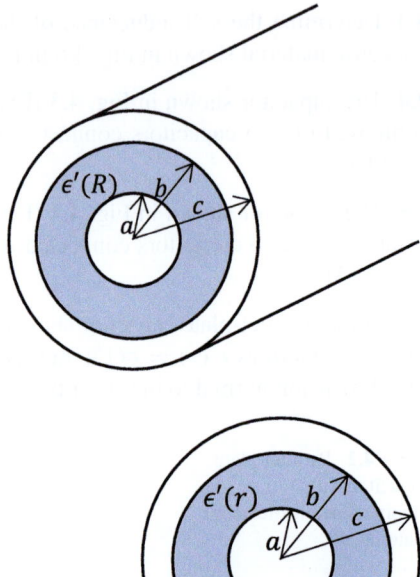

Fig. E4.5 Spherical capacitor with dielectric material with a dielectric constant that varies radially

4.10. Determine the capacitance of a capacitor in which a quarter of the space between the electrodes is occupied by a dielectric material with dielectric constant ϵ, as shown in Fig. E4.6. The dimension of the electrodes normal to the sheet is w.

4.11. Determine the polarization charge density on the interface between different dielectric materials for the refraction discussed in Sect. 4.3 (see Fig. 4.2).

4.12. A spherical conductor of radius a with electric charge Q is surrounded by dielectric material with dielectric constant ϵ. Determine the electric flux density, electric field, and electric polarization in the dielectric material, and the polarization charge density on the conductor surface.

Fig. E4.6 Parallel-plate capacitor in which a quarter of the space is occupied by a dielectric material

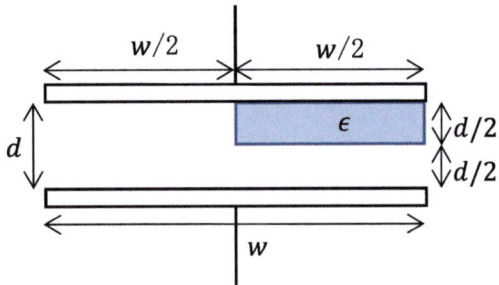

4.5 Exercises

Fig. E4.7 Hollow dielectric sphere in a uniform electric field

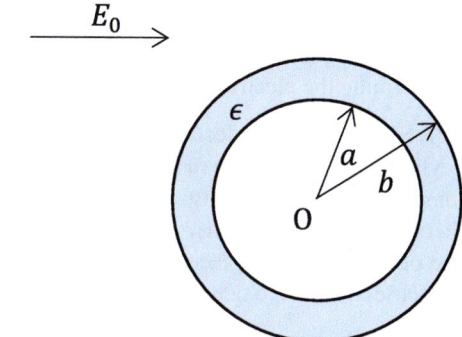

4.13. Solve the problem of Example 4.5 using the electric potential.

4.14. A hollow dielectric sphere is placed in a uniform electric field E_0, as shown in Fig. E4.7. Determine the electric field in each region.

4.15. A long, hollow dielectric cylinder is placed in a uniform perpendicular electric field E_0, as shown in Fig. E4.8. Determine the electric field in each region.

4.16. Suppose that an electric dipole moment p is placed at the center of a spherical hollow with radius a that is surrounded by a dielectric material with dielectric constant ϵ, as shown in Fig. E4.9. Determine the electric field and the polarization charge

Fig. E4.8 Hollow dielectric cylinder in a uniform electric field

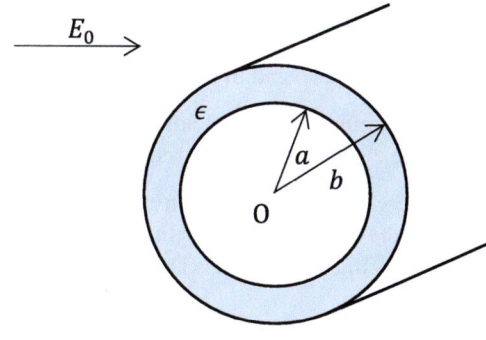

Fig. E4.9 Electric dipole moment p placed at the center of a spherical hollow in a dielectric material

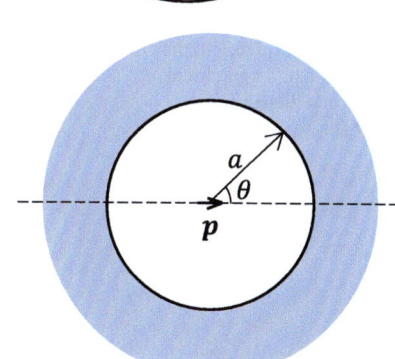

density on the inner surface of the dielectric material. Assume that an electric dipole moment p' is placed at the center after vacuum is replaced by the dielectric material to determine the electric field in the dielectric material.

4.17. Suppose that an electric dipole line of moment $\hat{p}$ is placed at the central axis of a cylindrical hollow with radius a surrounded by a dielectric material with dielectric constant ϵ, as shown in Fig. E4.10. Determine the electric field and the polarization charge density on the inner surface of the dielectric material. Assume that a dipole line of moment $\hat{p}'$ is placed at the central axis after replacing the vacuum by the dielectric material to determine the electric field in the dielectric material.

4.18. The electric potential is determined inside and outside the spherical dielectric material in a uniform applied electric field in Example 4.6. Prove that the obtained electric potential satisfies Laplace's equation.

4.19. The electric potential is determined inside and outside the cylindrical dielectric material in a uniform transverse electric field in Exercise 4.13. Prove that the obtained electric potential satisfies Laplace's equation.

4.20. The electric phenomenon is treated for an electric dipole moment p placed at the center of a spherical hollow of radius a surrounded by a dielectric material, as in Exercise 4.16. Prove that the electric potential satisfies Laplace's equation both in the hollow space and in the dielectric material.

4.21. Electric phenomenon is treated for an electric dipole line of moment $\hat{p}$ placed at the central axis of a cylindrical hollow of radius a surrounded by a dielectric material, as in Exercise 4.17. Prove that the electric potential satisfies Laplace's equation both in the hollow space and in the dielectric material.

4.22. Determine the equipotential surface for the cylindrical dielectric material in a transverse electric field discussed in Exercise 4.13.

4.23. The electric potential is treated when an electric dipole line of moment $\hat{p}$ is placed at the central axis of a cylindrical hollow of radius a surrounded by a dielectric material, as in Exercise 4.21. Determine the equipotential surface.

Fig. E4.10 Electric dipole line moment $\hat{p}$ placed at the central axis of a cylindrical hollow in a dielectric material

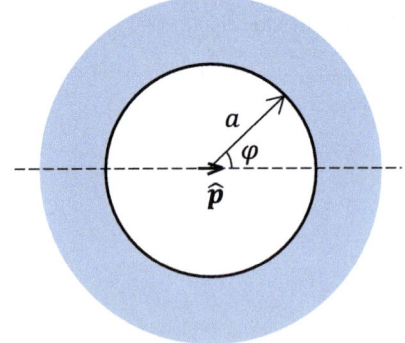

4.5 Exercises

4.24. Determine the polarization charge density on the surface of a dielectric material for the case where a point charge, q, is placed at a position of distance a from the surface, as treated in Example 4.4.

4.25. Determine the Coulomb force between the given charge and the polarization charge induced on the surface of the dielectric material in Exercise 4.24.

4.26. Determine the density of the polarization charge on the wide surface of the dielectric material facing a linear electric charge in Example 4.7.

4.27. Suppose that a straight-line charge with density λ is placed at a distance a from the surface of a dielectric material, as treated in Example 4.7. Determine the Coulomb force on the given line charge in a unit length due to the polarization charge induced on the surface of the dielectric material.

4.28. When a voltage V is applied to the parallel-plate capacitor treated in Exercise 4.6, determine the electrostatic energy of the capacitor.

4.29. When a voltage V is applied to the parallel-plate capacitor treated in Exercise 4.7, determine the electrostatic energy of the capacitor.

4.30. Suppose that electric charge $\pm Q'$ is given to the cylindrical capacitor in a unit length discussed in Exercise 4.2. Determine the electrostatic energy of the capacitor in a unit length using the electrostatic energy density.

4.31. When a voltage V is applied to the long cylindrical capacitor treated in Exercise 4.8, determine the electrostatic energy of the capacitor using the electrostatic energy density.

4.32. Suppose that electric charges $\pm Q'$ are given to the capacitor in a unit length treated in Exercise 4.8. Determine the electrostatic energy of the capacitor in a unit length using the electric potential difference.

4.33. When voltage V is applied to the concentric spherical capacitor treated in Exercise 4.9, determine the electrostatic energy of the capacitor.

4.34. Electrodes of a parallel-plate capacitor connected to a power source of voltage V are immersed in a liquid of dielectric constant ϵ and specific weight ρ, and a balanced situation is attained, as shown in Fig. E4.11. Determine the height h of the liquid level between the electrodes. The gravitational acceleration is denoted by g and neglect the effect of disorder near the edges of the capacitor.

4.35. Here, we treat the same problem as in Exercise 4.34 under the condition that the power source is removed and electric charges $\pm Q$ are given to each electrode. Derive the equation to determine the height of the liquid level. The length of the plate below the liquid level is l'.

Fig. E4.11 Parallel-plate capacitor immersed in a dielectric liquid

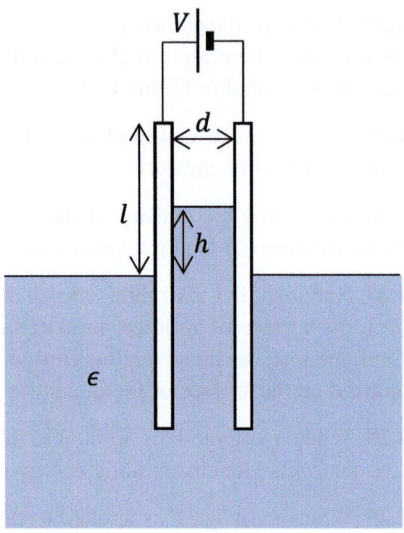

Answers to Exercises

4.1. We define cylindrical coordinates with the z-axis on the central axis of the dielectric cylinder and the azimuthal angle φ measured from the direction of the applied electric field. We denote the positive and negative electric charge densities and the relative displacement of these charges by $\pm\rho_\mathrm{p}$ and δ, respectively. Thus, the surface density of the polarization charge is given by $\sigma_\mathrm{p}(\varphi) = \rho_\mathrm{p}\delta\cos\varphi$. Since the magnitude of the electric polarization is equal to the amount of positive polarization charge that crosses the surface of a unit area, it is given by $P = \rho_\mathrm{p}\delta$. Thus, we have

$$\sigma_\mathrm{p}(\varphi) = P\cos\varphi.$$

4.2. It is assumed that electric charges of density σ_1 and σ_2 appear on the surface of the inner electrode facing the dielectric materials of ϵ_1 and ϵ_2, respectively. Thus, the total electric charge in a unit length on the inner electrode surface is $Q' = \pi a(\sigma_1 + \sigma_2)$. The values of the electric flux density in the dielectric materials 1 and 2 at distance R from the central axis are $D_1 = a\sigma_1/R$ and $D_2 = a\sigma_2/R$, respectively. Thus, the potential difference in each region is $(a\sigma_1/\epsilon_1)\log(b/a)$ and $(a\sigma_2/\epsilon_2)\log(b/a)$. Since these must be the same, the condition $\sigma_1/\epsilon_1 = \sigma_2/\epsilon_2$ should be satisfied, and the electric charge densities are determined to be

$$\sigma_1 = \frac{\epsilon_1 Q'}{\pi a(\epsilon_1 + \epsilon_2)}, \quad \sigma_2 = \frac{\epsilon_2 Q'}{\pi a(\epsilon_1 + \epsilon_2)}.$$

So, the potential difference is

$$V = \frac{Q'}{\pi(\epsilon_1 + \epsilon_2)}\log\frac{b}{a}.$$

4.5 Exercises

and the capacitance in a unit length is determined to be

$$C' = \frac{\pi(\epsilon_1 + \epsilon_2)}{\log(b/a)}.$$

4.3. We apply an electric charge Q' to the inner electrode in a unit length and then earth the outer electrode. Then, the electric field appears only in the region of the dielectric materials with the strength:

$$E(R) = \frac{Q'}{2\pi\epsilon_1 R}; \quad a < R < b,$$
$$= \frac{Q'}{2\pi\epsilon_2 R}; \quad b < R < c.$$

So, the potential difference between the electrodes is

$$V = \frac{Q'}{2\pi\epsilon_1}\int_a^b \frac{dR}{R} + \frac{Q'}{2\pi\epsilon_1}\int_b^c \frac{dR}{R} = \frac{Q'}{2\pi}\left(\frac{1}{\epsilon_1}\log\frac{b}{a} + \frac{1}{\epsilon_2}\log\frac{c}{b}\right).$$

The self-inductance in a unit length is determined to be

$$L' = \frac{Q'}{V} = \frac{2\pi\epsilon_1\epsilon_2}{\epsilon_2\log(b/c) + \epsilon_1\log(c/b)}.$$

4.4. The capacitance of the inner capacitor with the dielectric material 1 is

$$C_1 = \frac{4\pi\epsilon_1 ab}{b-a}$$

and the capacitance of the outer capacitor with the dielectric material 2 is

$$C_2 = \frac{4\pi\epsilon_2 bc}{c-b}.$$

Thus, the capacitance is determined to be

$$C = \frac{C_1 C_2}{C_1 + C_2} = \frac{4\pi\epsilon_1\epsilon_2 abc}{\epsilon_1 a(c-b) + \epsilon_2 c(b-a)}.$$

4.5. The capacitance of the right part of the capacitor with the dielectric material 1 is

$$C_1 = \frac{2\pi\epsilon_1 ab}{b-a}$$

and the capacitance of the left part of the capacitor with the dielectric material 2 is

$$C_2 = \frac{2\pi \epsilon_2 ab}{b-a}.$$

Thus, the capacitance is determined to be

$$C = C_1 + C_2 = \frac{2\pi ab(\epsilon_1 + \epsilon_2)}{b-a}.$$

4.6. We apply a voltage V to the capacitor. The electric field is V/d and the electric charge density on the electrode surface is $\sigma(x) = \epsilon(1+kx)V/d$. Then, the electric charge is

$$Q = l \int_0^w \sigma(x)\,\mathrm{d}x = \frac{lw\epsilon}{d}\left(1 + \frac{kw}{2}\right)V$$

and the capacitance is determined to be

$$C = \frac{lw\epsilon}{d}\left(1 + \frac{kw}{2}\right).$$

4.7. We apply a voltage V to the capacitor. The electric charge density that appears on the electrode surface is denoted by σ. Then, the electric field is $E(y) = \sigma/\epsilon'(y)$. So, the voltage is given by

$$V = \sigma \int_0^d \frac{\mathrm{d}y}{\epsilon'(y)} = \frac{\sigma}{\epsilon}\int_0^d \frac{\mathrm{d}y}{1+ky} = \frac{\sigma}{\epsilon k}\log(1+kd).$$

Since the electric charge is $Q = wl\sigma$, the capacitance is determined to be

$$C = \frac{lw\epsilon k}{\log(1+kd)}.$$

4.8. We apply electric charges $\pm Q'$ to the electrodes in a unit length. The electric flux density and electric field at radius R are $D(R) = Q'/2\pi R$ and $E(R) = Q'/2\pi \epsilon'(R)R$. Hence, the electric potential difference is

$$V = \frac{Q'}{2\pi\epsilon}\int_a^b \frac{\mathrm{d}R}{R(1+kR)} = \frac{Q'}{2\pi\epsilon}\log\frac{b(1+ka)}{a(1+kb)}.$$

4.5 Exercises

The capacitance in a unit length is determined to be

$$C' = \frac{2\pi\epsilon}{\log[b(1+ka)/a(1+kb)]}.$$

4.9. We apply electric charges $\pm Q$ to the electrodes. The electric field at radius r is $Q/[4\pi\epsilon'(r)r^2]$. Thus, the potential difference between the two electrodes is

$$V = \frac{Q}{4\pi\epsilon} \int_a^b \frac{dr}{(1+kr)r^2} = \frac{Q}{4\pi\epsilon} \int_a^b \left(\frac{1}{r^2} - \frac{k}{r} + \frac{k^2}{1+kr}\right) dr$$

$$= \frac{Q}{4\pi\epsilon} \left[\frac{1}{a} - \frac{1}{b} - k \log \frac{b(1+ka)}{a(1+kb)}\right].$$

The capacitance is determined to be

$$C = 4\pi\epsilon \left[\frac{1}{a} - \frac{1}{b} - k \log \frac{b(1+ka)}{a(1+kb)}\right]^{-1}.$$

4.10. This capacitor can be approximately replaced by the pair of capacitors, as shown in Fig. B4.1, although the condition in the region shown by the dashed circle is different from the original. If the thickness d is much smaller than the width w, this difference can be neglected. The capacitances of the left and right capacitors are $C_1 = \epsilon_0 w^2 / 2d$ and $C_2 = \epsilon_0 \epsilon w^2 / (\epsilon_0 + \epsilon) d$. So, the capacitance is determined to be

$$C = C_1 + C_2 = \frac{\epsilon_0(3\epsilon + \epsilon_0)w^2}{2(\epsilon_0 + \epsilon)d}.$$

4.11. The polarization charge density is equal to the difference between the normal components of the electric polarization:

$$\sigma_p = P_1 \cos\theta_1 - P_2 \cos\theta_2 = (\epsilon_1 - \epsilon_0)E_1 \cos\theta_1 - (\epsilon_2 - \epsilon_0)E_2 \cos\theta_2,$$

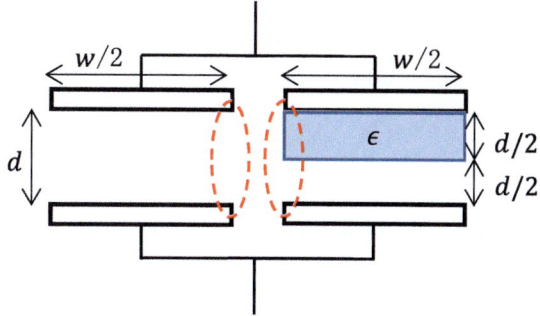

Fig. B4.1 Parallel connection of two capacitors as an approximation of the given capacitor

Using Eq. (4.16), this leads to

$$\sigma_p = \epsilon_0 \left(\frac{\epsilon_1}{\epsilon_2} - 1 \right) E_1 \cos \theta_1.$$

This result can also be directly derived from $\epsilon_0 (E_2 \cos \theta_2 - E_1 \cos \theta_1)$.

4.12. The electric flux density and electric field at a distance r from the center are

$$D(r) = \frac{Q}{4\pi r^2}, \quad E(r) = \frac{Q}{4\pi \epsilon r^2}$$

for $r > a$. So, the electric polarization is given by

$$P(r) = D(r) - \epsilon_0 E(r) = \frac{(\epsilon - \epsilon_0)Q}{4\pi \epsilon r^2}.$$

Since the electric polarization is normal to the surface, the polarization charge on the surface is

$$\sigma_p = P(a) = \frac{(\epsilon - \epsilon_0)Q}{4\pi \epsilon a^2}.$$

4.13. The electric potential outside the cylindrical dielectric material is given by

$$\phi(R, \varphi) = \left(-E_0 R + \frac{\hat{p}}{2\pi \epsilon_0 R} \right) \cos \varphi,$$

where $\hat{p}$ is the moment of the electric dipole line in a unit length on the central axis. If the uniform electric field in the dielectric material is denoted by E, the electric potential inside the dielectric cylinder is given by

$$\phi(R, \varphi) = -ER \cos \varphi.$$

From the condition given by Eq. (4.15) at $R = a$, we have

$$-E_0 + \frac{\hat{p}}{2\pi \epsilon_0 a^2} = -E, \tag{B4.1}$$

and Eq. (4.14) at $R = a$ leads to

$$\epsilon_0 \left(E_0 + \frac{\hat{p}}{2\pi \epsilon_0 a^2} \right) = \epsilon E. \tag{B4.2}$$

From Eqs. (B4.1) and (B4.2), we have

4.5 Exercises

$$\hat{p} = \frac{\epsilon - \epsilon_0}{\epsilon + \epsilon_0} \cdot 2\pi \epsilon_0 a^2 E_0, \quad E = \frac{2\epsilon_0}{\epsilon + \epsilon_0} E_0.$$

4.14. Positive and negative electric charges are displaced by the electric field and polarization charges appear on the inner and outer surfaces. Each surface polarization charge produces a uniform electric field inside it and a dipole electric field outside it. The uniform electric field in the hollow apace is denoted by E_1. The electric dipole moments equivalent to the polarization charges on the inner and outer surfaces are denoted by p and p', respectively. The polarization charge on the outer surface also produces a uniform electric field denoted by E_2 in the dielectric material. So, the electric field distribution is:

$$\begin{aligned} \mathbf{E} &= \mathbf{E}_1; & 0 \leq r < a, \\ &= \mathbf{E}_2 + \nabla \frac{p \cos\theta}{4\pi \epsilon_0 r^2}; & a < r < b, \\ &= \mathbf{E}_0 + \nabla \frac{(p+p')\cos\theta}{4\pi \epsilon_0 r^2}; & r > b, \end{aligned}$$

where θ is the zenithal angle. So, the boundary conditions at $r = a$ are

$$-E_1 \sin\theta = -E_2 \sin\theta + \frac{p \sin\theta}{4\pi \epsilon_0 a^3}, \quad \epsilon_0 E_1 \cos\theta = \epsilon \left(E_2 \cos\theta + \frac{p \cos\theta}{2\pi \epsilon_0 a^3} \right),$$

and the boundary conditions at $r = b$ are

$$-E_2 \sin\theta + \frac{p \sin\theta}{4\pi \epsilon_0 b^3} = -E_0 \sin\theta + \frac{(p+p')\sin\theta}{4\pi \epsilon_0 b^3},$$

$$\epsilon \left(E_2 \cos\theta + \frac{p \cos\theta}{2\pi \epsilon_0 b^3} \right) = \epsilon_0 \left[E_0 \cos\theta + \frac{(p+p')\cos\theta}{2\pi \epsilon_0 b^3} \right].$$

From these conditions, p, p', E_1, and E_2 are determined to be

$$p = -\frac{4\pi \epsilon_0 a^3}{3\alpha} \cdot \frac{\epsilon - \epsilon_0}{\epsilon} E_0, \quad p' = 4\pi \epsilon_0 b^3 \left(1 - \frac{2\epsilon + \epsilon_0}{3\alpha \epsilon} \right) E_0,$$

$$E_1 = \frac{1}{\alpha} E_0, \quad E_2 = \frac{2\epsilon + \epsilon_0}{3\alpha \epsilon} E_0,$$

where α is a constant given by

$$\alpha = \frac{(\epsilon + 2\epsilon_0)(2\epsilon + \epsilon_0)}{9\epsilon_0 \epsilon} - \frac{2a^3}{9b^3} \cdot \frac{(\epsilon - \epsilon_0)^2}{\epsilon_0 \epsilon}.$$

The electric field in each region is obtained from the above results.

4.15. Positive and negative electric charges are displaced by the electric field and polarization charges appear on the inner and outer surfaces. Each surface polarization charge produces a uniform electric field inside it and a dipole electric field outside it. The uniform electric field in the hollow space is denoted by $\boldsymbol{E}_1$. The two-dimensional electric dipole moments equivalent to the polarization charges on the inner and outer surfaces are denoted by $\hat{p}$ and $\hat{p}'$, respectively. The polarization charge on the outer surface also produces a uniform electric field denoted by $\boldsymbol{E}_2$ in the dielectric material. So, the electric field distribution is:

$$\boldsymbol{E} = \boldsymbol{E}_1; \qquad\qquad 0 \le R < a,$$
$$= \boldsymbol{E}_2 + \nabla \frac{\hat{p}\cos\varphi}{2\pi\epsilon_0 R^2}; \qquad a < R < b,$$
$$= \boldsymbol{E}_0 + \nabla \frac{(\hat{p}+\hat{p}')\cos\varphi}{2\pi\epsilon_0 R}; \qquad R > b,$$

where φ is the azimuthal angle. So, the boundary conditions at $R = a$ are

$$-E_1\sin\varphi = -E_2\sin\varphi + \frac{\hat{p}\sin\varphi}{2\pi\epsilon_0 a^2}, \qquad \epsilon_0 E_1 \cos\varphi = \epsilon\left(E_2\cos\varphi + \frac{\hat{p}\cos\varphi}{2\pi\epsilon_0 a^2}\right),$$

and the boundary conditions at $R = b$ are

$$-E_2\sin\varphi + \frac{\hat{p}\sin\varphi}{2\pi\epsilon_0 b^2} = -E_0\sin\varphi + \frac{(\hat{p}+\hat{p}')\sin\varphi}{2\pi\epsilon_0 b^2},$$
$$\epsilon\left(E_2\cos\varphi + \frac{\hat{p}\cos\varphi}{2\pi\epsilon_0 b^2}\right) = \epsilon_0\left[E_0\cos\varphi + \frac{(\hat{p}+\hat{p}')\cos\varphi}{2\pi\epsilon_0 b^2}\right].$$

From these conditions, $\hat{p}$, $\hat{p}'$, E_1, and E_2 are determined to be

$$\hat{p} = -\frac{\pi\epsilon_0 a^2}{\beta} \cdot \frac{\epsilon - \epsilon_0}{\epsilon} E_0, \qquad \hat{p}' = 2\pi\epsilon_0 b^2\left(1 - \frac{\epsilon + \epsilon_0}{2\beta\epsilon}\right) E_0,$$
$$E_1 = \frac{1}{\beta} E_0, \qquad E_2 = \frac{\epsilon + \epsilon_0}{2\beta\epsilon} E_0,$$

where β is a constant given by

$$\beta = \frac{(\epsilon + \epsilon_0)^2}{4\epsilon_0\epsilon} - \frac{a^2}{4b^2} \cdot \frac{(\epsilon - \epsilon_0)^2}{\epsilon_0\epsilon}.$$

The electric field in each region is obtained from the above results.

4.16. The electric field in the hollow space produced by the dipole moment p is

$$E_r = \frac{p}{2\pi\epsilon_0 r^3}\cos\theta, \qquad E_\theta = \frac{p}{4\pi\epsilon_0 r^3}\sin\theta.$$

4.5 Exercises

The uniform electric field in the hollow space produced by the polarization charge on the surface is denoted by E_0. Its radial and zenithal components are

$$E_0 \cos\theta, \quad -E_0 \sin\theta.$$

The electric field in the dielectric material is also influenced by the polarization charge. This can be reproduced by some dipole moment placed at the center. So, the assumed dipole moment p' includes this effect. After a virtual occupation of the hollow space by the dielectric material, the electric field in the dielectric material produced by the dipole moment p' is

$$E_r = \frac{p'}{2\pi\epsilon r^3}\cos\theta, \quad E_\theta = \frac{p'}{4\pi\epsilon r^3}\sin\theta.$$

So, the boundary conditions on the electric field and the electric flux density at $r = a$ are

$$\left(\frac{p}{4\pi\epsilon_0 a^3} - E_0\right)\sin\theta = \frac{p'}{4\pi\epsilon a^3}\sin\theta,$$

$$\left(\frac{p}{2\pi a^3} + \epsilon_0 E_0\right)\cos\theta = \frac{p'}{2\pi a^3}\cos\theta.$$

These yield

$$p' = \frac{3\epsilon}{2\epsilon + \epsilon_0}p, \quad E_0 = \frac{(\epsilon - \epsilon_0)}{2\pi\epsilon_0(2\epsilon + \epsilon_0)a^3}p.$$

Thus, we have

$$E_r = \frac{p}{2\pi\epsilon_0}\left[\frac{1}{r^3} + \frac{\epsilon - \epsilon_0}{(2\epsilon + \epsilon_0)a^3}\right]\cos\theta, \quad E_\theta = \frac{p}{4\pi\epsilon_0}\left[\frac{1}{r^3} - \frac{2(\epsilon - \epsilon_0)}{(2\epsilon + \epsilon_0)a^3}\right]\sin\theta$$

for $0 < r < a$ and

$$E_r = \frac{3}{2\pi(2\epsilon + \epsilon_0)r^3}p\cos\theta, \quad E_\theta = \frac{3}{4\pi(2\epsilon + \epsilon_0)r^3}p\sin\theta$$

for $r > a$. The polarization charge density is

$$\sigma_p = -\epsilon_0\bigl[E_r(r = a_{+0}) - E_r(r = a_{-0})\bigr] = -\frac{3(\epsilon - \epsilon_0)}{2\pi(2\epsilon + \epsilon_0)a^3}p\cos\theta.$$

4.17. The electric field produced in the hollow space by the electric moment $\hat{p}$ is

$$E_{mR} = \frac{\hat{p}}{2\pi\epsilon_0 R^2}\cos\varphi, \quad E_{m\varphi} = \frac{\hat{p}}{2\pi\epsilon_0 R^2}\sin\varphi.$$

The uniform electric field in the hollow space produced by the polarization charge on the surface is denoted by E_0. Its radial and azimuthal components are

$$E_0 \cos \varphi, \quad -E_0 \sin \varphi.$$

The electric field in the dielectric material is also influenced by the polarization charge. This can be reproduced by some linear dipole moment placed at the center. So, the assumed linear dipole moment $\hat{p}'$ includes this effect. After a virtual occupation of the hollow space by the dielectric material, the electric field in the dielectric material produced by the linear dipole moment $\hat{p}'$ is

$$E_R = \frac{\hat{p}'}{2\pi \epsilon R^2} \cos \varphi, \quad E_\varphi = \frac{\hat{p}'}{2\pi \epsilon R^2} \sin \varphi.$$

So, the boundary conditions on the electric field and the electric flux density at $R = a$ are

$$\left(\frac{\hat{p}}{2\pi \epsilon_0 a^2} - E_0 \right) \sin \varphi = \frac{\hat{p}'}{2\pi \epsilon a^2} \sin \varphi,$$

$$\left(\frac{\hat{p}}{2\pi a^2} + \epsilon_0 E_0 \right) \cos \varphi = \frac{\hat{p}'}{2\pi a^2} \cos \varphi.$$

These yield

$$\hat{p}' = \frac{2\epsilon}{\epsilon + \epsilon_0} \hat{p}, \quad E_0 = \frac{(\epsilon - \epsilon_0)}{2\pi \epsilon_0 (\epsilon + \epsilon_0) a^2} \hat{p}.$$

Thus, we have

$$E_R = \frac{\hat{p}}{2\pi \epsilon_0} \left[\frac{1}{R^2} + \frac{\epsilon - \epsilon_0}{(\epsilon + \epsilon_0) a^2} \right] \cos \varphi, \quad E_\varphi = \frac{\hat{p}}{2\pi \epsilon_0} \left[\frac{1}{R^2} - \frac{\epsilon - \epsilon_0}{(\epsilon + \epsilon_0) a^2} \right] \sin \varphi$$

for $0 < R < a$ and

$$E_R = \frac{\hat{p}}{\pi (\epsilon + \epsilon_0) R^2} \cos \varphi, \quad E_\varphi = \frac{\hat{p}}{\pi (\epsilon + \epsilon_0) R^2} \sin \varphi$$

for $R > a$. The polarization charge density on the surface is

$$\sigma_p = -\epsilon_0 \left[E_R(R = a_{+0}) - E_R(R = a_{-0}) \right] = -\frac{(\epsilon - \epsilon_0) \hat{p}}{\pi (\epsilon + \epsilon_0) a^2} \cos \varphi.$$

4.18. The electric potential inside the dielectric sphere is

$$\phi = -\frac{3\epsilon_0 E_0}{\epsilon + \epsilon_0} r \cos \theta.$$

4.5 Exercises

Using Eq. (19) in the Appendix, we have

$$\Delta\phi = \frac{1}{r^2} \cdot \frac{\partial}{\partial r}\left(r^2 \frac{\partial \phi}{\partial r}\right) + \frac{1}{r^2 \sin\theta} \cdot \frac{\partial}{\partial \theta}\left(\sin\theta \frac{\partial \phi}{\partial \theta}\right)$$

$$= -\frac{6\epsilon_0 E_0}{\epsilon + \epsilon_0} \cdot \frac{\cos\theta}{r} + \frac{6\epsilon_0 E_0}{\epsilon + \epsilon_0} \cdot \frac{\cos\theta}{r} = 0.$$

The electric potential outside the dielectric sphere is

$$\phi = E_0\left(-r + \frac{\epsilon - \epsilon_0}{\epsilon + \epsilon_0} \cdot \frac{a^3}{r^2}\right)\cos\theta.$$

Thus, we have

$$\Delta\phi = -2E_0\left(\frac{1}{r} - \frac{\epsilon - \epsilon_0}{\epsilon + \epsilon_0} \cdot \frac{a^3}{r^4}\right)\cos\theta + 2E_0\left(\frac{1}{r} - \frac{\epsilon - \epsilon_0}{\epsilon + \epsilon_0} \cdot \frac{a^3}{r^4}\right)\cos\theta = 0$$

and it is proved that Laplace's equation holds.

4.19. The electric potential inside the dielectric cylinder is given by

$$\phi(R, \varphi) = -\frac{2\epsilon_0}{\epsilon + \epsilon_0} E_0 R \cos\varphi.$$

Using Eq. (15) in the Appendix, we have

$$\Delta\phi = \frac{1}{R} \cdot \frac{\partial}{\partial R}\left(R\frac{\partial \phi}{\partial R}\right) + \frac{1}{R^2} \cdot \frac{\partial^2 \phi}{\partial \varphi^2}$$

$$= -\frac{2\epsilon_0 E_0}{\epsilon + \epsilon_0} \cdot \frac{\cos\varphi}{R} + \frac{2\epsilon_0 E_0}{\epsilon + \epsilon_0} \cdot \frac{\cos\varphi}{R} = 0.$$

Outside the dielectric material, the electric potential is given by

$$\phi(R, \varphi) = E_0\left(-R + \frac{\epsilon - \epsilon_0}{\epsilon + \epsilon_0} \cdot \frac{a^2}{R}\right)\cos\varphi,$$

and we have

$$\Delta\phi = -E_0\left(R - \frac{\epsilon - \epsilon_0}{\epsilon + \epsilon_0} \cdot \frac{a^2}{R}\right)\cos\varphi + E_0\left(R - \frac{\epsilon - \epsilon_0}{\epsilon + \epsilon_0} \cdot \frac{a^2}{R}\right)\cos\varphi = 0.$$

Thus, Laplace's equation holds for the electric potential inside and outside the dielectric material.

4.20. The electric potential in the hollow space $(0 < r < a)$ is

$$\phi = \frac{p}{4\pi \epsilon_0 r^2}\cos\theta - E_0 r \cos\theta.$$

In this case,

$$\nabla^2 \phi = \frac{1}{r^2} \cdot \frac{\partial}{\partial r}\left(r^2 \frac{\partial \phi}{\partial r}\right) + \frac{1}{r^2 \sin\theta} \cdot \frac{\partial}{\partial \theta}\left(\sin\theta \frac{\partial \phi}{\partial \theta}\right)$$

$$= \frac{p}{2\pi \epsilon_0 r^2}\left(\frac{\cos\theta}{r^2} - \frac{\cos\theta}{r^2}\right) - \frac{E_0}{r^2}(2r\cos\theta - 2r\cos\theta) = 0.$$

The electric potential in the dielectric material ($r > a$) is

$$\phi = \frac{p'}{4\pi \epsilon r^2} \cos\theta.$$

In this case, we have

$$\nabla^2 \phi = \frac{1}{r^2} \cdot \frac{\partial}{\partial r}\left(r^2 \frac{\partial \phi}{\partial r}\right) + \frac{1}{r^2 \sin\theta} \cdot \frac{\partial}{\partial \theta}\left(\sin\theta \frac{\partial \phi}{\partial \theta}\right)$$

$$= \frac{p'}{2\pi \epsilon r^2}\left(\frac{\cos\theta}{r^2} - \frac{\cos\theta}{r^2}\right) = 0.$$

Thus, Laplace's equation is satisfied for the electric potential in the hollow space and in the dielectric material.

4.21. The electric potential in the hollow space ($0 < R < a$) is

$$\phi = \frac{\hat{p}}{2\pi \epsilon_0 R} \cos\varphi - E_0 R \cos\varphi.$$

In this case,

$$\nabla^2 \phi = \frac{1}{R} \cdot \frac{\partial}{\partial R}\left(R \frac{\partial \phi}{\partial R}\right) + \frac{1}{R^2} \cdot \frac{\partial^2 \phi}{\partial \varphi^2}$$

$$= \frac{\hat{p}}{2\pi \epsilon_0 R}\left(\frac{\cos\varphi}{R^2} - \frac{\cos\varphi}{R^2}\right) - \frac{E_0}{R^2}(R\cos\varphi - R\cos\varphi) = 0.$$

The electric potential in the dielectric material ($R > a$) is

$$\phi = \frac{\hat{p}'}{2\pi \epsilon R} \cos\varphi.$$

In this case, we have

$$\nabla^2 \phi = \frac{1}{R} \cdot \frac{\partial}{\partial R}\left(R \frac{\partial \phi}{\partial R}\right) + \frac{1}{R^2} \cdot \frac{\partial^2 \phi}{\partial \varphi^2}$$

$$= \frac{\hat{p}'}{2\pi \epsilon R}\left(\frac{\cos\varphi}{R^2} - \frac{\cos\varphi}{R^2}\right) = 0.$$

4.5 Exercises

Thus, Laplace's equation is satisfied for the electric potential in the hollow space and in the dielectric material.

4.22. We define the x-axis in the direction of the applied electric field. The equipotential surface outside the dielectric material ($R > a$) is given by

$$E_0 \left(R - \frac{a'^2}{R} \right) \cos \varphi = K,$$

where $a' = [(\epsilon - \epsilon_0)/(\epsilon + \epsilon_0)]^{1/2} a$ and K is a constant. Using the relationships $R^2 = x^2 + y^2$ and $R \cos \varphi = x$, the above relation leads to

$$y = \pm a' \left(1 + \frac{K}{E_0 x - K} - \frac{x^2}{a'^2} \right)^{1/2}$$

in the x-y plane.

Inside the dielectric material ($R < a$), the equipotential surface is given by

$$ER \cos \varphi = K',$$

where $E = 2\epsilon_0 E_0/(\epsilon + \epsilon_0)$ and K' is a constant. This leads the equipotential surface:

$$x = \frac{K'}{E}.$$

4.23. The electric potential in the hollow space ($0 < R < a$) is

$$\phi = \frac{\hat{p}}{2\pi \epsilon_0 R} \cos \varphi - E_0 R \cos \varphi,$$

where E_0 is given by.

$$E_0 = \frac{(\epsilon - \epsilon_0)}{2\phi \epsilon_0 (\epsilon + \epsilon_0) a^2} \hat{p}.$$

Hence, the equipotential surface is given by

$$\left[\frac{1}{R} - \frac{(\epsilon - \epsilon_0) R}{(\epsilon + \epsilon_0) a^2} \right] \cos \varphi = K,$$

where K is a constant. Using $R^2 = x^2 + y^2$ and $R \cos \varphi = x$, the above equation leads to

$$y = \pm \left(\frac{x}{K + \alpha x} - x^2 \right)^{1/2},$$

where

$$\alpha = \frac{\epsilon - \epsilon_0}{(\epsilon + \epsilon_0)a^2}.$$

In the dielectric material ($R > a$), the electric potential is

$$\phi = \frac{\hat{p}'}{2\pi \epsilon R} \cos \varphi.$$

Hence, the equipotential surface is given by

$$\frac{\beta}{R} \sin \varphi = K',$$

where $\beta = \hat{p}'/2\pi\epsilon$ and K' is a constant. This leads the equipotential surface:

$$x^2 + \left(y - \frac{\beta}{2K'}\right)^2 = \left(\frac{\beta}{2K'}\right)^2.$$

4.24. The electric field on the vacuum side on the dielectric material surface is

$$E_v(0) = -\frac{q\epsilon a}{2\pi\epsilon_0(\epsilon + \epsilon_0)(x^2 + y^2 + a^2)^{3/2}},$$

and that on the dielectric material side on its surface is

$$E_d(0) = -\frac{qa}{2\pi(\epsilon + \epsilon_0)(x^2 + y^2 + a^2)^{3/2}}.$$

Hence, the polarization charge density is determined to be

$$\sigma_p = \epsilon_0[E_v(0) - E_d(0)] = -\frac{q(\epsilon - \epsilon_0)a}{2\pi(\epsilon + \epsilon_0)(x^2 + y^2 + a^2)^{3/2}}.$$

The total polarization charge is

$$Q_p = \int \sigma_p dS = -\frac{q(\epsilon - \epsilon_0)a}{2\pi(\epsilon + \epsilon_0)} \int_0^\infty \frac{2\pi r dr}{(r^2 + a^2)^{3/2}} = -\frac{\epsilon - \epsilon_0}{\epsilon + \epsilon_0} q,$$

which is equal to the image charge q' obtained in Example 4.4.

4.25. A perpendicular line is drawn from the point charge and the origin is defined at the foot of the perpendicular line on the surface. Assume a polarization charge in a narrow ring of radius from r to $r + dr$ on the surface. The electric field at the point charge produced by a polarization charge of a small area of azimuthal angle from φ to $\varphi + d\varphi$ is

4.5 Exercises

$$dE' = \frac{\sigma_p(r) r\,dr\,d\varphi}{4\pi\epsilon_0(r^2+a^2)}.$$

Only the vertical component, $dE'a/(r^2+a^2)^{1/2}$, remains from symmetry, and the contribution from this ring area to the electric field is

$$dE = \frac{a\sigma_p(r)r\,dr}{2\epsilon_0(r^2+a^2)^{3/2}}.$$

Using the result of $\sigma_p(r)$ obtained in Exercise 4.23, the electric field is determined to be

$$E = -\frac{q(\epsilon-\epsilon_0)a^2}{4\pi\epsilon_0(\epsilon+\epsilon_0)}\int_0^\infty \frac{r\,dr}{(r^2+a^2)^3} = \frac{q(\epsilon-\epsilon_0)a^2}{16\pi\epsilon_0(\epsilon+\epsilon_0)}\left[\frac{1}{(r^2+a^2)^2}\right]_0^\infty$$

$$= -\frac{q(\epsilon-\epsilon_0)}{16\pi\epsilon_0(\epsilon+\epsilon_0)a^2}.$$

Thus, the Coulomb force is

$$F = \frac{qq'}{16\pi\epsilon_0 a^2},$$

which is equal to the force between the given and image charges separated by distance $2a$.

4.26. The electric field on the surface is

$$(E_z)_{z=0+} = -\frac{a\epsilon\lambda}{\pi\epsilon_0(\epsilon+\epsilon_0)(x^2+a^2)}$$

on the vacuum side and

$$(E_z)_{z=0-} = -\frac{a\lambda}{\pi(\epsilon+\epsilon_0)(x^2+a^2)}$$

on the dielectric material side. So, the density of polarization charge on the surface is

$$\sigma_p = \epsilon_0\bigl[(E_z)_{z=0+} - (E_z)_{z=0-}\bigr] = -\frac{a(\epsilon-\epsilon_0)\lambda}{\pi(\epsilon+\epsilon_0)(x^2+a^2)}.$$

The total polarization charge on the surface is

$$\lambda_p = -\frac{a(\epsilon-\epsilon_0)\lambda}{\pi(\epsilon+\epsilon_0)}\int_{-\infty}^\infty \frac{dx}{(x^2+a^2)} = -\frac{(\epsilon-\epsilon_0)\lambda}{(\epsilon+\epsilon_0)},$$

which is equal to the image charge λ' obtained in Example 4.7.

4.27. A perpendicular plane is drawn from the given line charge to the dielectric material surface and the x-axis is defined normal to this plane with the origin on this plane. Assume a narrow region from x to $x + dx$. From the result of Exercise 4.26, the polarization charge in this region can be regarded as a line charge of density

$$\lambda_p dx = -\frac{a(\epsilon - \epsilon_0)\lambda dx}{\pi(\epsilon + \epsilon_0)(x^2 + a^2)}.$$

The Coulomb force on the given line charge from this line charge is $\lambda\lambda_p dx/2\pi(x^2+a^2)^{1/2}$. From symmetry, only its vertical component remains. Hence, the force in a unit length is

$$F' = -\frac{a^2(\epsilon-\epsilon_0)\lambda^2}{2\pi^2(\epsilon+\epsilon_0)}\int_{-\infty}^{\infty}\frac{dx}{(x^2+a^2)^2} = -\frac{(\epsilon-\epsilon_0)\lambda^2}{4\pi(\epsilon+\epsilon_0)a},$$

which is equal to the force between the given and image line charges separated by distance $2a$.

4.28. The electric field in the dielectric material is uniform and is given by $E = V/d$. So, the electrostatic energy is

$$U_e = dl \int_0^w \frac{\epsilon(1+kx)E^2 dx}{2} = \frac{l\epsilon V^2}{2d}\int_0^w (1+kx)dx$$

$$= \frac{lw\epsilon V^2}{2d}\left(1+\frac{kw}{2}\right) = \frac{1}{2}CV^2,$$

where C is the capacitance determined in Exercise 4.6.

4.29. We denote the electric field by $E(y)$. The electric flux density $D = \epsilon'(y)E(y)$ is uniform in the dielectric material and the D is determined by the condition:

$$V = D\int_0^d \frac{dy}{\epsilon'(y)} = \frac{D}{\epsilon}\int_0^d \frac{dy}{1+ky} = \frac{D}{\epsilon k}\log(1+kd).$$

The electrostatic energy is

$$U_e = lw\int_0^d \frac{D^2 dy}{2\epsilon'(y)} = \frac{lw\epsilon}{2}\left[\frac{kV}{\log(1+kd)}\right]^2 \int_0^d \frac{dy}{1+ky} = \frac{lwk\epsilon V^2}{2\log(1+kd)} = \frac{1}{2}CV^2,$$

where C is the capacitance determined in Exercise 4.7.

4.5 Exercises

4.30. According to the result of Exercise 4.2, the electric flux density in the dielectric materials 1 and 2 is, respectively, given by

$$D_1 = \frac{\epsilon_1 Q'}{\pi(\epsilon_1 + \epsilon_2) R}, \quad D_2 = \frac{\epsilon_2 Q'}{\pi(\epsilon_1 + \epsilon_2) R}.$$

Hence, the electrostatic energy density in each region is

$$u_{e1} = \frac{\epsilon_1 Q'^2}{2\pi^2 (\epsilon_1 + \epsilon_2)^2 R^2}, \quad u_{e2} = \frac{\epsilon_2 Q'^2}{2\pi^2 (\epsilon_1 + \epsilon_2)^2 R^2}.$$

So, the electrostatic energy in a unit length is determined to be

$$U'_e = \int_a^b (u_{e1} + u_{e2}) \pi R dR = \frac{Q'^2}{2\pi(\epsilon_1 + \epsilon_2)} \int_a^b \frac{dR}{R} = \frac{Q'^2}{2\pi(\epsilon_1 + \epsilon_2)} \log \frac{b}{a},$$

which is equal to $Q'^2/2C'$ with the capacitance in a unit length C' obtained in Exercise 4.2.

4.31. We apply electric charges $\pm Q'$ in a unit length. The electric flux density at radius R is $D(R) = Q'/2\pi R$. Hence, the electric charge Q' is determined by the condition:

$$V = \frac{Q'}{2\pi\epsilon} \int_a^b \frac{dR}{R(1+kR)} = \frac{Q'}{2\pi\epsilon} \log \frac{b(1+ka)}{a(1+kb)}.$$

The electrostatic energy in a unit length is

$$U'_e = \int_a^b \frac{1}{2\epsilon'(R)} \left(\frac{Q'}{2\pi R}\right)^2 2\pi R dR$$

$$= \pi\epsilon \left\{ \frac{V}{\log[b(1+ka)/a(1+kb)]} \right\}^2 \int_a^b \frac{dR}{R(1+kR)}$$

$$= \frac{\pi\epsilon V^2}{\log[b(1+ka)/a(1+kb)]} = \frac{1}{2} C' V^2,$$

where C' is the capacitance in a unit length determined in Exercise 4.8.

4.32. Since the electric flux density at distance R from the central axis is $D(R) = Q'/2\pi R$, the electric field there is $E(R) = Q'/2\pi\epsilon'(R) R$. Hence, the electric potential difference is

$$V = \frac{Q'}{2\pi\epsilon}\int_a^b \frac{dR}{R(1+kR)} = \frac{Q'}{2\pi\epsilon}\log\frac{b(1+ka)}{a(1+kb)}.$$

So, the electrostatic energy of the capacitor in a unit length is determined to be

$$U'_e = \frac{1}{2}Q'V = \frac{Q'^2}{4\pi\epsilon}\log\frac{b(1+ka)}{a(1+kb)}.$$

4.33. We apply electric charges $\pm Q$ to the electrodes. The electric field at radius r is $Q/[4\pi\epsilon'(r)r^2]$. Hence, the electric charge Q is determined by the condition:

$$V = \frac{Q}{4\pi\epsilon}\int_a^b \frac{dr}{(1+kr)r^2} = \frac{Q}{4\pi\epsilon}\left[\frac{1}{a}-\frac{1}{b}-k\log\frac{b(1+ka)}{a(1+kb)}\right].$$

The electrostatic energy is

$$U_e = \int_a^b \frac{1}{2\epsilon'(r)}\left(\frac{Q}{4\pi r^2}\right)^2 4\pi r^2 dr$$

$$= 2\pi\epsilon V^2 \left[\frac{1}{a}-\frac{1}{b}-k\log\frac{b(1+ka)}{a(1+kb)}\right]^{-2}\int_a^b \frac{dr}{(1+kr)r^2}$$

$$= 2\pi\epsilon V^2 \left[\frac{1}{a}-\frac{1}{b}-k\log\frac{b(1+ka)}{a(1+kb)}\right]^{-1} = \frac{1}{2}CV^2,$$

where C is the capacitance determined in Exercise 4.9.

4.34. We consider only the region above the liquid level outside the capacitor. The electric field strength is V/d in the whole region, and the electric energy density is $\epsilon_0 V^2/2d^2$ and $\epsilon V^2/2d^2$ in vacuum and in the dielectric liquid, respectively. Thus, the total energy is

$$U_e = \frac{wV^2[\epsilon_0(l-h)+\epsilon h]}{2d},$$

where w is the dimension normal to the sheet. The positive electric charge on the electrode is

$$Q = \frac{w[\epsilon_0(l-h)+\epsilon h]V}{d}.$$

So, when h is increased by Δh, U_e increases by

$$\Delta U_e = \frac{wV^2(\epsilon-\epsilon_0)}{2d}\Delta h.$$

4.5 Exercises

At the same time, the electric charge increases by

$$\Delta Q = \frac{wV(\epsilon - \epsilon_0)}{d}\Delta h.$$

This charge is fed by the power source. That is, the sum of the mechanical work to the outside, $F\Delta h$, and the energy to the power source, $-V\Delta Q$, must be equal to the decrease in the energy of the capacitor: $F\Delta h - V\Delta Q = -\Delta U_e$. Thus, we have

$$F = -\frac{\Delta U_e}{\Delta h} + V\frac{\Delta Q}{\Delta h} = \frac{wV^2(\epsilon - \epsilon_0)}{2d}.$$

This force is positive for increasing the height h, and is balanced with the gravitational force, $\rho wdhg$. Thus, we determine the height to be

$$h = \frac{V^2(\epsilon - \epsilon_0)}{2d^2 \rho g}.$$

4.35. We denote the electric potential difference between the two electrodes by V. Then, the electric charge density on the electrode is $\epsilon_0 V/d$ in the vacuum region and $\epsilon V/d$ in the region facing the dielectric liquid, respectively. Then, the total electric charge is

$$Q = \frac{wV\big[(l-h)\epsilon_0 + (l'+h)\epsilon\big]}{d}, \tag{B4.3}$$

and the electric energy is

$$U = \frac{1}{2}QV = \frac{Q^2 d}{2w[(l-h)\epsilon_0 + (l'+h)\epsilon]}.$$

Thus, the force on the dielectric liquid is

$$F = -\frac{\partial U}{\partial h} = \frac{Q^2 d(\epsilon - \epsilon_0)}{2w[(l-h)\epsilon_0 + (l'+h)\epsilon]^2},$$

which is directed to increasing h. From the force balance condition, $F = \rho wdhg$, the height h is given by the solution of the equation:

$$h = \frac{Q^2(\epsilon - \epsilon_0)}{2w^2 \rho g[(l-h)\epsilon_0 + (l'+h)\epsilon]^2}.$$

This is formally the same as the solution obtained in Exercise 4.34, if we use Eq. (B4.3). But, V is an unknown value to be determined. This cubic equation on h has a single real solution with a positive number.

Coffee Break: Validity of the Dipole

Various electric phenomena, such as electrostatic induction of conductors and electric polarization of dielectric materials, can be explained by using the electric dipole. The magnetic dipole also explains various magnetic phenomena such as magnetic shielding of superconductors and magnetization of magnetic materials. Here, we treat the case of a dielectric sphere in a uniform electric field, for example. Positive and negative electric charges are displaced in and against the direction of the applied electric field, respectively, as shown in Fig. C4.1a. This results in positive and negative polarization charges on the surface, as shown in Fig. C4.1b. The electric fields that displaced positive and negative charges produce outside the dielectric sphere are the same as those produced by the positive and negative electric charges concentrated on the respective centers, as shown in Fig. C4.1c. Hence, the resultant electric field outside the sphere is the same as that produced by some electric dipole placed on the center. The polarization charges on the surface produce a uniform electric field inside the dielectric material. Hence, the electric field inside the sphere is uniform with a uniform applied electric field. This supports the initial assumption that electric charges are displaced uniformly.

Since the magnetic phenomena for a superconducting sphere or magnetic sphere in a uniform magnetic flux density can be expressed as a uniform displacement of virtual magnetic charges, these can also be explained using the equivalent magnetic moment.

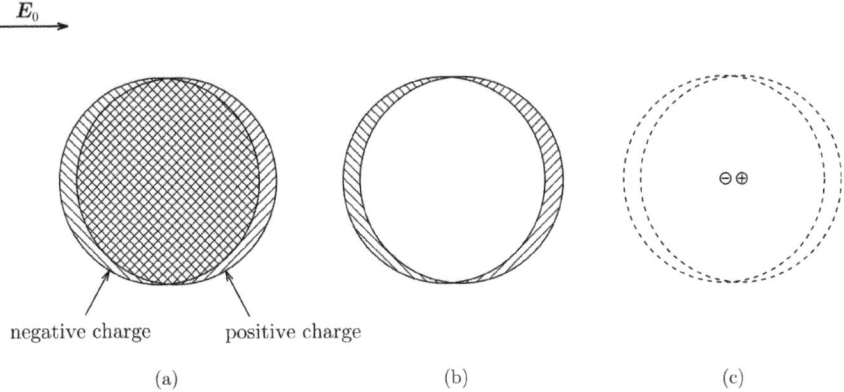

Fig. C4.1 Electric polarization in a dielectric sphere in a uniform electric field: **a** displacement of positive and negative electric charges driven by the electric field, **b** polarization charge that appears on the surface, and **c** electric dipole at the center

Chapter 5
Steady Current

5.1 Current

A conductor contains freely moving electric charge, and the Coulomb force can move the electric charge when an electric field is directly applied to the conductor. This movement of electric charge is current. We treat electric phenomena when a current flows that does not change with time.

The current is a vector with a magnitude and direction. When electric charge dQ passes through a cross-section within time, dt, the current is given by

$$I = \frac{dQ}{dt}. \tag{5.1}$$

Its unit is [C/s], which is denoted [A] (ampere).

Although the current is an amount of electric charge that passes through a certain cross-section in a unit time, it is not a quantity representing strength. We define current density as a quantity representing the strength of current. The current density i is also a vector. Its direction is the same as that of the current, and its magnitude is given by

$$i = \frac{dI}{dS}, \tag{5.2}$$

when current dI flows through a small normal cross-section of area dS. Its unit is [A/m^2].

Since the current is a flow of electric charge, we can describe it using the density and velocity of electric charge. Suppose that particles of electric charge q and density n move with velocity v. The current density is then given by

$$i = qnv. \tag{5.3}$$

© The Editor(s) (if applicable) and The Author(s), under exclusive license to Springer Nature Switzerland AG 2025
T. Matsushita, *Exercises in Electricity and Magnetism*,
https://doi.org/10.1007/978-3-031-67940-7_5

Since the electric charge density is given by $\rho = qn$, the current density is expressed as

$$i = \rho v. \tag{5.4}$$

The amount of electric charge is conserved. That is, the algebraic sum of positive and negative charges is conserved. We suppose a region V surrounded by a closed surface S and denote the electric charge density inside it by ρ. When current of density i flows across the surface, the electric charge that goes out of S in unit time is given by

$$-\frac{d}{dt}\int_V \rho dV = \int_S i \cdot dS. \tag{5.5}$$

Using Gauss's theorem for the right-hand side, Eq. (5.5) is written as

$$\int_V \left(\nabla \cdot i + \frac{\partial \rho}{\partial t}\right) dV = 0. \tag{5.6}$$

Since this relationship holds for arbitrary V, we have

$$\nabla \cdot i + \frac{\partial \rho}{\partial t} = 0. \tag{5.7}$$

This is called the continuity equation of current.

For a steady current when the electric charge distribution does not change with time, Eq. (5.7) reduces to

$$\nabla \cdot i = 0. \tag{5.8}$$

5.2 Ohm's Law

It is necessary to apply an electric potential difference to a material such as a metal to get a current. In many cases, it is empirically known that there is a proportional relationship between the electric potential difference V and the current I:

$$V = R_r I. \tag{5.9}$$

The proportional constant R_r is called the electric resistance or simply resistance. This constant is determined by the shape and property of the material that carries current. The unit of electric resistance is [V/A], which is denoted as [Ω] (ohm). Equation (5.9) is called Ohm's law. For a material of length l and uniform cross-sectional area S, the electric resistance is given by

$$R_r = \rho_r \frac{l}{S}, \quad (5.10)$$

where ρ_r is a constant inherent to the material and is called the resistivity or specific resistance. Its unit is [Ω m].

The relationship between the current and electric potential difference is also written as

$$I = GV. \quad (5.11)$$

In the above, the proportional constant $G = 1/R_r$ is called the conductance. Its unit is [S] (siemens). Using Eq. (5.10), the conductance is written as

$$G = \sigma_c \frac{S}{l}. \quad (5.12)$$

The constant $\sigma_c = 1/\rho_r$ is called the electric conductivity. Its unit is [S/m].

Using the relationships given by Eqs. (5.10) and (5.12), Ohm's law is written as

$$\boldsymbol{E} = \rho_r \boldsymbol{i} \quad (5.13)$$

or

$$\boldsymbol{i} = \sigma_c \boldsymbol{E}. \quad (5.14)$$

5.3 Fundamental Equations for Steady Electric Current

The fundamental physical property for the steady current is given by Eq. (5.8). This is the same as the electric flux density when the electric charge density is zero, as can be seen from Eq. (4.10). In addition, both the current density and the electric flux density are proportional to the electric field [see Eqs. (5.14) and (4.8)]. This indicates that the electric phenomenon is similar between these two cases. That is, the electric field $\boldsymbol{E}$ is common to the two cases, and the current density $\boldsymbol{i}$ corresponds to the electric flux density $\boldsymbol{D}$ and the electric conductivity σ_c corresponds to the dielectric constant ϵ.

One example is shown here. We suppose that electric charges $\pm Q$ are given to the outer and inner electrodes of a concentric spherical capacitor with a dielectric material of dielectric constant ϵ, as shown in Fig. 5.1. The electric flux density in the space between the electrodes is determined to be

Fig. 5.1 Concentric spherical capacitor

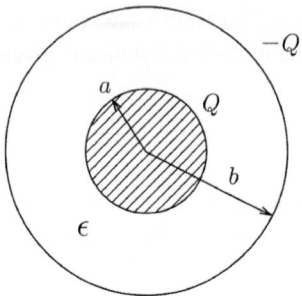

$$D = \frac{Q}{4\pi r^2} \tag{5.15}$$

at position r ($a < r < b$) from the center. Since the electric field is $E = D/\epsilon$, the electric potential difference is given by

$$V = \int_a^b \frac{Q}{4\pi \epsilon r^2} dr = \frac{Q}{4\pi \epsilon}\left(\frac{1}{a} - \frac{1}{b}\right). \tag{5.16}$$

Thus, the capacitance is

$$C = \frac{Q}{V} = \frac{4\pi \epsilon ab}{b-a}. \tag{5.17}$$

The dielectric material in the concentric spherical capacitor is replaced by a substance with electric conductivity σ_c. Now we determine the electric resistance between the two electrodes under the electric potential difference V. We denote the current by I. Since the current density does not have a divergence similarly to the electric flux density, Gauss's law gives the current density as

$$i = \frac{I}{4\pi r^2}, \tag{5.18}$$

corresponding to Eq. (5.15). Since the electric field is $E = i/\sigma_c$, the electric potential difference is given by

$$V = \int_a^b \frac{I}{4\pi \sigma_c r^2} dr = \frac{I}{4\pi \sigma_c}\left(\frac{1}{a} - \frac{1}{b}\right). \tag{5.19}$$

The electric resistance is

$$R_r = \frac{V}{I} = \frac{b-a}{4\pi \sigma_c ab}. \tag{5.20}$$

5.3 Fundamental Equations for Steady Electric Current

Thus, the above two problems are formally identical. Eliminating V common to each example from Eqs. (5.17) and (5.20), we have

$$CR_r = \frac{\epsilon}{\sigma_c}. \tag{5.21}$$

This quantity—the product of the capacitance and the electric resistance—does not depend on the shape of the capacitor or resistor and is given only by the dielectric constant and the electric conductivity. This relationship of Eq. (5.21) generally holds for a capacitor and resistor having electrodes of the same shape. However, this is limited to the case in which we can obtain a rigorous solution for the field.

Here, we discuss the boundary conditions to be satisfied for the steady current at an interface between substances with different electric conductivities, σ_{c1} and σ_{c2}, as shown in Fig. 5.2. Since the equation for the current density i is formally the same as that for the electric flux density D in the absence of electric charge, the boundary condition is also the same. That is, from Eq. (4.12) we have

$$\boldsymbol{n} \cdot (\boldsymbol{i}_1 - \boldsymbol{i}_2) = 0, \tag{5.22}$$

where $\boldsymbol{n}$ is a unit vector normal to the interface. This shows that the normal component of the current density is continuous at the interface. The continuity of the parallel component of the electric field is given by

$$\boldsymbol{n} \times (\sigma_{c1}\boldsymbol{i}_1 - \sigma_{c2}\boldsymbol{i}_2) = 0. \tag{5.23}$$

Here, we discuss the refraction of the current at an interface. Suppose an interface between two materials with electric conductivities σ_{c1} and σ_{c2}. Assume that a current of density i_1 is applied to material 1 in the direction of angle θ_1 from the normal direction to the interface, as shown in Fig. 5.3. The current density and the angle in material 2 are denoted by i_2 and θ_2, respectively. The continuity of the current on the interface is described as

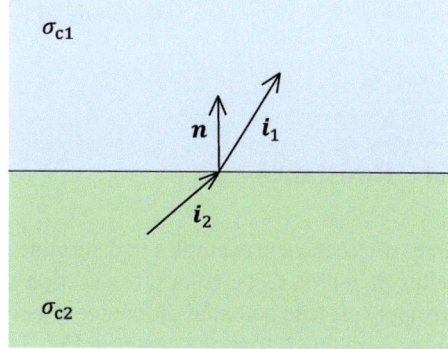

Fig. 5.2 Current densities at the interface between different materials

Fig. 5.3 Refraction of current at an interface

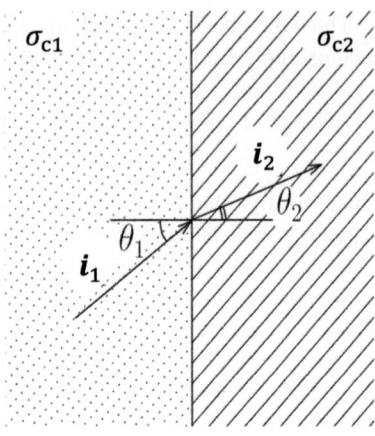

$$i_1 \cos \theta_1 = i_2 \cos \theta_2. \tag{5.24}$$

The continuity of the parallel component of the electric field is given by

$$\sigma_{c1} i_1 \sin \theta_1 = \sigma_{c2} i_2 \sin \theta_2. \tag{5.25}$$

These equations give

$$\frac{\tan \theta_1}{\tan \theta_2} = \frac{\sigma_{c2}}{\sigma_{c1}}. \tag{5.26}$$

This is the law of refraction of current. We obtain i_2 and θ_2 as

$$i_2 = i_1 \left[\left(\frac{\sigma_{c1}}{\sigma_{c2}} \right)^2 \sin^2 \theta_1 + \cos^2 \theta_1 \right]^{1/2}, \tag{5.27}$$

$$\theta_2 = \tan^{-1} \left(\frac{\sigma_{c1}}{\sigma_{c2}} \tan \theta_1 \right). \tag{5.28}$$

5.4 Resistance

Here, we treat some examples to determine the resistance. Suppose a quarter ring of radius R_0 with a rectangular cross-section and resistivity ρ_r, as shown in Fig. 5.4a. We apply electric potential difference V between the two edges. The electric field at an arc of radius R from the center in Fig. 5.4b is

5.4 Resistance

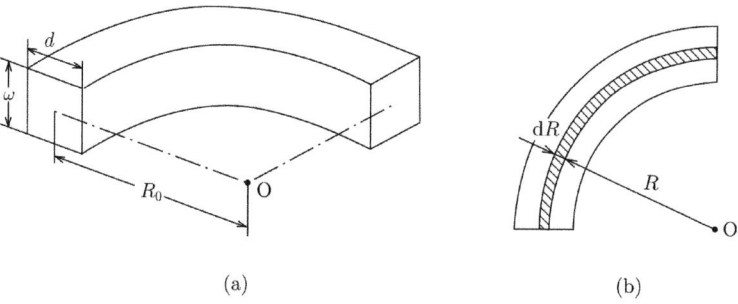

Fig. 5.4 **a** Quarter ring with rectangular cross-section and **b** part of a thin region from radius R to $R + dR$

$$E(R) = \frac{2V}{\pi R}, \tag{5.29}$$

and the current density is

$$i(R) = \frac{2V}{\pi \rho_r R}. \tag{5.30}$$

Then, the total current is

$$I = \int_{R_0 - d/2}^{R_0 + d/2} \frac{2wV}{\pi \rho_r R} dR = \frac{2wV}{\pi \rho_r} \log \frac{R_0 + d/2}{R_0 - d/2}. \tag{5.31}$$

Thus, the electric resistance is given by

$$R_r = \frac{\pi \rho_r}{2w \log[(R_0 + d/2)/(R_0 - d/2)]}. \tag{5.32}$$

Next, we discuss the electric resistance along the length of the truncated cone with resistivity ρ_r in Fig. 5.5. The cross-sectional area at position x from the bottom is

$$S(x) = \pi \left(b - \frac{b-a}{h} x \right)^2. \tag{5.33}$$

If the applied current is I, the current density at this position is $i(x) = I/S(x)$. Since the electric field is $E(x) = \rho_r i(x)$, the electric potential difference between the two edges is

$$V = \int_0^h \rho_r \frac{I}{S(x)} dx = \frac{\rho_r h I}{\pi a b}. \tag{5.34}$$

Fig. 5.5 Long truncated cone

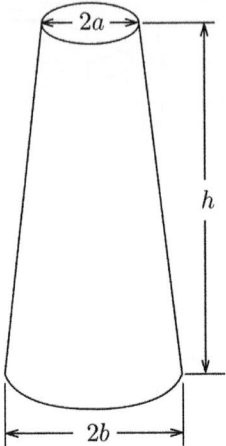

The electric resistance is

$$R_{\rm r} = \frac{\rho_{\rm r} h}{\pi ab}. \tag{5.35}$$

Example 5.1 Determine the electric resistance along the length of a quarter ring of radius R_0 with a circular cross-section of radius a, as shown in Fig. 5.6. The resistivity is $\rho_{\rm r}$.

Fig. 5.6 Quarter ring with circular cross-section

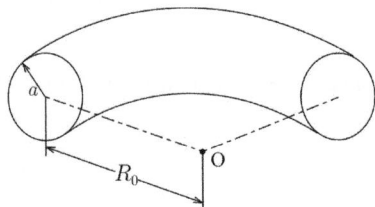

Solution 5.1 We apply voltage V between the two edges. The electric field along the circle of radius R from the center is $E(R) = 2V/(\pi R)$ (see Fig. 5.7). Hence, the current density at this point is $i(R) = 2V/(\pi \rho_{\rm r} R)$. Here, we define the angle θ as in the figure. Then, $R = R_0 + a\cos\theta$. The current flowing in the region between R and $R + {\rm d}R$ is $i(R) 2a\sin\theta {\rm d}R/[\pi \rho_{\rm r}(R_0 + a\cos\theta)]$. Hence, the total current is

$$I = \int_0^\pi \frac{4Va^2 \sin^2\theta\, {\rm d}\theta}{\pi \rho_{\rm r}(R_0 + a\cos\theta)}.$$

We transform the integrand as

5.4 Resistance

Fig. 5.7 Part in the region from R to $R + dR$, with respect to the center

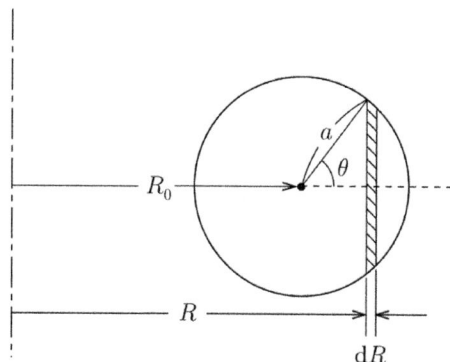

$$\frac{\sin^2\theta}{R_0 + a\cos\theta} = \frac{R_0}{a^2} - \frac{1}{a}\cos\theta - \left(\frac{R_0^2}{a^2} - 1\right)\frac{1}{R_0 + a\cos\theta}.$$

For integration of the third term, we use Eq. (A.23) in the Appendix. A simple calculation gives

$$I = \frac{4V}{\rho_r}\left[R_0 - \left(R_0^2 - a^2\right)^{1/2}\right].$$

Then, we obtain the electric resistance as

$$R_r = \frac{\rho_r}{4\left[R_0 - \left(R_0^2 - a^2\right)^{1/2}\right]}.$$

◆

Example 5.2 Determine the electric resistance along the length of a substance with resistivity ρ_r, as shown in Fig. 5.8.

Fig. 5.8 Long substance with rectangular cross-section

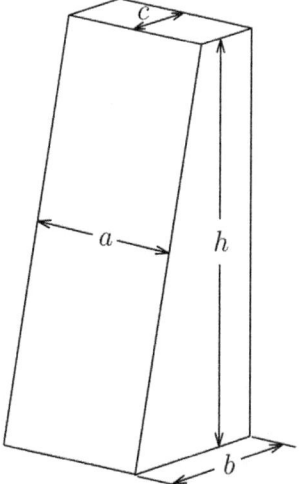

Solution 5.2 The cross-sectional area at height x from the bottom is $S(x) = a[b - (b-c)x/h]$, and the current density there is $i(x) = I/\{a[b - (b-c)x/h]\}$, when we apply current I. Since the electric field is $E(x) = \rho_r i(x)$, the voltage between the two edges is

$$V = \int_0^h \frac{\rho_r I}{a[b - (b-c)x/h]} dx = \frac{h\rho_r I}{a(b-c)} \log \frac{b}{c}.$$

The electric resistance is

$$R_r = \frac{h\rho_r}{a(b-c)} \log \frac{b}{c}.$$

♦

Example 5.3 The space between the electrodes in a concentric spherical resistor is occupied by two materials with resistivities ρ_{r1} and ρ_{r2}, as shown in Fig. 5.9. Determine the electric resistance between the two electrodes.

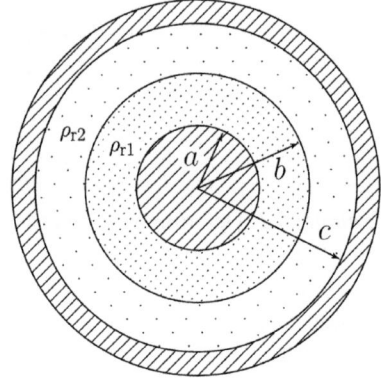

Fig. 5.9 Concentric spherical resistor with two materials

Solution 5.3 When current I is applied between the two electrodes, the current density at a position of radius r from the center is

$$i = \frac{I}{4\pi r^2}.$$

5.4 Resistance

The electric field strength is $E = \rho_{r1} i$ in the region $a < r < b$ and $E = \rho_{r2} i$ in the region $b < r < c$. The electric potential difference between the two electrodes is

$$V = \frac{\rho_{r1} I}{4\pi} \int_a^b \frac{dr}{r^2} + \frac{\rho_{r2} I}{4\pi} \int_b^c \frac{dr}{r^2} = \frac{\rho_{r1}(b-a)I}{4\pi ab} + \frac{\rho_{r2}(c-b)I}{4\pi bc}.$$

Thus, the electric resistance is determined to be

$$R_r = \frac{V}{I} = \frac{\rho_{r1}(b-a)}{4\pi ab} + \frac{\rho_{r2}(c-b)}{4\pi bc}.$$

◆

Example 5.4 The space between the electrodes in a concentric spherical resistor is occupied by two materials with resistivities ρ_{r1} and ρ_{r2}, as shown in Fig. 5.10. Determine the electric resistance between the two electrodes.

Fig. 5.10 Concentric spherical resistor with two materials

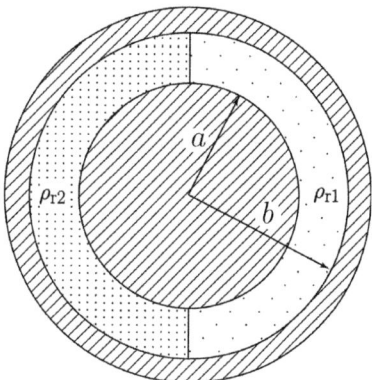

Solution 5.4 We use I_1 and I_2 to denote the currents flowing in the respective regions with the resistivities ρ_{r1} and ρ_{r2}, respectively, when we apply a voltage V between the electrodes. Then, the current densities at a distance r ($a \leq r \leq b$) in each region are $i_1 = I_1/2\pi r^2$ and $i_2 = I_2/2\pi r^2$, and the electric fields are $E_1 = \rho_{r1} I_1/2\pi r^2$ and $E_2 = \rho_{r2} I_2/2\pi r^2$. From the condition that the integrations of the electric fields from $r = a$ to $r = b$ are V, we have

$$I_1 = \frac{2\pi ab V}{(b-a)\rho_{r1}}, \quad I_2 = \frac{2\pi ab V}{(b-a)\rho_{r2}}.$$

Since $I_1 + I_2 = I$ is the total current, we have

$$R_{\mathrm{r}} = \frac{V}{I} = \frac{(b-a)\rho_{\mathrm{r}1}\rho_{\mathrm{r}2}}{2\pi ab(\rho_{\mathrm{r}1}+\rho_{\mathrm{r}2})}.$$

◆

Example 5.5 The space between two long coaxial electrodes is occupied by two substances with resistivities $\rho_{\mathrm{r}1}$ and $\rho_{\mathrm{r}2}$, as shown in Fig. 5.11. Determine the electric resistance between the two electrodes.

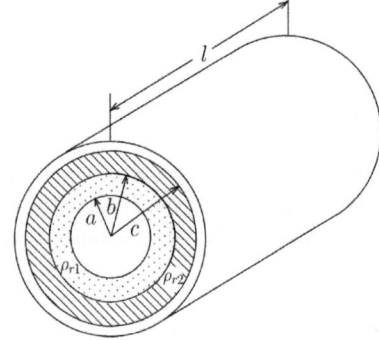

Fig. 5.11 Long coaxial resistor composed of two substances with different resistivities

Solution 5.5 When we apply current I, the current density at distance R from the central axis is

$$i(R) = \frac{I}{2\pi lR}.$$

The electric field is $E(R) = \rho_{\mathrm{r}1} i(R)$ for $a < R < b$ and $E(R) = \rho_{\mathrm{r}2} i(R)$ for $b < R < c$. Hence, the voltage between the two electrodes is

$$V = \int_a^b \frac{\rho_{\mathrm{r}1} I}{2\pi lR} dR + \int_b^c \frac{\rho_{\mathrm{r}2} I}{2\pi lR} dR = \frac{I}{2\pi l}\left(\rho_{\mathrm{r}1} \log \frac{b}{a} + \rho_{\mathrm{r}2} \log \frac{c}{b}\right).$$

The resistance is

$$R_{\mathrm{r}} = \frac{1}{2\pi l}\left(\rho_{\mathrm{r}1} \log \frac{b}{a} + \rho_{\mathrm{r}2} \log \frac{c}{b}\right).$$

◆

Example 5.6 The space between two long coaxial electrodes is occupied by two substances with resistivities $\rho_{\mathrm{r}1}$ and $\rho_{\mathrm{r}2}$, as shown in Fig. 5.12. Determine the electric resistance between the two electrodes.

Fig. 5.12 Long coaxial resistor composed of two substances with different resistivities

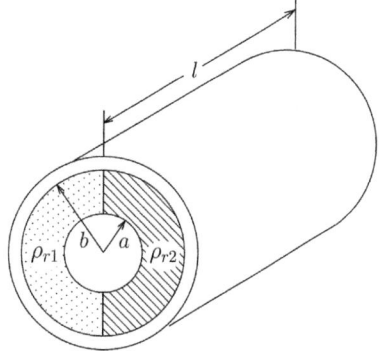

Solution 5.6 We use I_1 and I_2 to denote the currents flowing in the respective regions with the resistivities ρ_{r1} and ρ_{r2} when we apply voltage V between the electrodes. Then, the current densities at positions at distance R ($a \leq R \leq b$) in the respective regions are $i_1(R) = I_1/(\pi Rl)$ and $i_2(R) = I_2/(\pi Rl)$, and the electric fields are $E_1(R) = \rho_{r1}I_1/(\pi Rl)$ and $E_2(R) = \rho_{r2}I_2/(\pi Rl)$. From the condition that the integrations of the electric fields from $R = a$ to $R = b$ are V, we have

$$I_1 = \frac{\pi l V}{\rho_{r1} \log(b/a)}, \quad I_2 = \frac{\pi l V}{\rho_{r2} \log(b/a)}.$$

Since the total current is $I = I_1 + I_2$, we obtain the electric resistance as

$$R_r = \frac{\rho_{r1}\rho_{r2}}{\pi l(\rho_{r1} + \rho_{r2})} \log \frac{b}{a}.$$

♦

5.5 Electric Power

When we apply a current to a material with electric resistance, energy dissipation takes place. Suppose that a current I flows in a material under an electric potential difference V given by a power source. The work done by the power source in unit time is

$$P = VI. \tag{5.36}$$

This is called the electric power. Its unit is [VA], which is denoted as [W] (watt). Using Ohm's law, this is rewritten as

$$P = \rho_{\mathrm{r}} I^2 = \frac{V^2}{\rho_{\mathrm{r}}}, \tag{5.37}$$

and its value in a unit volume is

$$p = Ei = \sigma_{\mathrm{c}} E^2 = \rho_{\mathrm{r}} i^2. \tag{5.38}$$

Example 5.7 Suppose that, when we apply electric potential difference V between the two edges of the quarter ring shown in Fig. 5.4a, current I flows. Prove that the total electric power dissipated in this resistor is VI.

Solution 5.7 When we apply voltage V between the two edges, the electric field at a point of radius R is $E(R) = 2V/(\pi R)$. Hence, the electric power density is $p(R) = 4V^2/(\pi^2 \rho_{\mathrm{r}} R^2)$. The electric power in the region from R to $R + \mathrm{d}R$ is

$$\mathrm{d}P = \frac{\pi w R \mathrm{d}R p(R)}{2} = \frac{2wV^2}{\pi \rho_{\mathrm{r}}} \cdot \frac{\mathrm{d}R}{R}.$$

Thus, the total dissipated electric power is

$$P = \frac{2wV^2}{\pi \rho_{\mathrm{r}}} \int_{R_0 - d/2}^{R_0 + d/2} \frac{\mathrm{d}R}{R} = \frac{2wV^2}{\pi \rho_{\mathrm{r}}} \log \frac{R_0 + d/2}{R_0 - d/2}.$$

Using the current given by Eq. (5.31), this is equal to VI.

$\blacklozenge$

Example 5.8 Assume that current I flows through the long truncated cone shown in Fig. 5.5 when we apply electric potential difference V between its top and bottom. Prove that the dissipated total electric power is equal to VI.

Solution 5.8 The cross-sectional area of the resistor at a position x from the bottom is denoted by $S(x)$. The density of current flowing at this position is $i(x) = I/S(x)$ and the electric field strength is $E(x) = \rho_{\mathrm{r}} I / S(x)$. Thus, the dissipated electric power density is

$$p(x) = E(x) i(x) = \frac{\rho_{\mathrm{r}} I^2}{S(x)^2}.$$

The volume integral leads to the total dissipated power:

$$P = \int_0^h p(x)S(x)dx = I^2 \int_0^h \frac{\rho_r}{S(x)}dx.$$

In this equation

$$\int_0^h \frac{\rho_r}{S(x)}dx = R_r$$

is the electrical resistance. Thus, we have $P = R_r I^2 = VI$.

◆

5.6 Exercises

5.1. Currents $\pm I$ are applied to the electrodes of a parallel-plate capacitor, as shown in Fig. E5.1. Determine the current distribution in the electrode.

5.2. Determine the resistance of the cylindrical coaxial resistor shown in Fig. E5.2. The resistivity of the material is ρ_r.

Fig. E5.1 Parallel-plate capacitor

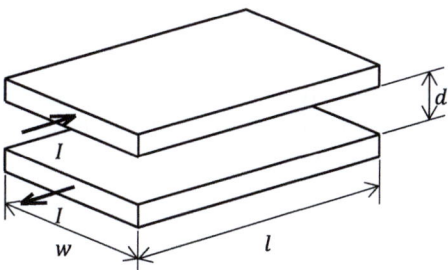

Fig. E5.2 Cylindrical coaxial resistor

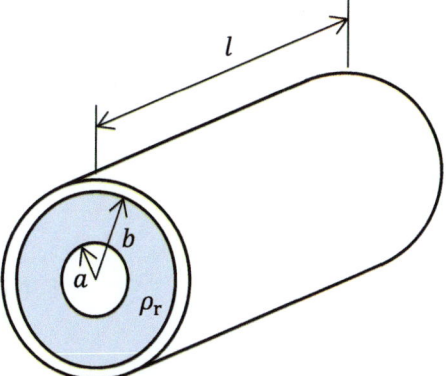

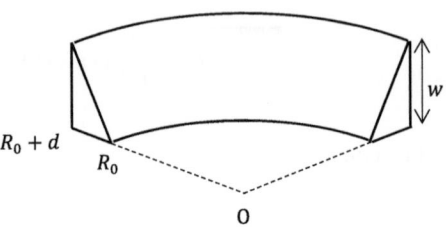

Fig. E5.3 Quarter circle with triangular cross section

5.3. Determine the electric resistance along the length of a quarter circle with a triangular cross-section, as shown in Fig. E5.3 The resistivity is ρ_r.

5.4. Determine the resistance of the resistor shown in Fig. 5.4a as a combined resistance for resistors connected in parallel.

5.5. The resistance of the resistor shown in Fig. 5.5 as a combined resistance for resistors connected in series.

5.6. The resistor treated in Example 5.3 can be regarded as a resistor composed of two resistors connected in series. Determine the resistance.

5.7. The resistor treated in Example 5.4 can be regarded as a resistor composed of two resistors connected in parallel. Determine the resistance.

5.8. Suppose that the resistivity of the resistor shown in Fig. 5.4a varies as $\rho_r(R) = \rho_{r0}(1 - \alpha R)$ with the radius R. Determine the resistance of this resistor.

5.9. Suppose that the resistivity of the resistor shown in Fig. 5.5 varies as $\rho_r(x) = \rho_{r0}(1 - \alpha x)$ with the distance x from the bottom. Determine the resistance of this resistor.

5.10. We apply a current to the material shown in Fig. E5.4 in the direction of the y-axis. When the resistivity varies as $\rho'_r = \rho_r(1 + ky)$ along the length, determine the resistance.

5.11. Assume that the resistivity varies as $\rho'_r = \rho_r(1 + kz)$ along the width in the material shown in Fig. E5.4 in Exercise 5.10. When we apply a current in the direction of the y-axis, determine the resistance.

5.12. In the long cylindrical coaxial resistor shown in Fig. E5.5, the resistivity varies radially as $\rho'_r(R) = \rho_r(1 + kR)$. Determine the radial resistance in a unit length.

5.13. In the spherical concentric resistor shown in Fig. E5.6, the resistivity varies radially as $\rho'_r(r) = \rho_r(1 + kr)$. Determine the resistance.

Fig. E5.4 Long resistor with resistivity that varies along the length

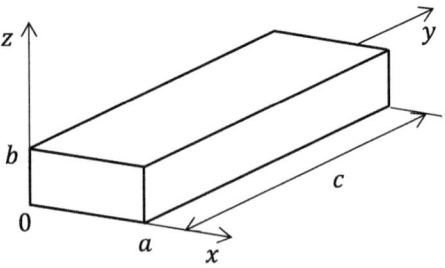

5.6 Exercises

Fig. E5.5 Cylindrical coaxial resistor with resistivity that varies radially

Fig. E5.6 Spherical concentric resistor with resistivity that varies radially

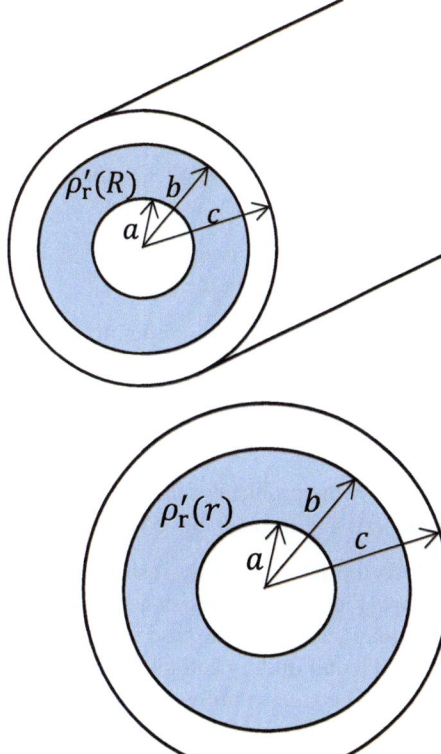

5.14. Suppose that voltage V is applied to a quarter ring of radius R_0 with a circular cross-section of radius a, as shown in Fig. 5.6 in Example 5.1. Determine the dissipated power in the resistor. The resistivity is ρ_r.

5.15. Determine the electric power dissipated in the resistor treated in Exercise 5.3, when we apply voltage V to the resistor.

5.16. Determine the electric power dissipated in the resistor treated in Example 5.2, when we apply current I to the resistor.

5.17. Determine the dissipated power when current I is applied to a spherical resistor with the structure shown in Fig. 5.1.

5.18. Determine the electric power dissipated in the spherical concentric resistor treated in Example 5.3, when we apply current I to the resistor.

5.19. Determine the electric power dissipated in the spherical concentric resistor treated in Example 5.4, when we apply voltage V to the resistor.

5.20. When voltage V is applied to the cylindrical resistor treated in Example 5.5, determine the dissipated power in the resistor.

5.21. When voltage V is applied to the cylindrical resistor treated in Example 5.6, determine the dissipated power in the resistor.

5.22. Determine the dissipated power when current I is applied to the cylindrical coaxial resistor discussed in Exercise 5.2. The resistivity of the material is ρ_r.

5.23. When voltage V is applied to the slab resistor treated in Exercise 5.10, determine the dissipated power in the resistor.

5.24. When voltage V is applied to the slab resistor treated in Exercise 5.11, determine the dissipated power in the resistor.

5.25. When voltage V is applied to the long cylindrical coaxial resistor treated in Exercise 5.12, determine the dissipated power in the resistor.

5.26. When voltage V is applied to the concentric spherical resistor treated in Exercise 5.13, determine the dissipated power in the resistor.

5.27. Solve the problem of the dissipated power in Example 5.7 with the fact that the current passing through any equipotential surface is equal to the given current.

5.28. We are determining the resistance between points P and Q in the resistor network shown in Fig. E5.7. The resistance of each resistor is R. For this purpose, we superpose two situations: One is the case in which the current source is connected between the terminal P and infinity and the other is the case in which the same current source is connected between infinity and the terminal Q. It is assumed that the electric potential is zero at infinity in this solution method. Prove that this assumption is valid from the viewpoint of electricity and determine the resistance.

5.29. The capacitance of a capacitor and the resistance of a resistor with the same concentric structure are treated in Example 4.1 and Example 5.3, respectively. A similar correspondence can be found between the capacitor treated in Example 4.2 and the resistor treated in Example 5.4. Discuss the general relationship between the capacitance and resistance for the capacitor and resistor with the same structure.

Fig. E5.7 Resistor network

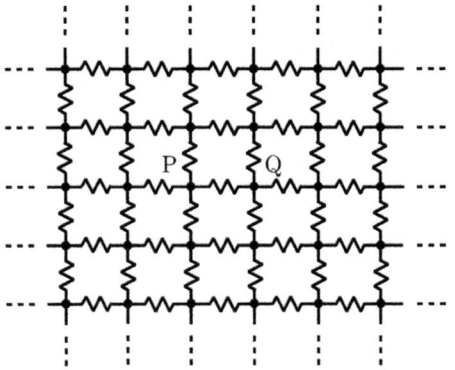

5.6 Exercises

5.30. Two parallel long cylindrical conductors with radius a are embedded in a uniform substance having electric conductivity σ_c, as shown in Fig. E5.8. The distance between the central axes of the two conductors is d ($> 2a$). When current I' is applied between the conductors in a unit length, the electric potential is given by

$$\phi(R, \varphi) = \frac{I'}{4\pi\sigma_c} \log \frac{R^2 + (d-h)^2 - 2R(d-h)\cos\varphi}{R^2 + h^2 - 2Rh\cos\varphi},$$

where R and φ are the distance from the central axis of the left conductor and the azimuthal angle measured from this axis, and $h = (d/2) - \sqrt{(d/2)^2 - a^2}$ (see Example 2.6: Note that the left region from the conductor surface in Example 2.6 corresponds to the left region in the present case.) Discuss the equipotential surface.

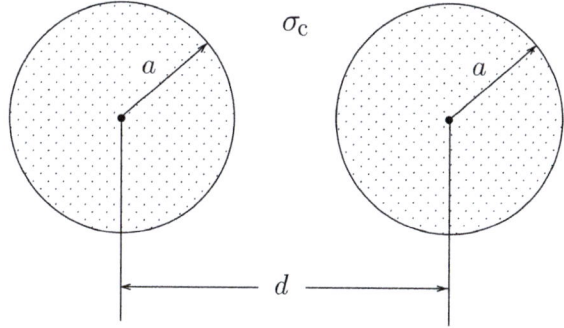

Fig. E5.8 Cross-sections of two parallel cylindrical conductors embedded in a uniform substance

5.31. A hollow sphere with electric conductivity σ_c is embedded in a material with electric conductivity σ_{c0}. A uniform current of density i_0 is applied, as shown in Fig. E5.9. Determine the current density in each region.

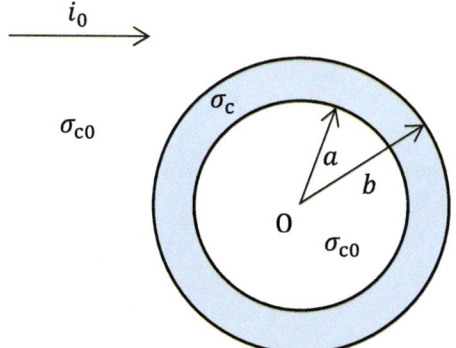

Fig. E5.9 Hollow sphere embedded in a material with different electric conductivity

5.32. A long hollow cylinder with electric conductivity σ_c is embedded in a material with electric conductivity σ_{c0}. A uniform current of density i_0 is applied, as shown in Fig. E5.10. Determine the current density in each region.

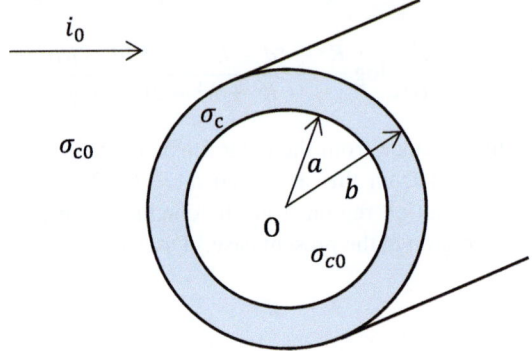

Fig. E5.10 Hollow cylinder embedded in a material with different electric conductivity

Answers to Exercises

5.1. The electric charge is uniformly distributed on the electrode surface. The increasing rate of the electric charge density is denoted by $\dot{\sigma}$. Then, the increasing rate of electric charge in the region from x to $x + dx$ is equal to $w\dot{\sigma}\,dx$ in the upper electrode, where x is the distance from the left edge of the electrode. This is supplied by the current. We denote by $I'(x)$ the current that flows at position x. Then, the increasing rate of the electric charge is given by $I'(x) - I'(x+dx) = -(\partial I'/\partial x)dx$. Thus, we have

$$\frac{\partial I'}{\partial x} = -w\dot{\sigma}.$$

Under the condition $I'(0) = I$, we have

$$I'(x) = I - w\dot{\sigma}x.$$

Another condition on the total electric charge is: $w\dot{\sigma}l = I$. So, the current distribution is determined to be

$$I'(x) = I\left(1 - \frac{x}{l}\right).$$

5.2. We apply current I. Then, the radial current density at distance R from the axis is

$$i(R) = \frac{I}{2\pi l R}$$

5.6 Exercises

and the electric field is $E(R) = \rho_r i(R)$. Thus, the electric potential between the two electrodes is

$$V = \int_a^b E(R)dR = \frac{\rho_r I}{2\pi l} \int_a^b \frac{1}{R} dR = \frac{\rho_r I}{2\pi l} \log \frac{b}{a}.$$

Thus, the resistance is determined to be

$$R_r = \frac{V}{I} = \frac{\rho_r}{2\pi l} \log \frac{b}{a}.$$

5.3. We apply electric potential difference V between the two edges. The electric field at distance R from the center is $E(R) = 2V/\pi R$ and the current density is $i(R) = 2V/\pi \rho_r R$. The height of this region is $w'(R) = w(R - R_0)/d$. Then, the total current is

$$I = \frac{2wV}{\pi \rho_r d} \int_{R_0}^{R_0+d} \frac{R - R_0}{R} dR = \frac{2wV}{\pi \rho_r d} \left[d - R_0 \log\left(1 + \frac{d}{R_0}\right) \right].$$

Thus, the electric resistance is given by

$$R_r = \frac{\pi \rho_r d}{2w[d - R_0 \log(1 + d/R_0)]}.$$

In the case $R_0 \gg d$, this reduces to $R_r = \pi \rho_r R_0 / 2S$ with $S = wd/2$ denoting the cross-sectional area.

5.4. The region from R to $R + dR$ is regarded as a thin resistor. Since the current flowing this region is

$$dI = i(R)wdR = \frac{2wVdR}{\pi \rho_r R},$$

the conductance of this region is

$$dG = \frac{dI}{V} = \frac{2wdR}{\pi \rho_r R}.$$

Hence, the total conductance of this resistor is

$$G = \int dG = \int_{R_0 - d/2}^{R_0 + d/2} \frac{2wdR}{\pi \rho_r R} = \frac{2w}{\pi \rho_r} \log \frac{R_0 + d/2}{R_0 - d/2},$$

and the resistance is determined to be

$$R_r = \frac{1}{G} = \frac{\pi \rho_r}{2w \log[(R_0 + d/2)/(R_0 - d/2)]}.$$

5.5. The region between x and $x + dx$ from the bottom is regarded as a thin resistor. The resistance of this region is given by

$$dR_r = \frac{\rho_r dx}{S(x)},$$

where $S(x)$ is the cross-sectional area of this region given by

$$S(x) = \pi \left(b - \frac{b-a}{h}x\right)^2.$$

So, the total resistance is determined to be

$$R_r = \int dR_r = \frac{\rho_r}{\pi} \int_0^h \frac{dx}{[b - (b-a)x/h]^2} = \frac{\rho_r h}{\pi ab}.$$

5.6. The resistance of the inner resistor with material 1 is

$$R_{r1} = \frac{\rho_{r1}(b-a)}{4\pi ab}$$

and the resistance of the outer resistor with material 2 is

$$R_{r2} = \frac{\rho_{r2}(c-b)}{4\pi bc}.$$

Thus, the resistance is determined to be

$$R_r = R_{r1} + R_{r2} = \frac{\rho_{r1}(b-a)}{4\pi ab} + \frac{\rho_{r2}(c-b)}{4\pi bc}.$$

5.7. The resistance of the right resistor with material 1 is

$$R_{r1} = \frac{\rho_{r1}(b-a)}{2\pi ab}$$

and the resistance of the left resistor with material 2 is

$$R_{r2} = \frac{\rho_{r2}(b-a)}{2\pi ab}.$$

5.6 Exercises

Thus, the resistance is determined to be

$$R_r = \frac{R_{r1} R_{r2}}{R_{r1} + R_{r2}} = \frac{(b-a)\rho_{r1}\rho_{r2}}{2\pi ab(\rho_{r1} + \rho_{r2})}.$$

5.8. We apply electric potential difference V between the two edges. Then, the electric field at an arc with radius R in Fig. 5.4b is $E(R) = 2V/\pi R$ and the current density is

$$i(R) = \frac{2V}{\pi \rho_{r0}} \cdot \frac{1}{R(1-\alpha R)} = \frac{2V}{\pi \rho_{r0}} \left(\frac{1}{R} - \frac{\alpha}{1-\alpha R} \right).$$

Then, the total current is

$$I = \frac{2wV}{\pi \rho_{r0}} \int_{R_0-d/2}^{R_0+d/2} \left(\frac{1}{R} - \frac{\alpha}{1-\alpha R} \right) dR = \frac{2wV}{\pi \rho_{r0}} \log \frac{(R_0+d/2)[1-\alpha(R_0+d/2)]}{(R_0-d/2)[1-\alpha(R_0-d/2)]},$$

and the resistance is determined to be

$$R_r = \frac{V}{I} = \frac{\pi \rho_{r0}}{2w} \left\{ \log \frac{(R_0+d/2)[1-\alpha(R_0+d/2)]}{(R_0-d/2)[1-\alpha(R_0-d/2)]} \right\}^{-1}.$$

5.9. When we apply current I, the current density at the position x from the bottom is $i(x) = I/S(x)$, where $S(x)$ is the cross-sectional area given by Eq. (5.33). Hence, the electric potential difference between the top and bottom edges is

$$V = \int_0^h \rho_r(x) \frac{I}{S(x)} dx = \frac{\rho_{r0} I}{\pi b^2} \int_0^h \frac{1-\alpha x}{(1-kx)^2} dx,$$

where $k = (b-a)/bh$. The integrand is written as

$$\frac{1-\alpha x}{(1-kx)^2} = \frac{\alpha/k}{1-kx} + \left(1 - \frac{\alpha}{k}\right) \frac{1}{(1-kx)^2},$$

and the resistance is determined to be

$$R_r = \frac{V}{I} = \frac{\rho_{r0}}{\pi ab} \left[1 - \frac{bh\alpha}{b-a} \left(1 - \frac{a}{b-a} \log \frac{b}{a} \right) \right].$$

5.10. The resistance of a small section from y to $y+dy$ is $dR_r = (\rho_r/ab)(1+ky)\,dy$. Hence, the resistance is determined to be

$$R_r = \frac{\rho_r}{ab} \int_0^c (1+ky)\,dy = \frac{c\rho_r}{ab} \left(1 + \frac{kc}{2} \right).$$

5.11. The conductance of a small section from z to $z+dz$ is $dG = [a/c\rho_r(1+kz)]dz$. Hence, the conductance is determined to be

$$G = \frac{a}{c\rho_r}\int_0^b \frac{dz}{1+kz} = \frac{a}{c\rho_r k}\log(1+kb).$$

The resistance is

$$R_r = \frac{c\rho_r k}{a\log(1+kb)}.$$

5.12. When we apply current I' in a unit length, the current density at radius R is

$$i(R) = \frac{I'}{2\pi R}$$

and the electric field there is

$$E(R) = \rho'_r(R)i(R) = \frac{\rho_r I'(1+kR)}{2\pi R}.$$

Hence, the potential difference between the two electrodes is

$$V = \frac{\rho_r I'}{2\pi}\int_a^b \frac{1+kR}{R}dR = \frac{\rho_r I'}{2\pi}\left[\log\frac{b}{a} + k(b-a)\right].$$

The resistance in a unit length is

$$R'_r = \frac{\rho_r}{2\pi}\left[\log\frac{b}{a} + k(b-a)\right].$$

5.13. When we apply current I, the current density at radius r is

$$i(r) = \frac{I}{4\pi r^2}$$

and the electric field there is

$$E(r) = \rho'_r(r)i(r) = \frac{\rho_r I}{4\pi}\left(\frac{1}{r^2} + \frac{k}{r}\right).$$

Hence, the potential difference between the two electrodes is

$$V = \frac{\rho_r I}{4\pi} \int_a^b \left(\frac{1}{r^2} + \frac{k}{r}\right) dr = \frac{\rho_r I}{4\pi}\left(\frac{b-a}{ab} + k \log \frac{b}{a}\right).$$

The resistance is

$$R_r = \frac{\rho_r}{4\pi}\left(\frac{b-a}{ab} + k \log \frac{b}{a}\right).$$

5.14. We define the distance R from the center and the angle θ, as shown in Fig. 5.7. The electric field in the hatched region, which is distant $R = R_0 + a\cos\theta$ from the center, is $E(R) = 2V/(\pi R)$. The width, thickness, and length of this part are $2a\sin\theta$, $a\sin\theta d\theta$, and $\pi R/2$. So, the dissipated power density there is

$$dP = \frac{4V^2}{\rho_r(\pi R)^2}\pi a^2 R \sin^2\theta d\theta = \frac{4V^2}{\pi\rho_r}\cdot\frac{a^2\sin^2\theta d\theta}{R_0 + a\cos\theta}.$$

Since we can transform as

$$\frac{a^2\sin^2\theta}{R_0 + a\cos\theta} = R_0 - a\cos\theta - \frac{R_0^2 - a^2}{R_0 + a\cos\theta},$$

The dissipated power is determined to be

$$P = \frac{4V^2}{\pi\rho_r}\int_0^\pi \frac{a^2\sin^2\theta d\theta}{R_0 + a\cos\theta} = \frac{4V^2}{\rho_r}\left[R_0 - (R_0^2 - a^2)^{1/2}\right],$$

where we have used Eq. (A.23) in the Appendix. Thus, the power is equal to V^2/R_r with the resistance R_r determined in Example 5.1.

5.15. The electric field and current density at distance R from the center are $E(R) = 2V/\pi R$ and $i(R) = 2V/\pi\rho_r R$, respectively. Then, the dissipated power density at this position is

$$p(R) = E(R)i(R) = \frac{4V^2}{\pi^2\rho_r R^2}.$$

So, the dissipated power in the region from R to $R + dR$ is

$$\frac{\pi R w'(R)p(R)}{2}dR = \frac{2wV^2}{\pi d\rho_r}\left(1 - \frac{R_0}{R}\right)dR.$$

The dissipated power is determined to be

$$P = \frac{2wV^2}{\pi d \rho_r} \int_{R_0}^{R_0+d} \left(1 - \frac{R_0}{R}\right) dR = \frac{2wV^2}{\pi d \rho_r} \left[d - R_0 \log\left(1 + \frac{d}{R_0}\right)\right],$$

which is equal to V^2/R_r with the resistance R_r obtained in Exercise 5.3.

5.16. The cross-sectional area at height x from the bottom is $S(x) = a[b - (b - c)x/h]$. Then, the current density at this position is $i(x) = I/S(x)$. The power density at this position is $p(x) = \rho_r i^2$, and the power dissipated in the region from x to $x + dx$ from the bottom is $p(x)S(x)dx$. Hence, the dissipated power is determined to be

$$P = \int_0^h p(x)S(x)dx = \frac{\rho_r I^2}{a} \int_0^h \frac{dx}{b - (b - c)x/h} = \frac{h\rho_r I^2}{a(b - c)} \log\frac{b}{c},$$

which is equal to $R_r I^2$ with the resistance R_r obtained in Example 5.2.

5.17. Current density at the position of distance r from the center is

$$i = \frac{I}{4\pi r^2}.$$

Hence, the total dissipated power is

$$P = \int_a^b \frac{i^2}{\sigma_c} 4\pi r^2 dr = \frac{I^2}{4\pi \sigma_c}\left(\frac{1}{a} - \frac{1}{b}\right) = \frac{(b-a)I^2}{4\pi \sigma_c ab}.$$

This is equal to $R_r I^2$ with the resistance given by Eq. (5.20).

5.18. The current density at distance r from the center is $i(r) = I/4\pi r^2$. Hence, the dissipated power density is given by $p(r) = \rho_{r1} i^2(r)$ for $a < r < b$ and $p(r) = \rho_{r2} i^2(r)$ for $b < r < c$. Thus, the dissipated power is

$$P = \rho_{r1} \int_a^b \left(\frac{I}{4\pi r^2}\right)^2 4\pi r^2 dr + \rho_{r1} \int_a^b \left(\frac{I}{4\pi r^2}\right)^2 4\pi r^2 dr$$

$$= \frac{\rho_{r1}(b - a)I^2}{4\pi ab} + \frac{\rho_{r2}(c - b)I^2}{4\pi bc},$$

which is equal to $R_r I^2$ with the resistance R_r obtained in Example 5.3.

5.19. The electric field is of the form of $E(r) = k/r^2$ with a constant k. From the condition

5.6 Exercises

$$V = \int_a^b \frac{k}{r^2} dr = \frac{(b-a)k}{ab},$$

we have $k = abV/(b-a)$. Then, the power density in material 1 with resistivity ρ_{r1} is

$$p_1(r) = \frac{E^2(r)}{\rho_{r1}} = \frac{(abV)^2}{(b-a)^2 \rho_{r1} r^4}.$$

Hence, the power in material 1 is given by

$$P_1 = \frac{(abV)^2}{(b-a)^2 \rho_{r1}} \int_a^b \frac{1}{r^4} 2\pi r^2 dr = \frac{2\pi abV^2}{(b-a)\rho_{r1}}.$$

The power in material 2 is similarly obtained, and the power is determined to be

$$P = P_1 + P_2 = \frac{2\pi ab(\rho_{r1} + \rho_{r2})V^2}{(b-a)\rho_{r1}\rho_{r2}},$$

which is equal to V^2/R_r with the resistance R_r obtained in Example 5.4.

5.20. We denote the current by I. Then, the current density is $i = I/2\pi lR$ at a distance R from the central axis. So, the electric power is

$$P = l\int_a^b \rho_{r1} i^2 2\pi R dR + l \int_b^c \rho_{r1} i^2 2\pi R dR$$

$$= \frac{I^2}{2\pi l}\left(\int_a^b \frac{\rho_{r1}}{R} dR + \int_b^c \frac{\rho_{r2}}{R} dR\right) = \frac{I^2}{2\pi l}\left(\rho_{r1} \log \frac{b}{a} + \rho_{r2} \log \frac{c}{b}\right).$$

This is given by $R_r I^2$ with the resistance R_r determined in Example 5.5.

5.21. We denote the electric field by $E = k/R$ as a function of radius R, where k is a constant. Then, the voltage is $V = k \log(b/a)$. So, we have $k = V/\log(b/a)$. Then, the electric power is

$$P = l \int_a^b \frac{E^2}{\rho_{r1}} \pi R dR + l \int_a^b \frac{E^2}{\rho_{r2}} \pi R dR = \frac{\pi l(\rho_{r1} + \rho_{r2})V^2}{\rho_{r1}\rho_{r2} \log(b/a)}.$$

This is given by V^2/R_r with the resistance R_r determined in Example 5.6.

5.22. The radial current density at distance R from the axis is

$$i(R) = \frac{I}{2\pi l R}.$$

Then, the power density is given by $\rho_r i^2(R)$. The total power is

$$P = l \int_a^b \rho_r \left(\frac{I}{2\pi l R}\right)^2 2\pi R dR = \frac{\rho_r I^2}{2\pi l} \log \frac{b}{a} = R_r I^2,$$

where $R_r = \rho_r \log(b/a)/2\pi l$ is the resistance.

5.23. We denote the current by I. Then, the current density $i = I/ab$ is uniform and the electric field is $E(y) = \rho_r(1 + ky)i = \rho_r(1 + ky)I/ab$. Thus, the voltage is

$$V = \int_0^c \frac{\rho_r I(1+ky) dy}{ab} = \frac{c\rho_r I}{ab}\left(1 + \frac{kc}{2}\right).$$

The dissipated power is

$$P = ab \int_0^c \rho_r(1+ky)\left(\frac{I}{ab}\right)^2 dy = \frac{c\rho_r I^2}{ab}\left(1 + \frac{kc}{2}\right)$$

$$= \frac{abV^2}{c\rho_r}\left(1 + \frac{kc}{2}\right)^{-1} = \frac{V^2}{R_r},$$

where R_r is the resistance obtained in Exercise 5.10.

5.24. The electric field $E = V/c$ is uniform and the dissipated power is given by

$$P = ac \int_0^b \frac{E^2 dx}{\rho_r(1+kx)} = \frac{aV^2}{c\rho_r k} \log(1+kb) = \frac{V^2}{R_r},$$

where R_r is the resistance obtained in Exercise 5.11.

5.25. We denote the current in a unit length by λ. Then, the current density is $i = \lambda/2\pi R$ and the electric field is $E(R) = \rho_r(1+kR)i = \rho_r(1+kR)\lambda/2\pi R$. Thus, the voltage is

$$V = \int_a^b \frac{\rho_r \lambda}{2\pi}\left(\frac{1}{R} + k\right) dR = \frac{\rho_r \lambda}{2\pi}\left[\log \frac{b}{a} + k(b-a)\right].$$

The dissipated power in a unit length is

$$P' = \int_a^b \rho_r(1+kR)\left(\frac{\lambda}{2\pi R}\right)^2 2\pi R\, dR = \int_a^b \frac{\rho_r \lambda^2}{2\pi}\left(\frac{1}{R}+k\right) dR$$

$$= \frac{\rho_r \lambda^2}{2\pi}\left[\log\frac{b}{a}+k(b-a)\right] = \frac{2\pi V^2}{\rho_r}\left[\log\frac{b}{a}+k(b-a)\right]^{-1} = \frac{V^2}{R'_r},$$

where R'_r is the resistance in a unit length obtained in Exercise 5.12.

5.26. We denote the current by I. Then, the current density is $i = I/4\pi r^2$ and the electric field is $E(r) = \rho_r(1+kr)i = \rho_r(1+kr)I/4\pi r^2$. Thus, the voltage is

$$V = \int_a^b \frac{\rho_r I}{4\pi}\left(\frac{1}{r^2}+\frac{k}{r}\right)dr = \frac{\rho_r I}{4\pi}\left(\frac{b-a}{ab}+k\log\frac{b}{a}\right).$$

The dissipated power is

$$P = \int_a^b \rho_r(1+kr)\left(\frac{I}{4\pi r^2}\right)^2 4\pi r^2 dr = \int_a^b \frac{\rho_r I^2}{4\pi}\left(\frac{1}{r^2}+\frac{k}{r}\right)dr$$

$$= \frac{\rho_r I^2}{4\pi}\left(\frac{b-a}{ab}+k\log\frac{b}{a}\right) = \frac{4\pi V^2}{\rho_r}\left(\frac{b-a}{ab}+k\log\frac{b}{a}\right)^{-1} = \frac{V^2}{R_r},$$

where R_r is the resistance obtained in Exercise 5.13.

5.27. The equipotential surface is a plane extending from the central axis, as shown in Fig. B5.1. We consider a narrow region between the two equipotential surfaces of angles θ and $\theta + d\theta$. The value of the dissipated power at position R between the two surfaces is $dP = wi(R)E(R)Rd\theta dR$. Using the relationship $E(R)Rd\theta = 2V d\theta/\pi$, this dissipated power is rewritten as $dP = (2wV/\pi)i(R)dRd\theta$. Integrating this along the direction of R, we have the total dissipated power:

Fig. B5.1 Two equipotential surfaces of angles θ and $\theta + d\theta$

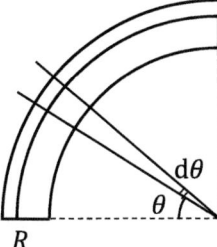

Fig. B5.2 Electric dipole line on the z-axis and observation point A

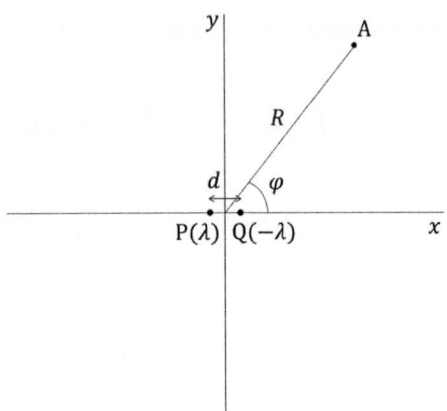

$$P = \frac{2VI}{\pi} \int_0^{\pi/2} d\theta = VI,$$

where we used $w \int i(R)dR = I$.

5.28. The given situation is equivalent to the two-dimensional electric dipole in which linear electric charges λ and $-\lambda$ are given on points P and Q, respectively. We define the x- and y-axes with the origin at the middle point between P and Q, as shown in Fig. B5.2. In this case, the electric potential at the observation point A of distance R from the origin and angle φ measured from the x-axis is given by

$$\phi(R, \varphi) = -\frac{\hat{p}}{2\pi \epsilon_0 R} \cos \varphi,$$

from Eq. (1.25) (note that the direction of the moment is opposite), where $\hat{p} = \lambda d$ is the dipole moment. So, the electric potential is zero at infinity independently of the angle φ, and the assumption is valid.

When DC current source of 1 A is connected between P and infinity, current 1/4 A flows directly from P to Q from symmetry. When DC current source of 1 A is connected between infinity and Q, current 1/4 A flows directly from P to Q from symmetry. From superposition of these cases, current 1/2 A flows from P to Q. Thus, the voltage between P and Q is $R/2$, and the resistance is determined to be $R/2$.

5.29. The capacitance of the concentric capacitor in Example 4.1 is

$$\frac{1}{C} = \frac{b-a}{4\pi \epsilon_1 ab} + \frac{c-b}{4\pi \epsilon_2 bc}$$

and the resistance of the resistor with the same structure in Example 5.3 is

$$R_r = \frac{\rho_{r1}(b-a)}{4\pi ab} + \frac{\rho_{r2}(c-b)}{4\pi bc}.$$

5.6 Exercises

Next, the capacitance of the capacitor in Example 4.2 is

$$\frac{1}{C} = \frac{b-a}{2\pi(\epsilon_1 + \epsilon_2)ab}$$

and the electric resistance of the resistor in Example 5.4 is

$$R_{\mathrm{r}} = \frac{b-a}{2\pi\left(\rho_{\mathrm{r}1}^{-1} + \rho_{\mathrm{r}2}^{-1}\right)ab}.$$

Such a relation is also found between the capacitance in Example 4.8 and the resistance in Example 5.5. These examples show that, if we respectively replace ϵ_1 and ϵ_2 with $\rho_{\mathrm{r}1}^{-1}$ and $\rho_{\mathrm{r}2}^{-1}$, $1/C$ and R_{r} have the same form.

The similarity can be expressed as follows:

$$CR = \frac{f(\epsilon_1, \epsilon_2, \ldots)}{f\left(\rho_{\mathrm{r}1}^{-1}, \rho_{\mathrm{r}2}^{-1}, \ldots\right)},$$

where $f(\epsilon_1, \epsilon_2, \ldots)$ is a function of ϵ_i ($i = 1, 2, \ldots$) and $f\left(\rho_{\mathrm{r}1}^{-1}, \rho_{\mathrm{r}2}^{-1}, \ldots\right)$ is a function in which ϵ_i is replaced with the corresponding $\rho_{\mathrm{r}i}^{-1}$. This is an extended form of Eq. (5.21).

5.30. From the given electric potential, the equipotential surface is given by

$$\frac{R^2 + (d-h)^2 - 2R(d-h)\cos\varphi}{R^2 + h^2 - 2Rh\cos\varphi} = \frac{1}{K},$$

where K is a constant. If we use Cartesian coordinates, we have $R\cos\varphi = x + d/2$ and $R\sin\varphi = y$. So, the above equation is rewritten as

$$\left[x - \frac{K+1}{2(K-1)}\sqrt{d^2 - 4a^2}\right]^2 + y^2$$

$$= \frac{(K+1)^2}{(K-1)^2}\left(\frac{d^2}{4} - a^2\right) - \frac{d^2}{2} + a^2 - \frac{K+1}{2(K-1)}d\sqrt{d^2 - 4a^2}$$

for $K \neq 1$. Thus, equipotential surfaces are cylindrical surfaces parallel to the cylindrical conductors. The surface of the left superconductor is an equipotential surface for $K = \left(d - \sqrt{d^2 - 4a^2}\right)^2/4a^2$. The plane $x = 0$ is also an equipotential surface for $K = 1$.

5.31. This problem is essentially the same as the determination of electric field around a hollow dielectric sphere placed in the uniform electric field treated in Exercise 4.14. The effect of each boundary at the outer and inner surfaces can be correctly expressed by a dipole moment placed at the origin for the current density outside the corresponding surface. On the other hand, a current with uniform density is produced inside the surface. We denote the uniform current density in the hollow space as i_1.

The dipole moments corresponding to the inner and outer surfaces are denoted by p and p', respectively. The outer surface also causes a uniform current density denoted as i_2 in the material with electric conductivity σ_c. So, the current distribution is:

$$i = i_1; \qquad 0 \le r < a,$$
$$= i_2 + \nabla \frac{\sigma_{c0} p \cos\theta}{4\pi r^2}; \qquad a < r < b,$$
$$= i_0 + \nabla \frac{\sigma_{c0}(p + p')\cos\theta}{4\pi r^2}; \qquad r > b,$$

where θ is the zenithal angle. So, the boundary conditions on continuity of the current and the parallel component of the electric field at $r = a$ are given by

$$-i_1 \sin\theta = -i_2 \sin\theta + \frac{\sigma_{c0} p \sin\theta}{4\pi a^3}, \qquad \frac{1}{\sigma_{c0}} i_1 \cos\theta = \frac{1}{\sigma_c}\left(i_2 \cos\theta + \frac{\sigma_{c0} p \cos\theta}{2\pi a^3}\right),$$

respectively, and the boundary conditions at $r = b$ are

$$-i_2 \sin\theta + \frac{\sigma_{c0} p \sin\theta}{4\pi b^3} = -i_0 \sin\theta + \frac{\sigma_{c0}(p+p') \sin\theta}{4\pi b^3},$$

$$\frac{1}{\sigma_c}\left(i_2 \cos\theta + \frac{\sigma_{c0} p \cos\theta}{2\pi b^3}\right) = \frac{1}{\sigma_{c0}}\left[i_0 \cos\theta + \frac{\sigma_{c0}(p+p')\cos\theta}{2\pi b^3}\right].$$

From these conditions, p, p', i_1, and i_2 are determined to be

$$p = -\frac{4\pi a^3}{3\alpha \sigma_{c0}} \cdot \frac{\sigma_{c0} - \sigma_c}{\sigma_{c0}} i_0, \qquad p' = \frac{4\pi b^3}{\sigma_{c0}}\left(1 - \frac{2\sigma_{c0} + \sigma_c}{3\alpha \sigma_{c0}}\right)i_0,$$

$$i_1 = \frac{1}{\alpha} i_0, \qquad i_2 = \frac{2\sigma_{c0} + \sigma_c}{3\alpha \sigma_{c0}} i_0,$$

where α is a constant given by

$$\alpha = \frac{(\sigma_c + 2\sigma_{c0})(2\sigma_c + \sigma_{c0})}{9\sigma_{c0}\sigma_0} - \frac{2a^3}{9b^3} \cdot \frac{(\sigma_{c0} - \sigma_0)^2}{\sigma_{c0}\sigma_0}.$$

5.32. This problem is essentially the same as the determination of electric field around a hollow dielectric cylinder placed in the uniform electric field treated in Exercise 4.15. The effect of each boundary at the outer and inner surfaces can be correctly expressed by a two-dimensional dipole moment placed at the origin for the current density outside the corresponding surface. On the other hand, a current with uniform density is produced inside the boundary. We denote the uniform current density in the hollow space as i_1. The linear dipole moments corresponding to the inner and outer surfaces are denoted by $\hat{p}$ and $\hat{p}'$, respectively. The outer surface also causes a

5.6 Exercises

uniform current density denoted as i_2 in the material with electric conductivity σ_c. So, the current distribution is:

$$i = i_1; \qquad\qquad\qquad 0 \leq R < a,$$
$$= i_2 + \nabla \frac{\sigma_{c0}\hat{p}\cos\varphi}{2\pi R}; \qquad a < R < b,$$
$$= i_0 + \nabla \frac{\sigma_{c0}(\hat{p}+\hat{p}')\cos\varphi}{2\pi R}; \qquad R > b,$$

where φ is the azimuthal angle. So, the boundary conditions on continuity of the current and the parallel component of the electric field at $R = a$ are given by

$$-i_1 \sin\varphi = -i_2 \sin\varphi + \frac{\sigma_{c0}\hat{p}\sin\varphi}{2\pi a^2}, \quad \frac{1}{\sigma_{c0}} i_1 \cos\varphi = \frac{1}{\sigma_c}\left(i_2 \cos\varphi + \frac{\sigma_{c0}\hat{p}\cos\varphi}{2\pi a^2}\right),$$

respectively, and the boundary conditions at $R = b$ are

$$-i_2 \sin\varphi + \frac{\sigma_{c0}\hat{p}\sin\varphi}{2\pi b^2} = -i_0 \sin\varphi + \frac{\sigma_{c0}(\hat{p}+\hat{p}')\sin\varphi}{2\pi b^2},$$

$$\frac{1}{\sigma_c}\left(i_2 \cos\varphi + \frac{\sigma_{c0}\hat{p}\cos\varphi}{2\pi b^2}\right) = \frac{1}{\sigma_{c0}}\left[i_0 \cos\varphi + \frac{\sigma_{c0}(\hat{p}+\hat{p}')\cos\varphi}{2\pi b^2}\right].$$

From these conditions, $\hat{p}, \hat{p}', i_1$, and i_2 are determined to be

$$\hat{p} = -\frac{\pi a^2(\sigma_{c0}-\sigma_c)}{\beta\sigma_{c0}^2}i_0, \quad \hat{p}' = \frac{2\pi b^2}{\sigma_{c0}}\left(1 - \frac{\sigma_{c0}+\sigma_c}{2\beta\sigma_{c0}}\right)i_0,$$

$$i_1 = \frac{1}{\beta}i_0, \quad i_2 = \frac{\sigma_{c0}+\sigma_c}{2\beta\sigma_{c0}}i_0,$$

where β is a constant given by

$$\beta = \frac{(\sigma_c+\sigma_{c0})^2}{4\sigma_{c0}\sigma_0} - \frac{a^2}{4b^2}\cdot\frac{(\sigma_{c0}-\sigma_0)^2}{\sigma_{c0}\sigma_0}.$$

Chapter 6
Current and Magnetic Flux Density

6.1 The Biot-Savart Law

A force works between currents. This force can be regarded as to be caused by a magnetic distortion of space, and the strength of this field is represented by the magnetic flux density. Suppose that current I flows along line C, as shown in Fig. 6.1. According to the Biot-Savart law, the magnetic flux density at point P produced by an elementary current $I \, d\mathbf{s}$ flowing in a small segment $d\mathbf{s}$ is given by

$$d\mathbf{B} = \frac{\mu_0 I \, d\mathbf{s} \times \mathbf{r}}{4\pi |\mathbf{r}|^3}, \tag{6.1}$$

where μ_0 is the magnetic permeability of vacuum given by

$$\mu_0 = 4\pi \times 10^{-7} \, \text{N/A}^2 \tag{6.2}$$

and $\mathbf{r}$ is the position vector of point P from the small segment. The unit of the magnetic flux density is [T] (tesla). If the angle between $d\mathbf{s}$ and $\mathbf{r}$ is θ (see Fig. 6.1), the magnitude of the magnetic flux density is

$$dB = \frac{\mu_0 I \, ds}{4\pi r^3} \sin\theta, \tag{6.3}$$

and the vector points along the motion of a screw when the screw driver is rotated from $d\mathbf{s}$ to $\mathbf{r}$.

Thus, the magnetic flux density produced at $\mathbf{r}$ by current I flowing through line C is given by

$$\mathbf{B}(\mathbf{r}) = \frac{\mu_0 I}{4\pi} \int_C \frac{d\mathbf{r}' \times (\mathbf{r} - \mathbf{r}')}{|\mathbf{r} - \mathbf{r}'|^3}. \tag{6.4}$$

© The Editor(s) (if applicable) and The Author(s), under exclusive license to Springer Nature Switzerland AG 2025
T. Matsushita, *Exercises in Electricity and Magnetism*,
https://doi.org/10.1007/978-3-031-67940-7_6

Fig. 6.1 Elementary current along C and observation point P

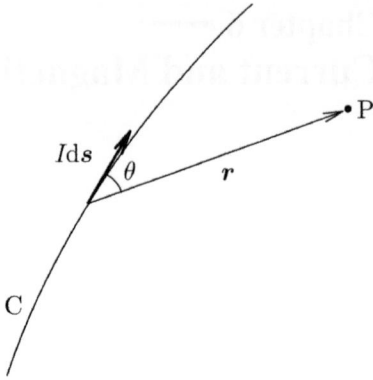

When a current flows with density *i* in space V, the magnetic flux density is given by

$$\boldsymbol{B}(\boldsymbol{r}) = \frac{\mu_0}{4\pi} \int_V \frac{\boldsymbol{i}(\boldsymbol{r}') \times (\boldsymbol{r} - \boldsymbol{r}')}{|\boldsymbol{r} - \boldsymbol{r}'|^3} dV'. \qquad (6.5)$$

Example 6.1 Current *I* flows in a circle of radius *a*, as shown in Fig. 6.2a. Determine the magnetic flux density at point P located at distance *z* in the normal direction from the center O of the circle.

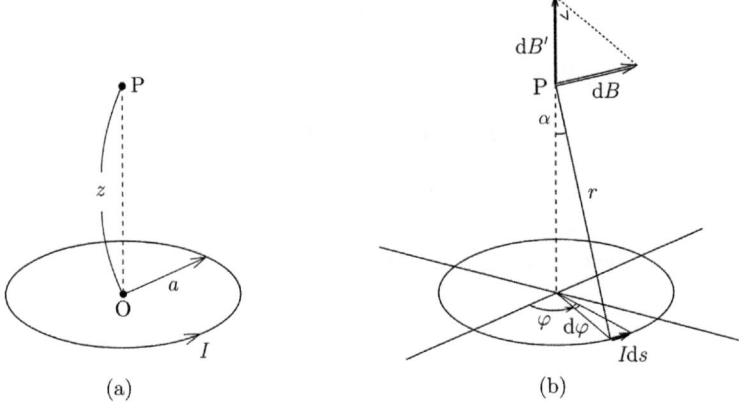

Fig. 6.2 a Point P on axis of circular current and **b** magnetic flux density at P produced by elementary current

Solution 6.1 We define the angle φ, as shown in Fig. 6.2b. The elementary current in the part from φ to $\varphi + d\varphi$ has a magnitude $Ids = Iad\varphi$ and is directed normally

6.1 The Biot-Savart Law

to the position vector from this segment to point P [$\theta = \pi/2$ in Eq. (6.3)], as shown in the figure. The magnetic flux density at P produced by this elementary current is

$$dB = \frac{\mu_0 I a d\varphi}{4\pi r^2}$$

with $r = (z^2 + a^2)^{1/2}$. From symmetry, only the component along the z-axis remains:

$$dB' = \frac{\mu_0 I a d\varphi}{4\pi r^2} \sin \alpha = \frac{\mu_0 I a^2 d\varphi}{4\pi r^3},$$

where angle α is defined in Fig. 6.2b. Integrating with respect to the angle φ, we have

$$B = \int_0^{2\pi} \frac{\mu_0 I a^2 d\varphi}{4\pi r^3} = \frac{\mu_0 I a^2}{2(z^2 + a^2)^{3/2}}.$$

◆

Example 6.2 Current I flows along a long straight line (see Fig. 6.3). Determine the magnetic flux density at point P at distance a from the line.

Fig. 6.3 Long straight line with current and observation point P

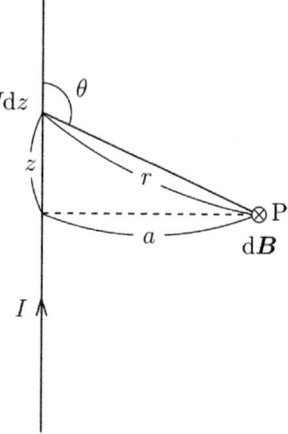

Solution 6.2 We define coordinates as shown in Fig. 6.3. The magnetic flux density produced at P by an elementary current $I dz$ in a small region from z to $z + dz$ is

$$dB = \frac{\mu_0 I dz}{4\pi r^2} \sin \theta,$$

where angle θ is defined as in the figure and $r = (z^2 + a^2)^{1/2} = a/\sin\theta$. The relationship $z = -a\cot\theta$ gives $dz = (a/\sin^2\theta)d\theta$. Thus, the elementary magnetic flux density is transformed to be

$$dB = \frac{\mu_0 I}{4\pi a}\sin\theta\, d\theta.$$

Since this vector is directed normally to this sheet, a simple superposition holds for summing the contribution from each small region. We obtain the magnetic flux density as

$$B = \int_0^\pi \frac{\mu_0 I}{4\pi a}\sin\theta\, d\theta = \frac{\mu_0 I}{2\pi a}. \tag{6.6}$$

◆

6.2 Force on Current

The force on an elementary part ds of a current I in a magnetic flux density $\boldsymbol{B}$ is given by

$$d\boldsymbol{F} = I d\boldsymbol{s} \times \boldsymbol{B} \tag{6.7}$$

(see Fig. 6.4). This is called the Lorentz force in a narrow sense. From the mathematical requirement that the force vector results from the product of two vectors, the vector product appears again. The force that line C with current I experiences in the magnetic flux density $\boldsymbol{B}$ is

$$\boldsymbol{F} = I \int_C d\boldsymbol{s} \times \boldsymbol{B}. \tag{6.8}$$

Fig. 6.4 Current I in magnetic flux density $\boldsymbol{B}$

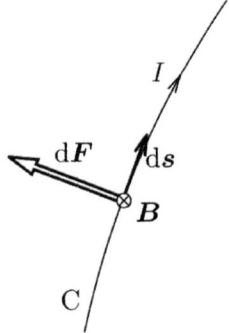

6.3 Magnetic Flux Lines

Since the current is composed of flowing electric charges, we can consider the force on the current to be a force on the electric charges. That is, the force on one electric charge q with velocity v in a magnetic flux density B is given by

$$F = qv \times B. \tag{6.9}$$

For the special case where the electric field E and the magnetic flux density B coexist, the force on the electric charge is

$$F = q(E + v \times B). \tag{6.10}$$

This is called the Lorentz force in a broad sense.

6.3 Magnetic Flux Lines

The magnetic flux density can be visualized by magnetic flux lines. The magnetic flux line is defined as follows: The direction of a tangential line at any point on a magnetic flux line is the same as the direction of B, and the line density of magnetic flux line is defined as equal to the magnitude of B. We define the magnetic flux that passes through arbitrary surface S as

$$\Phi = \int_S B \cdot dS. \tag{6.11}$$

The unit of magnetic flux is [T m^2], which is newly defined as [Wb] (weber).

The magnetic flux lines are closed lines. This is different from electric field lines, which start from positive electric charges and terminate at negative electric charges. Hence, the equation

$$\int_S B \cdot dS = 0 \tag{6.12}$$

holds for an arbitrary closed surface, S. Using Gauss's theorem on the left side, Eq. (6.12) gives

$$\nabla \cdot B = 0. \tag{6.13}$$

That is, the magnetic flux lines are closed lines. Equations (6.12) and (6.13) are called Gauss's law for magnetic flux and Gauss's divergence law for magnetic flux, respectively.

6.4 Ampere's Law

When we integrate the magnetic flux density along a closed line C around a current I, we have

$$\oint_C \mathbf{B} \cdot d\mathbf{s} = \mu_0 I. \tag{6.14}$$

When current flows with the density $\mathbf{i}$, the following equation holds:

$$\oint_C \mathbf{B} \cdot d\mathbf{s} = \mu_0 \int_S \mathbf{i} \cdot d\mathbf{S}, \tag{6.15}$$

where S is the surface surrounded by C. Equations (6.14) and (6.15) are called Ampere's law.

Applying Stokes' theorem to the left side of Eq. (6.15) yields

$$\nabla \times \mathbf{B} = \mu_0 \mathbf{i}. \tag{6.16}$$

This is called the differential form of Ampere's law. Thus, the current produces rotation of the magnetic flux density.

Example 6.3 Current I flows uniformly in a long cylinder of radius a. Determine the magnetic flux density inside and outside the cylinder.

Solution 6.3 We apply Ampere's law to a circle, C, of radius R from the central axis of the cylinder, as shown in Fig. 6.5. From symmetry, the magnetic flux density $\mathbf{B}$ is parallel to a line element, $d\mathbf{s}$, and its magnitude B is constant on C. Hence, the left side of Eq. (6.15) gives

$$\oint_C \mathbf{B} \cdot d\mathbf{s} = 2\pi R B.$$

Fig. 6.5 Circle C with the same central axis as cylinder for $R < a$

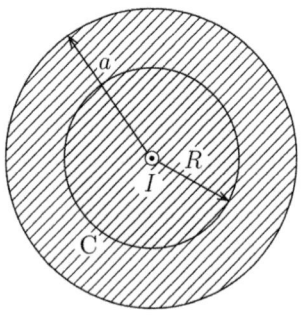

6.4 Ampère's Law

For $R > a$, the total current that flows inside C is I and the right side of Eq. (6.15) is equal to $\mu_0 I$. Thus, we have

$$B = \frac{\mu_0 I}{2\pi R}.$$

For $R < a$, the current inside C is $(R/a)^2 I$, and we have

$$B = \frac{\mu_0 I R}{2\pi a^2}.$$

◆

Example 6.4 Current flows uniformly with surface density τ on a wide plane. Determine the magnetic flux density at a position at distance h from the plane.

Solution 6.4 Suppose a rectangle, KLMN, normal to the direction of current with two sides (KL and MN) of length w parallel to the plane, as shown in Fig. 6.6. The other two sides (LM and NK) of length $2h$ are normal to the plane. We apply Ampère's law to this rectangle. From symmetry, the magnetic flux density has the same value on sides KL and MN at the same distance from the plane, and its vectors are parallel to these sides but opposite to each other. Hence, the contribution from these sides to the circular integral of the magnetic flux density gives $2wB$. On the other hand, the contribution from other two sides is zero. As a result, we have

$$\oint_C \mathbf{B} \cdot d\mathbf{s} = 2wB.$$

The total current inside the rectangle is $w\tau$. Thus, we obtain the magnetic flux density as

$$B = \frac{\mu_0 \tau}{2}. \tag{6.17}$$

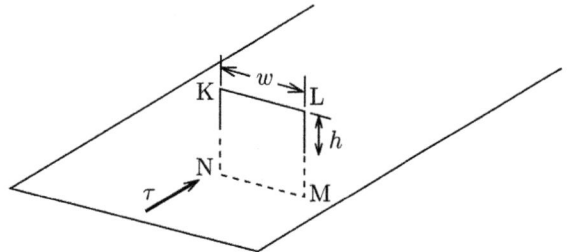

Fig. 6.6 Rectangle normal to the direction of current flowing uniformly on a plane

◆

6.5 Vector Potential

Magnetic flux density is a solenoidal field with no divergence, as shown by Eq. (6.13). Mathematically, the curl of a vector has no divergence. Hence, the magnetic flux density can be mathematically expressed as a curl of some vector:

$$\boldsymbol{B} = \nabla \times \boldsymbol{A}. \tag{6.18}$$

The quantity $\boldsymbol{A}$ is called the vector potential. We could add the gradient of any scalar function to the vector potential, and the vector potential would still correspond to the same magnetic flux density. This arbitrary gradient of some scalar function means that the vector potential cannot be uniquely determined without specifying some condition. In the case of a static magnetic phenomenon that does not change with time, the condition

$$\nabla \cdot \boldsymbol{A} = 0 \tag{6.19}$$

is usually used. This is called the Coulomb gauge.

The solution of the vector potential is given by

$$\boldsymbol{A}(\boldsymbol{r}) = \frac{\mu_0}{4\pi} \int_V \frac{\boldsymbol{i}(\boldsymbol{r}')}{|\boldsymbol{r} - \boldsymbol{r}'|} dV'. \tag{6.20}$$

Integrating the vector potential along C gives

$$\oint_C \boldsymbol{A} \cdot d\boldsymbol{s} = \int_S \nabla \times \boldsymbol{A} \cdot d\boldsymbol{S} = \int_S \boldsymbol{B} \cdot d\boldsymbol{S} = \Phi, \tag{6.21}$$

where S is a surface surrounded by C and Φ is a magnetic flux that penetrates C.

Substituting Eq. (6.18) into Eq. (6.16) gives

$$\nabla \times (\nabla \times \boldsymbol{A}) = \mu_0 \boldsymbol{i}. \tag{6.22}$$

With Eqs. (6.19) and (A.9) in the Appendix, Eq. (6.22) becomes

$$\Delta \boldsymbol{A} = -\mu_0 \boldsymbol{i}. \tag{6.23}$$

That is, each component of the vector potential satisfies Poisson's equation. In the region where current does not flow ($\boldsymbol{i} = 0$), this reduces to Laplace's equation,

$$\Delta \boldsymbol{A} = 0. \tag{6.24}$$

6.5 Vector Potential

Example 6.5 Determine the vector potential for the case discussed in Example 6.3.

Solution 6.5 We use cylindrical coordinates. Since the current flows only along the z-axis, the vector potential has only the z-component A_z, as indicated by Eq. (6.20). In addition, from symmetry, it does not depend on z or the azimuthal angle φ. The magnetic flux density has only the azimuthal component B_φ. Thus, we have

$$B_\varphi = -\frac{\partial A_z}{\partial R}.$$

The vector potential is given by

$$A_z = -\int_{R_0}^{R} B_\varphi \, dR,$$

where R_0 ($> a$) is the distance from the central axis to the reference point at which $A_z = 0$. The reason why infinity is not chosen as the reference point is that the vector potential diverges because of the requirement that the current flows over an infinitely long distance. We determine the vector potential as

$$A_z = \frac{\mu_0 I}{2\pi} \log \frac{R_0}{R}$$

from $B_\varphi = \mu_0 I/(2\pi R)$ for $R > a$ and as

$$A_z = \frac{\mu_0 I}{2\pi a^2}(a^2 - R^2) + \frac{\mu_0 I}{2\pi} \log \frac{R_0}{a}$$

from $B_\varphi = \mu_0 I R/(2\pi a^2)$ for $0 \leq R < a$. In the above, the vector potential takes on a constant value on the cylindrical surface with the radius R. Thus, we can define an equivector potential surface, which is similar to the equipotential surface. The vector potential is parallel to the equivector potential surface. ◆

Example 6.6 Current I is applied to an infinitely long solenoid coil of radius a with n turns in a unit length. Determine the vector potential.

Solution 6.6 Firstly, we determine the magnetic flux density. Suppose rectangles C_1 and C_2, as shown in Fig. 6.7. Applying Ampere's law to these rectangles, the

Fig. 6.7 Longitudinal cross-section of solenoid coil and rectangles C_1–C_3

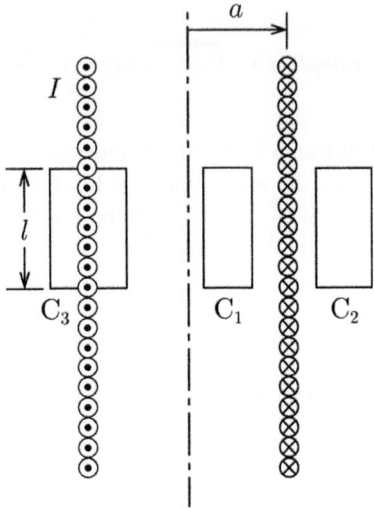

circular integral of the magnetic flux density is zero in each case. It shows that the magnetic flux density is constant inside and outside the coil. Since the magnetic flux density outside the coil must be uniform up to infinity and the total magnetic flux must be finite, we can show that the magnetic flux density must be zero outside the coil. Then, we apply Ampere's law to rectangle C_3 to determine the magnetic flux density B inside the coil. The left side of Eq. (6.15) is Bl with l denoting the axial length of C_3. Since the total current inside C_3 is nIl, we have

$$B = \mu_0 n I.$$

Then, we can determine the vector potential with Eq. (6.18), but we use Eq. (6.21) here. We apply this equation to a circle, C, of radius R from the central axis of the coil in Fig. 6.8. Since the current flows only in the azimuthal direction, the vector potential has only the azimuthal component A_φ. Hence, this is parallel to C and its magnitude is constant. We have

Fig. 6.8 Circle C of radius R from the central axis of the coil (for $R < a$)

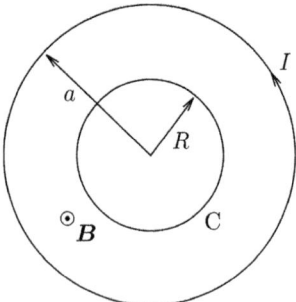

$$\oint_C \mathbf{A} \cdot d\mathbf{s} = 2\pi R A_\varphi(R).$$

On the other hand, the magnetic flux penetrating C is

$$\int_S \mathbf{B} \cdot d\mathbf{S} = B\pi R^2 = \pi \mu_0 n I R^2; \qquad 0 \leq R < a,$$
$$= B\pi a^2 = \pi \mu_0 n I a^2; \qquad R > a.$$

Thus, we determine the vector potential as

$$A_\varphi(R) = \frac{\mu_0 n I R}{2}; \qquad 0 \leq R < a,$$
$$= \frac{\mu_0 n I a^2}{2R}; \qquad R > a.$$

◆

6.6 Small Closed Current

Suppose that current I flows around a small square of side length d. The vector potential produced by this current is determined at a point P, sufficiently far from this square. We denote the direction vector from the center of the square to point P as $\mathbf{r}(|\mathbf{r}| \gg d)$. We define the origin of the coordinates at the center of the square placed on the x-y plane. Its sides are parallel to the x- or y-axis, as shown in Fig. 6.9. We also assume that P is on the y-z plane, and θ is the angle between $\mathbf{r}$ and the z-axis. Hence, the position of P is $\mathbf{r} = (0, r\sin\theta, r\cos\theta)$ in Cartesian coordinates.

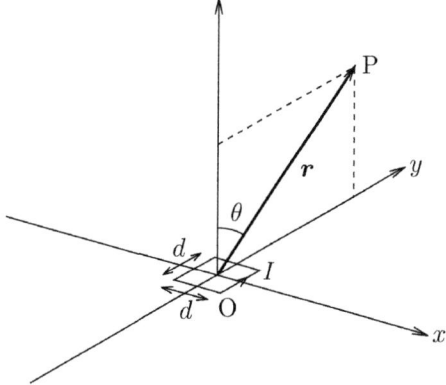

Fig. 6.9 Closed current flowing around a small square with its center at the origin and observation point P

The vector potential produced by this small closed current is given by

$$A(r) = \frac{\mu_0 (m \times r)}{4\pi r^3}, \qquad (6.25)$$

where m is the magnetic moment given by

$$m = IS \qquad (6.26)$$

with $S = d^2 i_z$. Practically, Eq. (6.25) reduces to

$$A_\varphi = \frac{\mu_0 m}{4\pi r^2} \sin\theta \qquad (6.27)$$

in spherical coordinates. In the above, $m = |m|$ is the magnitude of the magnetic moment. A_φ is constant on a surface on which the following condition holds:

$$\frac{\sin\theta}{r^2} = \text{const.} \qquad (6.28)$$

The magnetic flux density produced by the closed current is determined to be

$$\begin{aligned}
B_r &= \frac{1}{r\sin\theta} \cdot \frac{\partial}{\partial \theta}(\sin\theta A_\varphi) = \frac{\mu_0 m \cos\theta}{2\pi r^3}, \\
B_\theta &= -\frac{1}{r} \cdot \frac{\partial}{\partial r}(r A_\varphi) = \frac{\mu_0 m \sin\theta}{4\pi r^3}, \\
B_\varphi &= 0.
\end{aligned} \qquad (6.29)$$

Suppose that currents I and $-I$ flow along lines at $y = d/2$ and $y = -d/2$, respectively, on the y-z plane, as shown in Fig. 6.10. This pair of parallel opposite currents is called a magnetic dipole line. We define cylindrical coordinates, as shown in the figure, and measure the azimuthal angle φ from the x-axis. We denote the

Fig. 6.10 Magnetic dipole line composed of a pair of parallel opposite currents

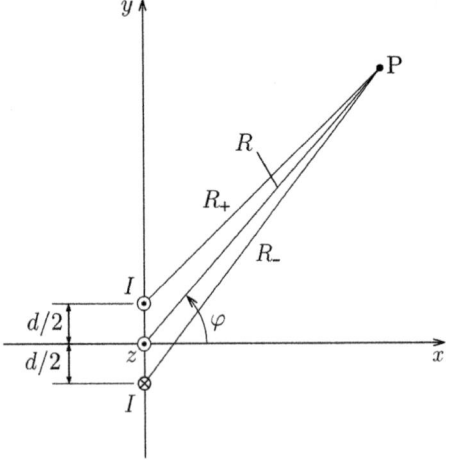

6.6 Small Closed Current

distances between observation point P and the lines at $y = d/2$ and $y = -d/2$ as R_+ and R_-, respectively. Then, the vector potential produced by these currents is

$$A_z(R, \varphi) = \frac{\mu_0 I}{2\pi} \log \frac{R_+}{R_-} \simeq \frac{\mu_0 \hat{m}}{2\pi R} \sin \varphi, \qquad (6.30)$$

where

$$\hat{m} = Id \qquad (6.31)$$

is the moment of a magnetic dipole line in a unit length. Hence, the vector potential is constant on a surface on which the following condition holds:

$$\frac{\sin \varphi}{R} = \text{const.} \qquad (6.32)$$

The general description of the vector potential is

$$\mathbf{A} = \frac{\mu_0 (\hat{m} \times \mathbf{R})}{2\pi R^2}. \qquad (6.33)$$

The magnetic flux density derived from Eq. (6.30) is

$$B_R = \frac{\mu_0 \hat{m}}{2\pi R^2} \cos \varphi,$$

$$B_\varphi = \frac{\mu_0 \hat{m}}{2\pi R^2} \sin \varphi,$$

$$B_z = 0. \qquad (6.34)$$

Example 6.7 Prove that the vector potential produced by a magnetic moment given by Eq. (6.27) satisfies Laplace's equation.

Solution 6.7 Using Eq. (A.18) in the Appendix, we have

$$\nabla \times \mathbf{A} = \mathbf{i}_r \frac{\mu_0 m}{2\pi} \cdot \frac{\cos \theta}{r^3} + \mathbf{i}_\theta \frac{\mu_0 m}{4\pi} \cdot \frac{\sin \theta}{r^3}$$

and

$$\nabla \times \nabla \times \mathbf{A} = \mathbf{i}_\varphi \frac{1}{r} \left(-\frac{\mu_0 m}{2\pi} \cdot \frac{\sin \theta}{r^3} + \frac{\mu_0 m}{2\pi} \cdot \frac{\sin \theta}{r^3} \right) = 0.$$

Since the vector potential satisfies the Coulomb gauge, Laplace's equation holds. ◆

6.7 Equivector Potential Surface

The equivector potential surface is characterized by the fact that the vector potential has the same value and direction on the surface in Cartesian coordinates. On the other hand, the vector potential obtained in Example 6.6 and that given by Eq. (6.30) do not have such equivector potential surfaces but have surfaces on which the vector potential has a single component that is constant in other coordinates. Namely, the direction of such a vector potential varies, while the value is the same on the surface. Although these are not equivector potential surfaces in the strict sense, we treat them as a kind of equivector potential surface.

There are two theorems on the equivector potential surface. Theorem I dictates that the vector potential on an equivector potential surface is parallel to the surface. Assume a closed equivector potential surface S. Integration of the vector potential on S leads to

$$\int_S \mathbf{A} \cdot d\mathbf{S} = \mathbf{A} \cdot \int_S d\mathbf{S} = A_n S, \tag{6.35}$$

where A_n is the normal component of the vector potential on S and S is the surface area of S. The left side of this equation leads to

$$\int_V \nabla \cdot \mathbf{A}\, dV, \tag{6.36}$$

where V is a region within S. Using the Coulomb gauge, $\nabla \cdot \mathbf{A} = 0$, this quantity is zero. Hence, we have

$$A_n = 0. \tag{6.37}$$

Thus, the vector potential is generally parallel to the equivector potential surface.

Theorem II dictates that the magnetic flux density on an equivector potential surface is parallel to the surface. Assume a closed line C on the equivector potential surface. Integration of the vector potential $\mathbf{A}$ on this line gives

$$\oint_C \mathbf{A} \cdot d\mathbf{s} = \mathbf{A} \cdot \oint_C d\mathbf{s} = 0. \tag{6.38}$$

The left-hand side of this equation leads to

$$\int_S \mathbf{B} \cdot d\mathbf{S} = \Phi, \tag{6.39}$$

6.7 Equivector Potential Surface

where S is the surface surrounded by C and Φ is the magnetic flux that penetrates S. Since this holds for arbitrary S, it can be said that the magnetic flux density is parallel to the equivector potential surface. Thus, if we denote the normal component of the magnetic flux density on the equivector potential surface by B_n, we have

$$B_n = 0. \qquad (6.40)$$

Thus, the magnetic flux density is parallel to equivector potential surface.

Theorem I holds generally. On the other hand, theorem II does not hold for all equivector potential surfaces. For example, the equivector potential surfaces for the solenoid coil in Example 6.6 and for the magnetic dipole line in Sect. 6.6 satisfy theorem II, while that for the magnetic moment does not. It is clear that theorem II holds for two-dimensional surface structures such as a long cylindrical surface. In the case of three-dimensional surface structures, Eq. (6.38) is not satisfied. We classify an equivector potential surface on which the magnetic flux density is not parallel to it as an equivector potential surface of the second kind.

***

Example 6.8 Show that an equivector potential surface is a cylindrical surface and that the magnetic flux density is also parallel to the cylindrical surface for the case of the magnetic dipole line discussed in Sect. 6.6.

***

Solution 6.8 The equivector potential surface is given by Eq. (6.32). Using the relationships $x = R\cos\varphi$ and $y = R\sin\varphi$, this is rewritten as

$$\frac{y}{x^2 + y^2} = c, \qquad (6.41)$$

where c is a constant. Thus, we have

$$x^2 + \left(y - \frac{1}{2c}\right)^2 = \frac{1}{4c^2}.$$

This shows a cylindrical surface along the z-axis passing through $x = 0$ and $y = 0$.

The magnetic flux density is given by Eq. (6.34) and we have

$$B_x = B_R \cos\varphi - B_\varphi \sin\varphi = \frac{\mu_0 \hat{m}}{2\pi R^2} \cos 2\varphi,$$

$$B_y = B_R \sin\varphi + B_\varphi \cos\varphi = \frac{\mu_0 \hat{m}}{2\pi R^2} \sin 2\varphi.$$

This leads to

$$\frac{B_y}{B_x} = \tan 2\varphi. \qquad (6.42)$$

This gives the slope of the magnetic flux density in the x-y plane. Here, we prove that the slope of the equivector-potential surface is equal to this. From Eq. (6.41) we have

$$\frac{dy}{dx} = -\frac{x}{y - (1/2c)}. \qquad (6.43)$$

The right-hand side of Eq. (6.42) leads to

$$\tan 2\varphi = \frac{2xy}{x^2 - y^2}.$$

By eliminating x^2 on the right-hand side using Eq. (6.41), we can show that the above equation reduces to the right-hand side of Eq. (6.43). Thus, the magnetic flux density is parallel to the equivector potential surface.

◆

6.8 Magnetic Charge

Magnets have north (N) and south (S) poles. The force between poles of the same kind is repulsive and the force between poles of the different kinds is attractive. This property is similar to that of electric charges. Hence, one can compare the magnetic interaction between magnetic poles to Coulomb's law for electric charges. A magnetic charge is an imaginary source that causes a magnetic interaction corresponding to the magnetic pole. In fact, it was assumed in the past that N and S poles had magnetic charges, q_m and $-q_m$, respectively, and that a force similar to the Coulomb force worked on magnetic charges. The force exerted by q'_m on q_m would then be given by

$$\boldsymbol{F}_m = \frac{\mu_0 q_m q'_m \boldsymbol{r}}{4\pi r^3} \qquad (6.44)$$

similarly to Eq. (1.1), where $\boldsymbol{r}$ is the position vector from q'_m to q_m and $r = |\boldsymbol{r}|$.

Since the magnetic force, Eq. (6.44), comes from some magnetic distortion in space, we can define the magnetic flux density $\boldsymbol{B}$ as the magnetic field that quantitatively expresses the strength of the magnetic distortion. When we express the magnetic force as

$$\boldsymbol{F}_m = q_m \boldsymbol{B}, \qquad (6.45)$$

the magnetic flux density is given by

$$\boldsymbol{B} = \frac{\mu_0 q'_m \boldsymbol{r}}{4\pi r^3}. \qquad (6.46)$$

It should be noted that this definition of magnetic charge is different by a factor of μ_0^{-1} from that used in other books, in which the magnetic field H defined in Chap. 9 was used for the magnetic interaction instead of the magnetic flux density B. One can also define a scalar potential, ϕ_m, called the magnetic potential similarly to the electric potential. Its relation to B is

$$B = -\nabla \phi_m. \tag{6.47}$$

Example 6.9 Assume magnetic charges with linear densities $\pm \lambda_m$ are positioned at $x = \pm d/2$, respectively, and define the moment of the magnetic dipole line by $\hat{m} = \lambda_m d$. This situation corresponds to the magnetic dipole line discussed in Sect. 6.6. Derive the magnetic potential and the magnetic flux density.

Solution 6.9 Using Eq. (1.25), the magnetic potential is given by

$$\phi_m = \frac{\mu_0 \hat{m} \cos \varphi}{2\pi R}.$$

Then, using Eq. (6.47), the magnetic flux density is determined to be

$$B_R = \frac{\mu_0 \hat{m}}{2\pi R^2} \cos \varphi, \quad B_\varphi = \frac{\mu_0 \hat{m}}{2\pi R^2} \sin \varphi, \quad B_z = 0,$$

which agrees with Eq. (6.34).

◆

6.9 Exercises

6.1. The straight current I and unknown straight currents I_x and I_y flow parallel to each other at positions located at each vertex of an equilateral triangle with side length a, as shown in Fig. E6.1. The magnetic flux density at point A, which is symmetric with the position of the current I_x, is zero. Determine the values of I_x and I_y.

Fig. E6.1 Current I and unknown currents I_x and I_y

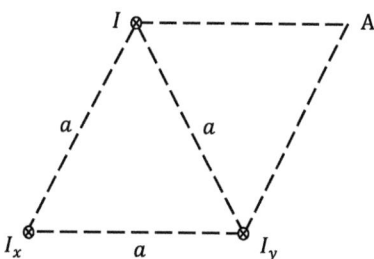

6.2. Current I flows on a closed triangle, as shown in Fig. E6.2. Determine the magnetic flux density at point A.

Fig. E6.2 Triangle on which current I flows and observation point A

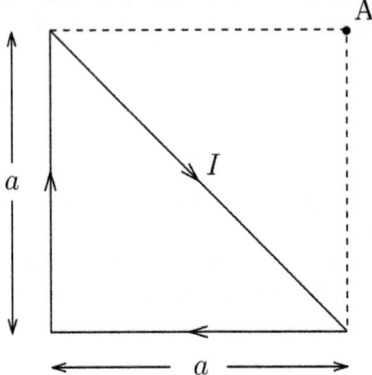

6.3. Suppose that current I flows on a closed circuit composed of two quarter circles and two straight lines on a common plane, as shown in Fig. E6.3. Determine the magnetic flux density at point O.

Fig. E6.3 Current I flowing in a closed circuit

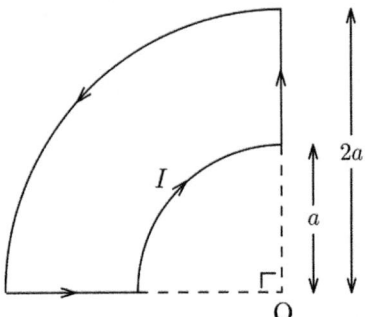

6.4. Two parallel currents of the same magnitude at distance $2a$ apart flow in opposite directions, as shown in Fig. E6.4. Determine the force on current I' equidistant from each current.

6.5. Two parallel currents of the same magnitude at distance $2a$ apart flow in the same direction, as shown in Fig. E6.5. Determine the magnetic flux density at point A at equidistance from each current. Determine also the force in a unit length on another current I' that flows at A and its direction.

6.6. Determine the Lorentz force between a current I_1 on a straight line and a current I_2 on a square, as shown in Fig. E6.6. These are placed on a common plane.

Fig. E6.4 Two parallel currents I flowing in opposite directions and current I' equidistant from each current

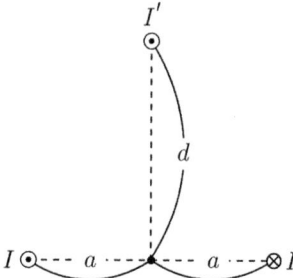

Fig. E6.5 Two parallel currents flowing in the same direction and point A at equidistance from each current

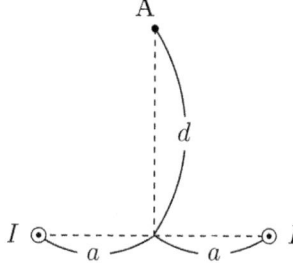

Fig. E6.6 Straight current and a square current

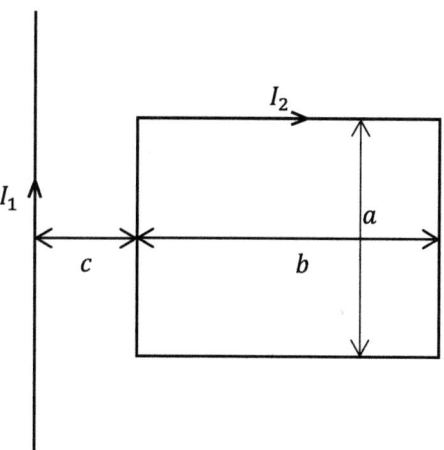

6.7. Determine the Lorentz force between a current I_1 on a straight line and a current I_2 on a circle of radius a. These are placed on a common plane and the distance between the line and the center of circle is d, as shown in Fig. E6.7.

6.8. Determine the Lorentz force between a current I_1 on a straight line and a current I_2 on a closed half-circle of radius a. These are placed on a common plane and the distance between the line and the center of circle is d, as shown in Fig. E6.8.

Fig. E6.7 Straight current and a circular current

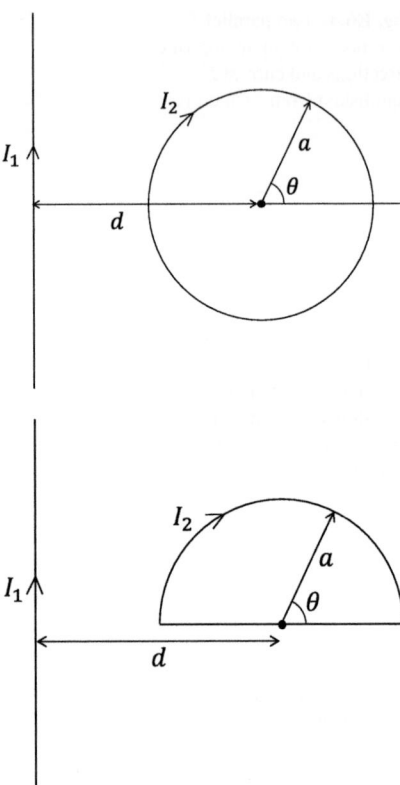

Fig. E6.8 Straight current and current on a closed half-circle

6.9. Suppose that current I_1 flows on a straight line and current I_2 flows on a closed loop on a common plane, as shown in Fig. E6.9. Prove that the component of the Lorentz force parallel to current I_1 is zero for arbitrary shape of the loop of current I_2.

Fig. E6.9 Straight current and a closed current loop of arbitrary shape

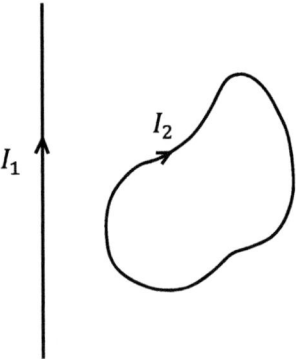

6.9 Exercises

6.10. Solve the problem of Example 6.4 with the Biot-Savart law.

6.11. Current I flows uniformly on a long thin plate of width $2w$, as shown in Fig. E6.10. Determine the magnetic flux density at point P at distance d from the center of the plane.

Fig. E6.10 Long thin plate with current and observation point P

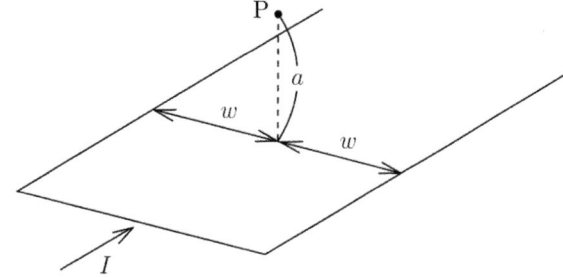

6.12. Current flows with surface density τ on a curved plane with a cross-section of a half circle with radius a, as shown in Fig. E6.11. Determine the magnetic flux density at the center of curvature of the half circle.

Fig. E6.11 Current flowing on a curved plane with a cross-section of a half circle

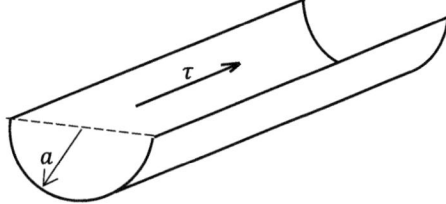

6.13. Suppose that a current flows uniformly along the y-axis with density i_0 inside a wide slab of thickness $2a$ parallel to the y-z plane, as shown in Fig. E6.12. Determine the magnetic flux density in each region using Ampere's law.

Fig. E6.12 Wide slab in which current flows uniformly

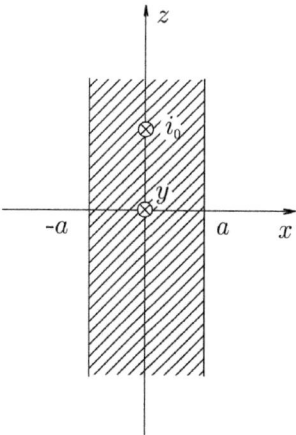

6.14. Straight currents I and $-I$ flow along the z-axis at $(a, 0)$ and $(-a, 0)$ in the x-y plane, respectively, as shown in Fig. E6.13. Determine the vector potential and magnetic flux density in the space.

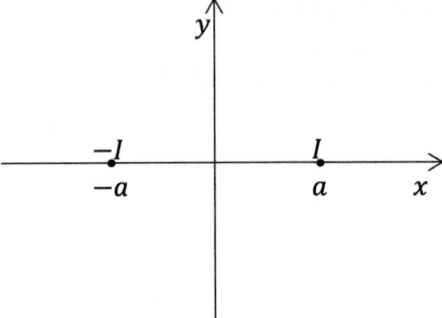

Fig. E6.13 Two straight currents

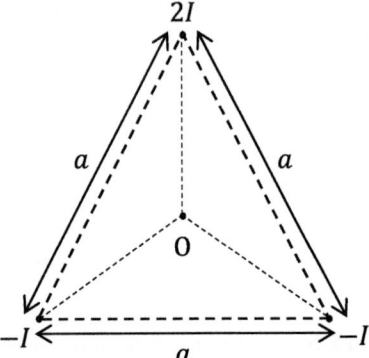

Fig. E6.14 Three straight currents at each vertex of an equilateral triangle

6.15. Three currents flow forward normal to the sheet at each vertex of an equilateral triangle with side length a, as shown in Fig. E6.14. Determine the vector potential and magnetic flux density field at the center of the equilateral triangle.

6.16. Assume that currents flow in four parallel straight lines, as shown in Fig. E6.15. Determine the magnetic flux density and vector potential at the center O.

6.17. Current I flows on a closed circle of diameter a. Determine the magnetic flux density and the vector potential at the center of the circle.

6.18. Current I flows on a closed equilateral triangle with side length a, as shown in Fig. E6.16. Determine the magnetic flux density and the vector potential at the center of the triangle.

6.19. Current I flows on a closed regular square with side length $2a$, as shown in Fig. E6.17. Determine the magnetic flux density and the vector potential at the center of the square.

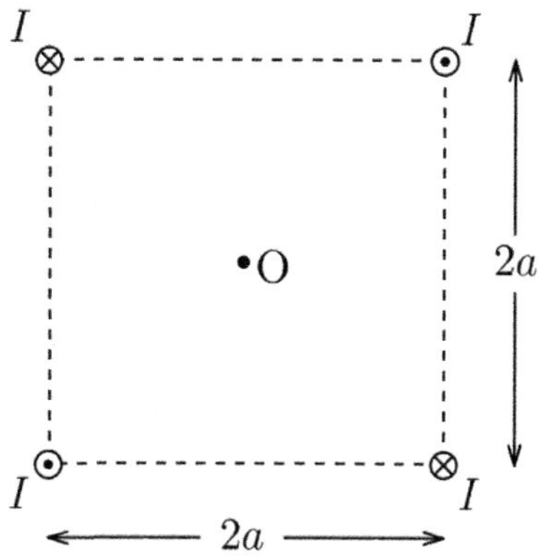

6.9 Exercises

Fig. E6.15 Currents flowing in four parallel straight lines

Fig. E6.16 Current flowing on a closed equilateral triangle and the center O

Fig. E6.17 Current flowing on a closed regular square and the center O

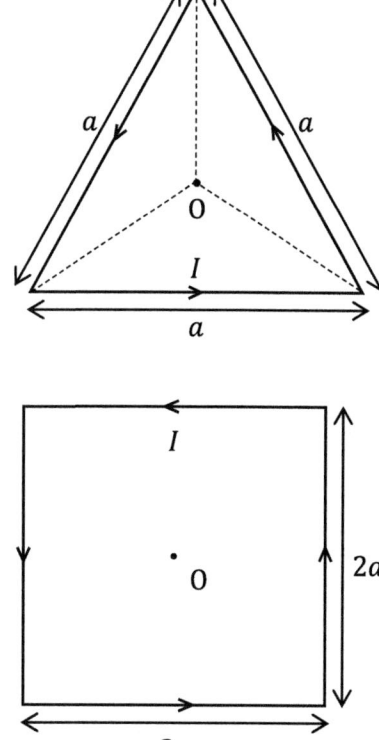

6.20. Determine the vector potential at point P for the circular current treated in Example 6.1.

6.21. Determine the vector potential produced by current I flowing on a circle of radius a.

6.22. Suppose that current flows along the z-axis as

$$i(R) = a - bR^2; \quad 0 \le R < (a/b)^{1/2};$$
$$= 0; \quad R > (a/b)^{1/2};$$

where R is the distance from the center. Determine the magnetic flux density and vector potential. The distance to the reference point of the vector potential is $R_0 \left[\gg (a/b)^{1/2} \right]$.

6.23. The vector potential described with cylindrical coordinates is given by

$$A = \frac{\mu_0 \gamma}{4} \left[(a^2 - R^2) + 2a^2 \log \frac{R_0}{a} \right] i_z; \quad 0 \le R < a,$$
$$= \frac{\mu_0 a^2 \gamma}{2} \log \frac{R_0}{R} i_z; \quad R > a.$$

Determine the current distribution.

6.24. The vector potential has a z-component in cylindrical coordinates and is given by

$$A_z(R) = \alpha \left(-R^2 + \frac{4}{9a} R^3 + \frac{5a^2}{9} \right) + \beta \log \frac{R_0}{a}; \quad 0 \le R < a,$$
$$= \beta \log \frac{R_0}{R}; \quad a < R < R_0,$$

where R_0 is the distance to the reference point of the vector potential. Determine the current distribution.

6.25. Current flows along the z-direction in cylindrical coordinates with the density given by

$$i_z(R) = a - bR^2; \quad 0 \le R < \left(\frac{2a}{b} \right)^{1/2},$$
$$= 0; \quad R > \left(\frac{2a}{b} \right)^{1/2}.$$

Determine the vector potential using Poisson's equation. Note that the total current is zero.

6.26. Determine the magnetic flux lines of the two-dimensional magnetic dipole line discussed in Sect. 6.6.

6.27. Discuss the magnetic flux lines of the magnetic dipole treated in Sect. 6.6.

6.28. Prove that the vector potential produced by a magnetic dipole line given by Eq. (6.30) satisfies Laplace's equation.

6.29. Determine the vector potential of a transverse magnetic quadrupole placed on the x-y plane ($z = 0$), as shown in Fig. E6.18.

Fig. E6.18 Transverse magnetic quadrupole composed of two small closed currents

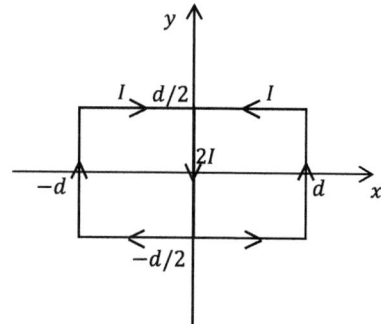

Fig. E6.19 Linear magnetic quadrupole composed of two small closed currents

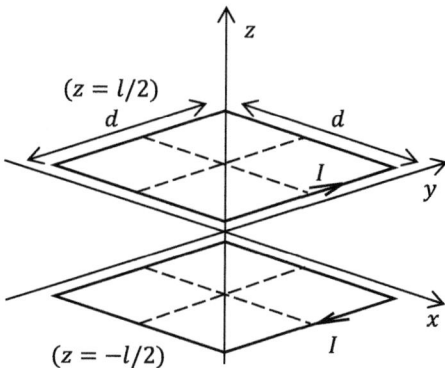

6.30. Determine the vector potential of a linear magnetic quadrupole placed on the x-y planes ($z = \pm l/2$), as shown in Fig. E6.19.

6.31. Prove that the vector potential of the linear magnetic quadrupole discussed in Exercise 6.30 satisfies Laplace's equation.

6.32. Express the magnetic flux density produced by a straight current I using the magnetic potential.

6.33. Determine the magnetic potential for the case of a current carrying cylinder treated in Examples 6.3 and 6.5. Compare it with the vector potential.

Answers to Exercises

6.1. Since the magnetic flux density at point A produced by I and I_y must be normal to the line connecting I_x and point A, we have $I_y = I$. Then, the magnetic flux density at A produced by two I's is

$$B = \frac{\sqrt{3}\mu_0 I}{2\pi a}$$

and the magnetic flux density at A produced by I_x is

$$B_x = \frac{\mu_0 I_x}{2\sqrt{3}\pi a}.$$

From the requirement $B + B_x = 0$, we have $I_x = -3I$.

6.2. First, we treat the contribution from the left side. The angle from an elementary line ds directed to observation point A is denoted by θ. Then, the distance from the elementary line to A is $r = a/\sin\theta$ and $|ds| = rd\theta/\sin\theta = ad\theta/\sin^2\theta$. The magnetic flux density due the elementary current is

$$dB = \frac{\mu_0 I ds}{4\pi r^2}\sin\theta = \frac{\mu_0 I}{4\pi a}\sin\theta\, d\theta.$$

Thus, the contribution from the left side is

$$B_1 = \frac{\mu_0 I}{4\pi a}\int_{\pi/4}^{\pi/2}\sin\theta\, d\theta = \frac{\mu_0 I}{4\sqrt{2}\pi a}.$$

Its direction is backward normal to the page. The contribution from the bottom side is the same.

Next, we treat the contribution from the inclined side. The distance between the line and A is $a/\sqrt{2}$, and the angle θ varies from $-3\pi/4$ to $-\pi/4$. Then, the magnetic flux density is directed forward normal to the page and its magnitude is

$$B_2 = -\frac{\mu_0 I}{2\sqrt{2}\pi a}\int_{-3\pi/4}^{-\pi/4}\sin\theta\, d\theta = \frac{\mu_0 I}{2\pi a}.$$

Thus, the total magnetic flux density is directed forward normal to the page and its magnitude is

$$B = -2B_1 + B_2 = \frac{(2-\sqrt{2})\mu_0 I}{4\pi a}.$$

6.3. Since the angle θ in Eq. (6.3) is 0 and π for the straight sections, there is no contribution to the magnetic flux density from these sections. On the quarter circle of radius a, θ is $\pi/2$ and the magnetic flux density is directed backward normal to the sheet. Its contribution is

$$B_a = \frac{\mu_0 I}{4\pi a^2}\int dr' = \frac{\mu_0 I}{8a}.$$

On the quarter circle of radius $2a$, θ is $-\pi/2$ and the magnetic flux density is directed forward normal to the sheet. Its contribution is

$$B_{2a} = \frac{\mu_0 I}{4\pi (2a)^2} \int dr' = \frac{\mu_0 I}{16a}.$$

Thus, the total magnetic flux density has a magnitude

$$B = B_a - B_{2a} = \frac{\mu_0 I}{16a}$$

and is directed backward normal to the page.

6.4. From the magnetic flux density obtained using Ampere's law, I' receives an attractive force of magnitude $\mu_0 II'/\left[2\pi \left(a^2 + d^2\right)^{1/2}\right]$ in a unit length from the current of the same direction and a repulsive force of the same magnitude from the current of the opposite direction. As a result, the combined force is $\mu_0 aII'/\left[\pi \left(a^2 + d^2\right)\right]$ in a unit length and is directed to the left, as shown in Fig. B6.1.

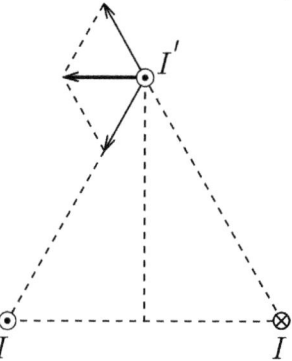

Fig. B6.1 Combined force produced by two currents

6.5. The magnetic flux density produced by one current is given by $B' = \mu_0 I/\left[2\pi \left(a^2 + d^2\right)^{1/2}\right]$. Since the vertical component is cancelled due to symmetry, as shown in Fig. B6.2, only the horizontal component remains. Hence, the magnetic flux density at point A is

$$B = 2B' \frac{d}{\left(a^2 + d^2\right)^{1/2}} = \frac{\mu_0 I d}{\pi \left(a^2 + d^2\right)}.$$

The force in a unit length on current I' flowing in the same direction at point A is given by

$$F' = I'B = \frac{\mu_0 II' d}{\pi \left(a^2 + d^2\right)}$$

and the force is directed vertically downward.

Fig. B6.2 Combined magnetic flux density due to two currents

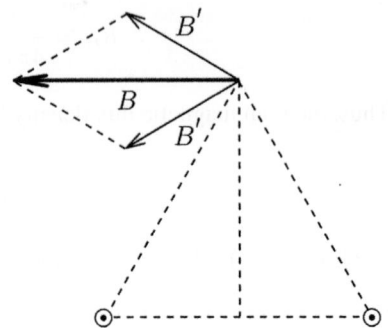

6.6. The Lorentz force between the long line current and a current of length dy on the left side of the square is $dF = \mu_0 I_1 I_2 dy/2\pi c$. Hence, the force on this side is

$$F_1 = \frac{\mu_0 I_1 I_2 a}{2\pi c}$$

directed to the left. The force on the right side is similarly obtained to be

$$F_2 = \frac{\mu_0 I_1 I_2 a}{2\pi (b+c)}$$

directed to the right. There is no force on other two sides. Hence, the total force is

$$F = F_1 - F_2 = \frac{\mu_0 I_1 I_2 a}{2\pi}\left(\frac{1}{c} - \frac{1}{b+c}\right) = \frac{\mu_0 I_1 I_2 ab}{2\pi (b+c)c}$$

and directed to the left.

6.7. The angle θ is defined in Fig. E6.7. The horizontal component of the Lorentz force only remains from symmetry. The force on the current in the region from θ to $\theta + d\theta$ is

$$dF = \frac{\mu_0 I_1 I_2 a \cos\theta \, d\theta}{2\pi (d + a\cos\theta)}.$$

So, the total Lorentz force is

$$F = \frac{\mu_0 I_1 I_2}{\pi} \int_0^\pi \left(1 - \frac{d}{d + a\cos\theta}\right) d\theta = -\mu_0 I_1 I_2 \left(\frac{d}{\sqrt{d^2 - a^2}} - 1\right)$$

and is attractive. We used Eq. (A.23) in the Appendix.

6.8. The angle θ is defined in Fig. E6.8. The horizontal component of the Lorentz force on the current on the half-circle in the region θ to $\theta + d\theta$ is

6.9 Exercises

$$dF_{h1} = \frac{\mu_0 I_1 I_2 a \cos\theta \, d\theta}{2\pi(d + a\cos\theta)}.$$

So, the total horizontal force is

$$F_{h1} = \frac{\mu_0 I_1 I_2}{2\pi} \int_0^\pi \left(1 - \frac{d}{d + a\cos\theta}\right) d\theta = -\frac{\mu_0 I_1 I_2}{2}\left(\frac{d}{\sqrt{d^2 - a^2}} - 1\right),$$

where we used Eq. (A.23) in the Appendix. The vertical component of the Lorentz force is given by

$$dF_{v1} = \frac{\mu_0 I_1 I_2 a \sin\theta \, d\theta}{2\pi(d + a\cos\theta)}.$$

So, the total vertical force is

$$F_{v1} = \frac{\mu_0 I_1 I_2 a}{2\pi} \int_0^\pi \frac{\sin\theta \, d\theta}{(d + a\cos\theta)} = \frac{\mu_0 I_1 I_2}{2\pi} \log \frac{d+a}{d-a}.$$

Next, the horizontal force on the straight section is 0, and the vertical force is

$$F_{v2} = -\frac{\mu_0 I_1 I_2}{2\pi} \int_{d-a}^{d+a} \frac{dR}{R} = -\frac{\mu_0 I_1 I_2}{2\pi} \log \frac{d+a}{d-a}.$$

Thus, the vertical force is totally cancelled and the horizontal force is

$$F_h = F_{h1} = -\frac{\mu_0 I_1 I_2}{2}\left(\frac{d}{\sqrt{d^2 - a^2}} - 1\right).$$

6.9. We divide the current loop I_2 into small segments composed of parallel and perpendicular parts to current I_1, as shown in Fig. B6.3a. The Lorentz force on the parallel segments is directed horizontal and that on the perpendicular segments is directed vertical. Since the force does not change even if these segments are moved with keeping the distance from current I_1 unchanged, the result is the same for the vertical component when the perpendicular segments are rearranged as shown in Fig. B6.3b. So, it is clear that the vertical component is totally cancelled.

6.10. Here, we define the x- and y- axes on the plane with the x-axis along the current, and the origin $(0, 0)$ is placed at the foot of a vertical line from the observation point. The current in the region from x to $x + dx$ is regarded as a line current of τdx that flows along the y-axis. If we use the result obtained with the Biot-Savart law in Example 6.2, the resultant magnetic flux density is $dB' = \mu_0 \tau dx / \left[2\pi \left(x^2 + h^2\right)^{1/2}\right]$.

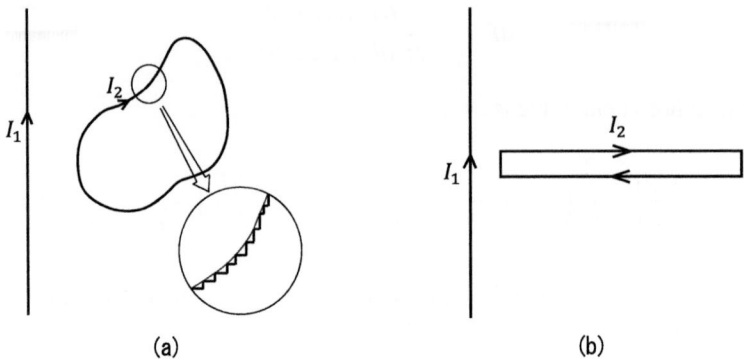

(a)   (b)

Fig. B6.3 **a** Division of current loop I_2 into small segments and **b** rearranged horizontal segments

From symmetry, only the horizontal component given by $dB = dB'h/(x^2 + h^2)^{1/2}$ remains (see Fig. B6.4). Thus, we have

$$B = \int_0^\infty \frac{\mu_0 \tau h dx}{\pi (x^2 + h^2)}.$$

If we put $x = h\tan\theta$, θ is the angle shown in Fig. B6.4. From the relationships $x^2 + h^2 = h^2/\cos^2\theta$ and $dx = hd\theta/\cos^2\theta$, we have

$$B = \frac{\mu_0 \tau}{\pi} \int_0^{\pi/2} d\theta = \frac{\mu_0 \tau}{2},$$

which agrees with the result in Example 6.4.

Fig. B6.4 Magnetic flux density produced by a current flowing in a narrow region and its vertical component

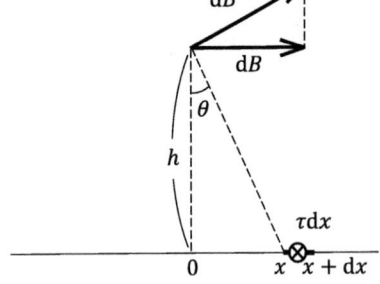

6.11. We define the x-axis in the direction of the width with the origin at the foot of the vertical line from point P on the plane, as shown in Fig. B6.5. Using Ampere's law, the magnetic flux density at point P produced by the current $(I/2w) dx$ flowing

Fig. B6.5 Magnetic flux density and its horizontal component produced by the current in the region x to $x + dx$

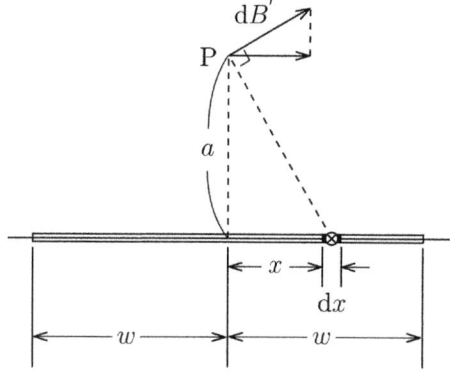

in a thin region from x to $x + dx$ is $dB' = \mu_0 I dx / \left[4\pi w (x^2 + a^2)^{1/2} \right]$. Only the x-component remains without cancellation, and integrating this, we have

$$B = 2 \int_0^w \frac{\mu_0 I a dx}{4\pi w (x^2 + a^2)}.$$

We put $x = a \tan\theta$. The magnetic flux density is determined to be

$$B = 2 \int_0^{\theta_w} \frac{\mu_0 I d\theta}{4\pi w} = \frac{\mu_0 I \theta_w}{2\pi w},$$

where $\theta_w = \tan^{-1}(w/a)$.

6.12. We define the angle φ on the cross-section, as shown in Fig. B6.6. The magnetic flux density produced by the current in a thin region of the angle from φ to $\varphi + d\varphi$ is

$$dB = \frac{\mu_0 \tau d\varphi}{2\pi},$$

and only the horizontal component, $dB' = dB \cos\varphi$, remains. Hence, the total magnetic flux density is

Fig. B6.6 Cross-section of the curved plane

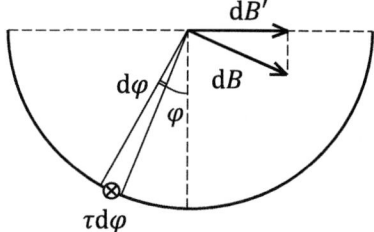

$$B = \frac{\mu_0 \tau}{2\pi} \int_{-\pi/2}^{\pi/2} \cos\varphi \, d\varphi = \frac{\mu_0 \tau}{\pi}$$

and is directed to the right.

6.13. We assume rectangle C with one side on the center, $x = 0$, as shown in Fig. B6.7. From symmetry, the magnetic flux density has only a z component (B_z), and its value must be zero at $x = 0$. The magnetic flux density is integrated along rectangle C, as shown by the arrow, so as to be consistent with the current flow along the positive y-axis. Since $\mathbf{B}$ is perpendicular to $d\mathbf{s}$ on the top and bottom sides of C, the integral is zero there. The left side of Eq. (6.15) is $-B_z l$, where l is the length of the side of C along the z-axis. The current penetrating C is lxi_0 for $0 \le x \le a$ and lai_0 for $x > a$. Thus, we have

$$\begin{aligned} B_z &= \mu_0 i_0 a; & x < -a, \\ &= -\mu_0 i_0 x; & -a \le x \le a, \\ &= -\mu_0 i_0 a; & x > a, \end{aligned}$$

where we have used the symmetric condition with respect to $x = 0$.

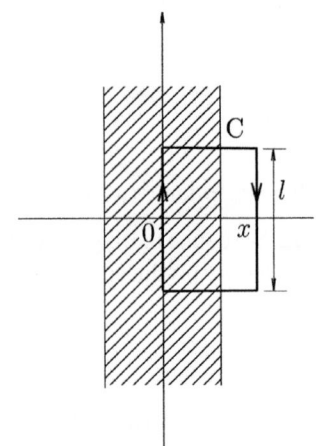

Fig. B6.7 Rectangle C for $x > a$

6.14. In this case, the vector potential is given by

$$A_z(x, y) = \frac{\mu_0 I}{4\pi} \log \frac{(x+a)^2 + y^2}{(x-a)^2 + y^2},$$

since the reference point can be chosen at infinity. The magnetic flux density is

$$B_x = \frac{\partial A_z}{\partial y} = \frac{\mu_0 I}{2\pi}\left[\frac{y}{(x+a)^2+y^2} - \frac{y}{(x-a)^2+y^2}\right],$$

$$B_y = -\frac{\partial A_z}{\partial x} = \frac{\mu_0 I}{2\pi}\left[\frac{x-a}{(x-a)^2+y^2} - \frac{x+a}{(x+a)^2+y^2}\right],$$

$$B_z = 0.$$

6.15. All the currents are placed at the same distance, $a/\sqrt{3}$, from the center, and the total current is zero. So, the vector potential at the center is zero. The magnetic flux density due to the current $2I$ is directed to the right and its value is $\sqrt{3}\mu_0 I/\pi a$. The vertical component of the magnetic flux density of the two currents $-I$ is cancelled, and only the horizontal component remains. Each horizontal component is $\sqrt{3}\mu_0 I/4\pi a$ and is directed to the right. The total magnetic flux density is directed to the right and its value is

$$\frac{\sqrt{3}\mu_0 I}{\pi a} + \frac{\sqrt{3}\mu_0 I}{4\pi a} \times 2 = \frac{3\sqrt{3}\mu_0 I}{2\pi a}.$$

6.16. The magnetic flux density at O produced by each current is given by $B = \mu_0 I/(2\sqrt{2}\pi a)$. But the magnetic flux density produced by pair of diagonal two currents is zero because of cancellation. So, the magnetic flux density is zero. The vector potential produced by the positive current (along the direction forward) is $A_z = (\mu_0 I/2\pi)\log(R_0/\sqrt{2}a)$ with R_0 denoting the distance to the reference point and that by the negative current is $A_z = -(\mu_0 I/2\pi)\log(R_0/\sqrt{2}a)$. As a result, the vector potential at O is zero.

6.17. We denote the azimuthal angle measured from the center by φ. The magnetic flux density at the center produced by a current element in the region from φ to $\varphi + d\varphi$ is

$$dB = \frac{\mu_0 I d\varphi}{4\pi a}.$$

Hence, the magnetic flux density at the center is determined to be

$$B = \int_0^{2\pi} \frac{\mu_0 I d\varphi}{4\pi a} = \frac{\mu_0 I}{2a}.$$

In the vicinity of the center, the magnetic flux density is almost uniform and is approximately given by the above value. So, it is reasonable to assume that the vector potential has an azimuthal component, A_φ. Integrating this over a small circle C with radius δ around the center on the plane that includes the center, we have

$$\int_C A_\varphi ds = 2\pi \delta A_\varphi.$$

This is equal to the magnetic flux penetrating C, $\pi\delta^2 B$. So, we have

$$A_\varphi = \frac{1}{2}\delta B.$$

This value goes to zero in the limit $\delta \to 0$. Thus, the vector potential at the center is zero.

6.18. We calculate the magnetic flux density produced by the current on the bottom side. We denote by x the distance of a point on this side from the foot of the vertical line from the center. The elementary current Idx produces the magnetic flux density forward normal to the sheet:

$$dB = \frac{a\mu_0 I dx}{8\sqrt{3}\pi \left(x^2 + a^2/12\right)^{3/2}}.$$

Putting $x = \left(a/2\sqrt{3}\right)\tan\theta$, the contribution from the bottom side leads to

$$B' = \frac{\sqrt{3}\mu_0 I}{\pi a} \int_0^{\pi/3} \cos\theta\, d\theta = \frac{3\mu_0 I}{2\pi a}.$$

The contributions are the same from other two side. Thus, we have

$$B = 3B' = \frac{9\mu_0 I}{2\pi a}.$$

In the vicinity of the center, the magnetic flux density is almost uniform and is approximately given by the above value. So, it is reasonable to assume that the vector potential has an azimuthal component, A_φ. Integrating this over a small circle C with radius δ around the center on the plane that includes the center, we have

$$\int_C A_\varphi ds = 2\pi \delta A_\varphi.$$

This is equal to the magnetic flux penetrating C, $\pi\delta^2 B$. So, we have

$$A_\varphi = \frac{1}{2}\delta B.$$

This value goes to zero in the limit $\delta \to 0$. Thus, the vector potential at the center is zero.

6.19. We calculate the magnetic flux density produced by the current on the bottom side. We denote by x the distance of a point on this side from the foot of the vertical line from the center. The elementary current $I\mathrm{d}x$ produces the magnetic flux density forward normal to the sheet:

$$\mathrm{d}B = \frac{a\mu_0 I \mathrm{d}x}{4\pi \left(x^2 + a^2\right)^{3/2}}.$$

Putting $x = a\tan\theta$, the contribution from the bottom side leads to

$$B' = \frac{\mu_0 I}{2\pi a} \int_0^{\pi/4} \cos\theta\, \mathrm{d}\theta = \frac{\mu_0 I}{2\sqrt{2}\pi a}.$$

The contributions are the same from other three side. Thus, we have

$$B = 4B' = \frac{\sqrt{2}\mu_0 I}{\pi a}.$$

In the vicinity of the center, the magnetic flux density is almost uniform and is approximately given by the above value. So, it is reasonable to assume that the vector potential has an azimuthal component, A_φ. Integrating this over a small circle C with radius δ around the center on the plane that includes the center, we have

$$\int_C A_\varphi\, \mathrm{d}s = 2\pi \delta A_\varphi.$$

This is equal to the magnetic flux penetrating C, $\pi\delta^2 B$. So, we have

$$A_\varphi = \frac{1}{2}\delta B.$$

This value goes to zero in the limit $\delta \to 0$. Thus, the vector potential at the center is zero.

6.20. Since the azimuthal current only flows, the vector potential has an azimuthal component, A_φ. From the result of Example 6.1, the magnetic flux density at point P is

$$B_z = \frac{\mu_0 I a^2}{2\left(z^2 + a^2\right)^{3/2}},$$

and the magnetic flux density does not change appreciably from this value. Hence, the vector potential around point P is approximately given by

$$A_\varphi = \frac{1}{R}\int B_z R \mathrm{d}R = \frac{\mu_0 I a^2 R}{4\left(z^2 + a^2\right)^{3/2}}.$$

Hence, we have $A_\varphi = 0$ in the limit of $R \to 0$. This result is supported by a more rigorous calculation in Exercise 6.21.

6.21. The circle is placed on the x-y plane with the center on the origin. The observation point P is assumed to be on the y-axis and at distance R from the axis, as shown Fig. B6.8. Then, the vector potential is directed along the x-axis at P. The elementary line vector for the current is $d\mathbf{r}' = a(-\mathbf{i}_x \sin\varphi + \mathbf{i}_y \cos\varphi) d\varphi$ and we have $d\mathbf{r}' \cdot \mathbf{i}_x = -a \sin\varphi \, d\varphi$. The vector potential is given by

$$A_\varphi = -\frac{\mu_0 I a}{4\pi} \int_0^{2\pi} \frac{\sin\varphi \, d\varphi}{\left(a^2 + R^2 + 2aR \sin\varphi + z^2\right)^{1/2}}.$$

Here, we put as $\varphi = 2\psi + \pi/2$. Then, the vector potential is rewritten as

$$A_\varphi = -\frac{\mu_0 I k}{2\pi} \left(\frac{a}{R}\right)^{1/2} \int_0^{\pi/2} \frac{1 - 2\sin^2\psi}{\left(1 - k^2 \sin^2\psi\right)^{1/2}} d\psi,$$

where $k = 2(aR)^{1/2}/[(a+R)^2 + z^2]^{1/2}$. In terms of complete elliptic integrals of the first and second kind,

$$F(k) = \int_0^{\pi/2} \frac{1}{\left(1 - k^2 \sin^2\psi\right)^{1/2}} d\psi,$$

$$E(k) = \int_0^{\pi/2} \left(1 - k^2 \sin^2\psi\right)^{1/2} d\psi,$$

Fig. B6.8 Circular current and the vector potential at observation point P

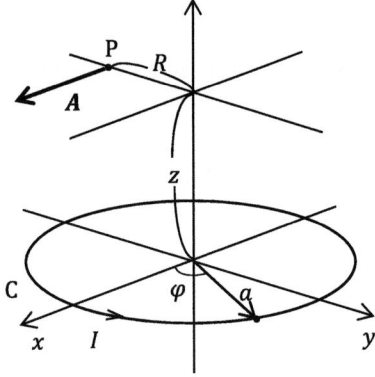

the vector potential is written as

$$A_\varphi = \frac{\mu_0 I}{2\pi k}\left(\frac{a}{R}\right)^{1/2}[(2-k^2)F(k) - 2E(k)].$$

In the limit $R \to 0 (k \to 0)$, the complete elliptic integrals reduce to

$$F(k) = \frac{\pi}{2}\left(1 + \frac{k^2}{4} + \frac{9k^4}{64}\right), \quad E(k) = \frac{\pi}{2}\left(1 - \frac{k^2}{4} - \frac{3k^4}{64}\right).$$

Then, we have

$$A_\varphi = \frac{\mu_0 I a^2 R}{4(a^2 + z^2)^{3/2}}. \tag{B6.1}$$

The vector potential in this limit can be simply determined by the following method. Using the result of Example 6.1, the magnetic flux density around the observation point is

$$B_z = \frac{\mu_0 I a^2}{2(a^2 + z^2)^{3/2}}. \tag{B6.2}$$

Then, substitution of Eq. (B6.2) into the relation,

$$B_z = \frac{1}{R}\frac{\partial(RA_\varphi)}{\partial R},$$

yields the same approximate solution (B6.1).

6.22. The magnetic flux density has an azimuthal component, B_φ. Equation (6.16) leads to

$$\frac{1}{R}\cdot\frac{\partial}{\partial R}(RB_\varphi) = \mu_0(a - bR^2)$$

for $0 \leq R < (a/b)^{1/2}$. After a simple calculation, we have

$$B_\varphi = \mu_0\left(\frac{a}{2}R - \frac{b}{4}R^3 + \frac{c}{R}\right),$$

where c is a constant. Since the magnetic flux density must be finite at $R = 0$, we have $c = 0$. The total current in the region $0 \leq R < (a/b)^{1/2}$ is

$$I = \int_0^{(a/b)^{1/2}} i(R) 2\pi R dR = \frac{\pi a^2}{2b}.$$

Thus, the magnetic flux density is determined to be

$$B_\varphi = \mu_0\left(\frac{a}{2}R - \frac{b}{4}R^3\right); \quad 0 \le R < \left(\frac{a}{b}\right)^{1/2},$$
$$= \frac{\mu_0 a^2}{4bR}; \quad R > \left(\frac{a}{b}\right)^{1/2}.$$

Then, the z-component of the vector potential is given by

$$A_z = \frac{\mu_0 a^2}{4b}\log\frac{R_0}{R}; \quad \left(\frac{a}{b}\right)^{1/2} < R < R_0,$$
$$= \frac{\mu_0}{16}\left(-4aR^2 + bR^4 + \frac{3a^2}{b}\right) + \frac{\mu_0 a^2}{8b}\log\frac{bR_0^2}{a}; \quad 0 \le R < \left(\frac{a}{b}\right)^{1/2}.$$

6.23. Using Eq. (6.23), the current distribution is determined to be

$$i(R) = -\frac{1}{\mu_0}\Delta A = -\frac{1}{\mu_0 R}\cdot\frac{\partial}{\partial R}\left(R\frac{\partial A_z}{\partial R}\right)i_z.$$

That is, the current flows along the z-axis and its distribution is

$$i(R) = \gamma; \quad 0 \le R < a,$$
$$= 0; \quad R > a.$$

6.24. Using Eq. (6.18), we obtain the magnetic flux density as

$$B_\varphi(R) = \alpha\left(2R - \frac{4}{3a}R^2\right); \quad 0 \le R < a,$$
$$= \frac{\beta}{R}; \quad a < R < R_0.$$

Then, the current distribution is determined to be

$$i(R) = \frac{4\alpha}{\mu_0}\left(1 - \frac{R}{a}\right); \quad 0 \le R < a,$$
$$= 0; \quad a < R < R_0.$$

6.25. It is clear that the vector potential has only a z-component, A_z, from Eq. (6.20). In addition, it is expected that A_z depends only on R. So, Poisson's equation is given by

$$-\frac{1}{R}\cdot\frac{\partial}{\partial R}\left(R\frac{\partial A_z}{\partial R}\right) = -\mu_0 i_z(R).$$

6.9 Exercises

Thus, we have

$$A_z(R) = \mu_0 \left(\frac{a}{4} R^2 - \frac{b}{16} R^4 \right) + \alpha \log R + \beta$$

for $0 \le R < (2a/b)^{1/2}$, where α and β are constants. Since the vector potential must be finite at $R = 0$, we have $\alpha = 0$. The other constant, β, is determined from the continuity of the vector potential at $R = (2a/b)^{1/2}$.

For $R > (2a/b)^{1/2}$, the vector potential is determined to be a constant. Thus, we have $A_z(R) = 0$, since the vector potential must be zero at infinity because of the zero total current.

Then, the continuity of the vector potential at $R = (2a/b)^{1/2}$ yields $\beta = -\mu_0(a^2/b)$. Thus, the vector potential is determined to be

$$A_z(R) = \frac{\mu_0}{4} \left(aR^2 - \frac{b}{4} R^4 - \frac{a^2}{b} \right); \quad 0 \le R < \left(\frac{2a}{b} \right)^{1/2},$$

$$= 0; \quad R > \left(\frac{2a}{b} \right)^{1/2}.$$

6.26. The equivector potential surface is given by Eq. (6.32):

$$A'_z(R, \varphi) = \frac{\sin \varphi}{R} = \text{const.}$$

Here, we assume the magnetic flux lines of the form:

$$\psi(R, \varphi) = \frac{\sin \varphi}{R^n} = \text{const.}$$

The exponent n is determined from the condition that they are parallel to each other:

$$\nabla A'_z \times \nabla \psi = 0.$$

Using

$$\nabla A'_z = \frac{\partial A'_z}{\partial R} i_R + \frac{1}{R} \cdot \frac{\partial A'_z}{\partial \varphi} i_\varphi = -\frac{\sin \varphi}{R^2} i_R + \frac{\cos \varphi}{R^2} i_\varphi,$$

$$\nabla \psi = \frac{\partial \psi}{\partial R} i_R + \frac{1}{R} \cdot \frac{\partial \psi}{\partial \varphi} i_\varphi = -\frac{n \sin \varphi}{R^{n+1}} i_R + \frac{\cos \varphi}{R^{n+1}} i_\varphi,$$

the above condition leads to

$$\frac{(1-n) \sin \varphi \cos \varphi}{R^{n+3}} i_z = 0.$$

Thus, we have $n = 1$, and the magnetic flux lines are given by

$$\psi(R, \varphi) = \frac{\sin \varphi}{R} = \text{const.} \qquad (B6.3)$$

This is the same form as the electric field lines of the two-dimensional electric dipole line discussed in Exercise 1.27. The solution is given by

$$x^2 + \left(y - \frac{1}{2c}\right)^2 = \frac{1}{4c^2}.$$

Equation (B6.3) is exactly of the same form as the equivector potential surface. So, it proves that the magnetic flux lines are parallel to the quivector potential surface. See Example 6.8.

6.27. The components of the magnetic flux density are given by Eq. (6.29). Then, its x- and y-components are given by

$$B_x = B_r \cos\theta - B_\theta \sin\theta, \quad B_y = B_r \sin\theta + B_\theta \cos\theta.$$

The magnetic flux lines are given by

$$\frac{B_x}{B_y} = \frac{dx}{dy},$$

which is reduced to

$$\frac{dx}{dy} = \frac{1}{3}\left(2\frac{x}{y} - \frac{y}{x}\right),$$

where we used $x/y = \cot\theta$. Here, we put $\Psi = x/y$. Then, the above equation leads to a form with separation of variables:

$$\int \frac{\Psi}{\Psi^2 + 1} d\Psi = -\frac{1}{3} \int \frac{1}{y'} dy'$$

with $y' = ky$, which leads to

$$\Psi^2 + 1 = c|y|^{-2/3}$$

with $c = k^{-2/3}$. So, the magnetic flux lines are expressed as

$$x^2 + y^2 = c|y|^{4/3}.$$

These lines pass through the origin and are symmetric with respect to $y = 0$.

On the other hand, the equivector potential surface is given by Eq. (6.28), which is expressed as

$$x^2 + y^2 = c'|y|^{2/3}.$$

Since this is different from the equation describing the magnetic field lines, it shows that the magnetic flux lines are not parallel to the equivector potential surface.

6.28. Using Eq. (A.14) in the Appendix, we have

$$\nabla \times \mathbf{A} = \mathbf{i}_R \frac{\mu_0 \hat{m}}{2\pi} \cdot \frac{\cos \varphi}{R^2} + \mathbf{i}_\varphi \frac{\mu_0 \hat{m}}{2\pi} \cdot \frac{\sin \varphi}{R^2}$$

and

$$\nabla \times \nabla \times \mathbf{A} = \mathbf{i}_z \frac{1}{R} \left(-\frac{\mu_0 \hat{m}}{2\pi} \cdot \frac{\sin \varphi}{R^2} + \frac{\mu_0 \hat{m}}{2\pi} \cdot \frac{\sin \varphi}{R^2} \right) = 0.$$

Since it is clear that the vector potential satisfies the Coulomb gauge, Laplace's equation holds.

6.29. The position of the observation point A of distance much larger than d from the origin is denoted by (r, θ, φ). Using Cartesian coordinates, this is expressed as $(r \sin \theta \cos \varphi, r \sin \theta \sin \varphi, r \cos \theta)$. First, we calculate the contribution from the current from $(d, d/2)$ to $(0, d/2)$. The distance between point P $(x, d/2)$ on this branch and point A is

$$r_P = \left[(r \sin \theta \cos \varphi - x)^2 + (r \sin \theta \sin \varphi - d/2)^2 + r^2 \cos^2 \theta \right]^{1/2}$$

$$= r \left[1 - \frac{1}{r} \sin \theta (2x \cos \varphi + d \sin \varphi) + \frac{1}{r^2} \left(x^2 + \frac{d^2}{4} \right) \right]^{1/2},$$

which leads to

$$\frac{1}{r_P} \simeq \frac{1}{r} + \frac{1}{2r^2} \sin \theta (2x \cos \varphi + d \sin \varphi) - \frac{1}{2r^3} \left(x^2 + \frac{d^2}{4} \right)$$

$$+ \frac{1}{8r^3} \sin^2 \theta (2x \cos \varphi + d \sin \varphi)^2.$$

Integrating this from $(d, d/2)$ to $(0, d/2)$, we have

$$-\frac{i_x}{r} \left[d + \frac{d^2 \sin \theta}{2r} (\cos \varphi + \sin \varphi) - \frac{d^3}{2r^2} \left(\frac{1}{3} + \frac{1}{4} \right) \right.$$

$$\left. + \frac{d^3 \sin^2 \theta}{8r^2} \left(\frac{4}{3} \cos^2 \varphi + 2 \cos \varphi \sin \varphi + \sin^2 \varphi \right) \right]$$

$$= -\frac{i_x}{r}\left[d + \frac{d^2 \sin\theta}{2r}(\cos\varphi + \sin\varphi) - \frac{7d^3}{24r^2}\right.$$
$$\left. + \frac{d^3 \sin^2\theta}{8r^2}\left(1 + \frac{1}{3}\cos^2\varphi + 2\cos\varphi\sin\varphi\right)\right].$$

A similar calculation for the contribution from $(0, -d/2)$ to $(d, -d/2)$ leads to

$$\frac{i_x}{r}\left[d + \frac{d^2 \sin\theta}{2r}(\cos\varphi - \sin\varphi) - \frac{7d^3}{24r^2}\right.$$
$$\left. + \frac{d^3 \sin^2\theta}{8r^2}\left(1 + \frac{1}{3}\cos^2\varphi - 2\cos\varphi\sin\varphi\right)\right].$$

From $(d, -d/2)$ to $(d, d/2)$ we have

$$\frac{i_y}{r}\left[d + \frac{d^2}{r}\sin\theta\cos\varphi - \frac{13d^3}{24r^2} + \frac{d^3 \sin^2\theta}{24r^2}(11\cos^2\varphi + 1)\right]$$

and from $(0, d/2)$ to $(0, -d/2)$ we have

$$-\frac{i_y}{r}\left[d - \frac{d^3}{24r^2}(1 - \sin^2\theta \sin^2\varphi)\right].$$

Thus, the vector potential produced by the right closed current is determined to be

$$\mathbf{A}_+ = \frac{\mu_0 I}{4\pi}\left\{-\mathbf{i}_x \frac{d^2}{r^2}\sin\theta\sin\varphi\left(1 + \frac{d}{2r}\sin\theta\cos\varphi\right)\right.$$
$$\left. + \mathbf{i}_y \frac{d^2}{r^2}\left[\sin\theta\cos\varphi - \frac{d}{2r}(1 - \sin^2\theta\cos^2\varphi)\right]\right\}.$$

The vector potential due to the left closed current is similarly obtained as

$$\mathbf{A}_- = \frac{\mu_0 I}{4\pi}\left\{\mathbf{i}_x \frac{d^2}{r^2}\sin\theta\sin\varphi\left(1 - \frac{d}{2r}\sin\theta\cos\varphi\right)\right.$$
$$\left. - \mathbf{i}_y \frac{d^2}{r^2}\left[\sin\theta\cos\varphi + \frac{d}{2r}(1 - \sin^2\theta\cos^2\varphi)\right]\right\}.$$

Then, the vector potential of the transverse magnetic quadrupole is given by

$$\mathbf{A} = \mathbf{A}_+ + \mathbf{A}_- = -\frac{\mu_0 I d^3}{4\pi r^3}[\mathbf{i}_x \sin^2\theta\sin\varphi\cos\varphi + \mathbf{i}_y(1 - \sin^2\theta\cos^2\varphi)].$$

6.30. The vector potential due to the upper closed current at $z = l/2$ is determined first. On the line from $(d/2, d/2, l/2)$ to $(-d/2, d/2, l/2)$, the distance between

6.9 Exercises

point P $(x, d/2, l/2)$ and observation point A $(r \sin\theta \cos\varphi, r \sin\theta \sin\varphi, r \cos\theta)$ is

$$r_P = \left[(r\sin\theta\cos\varphi - x)^2 + (r\sin\theta\sin\varphi - d/2)^2 + (r\cos\theta - l/2)^2\right]^{1/2}$$

$$= r\left[1 - \frac{1}{r}\sin\theta(2x\cos\varphi + d\sin\varphi) - \frac{1}{r}l\cos\theta + \frac{1}{r^2}\left(x^2 + \frac{d^2}{4} + \frac{l^2}{4}\right)\right]^{1/2},$$

which leads to

$$\frac{1}{r_P} \simeq \frac{1}{r} + \frac{1}{2r^2}\sin\theta(2x\cos\varphi + d\sin\varphi) + \frac{1}{2r^2}l\cos\theta - \frac{1}{2r^3}\left(x^2 + \frac{d^2}{4} + \frac{l^2}{4}\right)$$

$$+ \frac{1}{8r^3}[\sin\theta(2x\cos\varphi + d\sin\varphi) + l\cos\theta]^2.$$

Integrating this in the region from $x = -d/2$ to $x = d/2$, we have

$$-\frac{i_x}{r}\left\{d + \frac{d^2}{2r}\sin\theta\sin\varphi + \frac{1}{2r}ld\cos\theta - \frac{d}{2r^2}\left(\frac{d^2}{3} + \frac{l^2}{4}\right)\right.$$

$$\left. + \frac{d}{8r^2}\left[\frac{d^2}{3}\sin^2\theta(1 + 2\sin^2\varphi) + l^2\cos^2\theta + 2dl\sin\theta\cos\theta\sin\varphi\right]\right\}.$$

A similar calculation for the contribution from $(-d/2, -d/2, l/2)$ to $(d/2, -d/2, l/2)$ leads to

$$\frac{i_x}{r}\left\{d - \frac{d^2}{2r}\sin\theta\sin\varphi + \frac{1}{2r}ld\cos\theta - \frac{d}{2r^2}\left(\frac{d^2}{3} + \frac{l^2}{4}\right)\right.$$

$$\left. + \frac{d}{8r^2}\left[\frac{d^2}{3}\sin^2\theta(1 + 2\sin^2\varphi) + l^2\cos^2\theta - 2dl\sin\theta\cos\theta\sin\varphi\right]\right\}.$$

From $(d/2, -d/2, l/2)$ to $(d/2, d/2, l/2)$, we have

$$\frac{i_y}{r}\left\{d + \frac{d^2}{2r}\sin\theta\cos\varphi + \frac{ld}{2r}\cos\theta - \frac{d}{2r^2}\left(\frac{d^2}{3} + \frac{l^2}{4}\right)\right.$$

$$\left. + \frac{d}{8r^2}\left[\frac{d^2}{3}\sin^2\theta(3 - 2\sin^2\varphi) + l^2\cos^2\theta + 2dl\sin\theta\cos\theta\cos\varphi\right]\right\}$$

and from $(-d/2, d/2, l/2)$ to $(-d/2, -d/2, l/2)$, we have

$$-\frac{i_y}{r}\left\{d - \frac{d^2}{2r}\sin\theta\cos\varphi + \frac{ld}{2r}\cos\theta - \frac{d}{2r^2}\left(\frac{d^2}{3} + \frac{l^2}{4}\right)\right.$$

$$\left. + \frac{d}{8r^2}\left[\frac{d^2}{3}\sin^2\theta(3 - 2\sin^2\varphi) + l^2\cos^2\theta - 2dl\sin\theta\cos\theta\cos\varphi\right]\right\}.$$

Thus, the vector potential produced by the upper closed current is determined to be

$$A_+ = -\frac{\mu_0 I d^2}{4\pi r^2} \sin\theta \left[i_x \sin\varphi \left(1 + \frac{l}{2r}\cos\theta\right) - i_y \cos\varphi \left(1 + \frac{l}{2r}\cos\theta\right) \right].$$

The vector potential due to the lower closed current at $z = -l/2$ is similarly obtained as

$$A_- = \frac{\mu_0 I d^2}{4\pi r^2} \sin\theta \left[i_x \sin\varphi \left(1 - \frac{l}{2r}\cos\theta\right) - i_y \cos\varphi \left(1 - \frac{l}{2r}\cos\theta\right) \right].$$

Then, the vector potential of the linear magnetic quadrupole is given by

$$A = A_+ + A_- = -\frac{\mu_0 I d^2 l}{4\pi r^3} \sin\theta \cos\theta \left(i_x \sin\varphi - i_y \cos\varphi\right) = \frac{\mu_0 I (m \times r) \cos\theta}{4\pi r^4},$$

where $m = I d^2 i_z$ is the magnetic moment of the closed current at $z = l/2$.

6.31. The vector potential of the linear magnetic quadrupole is given by

$$A = \frac{\mu_0 I(m \times r) \cos\theta}{4\pi r^4},$$

which is written as

$$A = \frac{\mu_0 I m}{4\pi r^3} i_\varphi \sin\theta \cos\theta.$$

So, we have

$$\nabla \times A = \frac{\mu_0 I m}{4\pi r^4}\left[i_r(2 - 3\sin^2\theta) + i_\theta 2\sin\theta\cos\theta\right]$$

and

$$\nabla \times \nabla \times A = \frac{\mu_0 I m}{4\pi} i_\varphi \left[-\frac{1}{r^5} \cdot \frac{\partial}{\partial\theta}(2 - 3\sin^2\theta) + \frac{1}{r}\cdot\frac{\partial}{\partial r}\left(\frac{2}{r^3}\sin\theta\cos\theta\right)\right]$$

$$= \frac{\mu_0 I m}{4\pi r^5} i_\varphi (6\sin\theta\cos\theta - 6\sin\theta\cos\theta) = 0.$$

Thus, Laplace's equation is satisfied.

6.32. The magnetic flux density produced by the current I is

$$B_\varphi = \frac{\mu_0 I}{2\pi R},$$

6.9 Exercises

where R is the distance from the current. The magnetic potential is denoted by ϕ_m, and the relation, $B_\varphi = -\nabla \phi_m$, leads to

$$\frac{1}{R} \cdot \frac{\partial \phi_m}{\partial \varphi} = -\frac{\mu_0 I}{2\pi R}.$$

Thus, we have

$$\phi_m = -\frac{\mu_0 I}{2\pi} \varphi.$$

The magnetic potential is a multi-valued function of the azimuthal angle.

6.33. The magnetic flux density has an azimuthal component:

$$B_\varphi = \frac{\mu_0 I}{2\pi R}; \quad R > a,$$
$$= \frac{\mu_0 I R}{2\pi a^2}; \quad 0 \leq R < a.$$

Equation (6.47) is described as

$$\frac{1}{R} \cdot \frac{\partial \phi_m}{\partial \varphi} = -\frac{\mu_0 I}{2\pi R},$$

and we have

$$\phi_m = -\frac{\mu_0 I}{2\pi} \varphi; \quad R > a,$$
$$= -\frac{\mu_0 I R^2}{2\pi a^2} \varphi; \quad 0 \leq R < a.$$

Thus, the magnetic potential is a multivalued function. In addition, it predicts a radial component of the magnetic flux density inside the cylinder ($0 \leq R < a$). So, the magnetic potential cannot be used in the region in which current flows.

On the other hand, the vector potential is given by

$$A_z = \frac{\mu_0 I}{2\pi} \log \frac{R_0}{R}; \quad R > a,$$
$$= \frac{\mu_0 I}{4\pi a^2}(a^2 - R^2) + \frac{\mu_0 I}{2\pi} \log \frac{R_0}{R}; \quad 0 \leq R < a,$$

where R_0 is the distance to the reference point. The vector potential describes correctly the magnetic flux density and is applicable to any region.

Coffee Break: Does the Lorentz Force Do Mechanical Work?

It is stated in textbooks on electromagnetism that the Lorentz force does not do mechanical work, since it only changes the direction of motion of an electric charge in a magnetic field but does not influence the kinetic energy. On the other hand, when a current is applied to a conductor in a magnetic field, it can happen that the conductor is driven by the Lorentz force. In this case the mechanical work is done. How can we understand this phenomenon?

Suppose that there is an electron travelling in a conductor. The electron tends to move in the normal direction to the propagation direction in the conductor due to the Lorentz force. In practice, however, the electron moves along the conductor. This is caused by the Coulomb force of cations (and if the Hall effect is appreciable, the influence of the distributed Hall electrons is also included). That is, the transverse movement of the traveling electron is strongly suppressed by the cations. This means that a reaction, which is also the Coulomb force, is exerted on the cations. Hence, this reaction pushes the conductor in the direction of the Lorentz force. Thus, a Coulomb force of the same strength as the Lorentz force is exerted on the conductor. Hence, the statement that the Lorentz force moves the conductor is not correct.

In fact, the Coulomb force moves the conductor, and it is false to think that the Lorentz force moves it. The Coulomb force is not responsible for this, however, but the origin of this movement is electric power source. An electromotive force is induced when the conductor is forced to move, and the electromotive force works to stop the motion of the conductor. That is, if the work done by the electromotive force is realized, it cancels out the mechanical work that causes a loss energy. Nevertheless, the electric power source supplies additional energy to the circuit to keep the current constant. This causes the loss energy.

Someone is walking while revolving a ball by pulling a string connected to the ball. Assume that the ball approaches an object and the object falls down due to the wind produced by the moving ball and is damaged. The force that destroyed the object is the wind pressure, but it is not the centripetal force, i.e., the line tension of the string. The responsibility for the destruction lies with the person who revolved the ball and added a force to prevent the ball from slowing down by the reaction of the wind pressure. If he would not have added the force, the speed of the ball might have been reduced and the object might not have fallen down.

Chapter 7
Superconductors

7.1 Magnetic Properties of Superconductors

A superconductor is a material that loses its electric resistance when cooled below a characteristic temperature called the critical temperature. A superconductor also has perfect diamagnetism. That is, when a magnetic flux density is applied to a superconductor, the interior magnetic flux density is zero:

$$\boldsymbol{B} = 0. \tag{7.1}$$

This state is called the Meissner state. Since Eq. (7.1) holds inside the superconductor, from Eq. (6.16), we have

$$\boldsymbol{i} = 0. \tag{7.2}$$

Hence, current flows only on the surface of the superconductor. Equation (6.18) generally gives

$$\boldsymbol{A} = \nabla \alpha \tag{7.3}$$

with α denoting a scalar function. It is no problem, however, to assume as

$$\boldsymbol{A} = \text{const.} \tag{7.4}$$

in most cases. This is a special case of Eq. (7.3). If the superconducting region is not simply connected and a magnetic flux penetrates a space surrounded by the superconductor, the vector potential in the superconductor is not a constant.

The magnetic flux density is parallel to the superconductor surface, and there is a relationship between the magnetic flux density and the surface current density τ:

$$B = \mu_0 \tau. \qquad (7.5)$$

Example 7.1 Suppose that current I is applied to a long cylindrical superconductor with radius a. Determine the magnetic flux density and vector potential inside and outside the superconductor.

Solution 7.1 We define cylindrical coordinates with the z-axis on the central axis of the superconductor. The current flows uniformly only on the superconductor surface and the magnetic flux density does not appear inside the superconductor. Thus, the surface current density is $\tau = I/(2\pi a)$.

Suppose a circle, C, of radius R from the central axis on a plane normal to the axis (see Fig. 7.1). We apply Ampere's law to C. Since the current distribution is cylindrically symmetric, we can assume the magnetic flux density also has cylindrical symmetry. Hence, the magnetic flux density is parallel to C and its magnitude B is constant on C. Thus, the left side of Eq. (6.15) is $2\pi R B(R)$. All the current I flows through C and the right side of Eq. (6.15) is $\mu_0 I$ for $R > a$. We obtain the magnetic flux density as

$$B(R) = \frac{\mu_0 I}{2\pi R}; \quad R > a.$$

For $R < a$, the right side of Eq. (6.15) is zero. This gives

$$B(R) = 0; \quad 0 \leq R < a.$$

Thus, Eq. (7.1) is fulfilled inside the superconductor.

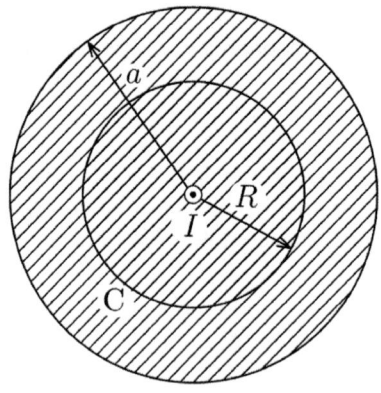

Fig. 7.1 Cross-section of cylindrical superconductor and closed circle C (for $R < a$)

7.1 Magnetic Properties of Superconductors

Now we determine the vector potential using the above results. From Eq. (6.20), we find that the vector potential has only the z-component, A_z, which is given by

$$A_z(R) = -\int B(R)dR.$$

We choose the reference point for zero vector potential at distance $R_0 (> a)$ from the central axis. Then, the vector potential is determined to be

$$A_z(R) = \frac{\mu_0 I}{2\pi} \log \frac{R_0}{R}; \quad R > a,$$

$$= \frac{\mu_0 I}{2\pi} \log \frac{R_0}{a}; \quad 0 \leq R < a.$$

◆

Example 7.2 Determine the magnetic flux density and vector potential when we apply currents I_1 and I_2 to the inner and outer superconductors, respectively, for the coaxial transmission line in Fig. 7.2.

Fig. 7.2 Long superconducting coaxial transmission line

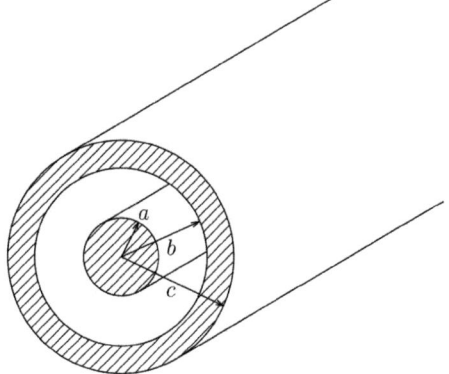

Solution 7.2 Current I_1 flows uniformly on the surface of the inner superconductor ($R = a$) and the induced current $-I_1$ flows uniformly on the inner surface of the outer superconductor ($R = b$). Current $I_1 + I_2$ flows on the outer surface of the outer superconductor ($R = c$), following the conservation law of current. The resultant magnetic flux density has the azimuthal component and its value is

$$B_\varphi = 0; \qquad 0 \le R < a,$$
$$= \frac{\mu_0 I_1}{2\pi R}; \qquad a < R < b,$$
$$= 0; \qquad b < R < c,$$
$$= \frac{\mu_0(I_1 + I_2)}{2\pi R}; \qquad R > c.$$

The vector potential has the z-component and its value is

$$A_z = \frac{\mu_0(I_1 + I_2)}{2\pi} \log \frac{R_0}{R}; \qquad R > c,$$
$$= \frac{\mu_0(I_1 + I_2)}{2\pi} \log \frac{R_0}{c}; \qquad b < R < c,$$
$$= \frac{\mu_0 I_1}{2\pi} \log \frac{bR_0}{cR} + \frac{\mu_0 I_2}{2\pi} \log \frac{R_0}{a}; \qquad a < R < b,$$
$$= \frac{\mu_0 I_1}{2\pi} \log \frac{bR_0}{ac} + \frac{\mu_0 I_2}{2\pi} \log \frac{R_0}{a}; \qquad 0 \le R < a,$$

where $R = R_0(> c)$ is the position of the reference point.

◆

Example 7.3 Two wide slab superconductors are parallel to each other, as shown in Fig. 7.3, and currents I and $-I$ are applied along the y-axis to the left and right superconductors, respectively. The length along the z-axis of each slab is l. Determine

Fig. 7.3 Two parallel slab superconductors

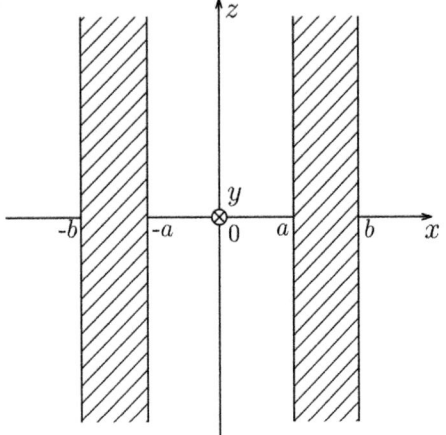

the current that appears on each superconductor surface, the magnetic flux density, and the vector potential inside and outside the superconductors.

Solution 7.3 We denote the current flowing on the surface at $x = b$ by I_b. Then, the current at $x = a$ is $-I - I_b$. So that the magnetic flux density does not penetrate the right superconductor, the current at $x = b$ (I_b) must be the same as the total current in the region $x \leq a$, i.e., $I - I - I_b = -I_b$. Thus, we have $I_b = 0$ and the current at $x = a$ is $-I$. The total current in the region of $-b < x < b$ must be zero from Eq. (6.15) so that the magnetic flux density is zero in the two superconductors. Hence, the current at $x = -a$ is I, and that at $x = -b$ is 0. Thus, the magnetic flux density along the z-axis is

$$B = 0; \quad x < -a,$$
$$= -\frac{\mu_0 I}{l}; \quad -a < x < a,$$
$$= 0; \quad x > a.$$

The vector potential has a y-component, A_y, and is given by

$$A_y = \frac{\mu_0 I a}{l}; \quad x < -a,$$
$$= -\frac{\mu_0 I x}{l}; \quad -a < x < a,$$
$$= -\frac{\mu_0 I a}{l}; \quad x > a.$$

◆

7.2 Special Solution Method for Magnetic Flux Density

Suppose that a straight current, I, flows at a distance a from a flat infinite superconductor surface, as shown in Fig. 7.4a. A current in the opposite direction appears on the superconductor surface to shield the superconductor, and it exerts an attractive force on I. The magnetic condition in vacuum can be determined using the method of images. The x-y plane is defined on the superconductor surface. Suppose that current I flows on the line at $x = 0$ and $z = a$ along the positive y-axis. The magnetic flux density on the surface ($z = 0$) must be parallel to the superconductor. We can realize this situation by virtually removing the superconductor and then applying a straight current with the same magnitude and opposite direction at the symmetric position ($x = 0$, $z = -a$) with respect to the superconductor surface. This will be confirmed below. The vector potential in the vacuum region ($z > 0$) produced by the two straight currents has only its y-component parallel to the currents, and we

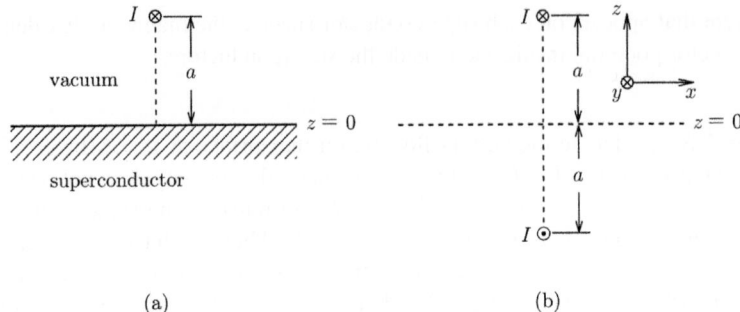

Fig. 7.4 **a** Thin straight current I placed at distance a from the surface of a wide, flat superconductor, and **b** imaginary current placed at a mirror position with respect to the superconductor surface

calculate it as

$$A_y(x, y, z) = \frac{\mu_0 I}{2\pi} \left\{ \log \frac{R_0}{[x^2 + (z-a)^2]^{1/2}} - \log \frac{R_0}{[x^2 + (z+a)^2]^{1/2}} \right\}$$

$$= \frac{\mu_0 I}{4\pi} \log \frac{x^2 + (z+a)^2}{x^2 + (z-a)^2}, \tag{7.6}$$

where R_0 is the distance from the current to the reference point. We can easily show that $A_y = 0$ on the superconductor surface ($z = 0$). Thus, the condition, Eq. (7.4), is satisfied. Hence, we conclude that Eq. (7.6) is the solution for the vector potential. The vector potential inside the superconductor ($z < 0$) is $A_y = 0$. Thus, the method of images is useful also for superconductors. The imaginary current placed at the mirror position is called an image current.

Then, the magnetic flux density in the vacuum region is

$$B_x = -\frac{\mu_0 I}{2\pi} \left[\frac{z+a}{x^2 + (z+a)^2} - \frac{z-a}{x^2 + (z-a)^2} \right],$$

$$B_y = 0,$$

$$B_z = -\frac{\mu_0 I x}{2\pi} \left[\frac{1}{x^2 + (z+a)^2} - \frac{1}{x^2 + (z-a)^2} \right]. \tag{7.7}$$

On the superconductor surface, it reduces to

$$B_x(x, y, 0) = -\frac{\mu_0 I a}{\pi (x^2 + a^2)}, \quad B_y(x, y, 0) = B_z(x, y, 0) = 0, \tag{7.8}$$

showing that the magnetic flux density is parallel to the surface. Figure 7.5 shows the magnetic flux lines. Then, from Eq. (7.8) we determine the density of the current induced on the superconductor surface to be

7.2 Special Solution Method for Magnetic Flux Density

Fig. 7.5 Magnetic flux lines produced by the given current and the current induced on the superconductor surface

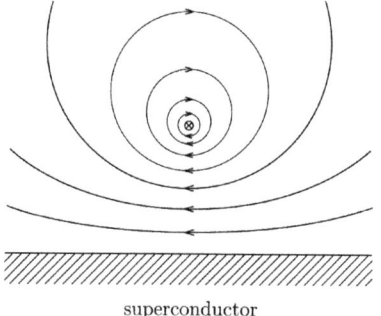

$$\tau = -\frac{Ia}{\pi\left(x^2 + a^2\right)}. \tag{7.9}$$

Example 7.4 Straight current I is placed at distances a and b from two flat superconductor surfaces that are perpendicular to each other, as shown in Fig. 7.6. Determine the vector potential and magnetic flux density in the vacuum.

Fig. 7.6 Two perpendicular flat superconductor surfaces and straight current I

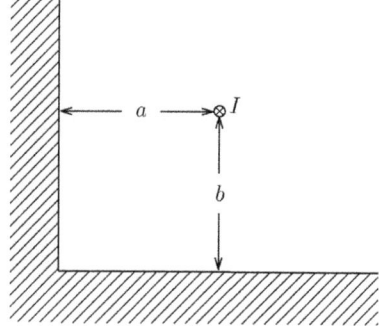

Solution 7.4 We denote the two superconductor surfaces that are perpendicular to each other by the x-y and y-z planes, as shown in Fig. 7.7. Suppose that the given current I is located at (a, b) on the x-z plane. We virtually remove the superconductor and place three image currents, $-I$, $-I$, and I, at $(a, -b)$, $(-a, b)$, and $(-a, -b)$, respectively. Then, the vector potential in the vacuum region $(x > 0, z > 0)$ is

$$A_y(x, z) = \frac{\mu_0 I}{4\pi} \log \frac{[(x-a)^2 + (z+b)^2][(x+a)^2 + (z-b)^2]}{[(x-a)^2 + (z-b)^2][(x+a)^2 + (z+b)^2]},$$

Fig. 7.7 Current I and three image currents

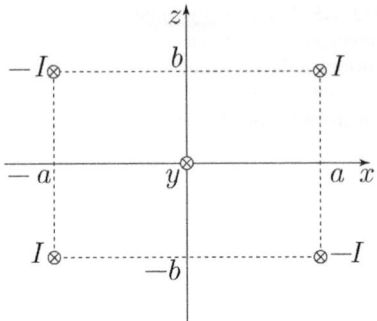

which satisfies $A_y = 0$ at infinity. This fulfills $A_y = 0$ on the surfaces $x = 0$ and $z = 0$, and hence, this gives the correct vector potential. We determine the current densities on the x-y and y-z planes to be

$$\tau(x, y, 0) = -\frac{1}{\mu_0}\left(\frac{\partial A_y}{\partial z}\right)_{z=0} = -\frac{4Iabx}{\pi\left[(x-a)^2 + b^2\right]\left[(x+a)^2 + b^2\right]},$$

$$\tau(0, y, z) = \frac{1}{\mu_0}\left(\frac{\partial A_y}{\partial x}\right)_{x=0} = -\frac{4Iabz}{\pi\left[(z-b)^2 + a^2\right]\left[(z+b)^2 + a^2\right]}.$$

◆

Suppose that we apply current I through a line A, separated by distance d from the central axis O of a grounded parallel long superconducting cylinder of radius a, as shown in Fig. 7.8a. Now we determine the vector potential outside the superconductor. We virtually remove the superconductor and apply an image current I' through a line B, separated by h from the central axis of the superconducting cylinder, as shown in Fig. 7.8b. The vector potential at point P on the superconductor surface is given by

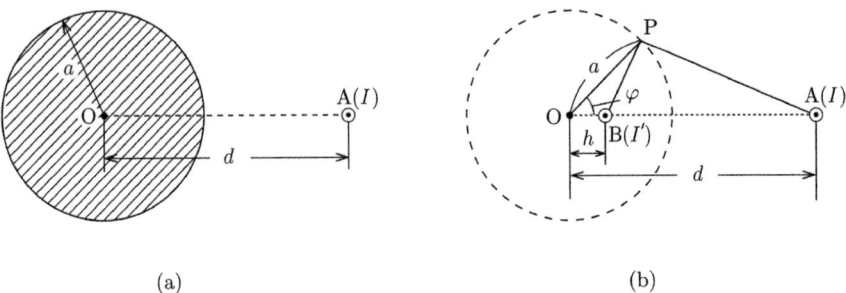

(a)   (b)

Fig. 7.8 a Long superconducting cylinder and straight current parallel to it, and **b** image current placed on line B after removing the superconductor

7.2 Special Solution Method for Magnetic Flux Density

$$A_z = \frac{\mu_0 I}{2\pi} \log \frac{R_0}{(a^2 + d^2 - 2ad\cos\varphi)^{1/2}} + \frac{\mu_0 I'}{2\pi} \log \frac{R'_0}{(a^2 + h^2 - 2ah\cos\varphi)^{1/2}}, \quad (7.10)$$

where φ is the angle POA. In the above, R_0 and R'_0 are the distances to a suitable reference point and are not important quantities. For the vector potential to be constant and independent of φ, the following conditions should be satisfied:

$$I' = -I, \quad (7.11)$$

$$h = \frac{a^2}{d}. \quad (7.12)$$

In this case, the current-carrying wire and superconductor are infinitely long and hence, the current induced in the superconductor has the same magnitude.

The above results give the vector potential at point (R, φ),

$$A_z(R, \varphi) = \frac{\mu_0 I}{2\pi} \log \frac{d\left[R^2 + (a^2/d)^2 - 2(a^2 R/d)\cos\varphi\right]^{1/2}}{a(R^2 + d^2 - 2Rd\cos\varphi)^{1/2}}, \quad (7.13)$$

which reduces to zero on the superconductor surface ($R = a$). That is, $R'_0 = (a/d)R_0$. We can calculate the magnetic flux density outside the superconductor, and Fig. 7.9 shows the magnetic flux lines. The current density on the superconductor surface is obtained as

$$\tau = \frac{1}{\mu_0} B_\varphi(R = a) = -\frac{1}{\mu_0}\left(\frac{\partial A_z}{\partial R}\right)_{R=a} = -\frac{(d^2 - a^2)I}{2\pi a(a^2 + d^2 - 2ad\cos\varphi)}. \quad (7.14)$$

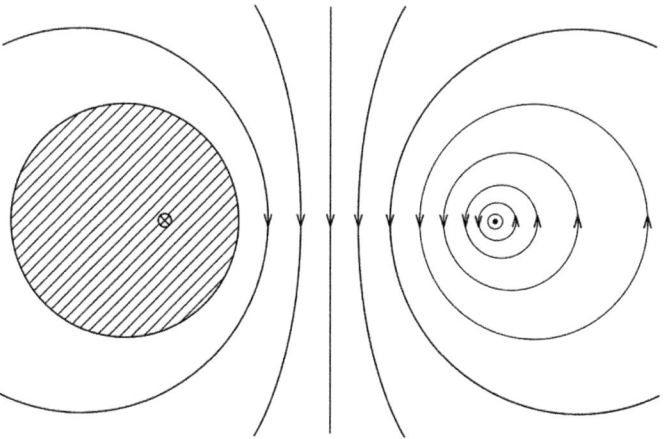

Fig. 7.9 Magnetic flux lines produced by the long superconducting cylinder and the straight current parallel to it

Example 7.5 A long superconducting cylinder with radius a is placed at distance $l(> a)$ from an infinite flat superconductor surface, as shown in Fig. 7.10, and a current I is applied to the superconducting cylinder. Determine the current density on the surfaces of the two superconductors.

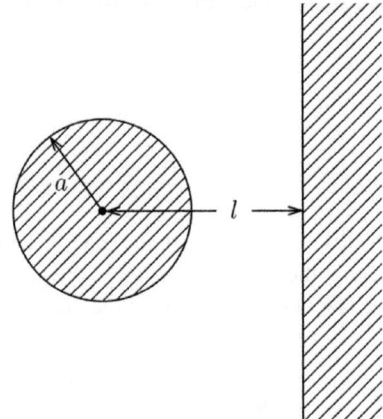

Fig. 7.10 Superconducting cylinder parallel to an infinite flat superconductor surface

Solution 7.5 We assume that the magnetic flux density in the vacuum region is the same as that when we place an image current I in the superconducting cylinder at distance h from the center of the cylinder and an image current $-I$ in the infinite superconductor at distance $l - h$ from its surface after virtually removing the two superconductors (see Fig. 7.11). In this case, the boundary condition on the infinite superconductor surface is satisfied. If the distance $2l - h$ between the image current

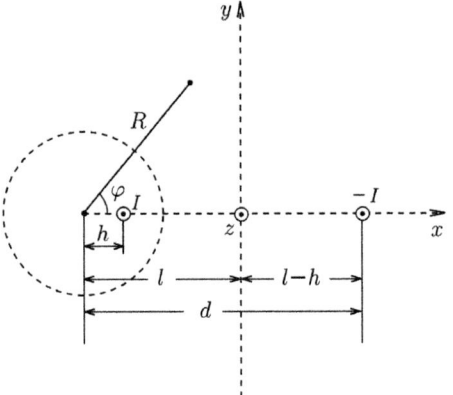

Fig. 7.11 Image currents placed in the two superconductors

7.2 Special Solution Method for Magnetic Flux Density

$-I$ and the cylinder center corresponds to d in Fig. 7.8b, the boundary condition on the cylinder surface is also satisfied. From the above relationship and Eq. (7.12), we obtain

$$d = l + \sqrt{l^2 - a^2}, \quad h = l - \sqrt{l^2 - a^2}.$$

Substituting these into Eq. (7.13) yields the vector potential outside the superconductors:

$$A_z(R, \varphi) = \frac{\mu_0 I}{4\pi} \log \frac{R^2 + \left(l + \sqrt{l^2 - a^2}\right)^2 - 2R\left(l + \sqrt{l^2 - a^2}\right)\cos\varphi}{R^2 + \left(l - \sqrt{l^2 - a^2}\right)^2 - 2R\left(l - \sqrt{l^2 - a^2}\right)\cos\varphi}.$$

We find that the vector potential on the surface, $A_z(a, \varphi) = [\mu_0 I/(2\pi)] \log\left[\left(l + \sqrt{l^2 - a^2}\right)/a\right]$, is constant. The current density on the cylinder surface is.

$$\tau = -\frac{1}{\mu_0}\left(\frac{\partial A_z}{\partial R}\right)_{R=a} = \frac{\sqrt{l^2 - a^2}\, I}{2\pi a(l - a\cos\varphi)}.$$

Next, we define Cartesian coordinates with the y-z plane ($x = 0$) on the infinite superconductor surface and the central axis of the cylindrical superconductor at $y = 0$. From the relationships $R\cos\varphi = x + l$ and $R\sin\varphi = y$, the vector potential is also expressed as

$$A_z(x, y) = \frac{\mu_0 I}{4\pi} \log \frac{\left(x - \sqrt{l^2 - a^2}\right)^2 + y^2}{\left(x + \sqrt{l^2 - a^2}\right)^2 + y^2}$$

for $x \leq 0$. Thus, we can easily confirm that $A_z(x = 0) = 0$ is satisfied. The density of the current (along the z-axis) on the infinite superconductor surface, which is equal to $-B_y(x=0)/\mu_0$, is

$$\tau = \frac{1}{\mu_0}\left(\frac{\partial A_z}{\partial x}\right)_{x=0} = -\frac{\sqrt{l^2 - a^2}\, I}{\pi(y^2 + l^2 - a^2)}.$$

◆

Example 7.6 Straight current I is placed at distance h from the central axis, O, of a hollow cylindrical superconductor, as shown in Fig. 7.12. Determine the vector potential in the vacuum and the current density on the inner surface of the superconductor.

Fig. 7.12 Hollow cylindrical superconductor and straight current inside the hollow space

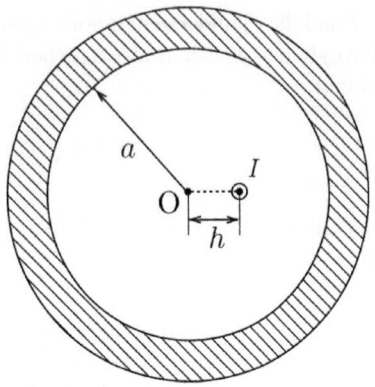

Solution 7.6 We virtually remove the superconductor and place an image current I' parallel to the current I on a plane including the central axis and the current I (see Fig. 7.13). We denote the distance between the central axis and the image current by d. The vector potential at point P on the inner surface of the superconductor is

$$A_z(a, \varphi) = \frac{\mu_0 I'}{2\pi} \log \frac{R'_0}{(a^2 + d^2 - 2ad \cos \varphi)^{1/2}} + \frac{\mu_0 I}{2\pi} \log \frac{R_0}{(a^2 + h^2 - 2ah \cos \varphi)^{1/2}}.$$

The conditions that satisfy $A_z(a, \varphi) = $ const. give

$$I' = -I, \quad d = \frac{a^2}{h}.$$

Since the total current is zero ($I + I' = 0$), we can choose infinity as the reference point of the vector potential, and we have $R_0 = R'_0$. The vector potential in the hollow interior is

$$A_z(R, \varphi) = \frac{\mu_0 I}{4\pi} \log \frac{R^2 + (a^2/h)^2 - 2(a^2 R/h) \cos \varphi}{R^2 + h^2 - 2Rh \cos \varphi}.$$

Fig. 7.13 Image current I'

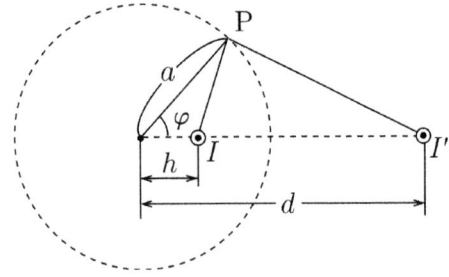

The current density on the inner surface is

$$\tau = \frac{1}{\mu_0} B_\varphi(R=a) = -\frac{1}{\mu_0}\left(\frac{\partial A_z}{\partial R}\right)_{R=a} = -\frac{(a^2 - h^2)I}{2\pi a(a^2 + h^2 - 2ah\cos\varphi)}.$$

◆

7.3 The Meissner State

Suppose that a superconducting sphere of radius a is placed in a uniform magnetic flux density, B_0 (see Fig. 7.14). The superconducting current flows on the surface to cancel the applied magnetic flux density inside the superconductor. Here, we determine the density of this current and the magnetic flux density outside the superconductor. We use spherical coordinates: The origin is at the center of the superconductor, and the z-axis is parallel to the direction of the applied magnetic flux density.

The effect of the surface current can be realized by a magnetic dipole placed at the center of the superconductor after virtually removing the superconductor. Then, the vector potential outside the superconductor has only the azimuthal component, A_φ, corresponding to the current. This component consists of $A_{f\varphi}$ caused by the uniform magnetic flux density B_0 and $A_{d\varphi}$ caused by the magnetic moment m. The former component is given by

$$A_{f\varphi} = \frac{B_0 r}{2}\sin\theta, \tag{7.15}$$

and the latter component is given by Eq. (6.27),

$$A_{d\varphi} = \frac{\mu_0 m}{4\pi r^2}\sin\theta. \tag{7.16}$$

Fig. 7.14 Superconducting sphere in a uniform magnetic flux density

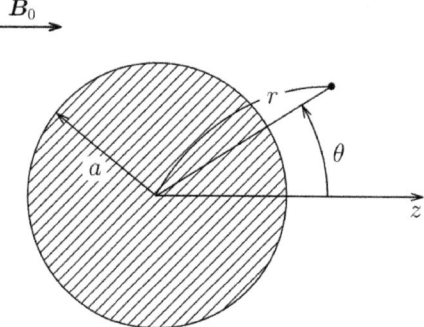

Thus, the vector potential is

$$A_\varphi = A_{f\varphi} + A_{d\varphi} = \left(\frac{B_0 r}{2} + \frac{\mu_0 m}{4\pi r^2}\right)\sin\theta. \tag{7.17}$$

The requirement that the vector potential is zero on the superconductor surface ($r = a$) gives

$$m = -\frac{2\pi a^3 B_0}{\mu_0}. \tag{7.18}$$

Hence, we determine the vector potential to be

$$A_\varphi = \frac{B_0}{2}\left(r - \frac{a^3}{r^2}\right)\sin\theta. \tag{7.19}$$

Inside the superconductor, the solution is $A_\varphi = 0$.

We obtain the magnetic flux density outside the superconductor as

$$B_r = \frac{1}{r\sin\theta}\cdot\frac{\partial}{\partial\theta}(\sin\theta A_\varphi) = B_0\left(1 - \frac{a^3}{r^3}\right)\cos\theta,$$

$$B_\theta = -\frac{1}{r}\cdot\frac{\partial}{\partial r}(rA_\varphi) = -B_0\left(1 + \frac{a^3}{2r^3}\right)\sin\theta,$$

$$B_\varphi = 0. \tag{7.20}$$

The current density in the azimuthal direction on the superconductor surface is

$$\tau = \frac{1}{\mu_0}B_\theta(r = a) = -\frac{3B_0}{2\mu_0}\sin\theta. \tag{7.21}$$

Example 7.7 An infinitely long superconducting cylinder of radius a is in a uniform perpendicular magnetic flux density of B_0. Determine the vector potential and magnetic flux density outside the superconductor and the current density on the superconductor surface.

Solution 7.7 We use cylindrical coordinates with the z-axis at the central axis of the superconductor and the azimuthal angle measured from the direction of the applied magnetic flux density. We virtually remove the superconductor and place a magnetic dipole line produced by a pair of anti-parallel straight currents at the axis. We denote the magnitude of the magnetic moment in a unit length along the z-axis by $\hat{m}$. From Eq. (6.30), the vector potential is given by

7.3 The Meissner State

$$A_z(R, \varphi) = \left(B_0 R - \frac{\mu_0 \hat{m}}{2\pi R}\right) \sin \varphi,$$

where the first and second terms are components of the applied magnetic flux density and magnetic moment, respectively. The requirement $A_z(R = a) = 0$ gives

$$\hat{m} = -\frac{2\pi a^2 B_0}{\mu_0}.$$

Thus, the vector potential outside the superconductor is

$$A_z(R, \varphi) = B_0\left(R - \frac{a^2}{R}\right) \sin \varphi.$$

We obtain the magnetic flux density as

$$B_R = \frac{1}{R} \cdot \frac{\partial A_z}{\partial \varphi} = B_0\left(1 - \frac{a^2}{R^2}\right) \cos \varphi,$$

$$B_\varphi = -\frac{\partial A_z}{\partial R} = -B_0\left(1 + \frac{a^2}{R^2}\right) \sin \varphi,$$

$$B_z = 0.$$

The surface current density is

$$\tau = \frac{1}{\mu_0} B_\varphi(R = a) = -\frac{2B_0}{\mu_0} \sin \varphi.$$

◆

Example 7.8 Derive the magnetic flux density outside the superconductor in Example 7.7 using the magnetic potential.

Solution 7.8 The value of the magnetic moment in a unit length on the central axis is denoted by $\hat{m}$. Then, the magnetic potential is given by

$$\phi_c = \frac{\mu_0 \hat{m}}{2\pi R} \cos \varphi$$

(see Example 6.9). The magnetic potential of the applied uniform magnetic flux density B_0 is given by

$$\phi_f = -B_0 R \cos \varphi.$$

Thus, the magnetic potential in the space outside the superconductor is

$$\phi_m = \phi_c + \phi_f = -\left(B_0 R - \frac{\mu_0 \hat{m}}{2\pi R}\right) \cos \varphi.$$

Since the magnetic flux density is normal to the superconductor surface, the following condition must be satisfied:

$$-\left(\frac{\partial \phi_m}{\partial R}\right)_{R=a} = -\left(B_0 + \frac{\mu_0 \hat{m}}{2\pi a^2}\right) \cos \varphi = 0.$$

Thus, we have $\hat{m} = -2\pi a^2 B_0 / \mu_0$, and the magnetic potential is

$$\phi_m = -B_0\left(R + \frac{a^2}{R}\right) \cos \varphi.$$

The magnetic flux density outside the superconductor is

$$B_R = -\frac{\partial \phi_m}{\partial R} = B_0\left(1 - \frac{a^2}{R^2}\right) \cos \varphi,$$

$$B_\varphi = -\frac{1}{R} \cdot \frac{\partial \phi_m}{\partial \varphi} = -B_0\left(1 + \frac{a^2}{R^2}\right) \sin \varphi,$$

$$B_z = -\frac{\partial \phi_m}{\partial z} = 0.$$

Thus, the same result is obtained.

◆

7.4 Exercises

7.1. Prove that Eq. (7.5) is fulfilled on the superconductor surfaces in Example 7.2.

7.2. Currents I and I_x are given to the left and right slab superconductors shown in Fig. 7.3 in Example 7.3. The magnetic flux density along the z-axis is $\mu_0 I / 2l$ in the region $-a < x < a$, where l is the length of each slab along the z-axis. Determine I_x, the current on each surface, and the magnetic flux density in each space.

7.3. Currents $2I$, $-I$, and I are applied along the y-axis to the left, middle, and right slab superconductors shown in Fig. E7.1. Determine the current on each surface, and the magnetic flux density in each space. The length of each slab along the z-axis is l.

7.4. Currents $2I$ and $-I$ are given to the left and right slab superconductors shown in Fig. E7.1 in Exercise 7.3. When we give some current to the middle slab superconductor, the magnetic flux density in the space $d < x < e$ is zero. Determine the current on each surface.

Fig. E7.1 Three slab superconductors

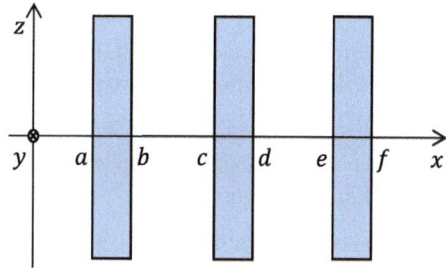

7.5. A long rectangular circuit made of a superconductor is placed near a straight line, as shown in Fig. E7.2. Current I is applied to the straight line, and the rectangular circuit is changed to the superconducting state by reducing the temperature below the critical temperature. Then, the current is reduced to zero. Determine the current induced in the rectangular circuit in the direction ABCD. The radius of the superconductor, δ, is assumed to be sufficiently small.

7.6. We assume two parallel superconducting slabs, as shown in Fig. 7.3 in Example 7.3, and apply currents I and $-I$ along the y-axis to the right and left superconductors, respectively. Derive the vector potential and the magnetic potential in this case and compare them. The size of the slabs along the z-axis is w.

7.7. The current density induced on the superconductor surface was determined using the image method in Sect. 7.2. Prove that the interior of the superconductor is completely shielded by the induced current given by Eq. (7.9).

7.8. Determine the force on the given current exerted by the current induced on the superconductor surfaces in Example 7.4.

7.9. The image method is employed in solving the problem of the two superconductor surfaces with a 90° angle between them in Example 7.4. Discuss the applicability of this method for other angles between the two surfaces.

7.10. The problem of a straight current placed in a long cylindrical hollow interior of a superconductor is treated, and the current induced on the inner surface is derived in Example 7.6. Determine the total current on the inner surface.

Fig. E7.2 Superconducting rectangular circuit and a straight current

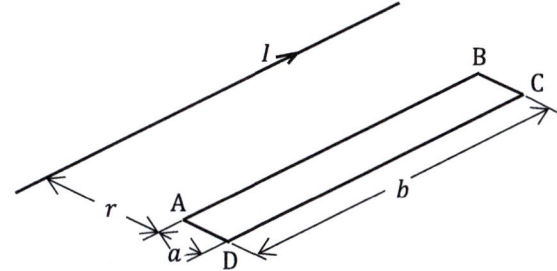

7.11. The problem of a long grounded cylindrical superconductor and a parallel straight current is treated in Sect. 7.2. When the superconductor is not grounded, determine the vector potential outside the superconductor.

7.12. The vector potential in the vacuum region is given by Eq. (7.6), when a straight current is put above an infinite superconductor surface. Prove that the vector potential satisfies Laplace's equation.

7.13. The vector potential in the vacuum region is given by Eq. (7.13), when a straight current is placed near a grounded cylindrical superconductor. Prove that the vector potential satisfies Laplace's equation.

7.14. The magnetic condition due to a straight current in a long cylindrical hollow interior of a superconductor is treated in Exercise 7.10. Prove that the vector potential satisfies Laplace's equation.

7.15. The magnetic condition for a long cylindrical superconductor with a current and a flat superconductor surface is treated in Example 7.5. Prove that the vector potential satisfies Laplace's equation.

7.16. The vector potential is given by Eq. (7.13) for the case of a grounded long superconducting cylinder and a straight current. Discuss the equivector potential surface.

7.17. The magnetic condition due to a straight current in a long cylindrical hollow interior of a superconductor is treated in Exercise 7.10. Discuss the equivector potential surface in the hollow region.

7.18. A magnetic phenomenon is treated in Example 7.5 for a long cylindrical superconductor with a current and a flat superconductor surface. Determine the equivector potential surface in the space between the cylindrical superconductor and the superconductor surface.

7.19. Determine the induced current density on the inner surface of a superconductor with a spherical hollow interior of radius a, when a magnetic moment m is placed at the center of the hollow space.

7.20. Determine the induced current density on the inner surface of a long superconductor with a cylindrical hollow interior of radius a, when a magnetic dipole line of moment $\hat{m}$ is placed at the central axis of the hollow space.

7.21. The vector potential is determined inside and outside of a spherical superconductor in a uniform magnetic flux density in Sect. 7.3. Prove that the obtained vector potential satisfies Laplace's equation.

7.22. The vector potential is determined inside and outside of a cylindrical superconductor in a uniform transverse magnetic flux density in Example 7.7. Prove that the obtained vector potential satisfies Laplace's equation.

7.4 Exercises

7.23. The magnetic condition in the spherical hollow interior of a superconductor is investigated when a magnetic moment m is placed at its center in Exercise 7.19. Prove that the vector potential satisfies Laplace's equation in the hollow space.

7.24. The magnetic condition is investigated in Exercise 7.20 when a magnetic dipole line of moment $\hat{m}$ in a unit length is placed at the central axis of the long cylindrical hollow interior of a superconductor. Prove that the vector potential satisfies Laplace's equation in the hollow space.

7.25. Determine the equivector potential surface for the cylindrical superconductor in a transverse magnetic flux density discussed in Example 7.7.

7.26. The magnetic condition is treated in Exercise 7.20 when a magnetic dipole line of moment $\hat{m}$ in a unit length is placed at the central axis of the long cylindrical hollow interior of a superconductor. Determine the equivector potential surface in the hollow space.

7.27. The magnetic condition around a cylindrical superconductor near a parallel straight current is discussed in Sect. 7.2. Solve the same problem using the boundary condition that the magnetic flux density is parallel to the superconductor surface.

7.28. The magnetic condition around a straight current within a cylindrical hollow interior of a superconductor is discussed in Example 7.6. Solve the same problem using the boundary condition that the magnetic flux density is parallel to the superconductor surface.

7.29. The Meissner state for a cylindrical superconductor in a normal magnetic flux density is discussed in Example 7.7. Solve the same problem using the boundary condition that the magnetic flux density is parallel to the superconductor surface.

7.30. Solve the problem of Example 7.5 with the boundary condition that the magnetic flux density must be parallel to the superconductor surface with Eq. (7.5).

7.31. Solve the problem of Exercise 7.19 with the boundary condition that the parallel component of the magnetic flux density on the superconductor surface satisfies Eq. (7.5).

7.32. Solve the problem of Exercise 7.20 with the boundary condition that the parallel component of the magnetic flux density on the superconductor surface satisfies Eq. (7.5).

7.33. The problem of determining the magnetic condition for a cylindrical superconductor and a parallel straight current is treated in Sect. 7.2 (see Fig. 7.8a) using the vector potential. Discuss if it can be solved using the magnetic potential.

7.34. Solve the problem of 7.19 using the magnetic potential.

7.35. Solve the problem of 7.20 using the magnetic potential.

Answers to Exercises

7.1. On the surface at $r = a$, the current density is $\tau_a = I_1/(2\pi a)$, and the magnetic flux density is $B_\varphi(a) = \mu_0 I_1/(2\pi a)$. Thus, the relationship of (7.5), $B_\varphi(a) = \mu_0 \tau_a$, holds. On the surface at $r = c$, the current density and magnetic flux density are $\tau_c = (I_1 + I_2)/(2\pi c)$ and $B_\varphi(c) = \mu_0(I_1 + I_2)/(2\pi c)$, respectively. So, the same relationship holds. On the surface at $r = b$, the current density is $\tau_b = -I_1/(2\pi b)$ and the magnetic flux density is $B(b) = \mu_0 I_1/(2\pi b)$. In this case, the normal vector on this surface is directed inward. Thus, Eq. (7.5) holds taking account of the geometry and the definition of curl in cylindrical coordinates.

7.2. From the value of the magnetic flux density in the region $-a < x < a$, the currents on the surfaces at $x = a$ and $x = -a$ are $I/2$ and $-I/2$, respectively. So, the currents on the surfaces at $x = b$ and $x = -b$ are $I_x - I/2$ and $3I/2$, respectively. From the requirement that the magnetic flux density is zero in the left superconductor, the current on the left surface of this superconductor must be equal to the total current in the region $x \geq -a$. This condition leads to $I_x = 2I$. So, the current at $x = b$ is $3I/2$.

The obtained magnetic flux density in each space is

$$B = \frac{3\mu_0 I}{2l}; \quad x < -b,$$

$$= \frac{\mu_0 I}{2l}; \quad -a < x < a,$$

$$= -\frac{3\mu_0 I}{2l}; \quad x > b.$$

7.3. We denote the currents on the surfaces at $x = a$, $x = b$, $x = c$, $x = d$, $x = e$, and $x = f$ by I_a, I_b, I_c, I_d, I_e, and I_f, respectively. Then, the condition that the magnetic flux density is zero in the left superconductor is $I_a = 2I - I_a$, which leads to $I_a = I$ and we have $I_b = I$. The condition for the middle superconductor is $I_c + 2I = -I_c$, which leads to $I_c = -I$ and we have $I_d = 0$. The condition for the right superconductor is $I_e + I = -I_e + I$, which leads to $I_e = 0$ and we have $I_f = I$.

The magnetic flux density along the z-axis in each space is determined to be

$$B = -\frac{\mu_0 I}{l}; \quad x < a,$$

$$= \frac{\mu_0 I}{l}; \quad b < x < c,$$

$$= 0; \quad d < x < e,$$

$$= \frac{\mu_0 I}{l}; \quad x > f.$$

7.4. The current flowing in the region $x \leq d$ and that flowing in the region $x \geq e$ must be the same. So, the current in the middle superconductor is determined to be $3I$. Using the method shown in Example 7.3 and Exercises 7.2 and 7.3, the current on each surface can be determined to be $-I(x = a)$, $3I(x = b)$, $-3I(x = c)$, $0(x = d)$, $0(x = e)$, and $-I(x = f)$.

7.5. The magnetic flux produced by the current penetrates the circuit downward. The magnetic flux is

$$\Phi = b \int_r^{r+a} \frac{\mu_0 I}{2\pi x} dx = \frac{\mu_0 b I}{2\pi} \log \frac{r+b}{r}.$$

When the circuit becomes superconducting by cooling down, the magnetic flux is expelled out of the superconductor. Since the superconductor is sufficiently thin, half of the trapped flux is expelled to the interior space. So, the above estimation of the trapped magnetic flux is reasonable. When the current is removed, the shielding current is induced in the circuit to keep this flux. The current induced along the direction ABCD is denoted as I_s. The magnetic flux produced by this current is

$$2b \int_\delta^{a-\delta} \frac{\mu_0 I_s}{2\pi x} dx = \frac{\mu_0 b I_s}{\pi} \log \frac{a-\delta}{\delta}.$$

This is equal to Φ, and the superconducting current is determined to be

$$I_s = 2 \frac{\log[(r+b)/r]}{\log[(a-\delta)/\delta]} I.$$

7.6. The current flows with the density I/w and $-I/w$ along the y-axis on the surfaces at $x = a$ and at $x = -a$, respectively. Then, the magnetic flux density appears only in the region $-a < x < a$ along the z-axis, and its value is $B = \mu_0 I/w$. The vector potential has a y-component and its distribution is

$$A_y = -\frac{\mu_0 I a}{w}; \quad x < -a,$$

$$= \frac{\mu_0 I x}{w}; \quad -a < x < a,$$

$$= \frac{\mu_0 I a}{w}; \quad x > a.$$

So, the equivector potential surfaces are parallel to the slab surfaces, and the superconductors are equivector potential.

On the other hand, if the magnetic potential is used, it is given by

$$\phi_m = -\frac{\mu_0 I z}{w}; \quad -a < x < a,$$

$$= 0; \quad \text{otherwise}.$$

The equipotential surfaces are normal to the z-axis and the superconductor surfaces are not equipotential. In addition, there is a gap in the value of the potential on the superconductor surfaces on which the current flows. Thus, the magnetic potential does not help in understanding the magnetic phenomena in this case.

7.7. The magnetic flux density produced by the given current and that produced by the surface current, Eq. (7.9), coexist inside the superconductor. The latter magnetic flux density is symmetric with that in the vacuum with respect to the superconductor surface ($z = 0$), and the magnetic flux density produced in the vacuum region is reproduced by the image current, $-I$. This means that the magnetic flux density produced inside the superconductor is reproduced by the current $-I$ placed at the position of the current I.

Hence, the magnetic flux density inside the superconductor is reproduced by the currents I and $-I$ placed at the same position, i.e., when no current is given. So, we can prove $B = 0$ inside the superconductor.

7.8. The force on the given current in a unit length due to the image current at $(x = -a, z = b)$ is

$$F'_{-a,b} = \frac{\mu_0 I^2}{8\pi a^2},$$

and it is directed to the right (the positive x-axis). The force on the given current in a unit length due to the image current at $(x = a, z = -b)$ is

$$F'_{a,-b} = \frac{\mu_0 I^2}{8\pi b^2},$$

and it is directed to upward (the positive z-axis). The force on the given current in a unit length due to the image current at $(x = -a, z = -b)$ is

$$F'_{-a,-b} = \frac{\mu_0 I^2}{8\pi (a^2 + b^2)},$$

and it is directed to the negative x- and z-axes along the direction connecting the given and image currents. So, the x- and z-axis components of the total force in a unit length are

$$F'_x = \frac{\mu_0 I^2}{8\pi}\left[\frac{1}{a^2} - \frac{a}{(a^2+b^2)^{3/2}}\right], \quad F'_z = \frac{\mu_0 I^2}{8\pi}\left[\frac{1}{b^2} - \frac{b}{(a^2+b^2)^{3/2}}\right].$$

7.4 Exercises

7.9. This method is useful only for special angles. This is the case in which the boundary conditions of the two surfaces are satisfied. If we say a conclusion, it is useful only when the angle between the two surfaces is π/n with n being an integer. That is, $n = 1$ in the problem in Sect. 7.2 and $n = 2$ in Example 7.4.

The case of $n = 3$ is shown in Fig. B7.1. The couple of the given current $I_1(I)$ and the image current $I_2(-I)$, that of $I_3(I)$ and $I_6(-I)$, and that of $I_4(-I)$ and $I_5(I)$ satisfy the boundary condition on the surface S_1. The couple of $I_1(I)$ and $I_6(-I)$, that of $I_2(-I)$ and $I_5(I)$, and that of $I_3(I)$ and $I_4(-I)$ satisfy the boundary on the surface S_2.

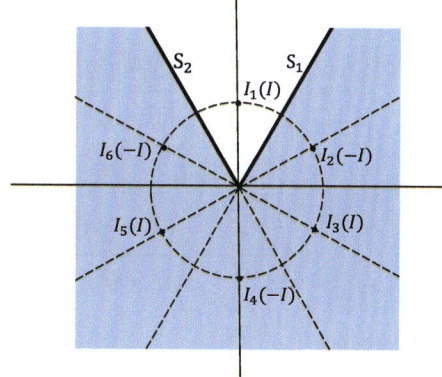

Fig. B7.1 Two superconductor surfaces with angle $\pi/3(n = 3)$

In the case of $n = 3/2$, it may seem to be possible. But there are only three points on which we can put image currents, and the image method cannot be applied.

7.10. According to the result in Example 7.6, the current density on the inner surface is

$$\tau = \frac{1}{\mu_0}\left(\frac{\partial A_z}{\partial R}\right)_{R=a} = -\frac{I(a^2 - h^2)}{2\pi a(a^2 + h^2 - 2ah\cos\varphi)}.$$

The total current is determined to be

$$\int_0^{2\pi} \tau a d\varphi = -\frac{I(a^2 - h^2)}{\pi}\int_0^{\pi}\frac{d\varphi}{a^2 + h^2 - 2ah\cos\varphi} = -I,$$

which is equal to the image current, where Eq. (A.23) in the Appendix was used.

7.11. The total current on the superconductor surface must be zero under the condition that it shields the interior of the superconductor. This situation can be realized by superposing two current distributions: One is the distribution determined in Sect. 7.2, which shields the magnetic flux density produced by the straight current, and the other is a uniform current distribution that does not produce magnetic flux density inside the superconductor. The latter current must be I from the above condition. So, the vector potential outside the cylindrical superconductor is

$$A_z(R, \varphi) = A'_z(R, \varphi) + \frac{\mu_0 I}{2\pi} \log \frac{R_0}{R},$$

where $A'_z(R, \varphi)$ is the vector potential obtained in Sect. 7.2 and R_0 is the distance to the reference point with zero vector potential.

7.12. For simplicity, we denote as $K = 4\pi/\mu_0 I$. Simple calculation leads to

$$K\frac{\partial^2 A_y}{\partial x^2} = \frac{-2x^2 + 2(z+a)^2}{[x^2 + (z+a)^2]^2} - \frac{-2x^2 + 2(z-a)^2}{[x^2 + (z-a)^2]^2},$$

$$K\frac{\partial^2 A_y}{\partial y^2} = 0,$$

$$K\frac{\partial^2 A_y}{\partial z^2} = \frac{2x^2 - 2(z+a)^2}{[x^2 + (z+a)^2]^2} - \frac{2x^2 - 2(z-a)^2}{[x^2 + (z-a)^2]^2}.$$

Thus, we have

$$K\left(\frac{\partial^2 A_y}{\partial x^2} + \frac{\partial^2 A_y}{\partial y^2} + \frac{\partial^2 A_y}{\partial z^2}\right) = 0$$

and Laplace's equation is satisfied.

7.13. The vector potential is given by

$$A_z(R, \varphi) = \frac{\mu_0 I}{4\pi}(A_1 - A_2 + \text{const.}) :$$

$$A_1 = \log\left[R^2 + \left(\frac{a^2}{d}\right)^2 - 2\left(\frac{a^2 R}{d}\right)\cos\varphi\right],$$

$$A_2 = \log(R^2 + d^2 - 2dR\cos\varphi).$$

Since A_1 and A_2 are functions of the same form, it is enough to show that A_2 satisfies Laplace's equation. $\nabla^2 A_2$ is given by:

$$\nabla^2 A_2 = \frac{1}{R} \cdot \frac{\partial}{\partial R}\left(R\frac{\partial A_2}{\partial R}\right) + \frac{1}{R^2} \cdot \frac{\partial^2 A_2}{\partial \varphi^2}.$$

After calculation we have

7.4 Exercises

$$\frac{1}{R} \cdot \frac{\partial}{\partial R}\left(R \frac{\partial A_2}{\partial R}\right) = -\frac{2d\left[(R^2 + d^2)\cos\varphi - 2dR\right]}{R(R^2 + d^2 - 2dR\cos\varphi)^2},$$

$$\frac{1}{R^2} \cdot \frac{\partial^2 A_2}{\partial \varphi^2} = \frac{2d\left[(R^2 + d^2)\cos\varphi - 2dR\right]}{R(R^2 + d^2 - 2dR\cos\varphi)^2}.$$

So, the equation, $\nabla^2 A_2 = 0$, holds. Hence, it is concluded that the vector potential satisfies Laplace's equation.

7.14. The vector potential is given by

$$A_z(R, \varphi) = \frac{\mu_0 I}{4\pi}(A_1 - A_2):$$

$$A_1 = \log\left[R^2 + \left(\frac{a^2}{h}\right)^2 - 2\left(\frac{a^2 R}{h}\right)\cos\varphi\right],$$

$$A_2 = \log(R^2 + h^2 - 2Rh\cos\varphi).$$

Since A_1 and A_2 are functions of the same form, it is enough to show that A_2 satisfies Laplace's equation. ΔA_2 is given by:

$$\Delta A_2 = \frac{1}{R} \cdot \frac{\partial}{\partial R}\left(R \frac{\partial A_2}{\partial R}\right) + \frac{1}{R^2} \cdot \frac{\partial^2 A_2}{\partial \varphi^2}.$$

After calculation we have

$$\frac{1}{R} \cdot \frac{\partial}{\partial R}\left(R \frac{\partial A_2}{\partial R}\right) = -\frac{2h\left[(R^2 + h^2)\cos\varphi - 2Rh\right]}{R(R^2 + h^2 - 2Rh\cos\varphi)^2},$$

$$\frac{1}{R^2} \cdot \frac{\partial^2 A_2}{\partial \varphi^2} = \frac{2h\left[(R^2 + h^2)\cos\varphi - 2Rh\right]}{R(R^2 + h^2 - 2Rh\cos\varphi)^2}.$$

So, the equation, $\Delta A_2 = 0$, holds. Hence, it is concluded that the vector potential satisfies Laplace's equation.

7.15. The vector potential is given by

$$A_z(x, y) = \frac{\mu_0 I}{4\pi} \log \frac{\left(x - \sqrt{l^2 - a^2}\right)^2 + y^2}{\left(x + \sqrt{l^2 - a^2}\right)^2 + y^2}$$

for $x < 0$, where I is the current given to the cylindrical superconductor. So, a simple calculation leads to

$$\frac{\partial^2 A_z}{\partial x^2} = \frac{\mu_0 I}{4\pi} \left\{ \frac{-\left(x - \sqrt{l^2 - a^2}\right)^2 + y^2}{\left[\left(x - \sqrt{l^2 - a^2}\right)^2 + y^2\right]^2} - \frac{-\left(x + \sqrt{l^2 - a^2}\right)^2 + y^2}{\left[\left(x + \sqrt{l^2 - a^2}\right)^2 + y^2\right]^2} \right\},$$

$$\frac{\partial^2 A_z}{\partial y^2} = \frac{\mu_0 I}{4\pi} \left\{ \frac{\left(x - \sqrt{l^2 - a^2}\right)^2 - y^2}{\left[\left(x - \sqrt{l^2 - a^2}\right)^2 + y^2\right]^2} - \frac{\left(x + \sqrt{l^2 - a^2}\right)^2 - y^2}{\left[\left(x + \sqrt{l^2 - a^2}\right)^2 + y^2\right]^2} \right\}.$$

Thus, we have

$$\frac{\partial^2 A_z}{\partial x^2} + \frac{\partial^2 A_z}{\partial y^2} = 0,$$

and Laplace's equation is satisfied.

7.16. The vector potential outside the superconductor is given by

$$A_z(R, \varphi) = \frac{\mu_0 I}{2\pi} \log \frac{d\left[R^2 + (a^2/d)^2 - 2(a^2 R/d) \cos \varphi\right]^{1/2}}{a\left(R^2 + d^2 - 2Rd \cos \varphi\right)^{1/2}},$$

where R and φ are the distance and azimuthal angle from the central axis of the superconductor. So, the equivector potential surface is given by

$$\frac{R^2 + (a^2/d)^2 - 2(a^2 R/d) \cos \varphi}{R^2 + d^2 - 2Rd \cos \varphi} = \frac{1}{K},$$

where K is a constant. If we use Cartesian coordinates, we have $R^2 = x^2 + y^2$ and $R \cos \varphi = x$. So, the above equation is rewritten as

$$\left[x + \frac{1}{K-1}\left(d - \frac{Ka^2}{d}\right)\right]^2 + y^2$$
$$= \frac{1}{(K-1)^2}\left(d - \frac{Ka^2}{d}\right)^2 + \frac{1}{K-1}\left[d^2 - K\left(\frac{a^2}{d}\right)^2\right].$$

Thus, the equivector potential surfaces are cylindrical surfaces parallel to the cylindrical superconductor. Especially, the superconductor surface ($x^2 + y^2 = a^2$) is an equivector potential surface, which is obtained for $K = d^2/a^2$. The minimum value of K is 0 and corresponds to the equivector potential surface just around the given current. So, the above solution holds for $0 < K \leq d^2/a^2$.

7.4 Exercises

7.17. The vector potential is given by

$$A_z(R, \varphi) = \frac{\mu_0 I}{4\pi} \log \frac{R^2 + (a^2/h)^2 - 2(a^2 R/h)\cos\varphi}{R^2 + h^2 - 2Rh\cos\varphi}.$$

Hence, the equivector potential surface is given by

$$\frac{R^2 + (a^2/h)^2 - 2(a^2 R/h)\cos\varphi}{R^2 + h^2 - 2Rh\cos\varphi} = K,$$

where K is a constant. Using Cartesian coordinates, we have $R^2 = x^2 + y^2$ and $R\cos\varphi = x$. So, the above equation leads to

$$\left[x - \frac{1}{K-1}\left(Kh - \frac{a^2}{h}\right)\right]^2 + y^2$$

$$= \frac{1}{K-1}\left[\left(\frac{a^2}{h}\right)^2 - Kh^2\right] + \frac{1}{(K-1)^2}\left(Kh - \frac{a^2}{h}\right)^2.$$

Thus, the equivector potential surfaces are cylindrical surfaces parallel to the cylindrical superconductor. Especially, the superconductor surface ($x^2 + y^2 = a^2$) is an equivector potential surface, which is obtained for $K = a^2/h^2$. The maximum value of K is infinity and corresponds to the equivector potential surface just around the given current. So, the above solution holds for $K \geq a^2/h^2$.

7.18. The vector potential is given by

$$A_z(x, y) = \frac{\mu_0 I}{4\pi} \log \frac{\left(x - \sqrt{l^2 - a^2}\right)^2 + y^2}{\left(x + \sqrt{l^2 - a^2}\right)^2 + y^2}$$

for $x < 0$, where I is the current given to the cylindrical superconductor. So, the equivector potential surface is given by

$$\frac{\left(x - \sqrt{l^2 - a^2}\right)^2 + y^2}{\left(x + \sqrt{l^2 - a^2}\right)^2 + y^2} = K,$$

where K is a positive constant. This leads to

$$\left(x + \frac{K+1}{K-1}\sqrt{l^2 - a^2}\right)^2 + y^2 = \frac{4K(l^2 - a^2)}{(K-1)^2}. \tag{B7.1}$$

Thus, the equivector potential surfaces are cylindrical surfaces parallel to the cylindrical superconductor. The surface of the cylindrical superconductor, which is an equivector potential surface expressed as

$$(x+l)^2 + y^2 = a^2,$$

is obtained for $K = \left[\left(l + \sqrt{l^2 - a^2}\right)/a\right]^2$. In the limit $K \to 1$, neglecting small terms such as x^2 and y^2, Eq. (B7.1) leads to

$$2\alpha x = 0,$$

where $\alpha = (K+1)\sqrt{l^2 - a^2}/(K-1)$. Thus, the wide superconductor surface ($x = 0$) is obtained. The range of K is $1 < K \leq \left[\left(l + \sqrt{l^2 - a^2}\right)/a\right]^2$.

7.19. The magnetic moment is assumed to be directed to the zenith ($\theta = 0$). The radial magnetic flux density must be cancelled in the superconducting region ($r > a$). So, a shielding current must be induced to produce a magnetic flux density of the same magnitude with the opposite sign. This means that the shielding current is equivalent to the magnetic moment $-m$ placed at the center for $r > a$. So, the shielding current density on the surface is given by

$$\tau_\varphi = -\frac{3m}{4\pi a^3} \sin\theta. \tag{B7.2}$$

On the other hand, the shielding current on the surface produces a uniform magnetic flux density in the hollow space. If we denote the uniform magnetic flux density by B_0, Eq. (7.18) gives

$$B_0 = -\frac{\mu_0 m}{2\pi a^3}.$$

The radial and zenithal components are

$$B_{0r} = -\frac{\mu_0 m}{2\pi a^3} \cos\theta, \quad B_{0\theta} = \frac{\mu_0 m}{2\pi a^3} \sin\theta.$$

The magnetic flux density produced in the hollow space by the magnetic moment m is

$$B_{mr} = \frac{\mu_0 m}{2\pi r^3} \cos\theta, \quad B_{m\theta} = \frac{\mu_0 m}{4\pi r^3} \sin\theta.$$

Thus, the magnetic flux density on the surface ($r = a$) is

$$B_r = 0, \quad B_\theta = \frac{3\mu_0 m}{4\pi a^3} \sin\theta.$$

Fig. B7.2 Zenithal magnetic flux density and azimuthal surface current density near the surface

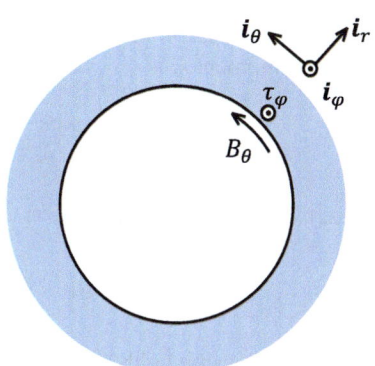

So, the relationship holds:

$$\tau_\varphi = -\frac{B_\theta(r=a)}{\mu_0},$$

and the boundary condition is satisfied. Note that the normal vector of the inner surface is opposite to the radial direction (see Fig. B7.2). This means that the induced current density on the surface is correctly given by Eq. (B7.2).

7.20. The moment of the magnetic dipole line is assumed to be directed to the azimuth ($\varphi = 0$). The radial magnetic flux density must be cancelled in the superconducting region ($R > a$). So, the shielding current must be induced to produce a magnetic flux density of the same magnitude with the opposite sign. This means that the shielding current is equivalent to the magnetic moment $-\hat{m}$ placed at the central axis for $R > a$. So, from the result in Example 7.7, the shielding current density on the surface is given by

$$\tau_z = -\frac{\hat{m}}{\pi a^2} \sin\varphi. \tag{B7.3}$$

On the other hand, the shielding current on the surface produces a uniform magnetic flux density in the hollow. If we denote the uniform magnetic flux density by B_0, it is given by

$$B_0 = -\frac{\mu_0 \hat{m}}{2\pi a^2}.$$

The radial and azimuthal components are

$$B_{0R} = -\frac{\mu_0 \hat{m}}{2\pi a^2}\cos\varphi, \quad B_{0\varphi} = \frac{\mu_0 \hat{m}}{2\pi a^2}\sin\varphi.$$

From Eq. (6.34), the magnetic flux density produced in the hollow space by the magnetic moment $\hat{m}$ is

$$B_{mR} = \frac{\mu_0 \hat{m}}{2\pi R^2} \cos\varphi, \quad B_{m\varphi} = \frac{\mu_0 \hat{m}}{2\pi R^2} \sin\varphi.$$

Thus, the magnetic flux density on the surface ($R = a$) is

$$B_R = 0, \quad B_\varphi = \frac{\mu_0 \hat{m}}{\pi a^2} \sin\varphi.$$

So, the relationship holds:

$$\tau_z = -\frac{B_\varphi(R=a)}{\mu_0},$$

and the boundary condition is satisfied. Note that the normal vector of the inner surface is opposite to the radial direction (see Fig. B7.3). This means that the induced current density on the surface is correctly given by Eq. (B7.3).

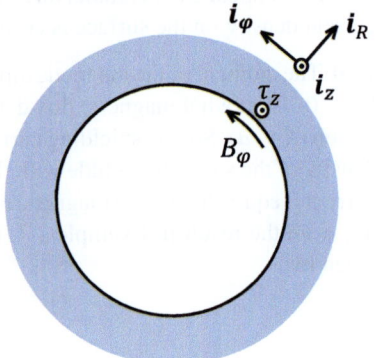

Fig. B7.3 Azimuthal magnetic flux density and axial surface current density near the surface

7.21. The vector potential inside the spherical superconductor is $A = 0$ and satisfies Laplace's equation. The vector potential outside the superconductor has an azimuthal component and is given by Eq. (7.19). Using Eq. (A.18) in the Appendix, we have

$$\nabla \times A = i_r B_0 \left(1 - \frac{a^3}{r^3}\right) \cos\theta - i_\theta B_0 \left(1 + \frac{a^3}{2r^3}\right) \sin\theta$$

and

$$\nabla \times \nabla \times A = i_\varphi \left[-\frac{B_0}{r}\left(1 - \frac{a^3}{r^3}\right)\sin\theta + \frac{B_0}{r}\left(1 - \frac{a^3}{r^3}\right)\sin\theta \right] = 0.$$

7.4 Exercises

Thus, Laplace's equation holds inside and outside the spherical superconductor.

7.22. The vector potential inside the cylindrical superconductor is $A = 0$ and satisfies Laplace's equation. The vector potential outside the superconductor is given by

$$A_z(R, \varphi) = B_0 \left(R - \frac{a^2}{R} \right) \sin \varphi.$$

Using Eq. (A.14) in the Appendix, we have

$$\nabla \times \nabla \times A = \frac{i_z}{R} \left[-\frac{\partial}{\partial R} \left(R \frac{\partial A_z}{\partial R} \right) - \frac{\partial}{\partial \varphi} \left(\frac{1}{R} \cdot \frac{\partial A_z}{\partial \varphi} \right) \right]$$

$$= -i_z \left[\frac{B_0}{R} \left(1 - \frac{a^2}{R^2} \right) \sin \theta - \frac{B_0}{R} \left(1 - \frac{a^2}{R^2} \right) \sin \theta \right] = 0.$$

Thus, Laplace's equation holds inside and outside the cylindrical superconductor.

7.23. The vector potential produced by the magnetic moment is

$$A_{m\varphi} = \frac{\mu_0 m}{4\pi r^2} \sin \theta,$$

and the vector potential produced by the induced current that produces a uniform magnetic flux density B_0 is

$$A_{f\varphi} = \frac{B_0 r}{2} \sin \theta = -\frac{\mu_0 m r}{4\pi a^3} \sin \theta.$$

Thus, the vector potential in the hollow space ($0 \leq r < a$) is given by

$$A_\varphi = A_{m\varphi} + A_{f\varphi} = \frac{\mu_0 m}{4\pi} \left(\frac{1}{r^2} - \frac{r}{a^3} \right) \sin \theta,$$

which leads to

$$\nabla \times A = i_r \frac{1}{r \sin \theta} \cdot \frac{\partial}{\partial \theta} (\sin \theta A_\varphi) - i_\theta \frac{1}{r} \cdot \frac{\partial}{\partial r} (r A_\varphi)$$

$$= \frac{\mu_0 m}{2\pi} \left[i_r \left(\frac{1}{r^3} - \frac{1}{a^3} \right) \cos \theta + i_\theta \left(\frac{1}{2r^3} - \frac{1}{a^3} \right) \sin \theta \right] \equiv i_r f_r + i_\theta f_\theta.$$

Since f_r and f_θ are independent of φ, we have

$$\nabla \times \nabla \times A = i_\varphi \frac{1}{r} \left[\frac{\partial}{\partial r} (r f_\theta) - \frac{\partial f_r}{\partial \theta} \right]$$

$$= i_\varphi \frac{\mu_0 m}{2\pi r} \left[\left(\frac{1}{r^3} - \frac{1}{a^3} \right) \sin \theta - \left(\frac{1}{r^3} - \frac{1}{a^3} \right) \sin \theta \right] = 0.$$

Thus, Laplace's equation is satisfied.

7.24. The vector potential produced by the magnetic moment is

$$A_{mz} = \frac{\mu_0 \hat{m}}{2\pi R} \sin \varphi,$$

and the vector potential produced by the induced current that produces a uniform magnetic flux density B_0 is

$$A_{fz} = B_0 R \sin \varphi = -\frac{\mu_0 \hat{m} R}{2\pi a^2} \sin \varphi.$$

The vector potential in the hollow space ($0 \leq R < a$) is given by

$$A_z = A_{mz} + A_{fz} = \frac{\mu_0 \hat{m}}{2\pi} \left(\frac{1}{R} - \frac{R}{a^2} \right) \sin \varphi,$$

which leads to

$$\nabla \times \mathbf{A} = \mathbf{i}_R \frac{1}{R} \cdot \frac{\partial A_z}{\partial \varphi} - \mathbf{i}_\varphi \frac{\partial A_z}{\partial R}$$

$$= \frac{\mu_0 \hat{m}}{2\pi} \left[\mathbf{i}_R \left(\frac{1}{R^2} - \frac{1}{a^2} \right) \cos \varphi + \mathbf{i}_\varphi \left(\frac{1}{R^2} + \frac{1}{a^2} \right) \sin \varphi \right] \equiv \mathbf{i}_R f_R + \mathbf{i}_\varphi f_\varphi.$$

Since f_R and f_φ are independent of z, we have

$$\nabla \times \nabla \times \mathbf{A} = \mathbf{i}_z \frac{1}{R} \left[\frac{\partial}{\partial R} (R f_\varphi) - \frac{\partial f_R}{\partial \varphi} \right]$$

$$= \mathbf{i}_z \frac{\mu_0 \hat{m}}{2\pi R} \left[-\left(\frac{1}{R^2} + \frac{1}{a^2} \right) \sin \varphi + \left(\frac{1}{R^2} + \frac{1}{a^2} \right) \sin \varphi \right] = 0.$$

Thus, Laplace's equation is satisfied.

7.25. We define the x-axis in the direction of the applied electric field with $x = 0$ on the central axis. The equivector potential surface is given by

$$B_0 \left(R - \frac{a^2}{R} \right) \sin \varphi = K,$$

where K is a constant. Using the relationships $R^2 = x^2 + y^2$ and $R \sin \varphi = y$, the above relation leads to

$$\left(1 - \frac{a^2}{x^2 + y^2} \right) y = \frac{K}{B_0}.$$

7.4 Exercises

So, we have

$$x = \pm a\left(1 + \frac{K}{B_0 y - K} - \frac{y^2}{a^2}\right)^{1/2}$$

in the x-y plane. The surface of the superconductor ($x^2 + y^2 = a^2$) is an equivector potential surface obtained for $K = 0$.

7.26. The vector potential has a z-component and is

$$A_z = \frac{\mu_0 \hat{m}}{2\pi}\left(\frac{1}{R} - \frac{R}{a^2}\right)\sin\varphi.$$

So, the equivector potential surface is given by

$$\left(\frac{1}{R} - \frac{R}{a^2}\right)\sin\varphi = K$$

with K denoting a constant. Noting $R^2 = x^2 + y^2$ and $R\sin\varphi = y$, the above equation is written as

$$\left(\frac{1}{x^2 + y^2} - \frac{1}{a^2}\right)y = K. \tag{B7.4}$$

So, we have

$$x = \pm\left(a^2 - \frac{Ka^4}{y + Ka^2} - y^2\right)^{1/2}.$$

The surface of the superconductor ($x^2 + y^2 = a^2$) is an equivector potential surface obtained for $K = 0$. When $|K|$ is very large, the second term on the left-hand side in Eq. (B7.4) can be neglected, and the equivector potential surface approaches

$$x^2 + \left(y - \frac{1}{2K}\right)^2 = \frac{1}{4K^2}.$$

7.27. The vector potential due to the given current I and the image current I' inside the superconductor is given by

$$A_z = \frac{\mu_0 I}{2\pi}\log\frac{R_0}{(R^2 + d^2 - 2Rd\cos\varphi)^{1/2}}$$
$$+ \frac{\mu_0 I'}{2\pi}\log\frac{R_0'}{(R^2 + h^2 - 2Rh\cos\varphi)^{1/2}},$$

where R_0 and R'_0 are constants denoting distances to a reference point of the vector potential. The normal component of the magnetic flux density at the superconductor surface ($R = a$) is

$$B_R(R=a) = \frac{1}{a} \cdot \left(\frac{\partial A_z}{\partial \varphi}\right)_{R=a}$$

$$= \frac{\mu_0}{2\pi} \sin\varphi \left(\frac{Id}{a^2 + d^2 - 2ad\cos\varphi} + \frac{I'h}{a^2 + h^2 - 2ah\cos\varphi}\right)$$

$$= \frac{\mu_0}{4\pi a} \sin\varphi \left[\frac{I}{(a^2 + d^2)/2ad - \cos\varphi} + \frac{I'}{(a^2 + h^2)/2ah - \cos\varphi}\right].$$

The following conditions must be satisfied so that the above magnetic flux density is identically zero:

$$I' = -I, \quad \frac{a^2 + h^2}{2ah} = \frac{a^2 + d^2}{2ad}.$$

The latter condition is reduced to $h = a^2/d$. Thus, the obtained vector potential agrees with that in Eq. (1.23), and the correct result is obtained using the boundary condition.

7.28. The vector potential due to the given current I and the image current I' in the hollow interior is given by

$$A_z = \frac{\mu_0 I}{2\pi} \log \frac{R_0}{(R^2 + h^2 - 2Rh\cos\varphi)^{1/2}} + \frac{\mu_0 I'}{2\pi} \log \frac{R'_0}{(R^2 + d^2 - 2Rd\cos\varphi)^{1/2}},$$

where R_0 and R'_0 are constants denoting distances to a reference point of the vector potential. The normal component of the magnetic flux density at the superconductor surface ($R = a$) is

$$B_R(R=a) = \frac{1}{a} \cdot \left(\frac{\partial A_z}{\partial \varphi}\right)_{R=a}$$

$$= \frac{\mu_0}{2\pi} \sin\varphi \left(\frac{Ih}{a^2 + h^2 - 2ah\cos\varphi} + \frac{I'd}{a^2 + d^2 - 2ad\cos\varphi}\right)$$

$$= \frac{\mu_0}{4\pi a} \sin\varphi \left[\frac{I}{(a^2 + h^2)/2ah - \cos\varphi} + \frac{I'}{(a^2 + d^2)/2ad - \cos\varphi}\right].$$

The following conditions must be satisfied so that the above magnetic flux density is identically zero:

$$I' = -I, \quad \frac{a^2 + d^2}{2ad} = \frac{a^2 + h^2}{2ah}.$$

7.4 Exercises

The latter condition is reduced to $d = a^2/h$. Thus, the obtained vector potential agrees with the result in Example 7.6, and the correct result is obtained using the boundary condition.

7.29. Using the vector potential in the vacuum $(R > a)$,

$$A_z(R, \varphi) = \left(B_0 R + \frac{\mu_0 \hat{m}}{2\pi R}\right) \sin \varphi,$$

the radial component of the magnetic flux density in this region is given by

$$B_r = \frac{1}{R} \cdot \frac{\partial A_z}{\partial \varphi} = \left(B_0 + \frac{\mu_0 \hat{m}}{2\pi R^2}\right) \cos \varphi.$$

From the requirement that this component is zero on the superconductor surface $(R = a)$, we have

$$\hat{m} = -\frac{2\pi a^2 B_0}{\mu_0}.$$

Thus, the same result is obtained as in Example 7.7 using the boundary condition on the magnetic flux density.

7.30. We place an image current I in the superconducting cylinder at distance h from the center of the cylinder and an image current $-I$ in the infinite superconductor at distance $l - h$ from its surface after virtually removing the two superconductors, as shown in Fig. 7.11, so that the boundary condition on the infinite superconductor surface is satisfied. Then, the vector potential in the vacuum region is given by

$$A_z(R, \varphi) = \frac{\mu_0 I}{4\pi} \log \frac{R^2 + d^2 - 2Rd \cos \varphi}{R^2 + h^2 - 2Rh \cos \varphi},$$

where R and φ are the distance and azimuthal angle from the central axis of the superconducting cylinder. Hence, the radial magnetic flux density on the superconductor surface $(R = a)$ is

$$B_R(a, \varphi) = \frac{\mu_0 I \sin \varphi}{2\pi} \left[\frac{1}{(a^2 + d^2)/d - 2a \cos \varphi} - \frac{1}{(a^2 + h^2)/h - 2a \cos \varphi}\right].$$

So that this magnetic flux density is identically zero, the following condition should be satisfied:

$$\frac{a^2 + d^2}{d} = \frac{a^2 + h^2}{h},$$

which leads to $h = a^2/d$, or

$$h = l - \sqrt{l^2 - a^2}.$$

Then, the azimuthal magnetic flux density on the superconductor surface is

$$B_\varphi(R=a) = -\frac{\mu_0 I}{4\pi}\left(\frac{a - d\cos\varphi}{a^2 + d^2 - 2ad\cos\varphi} - \frac{a - h\cos\varphi}{a^2 + h^2 - 2ah\cos\varphi}\right)$$

$$= \frac{\mu_0\sqrt{l^2 - a^2}\,I}{2\pi a(l - a\cos\varphi)}.$$

The surface current density is obtained from Eq. (7.5) as

$$\tau(\varphi) = \frac{1}{\mu_0} B_\varphi(R=a) = \frac{\mu_0\sqrt{l^2 - a^2}\,I}{2\pi a(l - a\cos\varphi)}.$$

These agree with the results obtained in Example 7.5.

7.31. The azimuthal superconducting current of density

$$\tau(\theta) = \tau_0 \sin\theta$$

is assumed on the superconductor surface ($r = a$.). From Eq. (7.21), this current produces a uniform magnetic flux density in the direction of $\theta = 0$:

$$B_0 = \frac{2\mu_0}{3}\tau_0.$$

Hence, the magnetic flux density inside the hollow space is

$$B_r = \left(\frac{\mu_0 m}{2\pi r^3} + \frac{2\mu_0}{3}\tau_0\right)\cos\theta, \quad B_\theta = \left(\frac{\mu_0 m}{4\pi r^3} - \frac{2\mu_0}{3}\tau_0\right)\sin\theta.$$

Thus, the zenithal magnetic flux density on the superconductor surface is

$$B_\theta(r=a) = \left(\frac{\mu_0 m}{4\pi a^3} - \frac{2\mu_0}{3}\tau_0\right)\sin\theta.$$

The boundary condition, $B_\theta(r=a) = -\mu_0\tau(\theta)$, leads to

$$\tau_0 = -\frac{3m}{4\pi a^3},$$

(see Fig. B7.2). Then, the radial component of the magnetic flux density on the superconductor surface is

$$B_r(r=a) = 0.$$

Hence, from the above two conditions, the obtained result is concluded to be valid.

7.4 Exercises

7.32. The axial superconducting current of density

$$\tau(\varphi) = \tau_0 \sin \varphi$$

is assumed on the superconductor surface ($R = a$). From the result in Example 7.7, this current produces a uniform magnetic flux density in the direction of $\varphi = 0$:

$$B_0 = \frac{\mu_0}{2}\tau_0.$$

Hence, the magnetic flux density inside the hollow space is

$$B_R = \left(\frac{\mu_0 \hat{m}}{2\pi R^2} + \frac{\mu_0}{2}\tau_0\right)\cos\varphi, \quad B_\varphi = \left(\frac{\mu_0 \hat{m}}{2\pi R^2} - \frac{\mu_0}{2}\tau_0\right)\sin\varphi.$$

Thus, the zenithal magnetic flux density on the superconductor surface is

$$B_\varphi(R = a) = \left(\frac{\mu_0 \hat{m}}{2\pi a^2} - \frac{\mu_0}{2}\tau_0\right)\sin\varphi.$$

The boundary condition, $B_\varphi(R = a) = -\mu_0 \tau(\varphi)$, leads to

$$\tau_0 = -\frac{\hat{m}}{\pi a^2},$$

(see Fig. B7.3). Then, the radial component of the magnetic flux density on the superconductor surface is

$$B_R(R = a) = 0.$$

Hence, from the above two conditions, the obtained result is concluded to be valid.

7.33. We assume the image current inside the cylindrical superconductor, as shown in Fig. 7.8b. Then, the magnetic potential is given by

$$\phi_m = \frac{\mu_0 I}{2\pi} \log \frac{R_0}{(R^2 + d^2 - 2Rd\cos\varphi)^{1/2}} + \frac{\mu_0 I'}{2\pi} \log \frac{R_0'}{(R^2 + h^2 - 2Rh\cos\varphi)^{1/2}},$$

where R and φ are the distance and azimuthal angle measured from the central axis of the superconductor, and R_0 and R_0' are the distances to the reference points. So, the radial magnetic flux density on the superconductor surface ($R = a$) is

$$B_R(R = a) = -\left(\frac{\partial \phi_m}{\partial R}\right)_{R=a}$$

$$= \frac{\mu_0}{2\pi}\left[\frac{(a - h\cos\varphi)I'}{a^2 + h^2 - 2ah\cos\varphi} + \frac{(a - d\cos\varphi)I}{a^2 + d^2 - 2ad\cos\varphi}\right].$$

From the condition that this magnetic flux density must be identically zero, we have

$$\frac{a^2 + d^2}{d} = \frac{a^2 + h^2}{h}, \quad I' = -I.$$

The former condition leads to $h = a^2/d$, or

$$h = l - \sqrt{l^2 - a^2}.$$

This result agrees with the result obtained using the vector potential. However, the azimuthal magnetic flux density cannot be correctly derived with the magnetic potential. Thus, it can be concluded that the magnetic potential cannot be used for the magnetic fields produced by currents, while it is useful for those produced by magnetic moments.

7.34. From Eq. (1.22), the magnetic potential for the magnetic moment m is given by

$$\phi_m = \frac{\mu_0 m}{4\pi r^2} \cos\theta.$$

The magnetic potential that produces a uniform magnetic flux density B_0 is given by

$$\phi_f = -B_0 r \cos\theta.$$

So, the normal component of the magnetic flux density on the surface ($r = a$) is

$$B_r(r = a) = -\left[\frac{\partial}{\partial r}(\phi_m + \phi_f)\right]_{r=a} = \left(\frac{\mu_0 m}{2\pi a^3} + B_0\right)\cos\theta.$$

From the requirement that this component is zero on the superconductor surface, we have

$$B_0 = -\frac{\mu_0 m}{2\pi a^3}.$$

Thus, the radial and zenithal components of the magnetic flux density in the hollow space are

$$B_r = \frac{\mu_0 m}{2\pi}\left(\frac{1}{r^3} - \frac{1}{a^3}\right)\cos\theta, \quad B_\theta = \frac{\mu_0 m}{2\pi}\left(\frac{1}{2r^3} + \frac{1}{a^3}\right)\sin\theta.$$

Then, the zenithal component of the magnetic flux density on the surface is

$$B_\theta(r = a) = \frac{3\mu_0 m}{4\pi a^3}\sin\theta.$$

7.4 Exercises

Hence, the shielding current density on the surface is

$$\tau_\varphi = -\frac{B_\theta(r=a)}{\mu_0} = -\frac{3m}{4\pi a^3}\sin\theta.$$

7.35. From Eq. (1.25), the magnetic potential for the magnetic moment $\hat{m}$ is given by

$$\phi_m = \frac{\mu_0 \hat{m}}{2\pi R}\cos\varphi.$$

The magnetic potential that produces a uniform magnetic flux density B_0 is given by

$$\phi_f = -B_0 R \cos\varphi.$$

So, the normal component of the magnetic flux density on the surface ($R=a$) is

$$B_R(R=a) = -\left[\frac{\partial}{\partial R}(\phi_m + \phi_f)\right]_{R=a} = \left(\frac{\mu_0 \hat{m}}{2\pi a^2} + B_0\right)\cos\varphi.$$

From the requirement that this component is zero on the superconductor surface, we have

$$B_0 = -\frac{\mu_0 \hat{m}}{2\pi a^2}.$$

Thus, the radial and azimuthal components of the magnetic flux density in the hollow space are

$$B_R = \frac{\mu_0 \hat{m}}{2\pi}\left(\frac{1}{R^2} - \frac{1}{a^2}\right)\cos\varphi, \quad B_\varphi = \frac{\mu_0 \hat{m}}{2\pi}\left(\frac{1}{R^2} + \frac{1}{a^2}\right)\sin\varphi.$$

Then, the azimuthal component of the magnetic flux density on the surface is

$$B_\varphi(R=a) = \frac{\mu_0 \hat{m}}{\pi a^2}\sin\varphi.$$

Hence, the shielding current density on the surface is

$$\tau_z = -\frac{B_\varphi(R=a)}{\mu_0} = -\frac{\hat{m}}{\pi a^2}\sin\varphi.$$

Chapter 8
Current Systems

8.1 Inductance

When a current I flows in a closed circuit C, the magnetic flux penetrating C is proportional to I:

$$\Phi = LI. \tag{8.1}$$

The proportional constant L is called the self-inductance. The unit of self-inductance is [Wb/A] and is newly defined as [H] (henry). The self-inductance is determined only by the shape of C and is defined as a positive quantity. That is, the directions of the current and magnetic flux follow the right-hand rule.

Suppose a system composed of n electric circuits, and assume that current I_i flows in the i-th circuit C_i ($i = 1, 2, \ldots, n$). We express the magnetic flux Φ_i penetrating C_i as

$$\Phi_i = \sum_{j=1}^{n} L_{ij} I_j. \tag{8.2}$$

In the above, the L_{ij}'s are inductance coefficients. The L_{ii}'s are self-inductances, and L_{ij}'s ($i \neq j$) are mutual inductances. The reciprocity theorem

$$L_{ij} = L_{ji} \tag{8.3}$$

holds generally.

Example 8.1 Determine the self-inductance of a unit length of the parallel-wire transmission line of radius a separated by distance d in Fig. 8.1. Assume that d is much larger than a and we can neglect the magnetic flux inside the conductors.

Fig. 8.1 Parallel-wire transmission line

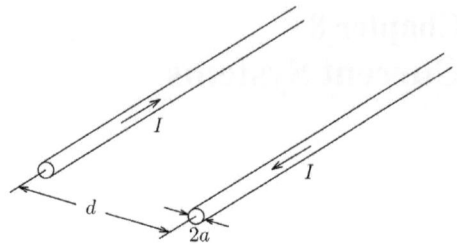

Solution 8.1 We suppose that current I flows as shown in the figure. We assume a plane that includes the axes of the two cylindrical conductors and calculate the magnetic flux penetrating the plane between the two conductors. The magnetic flux density produced by current I flowing along the left conductor at distance $x(> a)$ from its central axis is

$$B_1 = \frac{\mu_0 I}{2\pi x}$$

and is directed downwards. Hence, the magnetic flux in a unit length produced by this current is

$$\Phi'_1 = \int_a^{d-a} \frac{\mu_0 I}{2\pi x} dx = \frac{\mu_0 I}{2\pi} \log \frac{d-a}{a}.$$

The magnetic flux produced by the current along the right conductor is the same, and we have the self-inductance of a unit length as

$$L' = \frac{2\Phi'_1}{I} = \frac{\mu_0}{\pi} \log \frac{d-a}{a} \simeq \frac{\mu_0}{\pi} \log \frac{d}{a}.$$

◆

8.2 Coils

The component used to store magnetic flux is a coil. Coils are also used for other purposes, such as producing various magnetic flux densities or generating electric power, which will be covered in Chap. 10. Here, we introduce the magnetic property of coils.

A coil used to produce a uniform magnetic flux density is called a solenoid coil. For example, when we apply current I to a long solenoid coil with a winding of n turns in a unit length, the interior magnetic flux density is uniform with the value

$$B = \mu_0 n I, \tag{8.4}$$

8.2 Coils

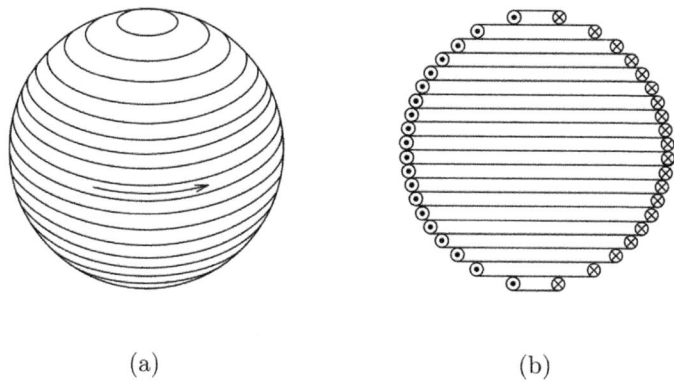

Fig. 8.2 Spherical coil: **a** geometry and **b** windings

as shown in Example 6.6. This value does not depend on the radius or length of the coil. For a coil of radius a, the magnetic flux that penetrates one turn of the coil is

$$\phi = \pi \mu_0 n a^2 I. \tag{8.5}$$

Thus, the magnetic flux penetrating the coil with a unit length is

$$\Phi' = n\phi = \pi \mu_0 n^2 a^2 I. \tag{8.6}$$

Hence, the self-inductance in a unit length is given by

$$L' = \frac{\Phi'}{I} = \pi \mu_0 n^2 a^2. \tag{8.7}$$

A spherical coil is a special coil for producing a uniform magnetic flux density in a limited space. That is, when current flows on the surface of a sphere, as given by Eq. (7.21), the interior magnetic flux density is uniform. Assume a spherical coil of radius a with N turns, as in Fig. 8.2a. We take the number of turns in a unit zenithal length to be $[N/(2a)] \sin \theta$, where θ is the zenithal angle. When we apply current I, the surface current density on the sphere is

$$\tau = \frac{NI}{2a} \sin \theta \tag{8.8}$$

and from Eq. (7.21), we obtain the uniform interior magnetic flux density as

$$B_0 = \frac{\mu_0 NI}{3a}. \tag{8.9}$$

To realize such a winding, the number of turns in a unit length along the axis is $N/(2a)$.

Example 8.2 Determine the self-inductance of the spherical coil with diameter a and total number of windings N shown in Fig. 8.2.

Solution 8.2 When current I is applied to the coil, the magnetic flux that penetrates one turn of the coil at zenithal angle θ is

$$\phi(\theta) = B_0 \pi (a \sin \theta)^2 = \frac{\mu_0}{3} \pi a N I \sin^2 \theta,$$

where $B_0 = \mu_0 N I / 3a$ is the uniform magnetic flux density in the coil given by Eq. (8.9). Hence, the total magnetic flux penetrating the coil is

$$\Phi = \int_0^\pi \frac{1}{2} N \phi(\theta) \sin \theta \, d\theta = \frac{\mu_0}{6} \pi a N^2 I \int_0^\pi \sin^3 \theta \, d\theta = \frac{2\mu_0}{9} \pi a N^2 I.$$

The self-inductance is

$$L = \frac{\Phi}{I} = \frac{2\pi}{9} \mu_0 a N^2.$$

◆

8.3 Magnetic Energy

The electric field fills the space between two electrodes in a charged capacitor, and we can regard the space as filled with electric energy. For a coil that stores magnetic flux, we can also regard the interior space as filled with magnetic energy.

Here, we suppose two superconductors in electrical contact with each other, as shown in Fig. 8.3. One of them is a movable plate. Assume that magnetic flux Φ is trapped within the space surrounded by the superconductors. A current flows on the inner surface to shield the superconductors from the magnetic flux. Hence, a repulsive force given by Eq. (6.7) is exerted on the movable superconducting plate. If the plate is displaced by distance x, the interior magnetic flux density changes to

$$B = \frac{\Phi}{(a+x)b}, \tag{8.10}$$

and the density of current flowing on the inner surface changes to

$$\tau = \frac{B}{\mu_0} = \frac{\Phi}{\mu_0(a+x)b}. \tag{8.11}$$

8.3 Magnetic Energy

Fig. 8.3 Magnetic flux trapped in closed circuit composed of two superconductors. The superconducting plate is movable as shown by the arrow and in electrical contact with the fixed piece

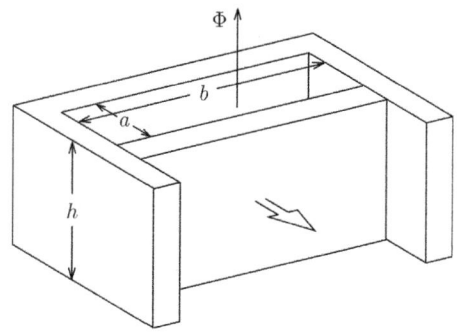

We assume that a is much smaller than b. Since the magnetic flux density produced by the fixed superconductor is half of the value given by Eq. (8.10), we estimate the force on the movable plate to be

$$F = \frac{1}{2}\tau Bbh = \frac{\Phi^2 h}{2\mu_0(a+x)^2 b}, \tag{8.12}$$

which is directed along increasing x. This is an isolated system, and there is no electromagnetic interaction with the surroundings after the initial condition is established. Thus, this force is attributed to the variation in the magnetic energy U_m of this system. From the relationship

$$F = -\frac{\partial U_m}{\partial x}, \tag{8.13}$$

we estimate the magnetic energy as

$$U_m = \frac{\Phi^2 h}{2\mu_0(a+x)b} = \frac{1}{2\mu_0}B^2(a+x)bh. \tag{8.14}$$

In the above, $(a+x)bh$ is the volume of the space in which the uniform magnetic flux is trapped. Hence, the magnetic energy density is given by

$$u_m = \frac{1}{2\mu_0}B^2. \tag{8.15}$$

Since the total current flowing in the closed circuit is $I = \tau h$, the self-inductance of the system is

$$L = \frac{\Phi}{I} = \frac{\mu_0(a+x)b}{h}. \tag{8.16}$$

In terms of the self-inductance, we rewrite the magnetic energy, Eq. (8.14), as

$$U_m = \frac{1}{2}LI^2 = \frac{1}{2}\Phi I = \frac{1}{2L}\Phi^2. \tag{8.17}$$

We consider a system composed of n closed electric circuits. Suppose that current I_i flows in the i-th circuit and that magnetic flux Φ_i penetrates it ($i = 1, 2, \ldots, n$). Then, extending the result of Eq. (8.17) to this case, the magnetic energy of this system is given by

$$U_m = \frac{1}{2}\sum_{i=1}^{n}\Phi_i I_i = \frac{1}{2}\sum_{i=1}^{n}\sum_{j=1}^{n}L_{ij}I_i I_j. \tag{8.18}$$

When current flows with density $\boldsymbol{i}$ in volume V, the magnetic energy in volume V is given by

$$U_m = \frac{1}{2}\int_V \boldsymbol{A} \cdot \boldsymbol{i}\, dV. \tag{8.19}$$

Using the magnetic energy density, Eq. (8.15), the magnetic energy in volume V is also given by

$$U_m = \frac{1}{2\mu_0}\int_V B^2\, dV. \tag{8.20}$$

Example 8.3 Current I is applied to the spherical coil with diameter a and total number of windings N shown in Fig. 8.2. Determine the total magnetic energy using Eq. (8.20).

Solution 8.3 Using the uniform magnetic flux density $B_0 = \mu_0 NI/3a$, the magnetic energy inside the coil is

$$U_{m1} = \frac{B_0^2}{2\mu_0} \cdot \frac{4\pi}{3}a^3 = \frac{2\pi}{27}\mu_0 aN^2 I^2.$$

The magnetic flux density outside the coil is equal to that given by the magnetic moment, which is given by Eq. (6.29) with $m = (2\pi/\mu_0)B_0 a^3$, placed on the coil center. Thus, the magnetic energy density outside the coil is

$$u_{m2} = \frac{1}{2\mu_0}(B_r^2 + B_\theta^2) = \frac{B_0^2 a^6}{2\mu_0 r^6}\left(\cos^2\theta + \frac{\sin^2\theta}{4}\right)$$
$$= \frac{\mu_0 a^4 N^2 I^2}{18r^6}\left(\cos^2\theta + \frac{\sin^2\theta}{4}\right).$$

8.3 Magnetic Energy

Integrating this in the space outside the coil, the magnetic energy in this region is obtained as

$$U_{m2} = \frac{\mu_0 a^4 N^2 I^2}{18} \int_a^\infty \frac{1}{r^6} 2\pi r^2 \, dr \int_0^\pi \left(\cos^2\theta + \frac{\sin^2\theta}{4} \right) \sin\theta \, d\theta$$

$$= \frac{\pi \mu_0 a^4 N^2 I^2}{9} \cdot \frac{1}{3a^3} \cdot \left(\frac{2}{3} + \frac{1}{3} \right) = \frac{\pi}{27} \mu_0 a N^2 I^2.$$

The total magnetic energy is determined to be

$$U_m = U_{m1} + U_{m2} = \frac{\pi}{9} \mu_0 a N^2 I^2.$$

◆

Example 8.4 We apply current I to the superconducting coaxial transmission line in Fig. 7.2 in Example 7.2. Determine the magnetic energy stored in a unit length of the transmission line and derive the self-inductance with this result.

Solution 8.4 Currents flow only on the surfaces $R = a$ and $R = b$ so that the magnetic flux does not penetrate the superconductors. The magnetic flux density is $B = \mu_0 I/(2\pi R)$ only in the region $a < R < b$ and is zero in other regions. Hence, the magnetic energy is non-zero only in the region $a < R < b$ and its density is

$$\frac{B^2}{2\mu_0} = \frac{\mu_0 I^2}{8\pi^2 R^2}.$$

Integrating this over the volume in a unit length, we have

$$U'_m = \frac{\mu_0 I^2}{8\pi^2} \int_a^b \frac{1}{R^2} 2\pi R \, dR = \frac{\mu_0 I^2}{4\pi} \log \frac{b}{a}.$$

We can also obtain the magnetic energy from Eq. (8.19). The vector potential has only the axial component, A_z, similarly to the current. From the relationship $\partial A_z/\partial R = -B$ with $A_z(b) = 0$, we obtain the vector potential as

$$A_z(R) = \frac{\mu_0 I}{2\pi} \log \frac{b}{R}$$

in the region $a < R < b$. Thus, the magnetic energy is

$$U'_m = \frac{1}{2}A_z(a)I = \frac{\mu_0 I^2}{4\pi}\log\frac{b}{a}.$$

which agrees with the above result.

Using this result, the self-inductance in a unit length is

$$L' = \frac{2U'_m}{I^2} = \frac{\mu_0}{2\pi}\log\frac{b}{a}.$$

♦

Example 8.5 Suppose that the coaxial transmission line in Example 8.4 is not made of a superconductor but of a usual conductor. Determine the self-inductance with the magnetic energy.

Solution 8.5 In the case of conductor, the current flows uniformly inside the conductor, and the magnetic flux densities in the regions $0 \leq R < a$ and $b < R < c$ are respectively given by

$$B(R) = \frac{\mu_0 I R}{2\pi a^2}; \qquad 0 \leq R < a,$$

$$= \frac{\mu_0 I}{2\pi(c^2 - b^2)}\left(\frac{c^2}{R} - R\right); \quad b < R < c.$$

Hence, in comparison with the case of a superconductor, the magnetic energy increases by

$$\Delta U'_m = \frac{1}{2\mu_0}\int_0^a \left(\frac{\mu_0 I R}{2\pi a^2}\right)^2 2\pi R\, dR + \frac{1}{2\mu_0}\int_b^c \left[\frac{\mu_0 I}{2\pi(c^2-b^2)}\left(\frac{c^2}{R}-R\right)\right]^2 2\pi R\, dR$$

$$= \frac{\mu_0 c^2 I^2}{8\pi(c^2-b^2)}\left(\frac{2c^2}{c^2-b^2}\log\frac{c}{b}-1\right).$$

Adding this contribution to the result in Example 8.4, we obtain the self-inductance in a unit length as

$$L' = \frac{\mu_0}{2\pi}\log\frac{b}{a} + \frac{\mu_0 c^2}{4\pi(c^2-b^2)}\left(\frac{2c^2}{c^2-b^2}\log\frac{c}{b}-1\right).$$

♦

8.3 Magnetic Energy

Example 8.6 We apply a current I to a coaxial transmission line with the cross-section shown in Fig. 8.4. Determine the magnetic energy in a unit length using Eq. (8.19) and the self-inductance. The thickness of the outer conductor is neglected.

Fig. 8.4 Cross-section of coaxial transmission line

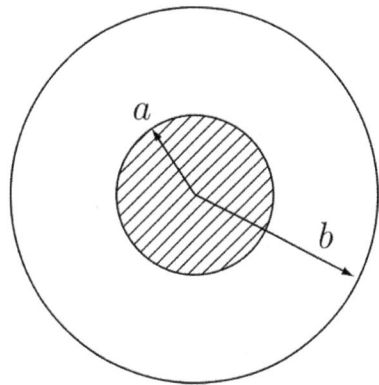

Solution 8.6 The magnetic flux density is

$$B = \frac{\mu_0 I R}{2\pi a^2}; \quad 0 < R < a,$$

$$= \frac{\mu_0 I}{2\pi R}; \quad a < R < b,$$

$$= 0; \quad R > b.$$

Since $B = 0$ for $R > b$, we obtain $A(R = b) = 0$. Thus, the vector potential is determined to be

$$A(R) = \int_R^b B(R)\,dR = \frac{\mu_0 I}{2\pi} \log \frac{b}{R}; \quad a < R < b,$$

$$= \frac{\mu_0 I}{2\pi} \log \frac{b}{a} + \frac{\mu_0 I}{4\pi a^2}(a^2 - R^2); \quad 0 < R < a.$$

The magnetic energy in a unit length is

$$U'_m = \frac{1}{2} \int_0^a A(R) \frac{I}{\pi a^2} 2\pi R\,dR = \frac{\mu_0 I^2}{4\pi}\left(\log \frac{b}{a} + \frac{1}{4}\right),$$

and the self-inductance in a unit length is

$$L' = \frac{\mu_0}{2\pi}\left(\log\frac{b}{a} + \frac{1}{4}\right).$$

The difference from the result of Example 8.4 is due to the contribution from the additional magnetic flux in the inner conductor.

◆

Example 8.7 We denote the inner, middle, and outer long coaxial superconducting cylinders in Fig. 8.5 as superconductors 1–3. (a) Determine the inductance coefficients assuming that the reference point for zero vector potential is at $R = R_\infty (> R_4)$, and (b) determine the magnetic energy in a unit length using the inductance coefficients when we apply currents I_1, I_2, and I_3 to superconductors 1, 2, and 3, respectively.

Fig. 8.5 Cross-section of coaxial superconducting cylinders

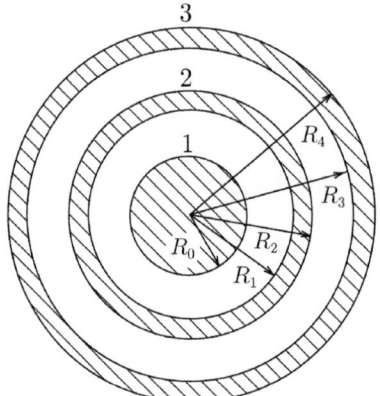

Solution 8.7 The inductance coefficients in a unit length are

$$L'_{11} = \frac{\mu_0}{2\pi} \log \frac{R_1 R_3 R_\infty}{R_0 R_2 R_4},$$

$$L'_{21} = L'_{12} = L'_{22} = \frac{\mu_0 I}{2\pi R} \log \frac{R_3 R_\infty}{R_2 R_4},$$

$$L'_{31} = L'_{32} = L'_{13} = L'_{23} = L'_{33} = \frac{\mu_0 I}{2\pi R} \log \frac{R_\infty}{R_4}.$$

From Eq. (8.18), we calculate the magnetic energy in a unit length as

$$U'_m = \frac{1}{2}L'_{11}I_1^2 + \frac{1}{2}L'_{22}I_2^2 + \frac{1}{2}L'_{33}I_3^2 + L'_{12}I_1 I_2 + L'_{23}I_2 I_3 + L'_{31}I_3 I_1$$

$$= \frac{\mu_0}{4\pi}\left[I_1^2 \log \frac{R_1 R_3 R_\infty}{R_0 R_2 R_4} + \left(I_2^2 + 2I_1 I_2\right) \log \frac{R_3 R_\infty}{R_2 R_4}\right.$$
$$\left. + \left(I_3^2 + 2I_2 I_3 + 2I_3 I_1\right) \log \frac{R_\infty}{R_4}\right]$$
$$= \frac{\mu_0}{4\pi}\left[(I_1 + I_2 + I_3)^2 \log \frac{R_\infty}{R_4} + (I_1 + I_2)^2 \log \frac{R_3}{R_2} + I_1^2 \log \frac{R_1}{R_0}\right].$$

◆

8.4 Mean Magnetic Flux

In the case where the path of the current is spread over a finite cross-sectional area, the determination of the inductance is not easy, since the path of integration to determine the magnetic flux is not clear. In this case, it is recommended to use the magnetic energy for determination of the inductance, as shown in Examples 8.5 and 8.6. We discuss here the meaning of this method.

We suppose a set of integral paths $\{S_i\}$ ($i = 1, 2, \ldots$) for the current that is spread over a finite cross-sectional area, and denote by ΔI_i and Φ_i the current flowing in the i-th path and the magnetic flux penetrating it, respectively. In this case, the mean magnetic flux is given by

$$\langle \Phi \rangle = \frac{\sum_i \Phi_i \Delta I_i}{\sum_i \Delta I_i} = \frac{\sum_i \Phi_i \Delta I_i}{I}, \qquad (8.21)$$

where I is the total current. In the case where the current density is uniform, the mean magnetic flux simply gives the mean value of the magnetic flux for each integral path.

The magnetic energy of this system is given by

$$U_m = \frac{1}{2} \sum_i \Phi_i \Delta I_i. \qquad (8.22)$$

Hence, the inductance determined using the mean magnetic flux is

$$L = \frac{\langle \Phi \rangle}{I} = \frac{2 U_m}{I^2}. \qquad (8.23)$$

Thus, it is found that the inductance determined using the magnetic energy is identical to that obtained using the mean magnetic flux.

Equation (8.21) is generally written as

$$\langle \Phi \rangle = \sum_i \Phi_i f_i, \qquad (8.24)$$

with the probability function that satisfies $\sum_i f_i = 1$. In a case in which the current is continuously distributed, this can be generalized to

$$\langle \Phi \rangle = \int \Phi(s) f(s)\, ds \qquad (8.25)$$

with

$$\int f(s)\, ds = 1, \qquad (8.26)$$

where $\Phi(s)$ is the magnetic flux in a closed circle represented by path s.

Example 8.8 Solve the problem of Example 8.6 using the mean magnetic flux.

Solution 8.8 The magnetic flux in the region from $R = R'(< a)$ to $R = b$ is

$$\Phi(R') = \frac{\mu_0 I}{2\pi a^2} \int_{R'}^{a} R\, dR + \frac{\mu_0 I}{2\pi} \int_{a}^{b} \frac{dR}{R}$$

$$= \frac{\mu_0 I}{4\pi a^2} (a^2 - R'^2) + \frac{\mu_0 I}{2\pi} \log \frac{b}{a}.$$

Thus, the mean magnetic flux is

$$\langle \Phi \rangle = \int_{0}^{a} \Phi(R') f(R')\, dR',$$

where $f(R')$ is a probability function. Since the portion of the current paths with the same distance from the central axis has a same weight, $f(R')$ is proportional to R'. So, we have

$$f(R') = \frac{2R'}{a^2}.$$

The mean magnetic flux is

$$\langle \Phi \rangle = \frac{\mu_0 I}{2\pi} \log \frac{b}{a} + \frac{\mu_0 I}{2\pi a^4} \int_{0}^{a} (a^2 R' - R'^3)\, dR' = \frac{\mu_0 I}{2\pi} \left(\log \frac{b}{a} + \frac{1}{4} \right).$$

The self-inductance in a unit length is determined to be

8.4 Mean Magnetic Flux

$$L' = \frac{\langle \Phi \rangle}{I} = \frac{\mu_0}{2\pi}\left(\log\frac{b}{a} + \frac{1}{4}\right).$$

♦

Example 8.9 We apply current I through the parallel-plate transmission line shown in Fig. 8.6. Determine the magnetic energy and self-inductance in a unit length. Discuss the difference between conducting and superconducting transmission lines.

Fig. 8.6 Parallel-plate transmission line

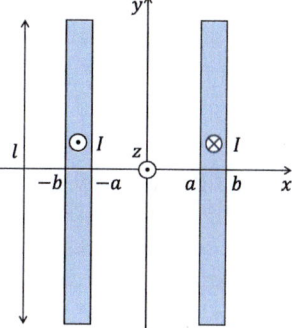

Solution 8.9 First, we treat the conducting transmission line. The magnetic flux density is

$$B_y = 0; \qquad x < -b, x > b,$$

$$= \frac{\mu_0 I(x+b)}{(b-a)l}; \quad -b < x < -a,$$

$$= \frac{\mu_0 I}{l}; \qquad -a < x < a,$$

$$= \frac{\mu_0 I(b-x)}{(b-a)l}; \quad a < x < b.$$

Hence, the magnetic energy in a unit length is

$$U'_m = \frac{1}{2}\mu_0 I^2 \left[\frac{1}{(b-a)^2 l}\int_{-b}^{-a}(x+b)^2\,dx + \frac{2a}{l} + \frac{1}{(b-a)^2 l}\int_{a}^{b}(b-x)^2\,dx\right]$$

$$= \frac{(2a+b)\mu_0 I^2}{3l}.$$

Thus, the self-inductance in a unit length is

$$L' = \frac{2(2a+b)\mu_0}{3l}.$$

For the superconducting transmission line, the magnetic flux density is zero in the regions where $-b < x < -a$ and $a < x < b$. The magnetic energy in a unit length is

$$U'_m = \frac{\mu_0 I^2 a}{l},$$

and the self-inductance in a unit length is

$$L' = \frac{2\mu_0 a}{l}.$$

The difference between the two cases comes from the magnetic energy inside the conducting regions. If we express b as $b = a(1 + \delta)$, the effective distance between the two conducting regions is $2a(1 + \delta/3)$ for the conducting case, while that for the superconducting case is $2a$. Note that this is not the mean distance, which is $2a(1 + \delta/2)$.

◆

8.5 Exercises

8.1. Prove that the magnetic flux is generally proportional to the current as stated in Eq. (8.1).

8.2. Determine the self-inductance in a unit length of the parallel superconducting transmission line shown in Fig. E8.1.

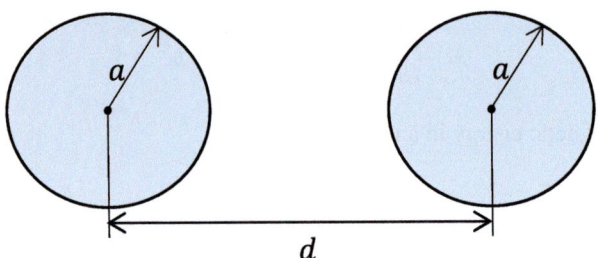

Fig. E8.1 Parallel superconducting transmission line

8.5 Exercises

Fig. E8.2 Superconducting parallel-plate transmission line with a tilted plate

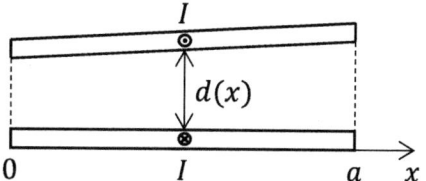

8.3. The shielding current flowing in a wide superconductor surface is calculated when a current is applied to a parallel superconducting cylinder in Example 7.5. Determine the self-inductance for the superconducting cylinder and superconductor surface.

8.4. The upper plate superconductor is slightly tilted in the parallel-plate transmission line shown in Fig. E8.2. The distance between the plates changes as $d(x) = d_0(1+kx)$. Determine the self-inductance in a unit length of this transmission line.

8.5. The upper plate superconductor is slightly tilted in a parallel-plate transmission line, as shown in Fig. E8.3. The distance between the plates changes as $d(y) = d_0(1 + ky)$ along the length. Determine the self-inductance of this transmission line.

8.6. The parallel-plate transmission line treated in Exercise 8.5 can be regarded as being composed of narrow transmission lines with different distances between the plates. Determine the self-inductance of this transmission line as an assembly of such narrow transmission lines connected in series.

8.7. Determine the mutual inductance between a parallel-wire transmission line and a closed rectangular circuit, as shown in Fig. E8.4. These are placed on a common plane, and the current direction shown by the arrows is defined as positive.

Fig. E8.3 Superconducting parallel-plate transmission line with a tilted plate

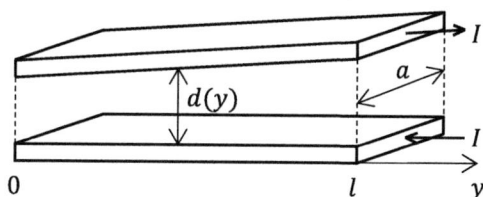

Fig. E8.4 Parallel-wire transmission line and a closed rectangular circuit

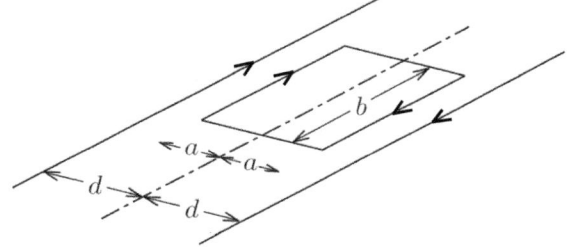

Fig. E8.5 Two parallel-wire transmission lines

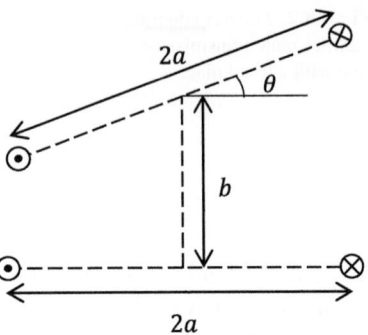

8.8. Two parallel-wire transmission lines are arranged as shown in Fig. E8.5. Determine the mutual inductance between these transmission lines in a unit length and the angle at which the mutual inductance is zero. The directions of the currents are shown in the figure.

8.9. The inductance coefficients of the long coaxial superconducting cylinders shown in Fig. 8.5 are determined in Example 8.7. We denote the inner, middle, and outer superconductors as superconductors 1, 2, and 3. Currents I_1, I_2, and I_3 are applied to superconductors 1, 2, and 3. Determine the value of I_2 when the magnetic flux between superconductor 2 and the reference point at $R = R_\infty (> R_4)$ is zero.

8.10. Design a magnet that produces a uniform transverse magnetic flux density B_0 in a long cylindrical space of radius a, as shown in Fig. E8.6.

8.11. Determine the self-inductance in a unit length of the long cylindrical coil of radius a and the total number of windings N discussed in Exercise 8.10.

8.12. Prove Eq. (8.18) by mathematical induction.

8.13. Determine the magnetic energy in a unit length when we give currents $\pm I$ to the parallel-wire transmission line discussed in Example 8.1. The radius a of the conductor is much smaller than the distance d.

8.14. Determine the magnetic energy of the parallel superconducting transmission line in a unit length discussed in Exercise 8.2, when current I is applied.

8.15. Determine the magnetic energy in a unit length when current I is applied to the superconducting transmission line in Exercise 8.4.

Fig. E8.6 Cross-sectional view of cylindrical magnet that produces a uniform transverse magnetic flux density

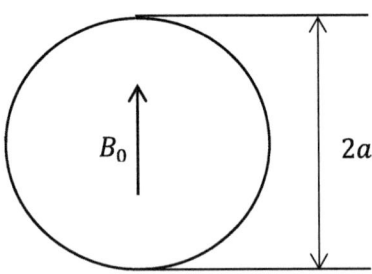

8.5 Exercises

8.16. Determine the magnetic energy when current I is applied to the superconducting transmission line in Exercise 8.5.

8.17. The magnetic energy of a parallel-plate transmission line carrying current I is treated in Example 8.9 using the magnetic energy density for conducting and superconducting transmission lines. Treat the same problem for the conducting transmission line using the vector potential.

8.18. Solve the problem in Exercise 8.17 for a superconducting transmission line using the vector potential.

8.19. Solve the problem in Exercise 8.9 using the vector potential.

8.20. We apply current I to the long cylindrical coil of radius a and total number of windings N discussed in Exercise 8.10. Determine the magnetic energy stored in a unit length of the coil using the vector potential.

8.21. Solve the same problem as in Exercise 8.20 using the magnetic energy density.

8.22. Suppose that we apply current I to the parallel superconducting transmission line treated in Exercise 8.2. Determine the magnetic energy in a unit length of the transmission line using the vector potential.

8.23. Solve the problem of Exercise 8.22 using the magnetic flux.

8.24. The magnetic energy of a coaxial transmission line with a thin outer conductor is discussed in Example 8.6 using the vector potential. Solve the same problem using the magnetic energy density.

8.25. We apply currents $\pm I$ to a coaxial transmission line made of a conductor with the cross-section shown in Fig. E8.7. Determine the self-inductance in a unit length using Eq. (8.20). The thickness of the inner conductor is neglected.

8.26. Determine the magnetic energy in a unit length of the coaxial transmission line discussed in Exercise 8.25 using the vector potential, when currents $\pm I$ are applied. The thickness of the inner conductor can be neglected.

8.27. Solve the problem of Exercise 8.25 using the mean magnetic flux.

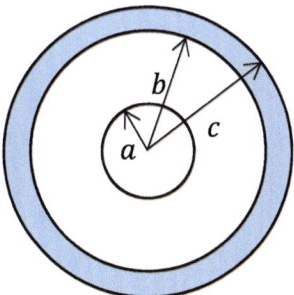

Fig. E8.7 Coaxial conducting transmission line with thin inner conductor

Fig. E8.8 Coaxial conducting transmission line

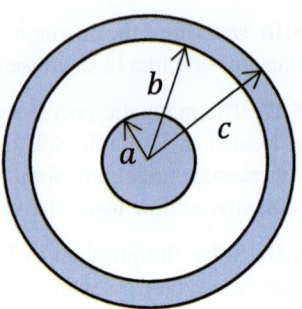

8.28. We apply currents $\pm I$ to a coaxial transmission line made of a conductor with the cross-section shown in Fig. E8.8. Determine the self-inductance in a unit length using the vector potential.

8.29. Solve the same problem as Exercise 8.28 using the mean magnetic flux.

8.30. Determine the self-inductance of a solenoid coil of inner radius a, wire thickness t, length l, and total number of windings N using the magnetic energy density when current I is applied. Assume that the current flows with a uniform density in the winding area.

8.31. Determine the magnetic energy of a solenoid coil of inner radius a, wire thickness t, length l, and total number of winding N, which is discussed in Exercise 8.30, when current I is applied. Use the vector potential.

8.32. Solve the problem of Exercise 8.30 with the mean magnetic flux.

8.33. The shielding current flowing in a wide superconducting surface is calculated when a current is applied to a parallel superconducting cylinder in Example 7.5. Determine the force between them. Discuss if it is directly derived from the magnetic energy.

Answers to Exercises

8.1. The magnetic flux density is given by Eq. (6.4) as

$$\mathbf{B}(\mathbf{r}) = \frac{\mu_0 I}{4\pi} \int_C \frac{d\mathbf{r}' \times (\mathbf{r} - \mathbf{r}')}{|\mathbf{r} - \mathbf{r}'|^3}.$$

So, the magnetic flux penetrating a surface S is given by

$$\Phi = \int_S \mathbf{B}(\mathbf{r}) \cdot d\mathbf{S} = \frac{\mu_0 I}{4\pi} \int_S \int_C \frac{d\mathbf{r}' \times (\mathbf{r} - \mathbf{r}')}{|\mathbf{r} - \mathbf{r}'|^3} \cdot d\mathbf{S},$$

where the surface integral is with respect to *r*. Thus, the magnetic flux is proved to be proportional to the current I and the proportional constant, the inductance, is determined only by the geometrical factor of the current flow.

8.2. We apply currents $\pm I$. The image currents are assumed as in the condition shown in Fig. B8.1. We define cylindrical coordinates with the z-axis on the central axis of the left superconductor. Using the result of Example 7.5, the vector potential in the space outside the superconductors is

$$A_z(R, \varphi) = \frac{\mu_0 I}{4\pi} \log \frac{R^2 + (d-h)^2 - 2R(d-h)\cos\varphi}{R^2 + h^2 - 2Rh\cos\varphi},$$

where φ is the azimuthal angle and h is given by

$$h = \frac{1}{2}\left(d - \sqrt{d^2 - 4a^2}\right).$$

Then, the azimuthal magnetic flux density is

$$B_\varphi(R, \varphi) = -\frac{\partial A_z}{\partial R}$$
$$= \frac{\mu_0 I}{2\pi}\left[\frac{R - h\cos\varphi}{R^2 + h^2 - 2Rh\cos\varphi} - \frac{R - (d-h)\cos\varphi}{R^2 + (d-h)^2 - 2R(d-h)\cos\varphi}\right].$$

On the plane including the central axes of the two superconductors ($\varphi = 0$), the magnetic flux density is directed vertical and the total magnetic flux between the two superconductors in a unit length is

$$\Phi' = \int_a^{d-a} B_\varphi(R, 0)\, dR = \frac{\mu_0 I}{2\pi} \int_a^{d-a} \left(\frac{1}{R-h} - \frac{1}{R-d+h}\right) dR$$
$$= \frac{\mu_0 I}{\pi} \log \frac{d-a-h}{a-h} = \frac{\mu_0 I}{\pi} \log \frac{d + \sqrt{d^2 - 4a^2}}{2a}.$$

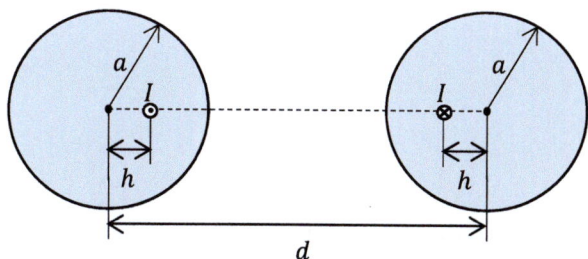

Fig. B8.1 Image current in each superconductor

Thus, the self-inductance in a unit length is determined to be

$$L' = \frac{\mu_0}{\pi} \log \frac{d + \sqrt{d^2 - 4a^2}}{2a}.$$

8.3. Using the obtained vector potential obtained in Example 7.5, the y-component of the magnetic flux density in the region $-(l-a) < x < 0$ is determined to be

$$B_y = -\frac{\partial A_z}{\partial x} = \frac{\mu_0 I}{2\pi} \left[\frac{x + \sqrt{l^2 - a^2}}{\left(x + \sqrt{l^2 - a^2}\right)^2 + y^2} - \frac{x - \sqrt{l^2 - a^2}}{\left(x - \sqrt{l^2 - a^2}\right)^2 + y^2} \right].$$

Hence, the magnetic flux between the two superconductors in a unit length is given by

$$\Phi' = \int_{a-l}^{0} B_y(y=0)\,dx = \frac{\mu_0 I}{2\pi} \int_{a-l}^{0} \left[\frac{1}{x - \sqrt{l^2 - a^2}} - \frac{1}{x + \sqrt{l^2 - a^2}} \right] dx$$

$$= \frac{\mu_0 I}{2\pi} \log \frac{\sqrt{l^2 - a^2} + l - a}{\sqrt{l^2 - a^2} - l + a} = \frac{\mu_0 I}{2\pi} \log \frac{l + \sqrt{l^2 - a^2}}{a}.$$

Thus, the self-inductance in a unit length is given by

$$L' = \frac{\Phi'}{I} = \frac{\mu_0}{2\pi} \log \frac{l + \sqrt{l^2 - a^2}}{a}.$$

8.4. We apply current I to this transmission line. Since the magnetic flux must be kept constant, the current density changes with x. So, the current density denoted by $\tau(x)$ changes as $\tau(x) = c/d(x)$ with c denoting a constant. The total current is given by

$$I = \int_0^a \tau(x)\,dx = \int_0^a \frac{c\,dx}{d_0(1 + kx)} = \frac{c}{d_0 k} \log(1 + ka).$$

Thus, we have

$$c = \frac{d_0 k I}{\log(1 + ka)}.$$

The magnetic field is $H(x) = \tau(x)$, and the magnetic flux in a unit length is

$$\Phi' = \mu_0 d_0 H(0) = \mu_0 c = \frac{\mu_0 d_0 k I}{\log(1 + ka)}.$$

8.5 Exercises

Hence, the self-inductance in a unit length is determine to be

$$L' = \frac{\Phi'}{I} = \frac{\mu_0 d_0 k}{\log(1+ka)}.$$

8.5. We apply current I to this transmission line. The magnetic field is uniform and is given by $H(y) = I/a$. So, the magnetic flux penetrating the transmission line is

$$\Phi = \int_0^l \mu_0 H(y) d(y) \, dy = \frac{\mu_0 I d_0}{a} \int_0^l (1+ky) \, dy = \frac{\mu_0 I d_0 l}{a}\left(1+\frac{kl}{2}\right).$$

Hence, the self-inductance is determined to be

$$L = \frac{\mu_0 d_0 l}{a}\left(1+\frac{kl}{2}\right).$$

8.6. When current I is applied, the magnetic flux in a narrow region from y to $y + dy$ is

$$d\Phi = \mu_0 H(y) d(y) \, dy = \frac{\mu_0 I d_0}{a}(1+ky) \, dy.$$

Hence, the self-inductance of this region is

$$dL = \frac{d\Phi}{I} = \frac{\mu_0 d_0}{a}(1+ky) \, dy.$$

Thus, the inductance is determined to be

$$L = \int dL = \frac{\mu_0 d_0}{a} \int_0^l (1+ky) \, dy = \frac{\mu_0 d_0 l}{a}\left(1+\frac{kl}{2}\right).$$

8.7. When we apply current I to the left line as shown in the arrow, the magnetic flux penetrating the rectangular circuit is directed downward and its direction is the same as that produced by the current flowing in the rectangular circuit. Its value is

$$\Phi_1 = b\frac{\mu_0 I}{2\pi} \int_{d-a}^{d+a} \frac{dr}{r} = \frac{\mu_0 I b}{2\pi} \log\frac{d+a}{d-a}.$$

Fig. B8.2 Distances between lines

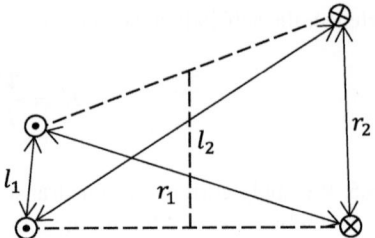

The magnetic flux produced by the right line is in the same direction and the magnitude is the same. Hence, the total magnetic flux is $\Phi = 2\Phi_1$ and the mutual inductance is determined to be

$$M = \frac{\Phi}{I} = \frac{\mu_0 b}{\pi} \log \frac{d+a}{d-a}.$$

8.8. The distances are defined as shown in Fig. B8.2. The magnetic flux in the region of distance from r_1 to r_2 from the lower right current and that in the region of distance from l_1 to l_2 from the left current penetrate the upper transmission line. So, the mutual inductance in a unit length is

$$M' = \frac{\mu_0}{2\pi} \log \frac{l_2 r_1}{l_1 r_2}$$
$$= \frac{\mu_0}{4\pi} \log \frac{[(b+a\sin\theta)^2 + (a+a\cos\theta)^2][(b-a\sin\theta)^2 + (a+a\cos\theta)^2]}{[(b-a\sin\theta)^2 + (a-a\cos\theta)^2][(b+a\sin\theta)^2 + (a-a\cos\theta)^2]}.$$

The condition of zero mutual inductance is given by

$$\frac{[(b+a\sin\theta)^2 + (a+a\cos\theta)^2][(b-a\sin\theta)^2 + (a+a\cos\theta)^2]}{[(b-a\sin\theta)^2 + (a-a\cos\theta)^2][(b+a\sin\theta)^2 + (a-a\cos\theta)^2]} = 1,$$

which is reduced to $\cos\theta = 0$, or

$$\theta = \frac{\pi}{2}.$$

8.9. The magnetic flux in a unit length between superconductor 2 and the reference point is given by

$$\Phi_2 = L_{21}I_1 + L_{22}I_2 + L_{23}I_3.$$

So, the condition $\Phi_2 = 0$ is obtained when current I_2 is equal to

$$I_2 = -\frac{L_{21}I_1 + L_{23}I_3}{L_{22}} = -I_1 - \frac{\log(R_\infty/R_4)}{\log(R_3 R_\infty/R_2 R_4)} I_3.$$

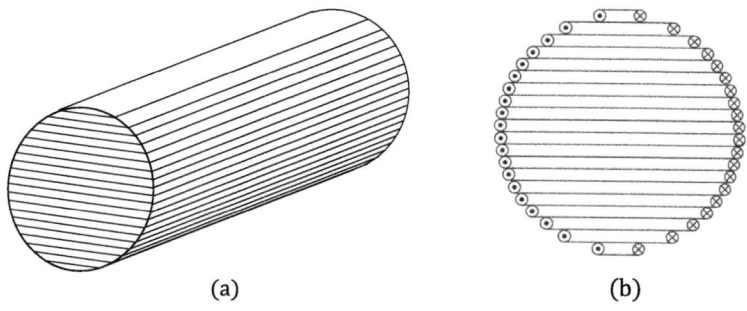

Fig. B8.3 **a** Overview of cylindrical magnet and **b** winding structure

8.10. The shielding current that flows on the cylindrical superconductor surface in a uniform transverse magnetic flux density discussed in Example 7.7 produces a uniform magnetic flux density inside the superconductor. So, if this current distribution is realized, the aimed magnet can be made in the inner space. The current density on the cylindrical superconductor surface is

$$\tau = \frac{2B_0}{\mu_0} \sin\varphi,$$

where φ is the azimuthal angle measured from the direction of the magnetic flux density B_0. To realize this current distribution, we propose a magnet shown in Fig. B8.3a. The winding structure is shown in Fig. B8.3b. If the total number of windings is N and the current is I, the current density is

$$\tau = \frac{NI}{2a} \sin\varphi,$$

and the produced uniform magnetic flux density is

$$B_0 = \frac{\mu_0 NI}{4a}.$$

8.11. When current I is applied to the coil, the magnetic flux that penetrates one turn of the coil in a unit length at the azimuthal angle φ is

$$\phi'(\varphi) = 2aB_0 \sin\varphi,$$

where $B_0 = \mu_0 NI/4a$ is the uniform magnetic flux density in the coil. Hence, the total magnetic flux penetrating the coil in a unit length is

$$\Phi' = \int_0^\pi \frac{1}{2} N\phi'(\varphi) \sin\varphi \, d\varphi = \frac{\pi}{8} \mu_0 N^2 I.$$

So, the self-inductance in a unit length is determined to be

$$L' = \frac{\Phi'}{I} = \frac{\pi}{8}\mu_0 N^2.$$

8.12. The result is apparent for $n = 1$, and the case of $n = 2$ is proved in Example 10.7. Then, we assume that the relationship holds for a system with n coils. If we can show that the relationship holds for a system with $n + 1$ coils under this condition, we can prove the relationship for any system. We denote by I_i the current flowing in the i-th coil in the n coils system. Then, we assume that the magnetic flux penetrating it and the total magnetic energy are, respectively, given by

$$\Phi_i = \sum_{j=1}^{n} L_{ij} I_j,$$

and

$$U_\mathrm{m}(n) = \frac{1}{2}\sum_{i=1}^{n}\sum_{j=1}^{n} L_{ij} I_i I_j.$$

Here, we apply current I_{n+1} to the $(n + 1)$-th coil. Then, the magnetic fluxes in other coils are given by

$$\Phi_i = \sum_{j=1}^{n+1} L_{ij} I_j$$

and the increase in the magnetic energy is given by

$$\Delta U'_\mathrm{m} = \frac{1}{2}\sum_{i=1}^{n}\sum_{j=1}^{n} L_{ij} I_i I_{n+1}.$$

The magnetic energy of the $(n + 1)$-th coil is

$$\Delta U''_\mathrm{m} = \frac{1}{2}L_{n+1\,n+1} I_{n+1}^2.$$

Hence, the total magnetic energy is given by

$$U_\mathrm{m}(n+1) = U_\mathrm{m}(n) + \Delta U'_\mathrm{m} + \Delta U''_\mathrm{m} = \frac{1}{2}\sum_{i=1}^{n+1}\sum_{j=1}^{n+1} L_{ij} I_i I_j = \frac{1}{2}\sum_{i=1}^{n+1} \Phi_i I_i.$$

Thus, the form of Eq. (8.18) holds for $n + 1$ coils. So, we can prove that Eq. (8.18) is valid generally.

8.5 Exercises

8.13. The vector potential has a component in the direction of the current and its value of the wire with current I is

$$A_{z+} = -\frac{\mu_0 I}{2\pi} \log a + \frac{\mu_0 I}{2\pi} \log(d-a) = \frac{\mu_0 I}{2\pi} \log \frac{d-a}{a}.$$

The vector potential of the wire with current $-I$ is similarly obtained as

$$A_{z-} = \frac{\mu_0 I}{2\pi} \log a - \frac{\mu_0 I}{2\pi} \log(d-a) = -\frac{\mu_0 I}{2\pi} \log \frac{d-a}{a}.$$

So, the magnetic energy in a unit length is determined to be

$$U'_m = \frac{1}{2}(A_{z+}I - A_{z-}I) = \frac{\mu_0 I^2}{2\pi} \log \frac{d-a}{a}.$$

This result can also be derived from the inductance determined in Example 8.1.

8.14. Using the result of Exercise 8.2, the vector potential of the left superconductor is

$$A_z(a, \varphi) = \frac{\mu_0 I}{4\pi} \log \frac{a^2 + (d-h)^2 - 2a(d-h)\cos\varphi}{a^2 + h^2 - 2ah\cos\varphi}$$

$$= \frac{\mu_0 I}{2\pi} \log \frac{d + \sqrt{d^2 - 4a^2}}{2a},$$

where

$$h = \frac{1}{2}\left(d - \sqrt{d^2 - 4a^2}\right).$$

So, the contribution from the left superconductor to the magnetic energy is

$$U_{m1} = \frac{1}{2} A_z(a, \varphi)I = \frac{\mu_0 I^2}{4\pi} \log \frac{d + \sqrt{d^2 - 4a^2}}{2a}.$$

The right superconductor has the same contribution, and we have

$$U_m = U_{m1} + U_{m2} = \frac{\mu_0 I^2}{2\pi} \log \frac{d + \sqrt{d^2 - 4a^2}}{2a}.$$

8.15. Since the current density on the superconductor surface is

$$\tau(x) = \frac{kI}{\log(1+ka)(1+kx)},$$

the magnetic flux density is

$$B(x) = \frac{\mu_0 kI}{\log(1+ka)(1+kx)}.$$

So, the magnetic energy in a unit length is given by

$$U'_m = \frac{1}{2\mu_0} \int_0^a B^2(x) d(x) \, dx = \frac{\mu_0 k^2 d_0 I^2}{2[\log(1+ka)]^2} \int_0^a \frac{dx}{1+kx}$$

$$= \frac{\mu_0 k d_0 I^2}{2\log(1+ka)}.$$

This is equal to $(1/2)L'I^2$ with the self-inductance L' in a unit length determined in Exercise 8.4.

8.16. Since the magnetic field in the space of the transmission line is $H(y) = I/a$, the magnetic energy is

$$U_m = \frac{\mu_0 a}{2} \int_0^l H^2(y) d(y) \, dy = \frac{\mu_0 d_0 I^2}{2a} \int_0^l (1+ky) \, dy$$

$$= \frac{\mu_0 d_0 l}{2a} \left(1 + \frac{kl}{2}\right) I^2.$$

This is equal to $(1/2)LI^2$ with the self-inductance L determined in Exercise 8.5.

8.17. Cartesian coordinates are defined in Fig. 8.6. In the case of a conducting transmission line, the distribution of the vector potential is given by

$$A_z = \frac{\mu_0 I}{2l}(a+b); \qquad\qquad x < -b,$$

$$= -\frac{\mu_0 I}{2(b-a)l}(x^2 + 2bx + a^2); \quad -b < x < -a,$$

$$= -\frac{\mu_0 I}{l}x; \qquad\qquad -a < x < a,$$

$$= \frac{\mu_0 I}{2(b-a)l}(x^2 - 2bx + a^2); \quad a < x < b,$$

$$= -\frac{\mu_0 I}{2l}(a+b); \qquad\qquad x > b.$$

Since the current flows with density $I/(b-a)l$ and $-I/(b-a)l$ for $-b < x < -a$ and $a < x < b$, respectively, the magnetic energy in a unit length is

$$U'_m = \frac{1}{2} \int_{-b}^{-a} A_z i \, dx + \frac{1}{2} \int_a^b A_z i \, dx = \frac{\mu_0 I^2 (2a+b)}{3l}.$$

8.18. In the case of a superconducting transmission line, the vector potential is given by

$$A_z = \frac{\mu_0 I}{l} a; \quad x < -a,$$
$$= -\frac{\mu_0 I}{l} x; \quad -a < x < a,$$
$$= -\frac{\mu_0 I}{l} a; \quad x > b.$$

That is, the vector potential is $\mu_0 I a/l$ for the left superconductor and $-\mu_0 I a/l$ for the right superconductor. Hence, the magnetic energy in a unit length is determined to be

$$U'_m = \frac{\mu_0 I^2 a}{2l} + \frac{\mu_0 I^2 a}{2l} = \frac{\mu_0 I^2 a}{l}.$$

8.19. The magnetic flux density is given by

$$B_\varphi = \frac{\mu_0 I_1}{2\pi R}; \quad R_0 < R < R_1,$$
$$= \frac{\mu_0 (I_1 + I_2)}{2\pi R}; \quad R_2 < R < R_3,$$
$$= \frac{\mu_0 (I_1 + I_2 + I_3)}{2\pi R}; \quad R > R_4,$$
$$= 0; \quad \text{otherwise}.$$

Hence, the vector potential of superconductors 1, 2, and 3 are:

$$A_{\varphi 3} = \frac{\mu_0 (I_1 + I_2 + I_3)}{2\pi} \log \frac{R_\infty}{R_4},$$
$$A_{\varphi 2} = \frac{\mu_0 (I_1 + I_2)}{2\pi} \log \frac{R_3}{R_2} + \frac{\mu_0 (I_1 + I_2 + I_3)}{2\pi} \log \frac{R_\infty}{R_4},$$
$$A_{\varphi 1} = \frac{\mu_0 I_1}{2\pi} \log \frac{R_1}{R_0} + \frac{\mu_0 (I_1 + I_2)}{2\pi} \log \frac{R_3}{R_2} + \frac{\mu_0 (I_1 + I_2 + I_3)}{2\pi} \log \frac{R_\infty}{R_4}.$$

Thus, the magnetic energy in a unit length is determined to be

$$U'_m = \frac{1}{2}(A_{\varphi 1} I_1 + A_{\varphi 2} I_2 + A_{\varphi 3} I_3)$$
$$= \frac{\mu_0}{4\pi} \left[(I_1 + I_2 + I_3)^2 \log \frac{R_\infty}{R_4} + (I_1 + I_2)^2 \log \frac{R_3}{R_2} + I_1^2 \log \frac{R_1}{R_0} \right].$$

8.20. The vector potential due to the current is equivalent to that due to the shielding current flowing on the surface of the long cylindrical superconductor in Example 7.7. This is given by

$$A_z(a, \varphi) = \frac{\mu_0 |\hat{m}|}{2\pi a} \sin \varphi = B_0 a \sin \varphi,$$

where $|\hat{m}|$ is the magnetic moment in a unit length produced by the current, and $B_0 = \mu_0 NI/4a$ is the uniform magnetic flux density in the coil. The current that flows in the region from φ to $\varphi + d\varphi$ in the azimuthal angle is

$$\tau a \, d\varphi = \frac{NI}{2} \sin \varphi \, d\varphi.$$

Hence, the magnetic energy in a unit length is

$$U'_m = \frac{1}{2} \int_0^{2\pi} A_z(a, \varphi) \tau a \, d\varphi = \frac{1}{16} \mu_0 N^2 I^2 \int_0^{2\pi} \sin^2 \varphi \, d\varphi = \frac{\pi}{16} \mu_0 N^2 I^2.$$

This result agrees with the magnetic energy determined with the self-inductance determined in Exercise 8.11.

8.21. When current I is applied to the long cylindrical coil, the magnetic flux density inside the coil is

$$B = B_0 = \frac{\mu_0 NI}{4a}.$$

The magnetic energy inside the coil in a unit length is

$$U'_{m1} = \frac{B_0^2}{2\mu_0} \pi a^2 = \frac{\pi}{32} \mu_0 N^2 I^2.$$

Outside the coil, the radial and azimuthal components of the magnetic flux density are

$$B_R = B_0 \frac{a^2}{R^2} \cos \varphi, \quad B_\varphi = B_0 \frac{a^2}{R^2} \sin \varphi.$$

Hence, the magnetic energy outside the coil in a unit length is

$$U'_{m2} = \frac{1}{2\mu_0} \int_a^\infty (B_R^2 + B_\varphi^2) 2\pi R \, dR = \frac{\pi}{\mu_0} B_0^2 a^4 \int_a^\infty \frac{dR}{R^3} = \frac{\pi}{32} \mu_0 N^2 I^2.$$

Thus, the magnetic energy in a unit length is

8.5 Exercises

$$U'_m = U'_{m1} + U'_{m2} = \frac{\pi}{16}\mu_0 N^2 I^2.$$

This agrees with the result of Exercise 8.20.

8.22. Using the result of Exercise 8.2, the vector potential of the left superconductor is

$$A_z(a, \varphi) = \frac{\mu_0 I}{4\pi} \log \frac{a^2 + (d-h)^2 + 2a(d-h)\cos\varphi}{a^2 + h^2 + 2ah\cos\varphi} = \frac{\mu_0 I}{2\pi} \log \frac{d + \sqrt{d^2 - 4a^2}}{2a}.$$

Hence, the magnetic energy of the left superconductor in a unit length is

$$U'_{m1} = \frac{1}{2} A_z(a, \varphi) I = \frac{\mu_0 I^2}{4\pi} \log \frac{d + \sqrt{d^2 - 4a^2}}{2a}.$$

The magnetic energy of the right superconductor, U'_{m2}, is the same. The total magnetic energy in a unit length is

$$U'_m = U'_{m1} + U'_{m2} = \frac{\mu_0 I^2}{2\pi} \log \frac{d + \sqrt{d^2 - 4a^2}}{2a}.$$

8.23. Using the result of Exercise 8.2, the magnetic flux between the two superconductors in a unit length when current I flows is

$$\Phi' = \frac{\mu_0 I}{\pi} \log \frac{d + \sqrt{d^2 - 4a^2}}{2a}.$$

Hence, the magnetic energy in a unit length is determined to be

$$U'_m = \frac{1}{2} \Phi' I = \frac{\mu_0 I^2}{2\pi} \log \frac{d + \sqrt{d^2 - 4a^2}}{2a}.$$

8.24. The magnetic flux density is given by

$$B_\varphi = \frac{\mu_0 I R}{2\pi a^2}; \quad 0 < R < a,$$

$$= \frac{\mu_0 I}{2\pi R}; \quad a < R < b,$$

$$= 0; \quad R > b.$$

Hence, the magnetic energy is a unit length is determined to be

$$U'_m = \frac{1}{2\mu_0} \int B_\varphi^2 2\pi R \, dR = \frac{\mu_0 I^2}{4\pi a^4} \int_0^a R^3 \, dR + \frac{\mu_0 I^2}{4\pi} \int_a^b \frac{dR}{R}$$

$$= \frac{\mu_0 I^2}{4\pi} \left(\log \frac{b}{a} + \frac{1}{4} \right).$$

8.25. The magnetic flux density is

$$B(R) = \frac{\mu_0 I}{2\pi R}; \qquad\qquad a < R < b,$$

$$= \frac{\mu_0 I}{2\pi (c^2 - b^2)} \left(\frac{c^2}{R} - R \right); \quad b < R < c,$$

$$= 0; \qquad\qquad \text{otherwise.}$$

The magnetic energy in a unit length is

$$U'_m = \frac{\mu_0 I^2}{4\pi} \left[\int_a^b \frac{dR}{R} + \frac{1}{(c^2 - b^2)^2} \int_b^c \left(\frac{c^2}{R} - R \right)^2 R \, dR \right]$$

$$= \frac{\mu_0 I^2}{4\pi} \left[\log \frac{b}{a} + \frac{c^4}{(c^2 - b^2)^2} \log \frac{c}{b} - \frac{3c^2 - b^2}{4(c^2 - b^2)} \right].$$

So, the self-inductance is determined to be

$$L' = \frac{\mu_0}{2\pi} \left[\log \frac{b}{a} + \frac{c^4}{(c^2 - b^2)^2} \log \frac{c}{b} - \frac{3c^2 - b^2}{4(c^2 - b^2)} \right].$$

8.26. The vector potential has a z-component given by

$$A_z(R) = \frac{\mu_0 I}{2\pi} \log \frac{b}{R} + \frac{\mu_0 I}{2\pi (c^2 - b^2)} \left[c^2 \log \frac{c}{b} - \frac{1}{2}(c^2 - b^2) \right]; \quad a < R < b,$$

$$= \frac{\mu_0 I}{2\pi (c^2 - b^2)} \left[c^2 \log \frac{c}{R} - \frac{1}{2}(c^2 - R^2) \right]; \qquad\qquad b < R < c.$$

The current density in the region from $R = b$ to $R = c$ is $i = -I/\pi(c^2 - b^2)$. Hence, the magnetic energy in a unit length is determined to be

8.5 Exercises

$$U'_m = \frac{1}{2}A_z(a)I - \frac{I}{2\pi(c^2-b^2)}\int_b^c A_z(R)2\pi R\,dR$$

$$= \frac{\mu_0 I^2}{4\pi}\log\frac{b}{a} + \frac{\mu_0 I^2}{4\pi(c^2-b^2)}\left[c^2\log\frac{c}{b} - \frac{1}{2}(c^2-b^2)\right]$$

$$- \frac{\mu_0 I^2}{2\pi(c^2-b^2)^2}\left[\frac{1}{8}(c^4-b^4) - \frac{b^2 c^2}{2}\log\frac{c}{b}\right]$$

$$= \frac{\mu_0 I^2}{4\pi}\left[\log\frac{b}{a} + \frac{c^4}{(c^2-b^2)^2}\log\frac{c}{b} - \frac{3c^2-b^2}{4(c^2-b^2)}\right].$$

8.27. The magnetic flux in the region from $R=a$ to $R=R'(>b)$ is

$$\Phi(R') = \frac{\mu_0 I}{2\pi}\int_a^b \frac{dR}{R} + \frac{\mu_0 I}{2\pi(c^2-b^2)}\int_b^{R'}\left(\frac{c^2}{R} - R\right)dR$$

$$= \frac{\mu_0 I}{2\pi}\left[\log\frac{b}{a} + \frac{1}{c^2-b^2}\left(c^2\log\frac{R'}{b} - \frac{R'^2-b^2}{2}\right)\right].$$

Thus, the mean magnetic flux is

$$\langle\Phi\rangle = \int_b^c \Phi(R')f(R')\,dR',$$

where $f(R')$ is a weight function. Since the portion of the current path with the same distance from the central axis has a same weight, $f(R')$ is proportional to R' within the range of R' from b to c. So, we have

$$f(R') = \frac{2R'}{c^2-b^2}.$$

The mean magnetic flux is

$$\langle\Phi\rangle = \frac{\mu_0 I}{2\pi}\log\frac{b}{a} + \frac{\mu_0 I}{2\pi(c^2-b^2)^2}\int_b^c \left(2c^2 R'\log\frac{R'}{b} - R'^3 + b^2 R'\right)dR'$$

$$= \frac{\mu_0 I}{2\pi}\left[\log\frac{b}{a} + \frac{c^4}{(c^2-b^2)^2}\log\frac{c}{b} - \frac{3c^2-b^2}{4(c^2-b^2)}\right].$$

The self-inductance is determined to be

$$L' = \frac{\langle \Phi \rangle}{I} = \frac{\mu_0}{2\pi}\left[\log\frac{b}{a} + \frac{c^4}{(c^2-b^2)^2}\log\frac{c}{b} - \frac{3c^2-b^2}{4(c^2-b^2)}\right].$$

8.28. The vector potential is

$$A_z(R) = \frac{\mu_0 I}{2\pi}\log\frac{b}{a} + \frac{\mu_0 I}{4\pi a^2}(a^2 - R^2); \qquad 0 < R < a,$$

$$= \frac{\mu_0 I}{2\pi}\log\frac{b}{R} + \frac{\mu_0 I}{2\pi(c^2-b^2)}\left[c^2\log\frac{c}{b} - \frac{1}{2}(c^2-b^2)\right]; \qquad a < R < b,$$

$$= \frac{\mu_0 I}{2\pi(c^2-b^2)}\left[c^2\log\frac{c}{R} - \frac{1}{2}(c^2 - R^2)\right]; \qquad b < R < c.$$

The current density is $i = I/\pi a^2$ in the region $0 \le R < a$ and $i' = -I/\pi(c^2-b^2)$ in the region $b < R < c$. Hence, the magnetic energy in a unit length is determined to be

$$U'_m = \frac{I}{2\pi a^2}\int_0^a A_z(R) 2\pi R \, dR - \frac{I}{2\pi(c^2-b^2)}\int_b^c A_z(R) 2\pi R\, dR$$

$$= \frac{\mu_0 I^2}{4\pi}\left(\log\frac{b}{a} + \frac{1}{4}\right) + \frac{\mu_0 I^2}{4\pi(c^2-b^2)}\left[c^2\log\frac{c}{b} - \frac{1}{2}(c^2-b^2)\right]$$

$$- \frac{\mu_0 I^2}{2\pi(c^2-b^2)^2}\left[\frac{1}{8}(c^4 - b^4) - \frac{b^2c^2}{2}\log\frac{c}{b}\right]$$

$$= \frac{\mu_0 I^2}{4\pi}\left[\log\frac{b}{a} + \frac{c^4}{(c^2-b^2)^2}\log\frac{c}{b} - \frac{c^2}{2(c^2-b^2)}\right].$$

8.29. The method of the mean magnetic flux is used for an inner thin conductor in Example 8.8, and for an outer thin conductor in Exercise 8.27. From these results, the mean magnetic flux in a unit length is determined to be

$$\Phi(R') = \frac{\mu_0 I}{2\pi}\left[\log\frac{b}{a} + \frac{1}{4} + \frac{c^4}{(c^2-b^2)^2}\log\frac{c}{b} - \frac{3c^2-b^2}{4(c^2-b^2)}\right].$$

Thus, the self-inductance in a unit length is

$$L' = \frac{\mu_0}{2\pi}\left[\log\frac{b}{a} + \frac{1}{4} + \frac{c^4}{(c^2-b^2)^2}\log\frac{c}{b} - \frac{3c^2-b^2}{4(c^2-b^2)}\right].$$

The obtained result agrees with the result of Example 8.5 using the magnetic energy density and the result of Exercise 8.28.

8.30. The magnetic flux density is

$$B = \mu_0 \frac{N}{l} I; \qquad 0 \leq R < a,$$

$$= \mu_0 \frac{N}{l} I \left(1 - \frac{R-a}{t}\right); \qquad a < R < a+t.$$

So, the magnetic energy in the magnet bore is

$$U_{m1} = \frac{1}{2\mu_0} \left(\mu_0 \frac{N}{l} I\right)^2 \pi a^2 l = \frac{\pi \mu_0}{2l} (NaI)^2$$

and that in the winding area is

$$U_{m2} = \frac{\mu_0 (NI)^2}{2l} \int_a^{a+t} \left(1 - \frac{R-a}{t}\right)^2 2\pi R\, dR = \frac{\pi \mu_0 (NI)^2}{3l} t \left(a + \frac{t}{4}\right).$$

Thus, the magnetic energy is

$$U_m = U_{m1} + U_{m2} = \frac{\pi \mu_0 (NI)^2}{2l} \left(a^2 + \frac{2at}{3} + \frac{t^2}{6}\right)$$

and the self-inductance is determined to be

$$L = \frac{\pi \mu_0 N^2}{l} \left(a^2 + \frac{2at}{3} + \frac{t^2}{6}\right).$$

8.31. The vector potential in the winding area ($a < R < a+t$) is given by

$$A_\varphi = \frac{1}{2} B_0 a + \frac{1}{tR} B_0 \left[-\frac{1}{3} R^3 + \frac{1}{2}(a+t)R^2 - \frac{1}{6} a^2 (a+3t)\right],$$

where $B_0 = \mu_0 NI/l$ is a uniform magnetic flux density in the bore. It can be easily shown that the above vector potential gives the magnetic flux density derived in Exercise 8.30 from $B = (\partial RA_\varphi / \partial R)/R$. The current density in this area is $i = NI/lt$. Hence, the magnetic energy is determined to be

$$U_m = \frac{l}{2} \int_a^{a+t} A_\varphi i 2\pi R\, dR = \frac{\pi \mu_0 (NI)^2}{2l} \left(a^2 + \frac{2at}{3} + \frac{t^2}{6}\right).$$

8.32. The magnetic flux density is

$$B = \mu_0 \frac{N}{l} I; \qquad 0 \leq R < a,$$

$$= \mu_0 \frac{N}{l} I \left(1 - \frac{R-a}{t}\right); \qquad a < R < a+t.$$

So, the magnetic flux in the magnet bore that interlinks the winding is

$$\Phi_0 = \frac{\pi \mu_0 N^2 a^2 I}{l}.$$

Next, the mean magnetic flux in the winding area is calculated. The magnetic flux in the region from a to $R (a < R < a+t)$ is given by

$$\Delta\Phi(R) = \frac{\mu_0 N^2 I}{l} \int_a^R \left(1 - \frac{R'-a}{t}\right) 2\pi R' \, dR'$$

$$= \frac{2\pi \mu_0 N^2 I}{l} \left[\frac{a+t}{2t}(R^2 - a^2) - \frac{1}{3t}(R^3 - a^3)\right].$$

Hence, the mean magnetic flux in the winding area is

$$\Delta\Phi = \frac{1}{t} \int_a^{a+t} \Delta\Phi(R) \, dR = \frac{2\pi \mu_0 N^2 I}{3l} t \left(a + \frac{t}{4}\right).$$

The total mean magnetic flux is

$$\langle\Phi\rangle = \Phi_0 + \Delta\Phi = \frac{\pi \mu_0 N^2 I}{l}\left(a^2 + \frac{2at}{3} + \frac{t^2}{6}\right)$$

and the self-inductance is determined to be

$$L = \frac{\pi \mu_0 N^2}{l}\left(a^2 + \frac{2at}{3} + \frac{t^2}{6}\right).$$

8.33. The self-inductance of this system is calculated in Exercise 8.3. The result shows that the inductance changes with the distance between the superconducting cylinder and the superconducting surface. It means that the magnetic flux comes in or goes out along the length. So, the electromagnetic induction is involved and the force cannot be calculated from the magnetic energy. The force can be simply calculated as a force between two image currents separated by $2(l-h)$. That is, the force is attractive and its value in a unit length is

$$F' = \frac{\mu_0 I^2}{4\pi(l-h)} = \frac{\mu_0 I^2}{4\pi\sqrt{l^2-a^2}}.$$

Coffee Break: Prediction of Superconductivity

Now, the *E*-*B* analogy is commonly used in electricity and magnetism. If this analogy was believed just after the formulation of Maxwell's theory in the nineteenth century, someone might have happened to predict a material showing perfect diamagnetism, $B = 0$, which was analogous to a conductor showing $E = 0$. Since it can be easily proved that such a material has zero electric resistivity, he might be a person who predicted a superconductor before 1911 in which the superconductor was discovered.

Here, the proof of the zero-resistivity is treated. We assume that a current is applied to this material. In this case, the current must flow on the surface of the material so that magnetic flux does not invade the material. We suppose that rectangle C is located inside the material in such a way that one side is placed on the surface and parallel to the current, as shown in Fig. 8.18. The current density is integrated on C. The integrated value is not zero on the side on the surface, while the contribution from other three sides is zero. That is, we have

$$\oint_C \boldsymbol{i} \cdot \mathrm{d}\boldsymbol{s} \neq 0.$$

If Ohm's law $\boldsymbol{E} = \rho_r \boldsymbol{i}$ holds, the above equation leads to

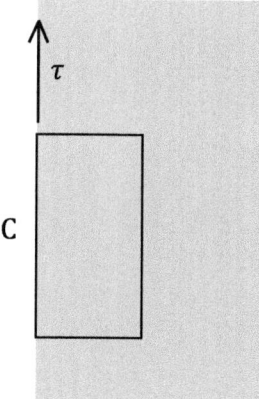

Fig. 8.18 Rectangle C in superconductor with one side on the surface on which current flows

$$\oint_C \boldsymbol{E} \cdot \mathrm{d}\boldsymbol{s} \neq 0.$$

The left-hand side must be zero for a static electric field, however. To be consistent with the first equation, the electric resistivity ρ_r of this material must be zero. So, the material that shows $\boldsymbol{B} = 0$ under any conditions is nothing else than a superconductor. Thus, the prediction of a diamagnetic material is equivalent to the prediction of a superconductor.

Chapter 9
Magnetic Materials

9.1 Magnetization

Some materials possess a magnetic moment that causes magnetic phenomena when an external magnetic flux density is applied to these materials. These materials are classified as magnetic materials.

The appearance of the magnetic moment in magnetic materials is analogous to the way in which electric polarization arises because of a relative displacement between electric charges of different signs in an external electric field. In fact, it looks as if the magnetic moment appears because of a relative displacement of magnetic charges of different signs in an external magnetic flux density, as shown in Eq. (6.45). Magnetic charges do not exist, however, as discussed in Sect. 6.8. On the other hand, we have learned that the magnetic moment can be equivalently expressed by closed currents. Such an equivalent current that produces the magnetic moment is called a magnetizing current, although true currents produce a static magnetic moment only in superconductors. The magnetizing current in a magnetic material cannot be taken outside. This is also similar to a polarization charge, which cannot be taken out of a dielectric material. Thus, the magnetizing current is a virtual substance like the magnetic charge. The magnetic moment is caused by electron spins, electron orbital motion, nuclear magnetic moments, etc.

The resultant magnetic moment in a unit volume of a material is called the magnetization and is represented by M. Its unit is [A/m]. This definition of magnetization is the same as that for superconductors. The magnetization M in a magnetic material is usually directed parallel to an applied magnetic flux density, B, and is proportional to $B = |B|$. Thus, it can be expressed as

$$M = \frac{\chi_m}{\mu_0} B, \qquad (9.1)$$

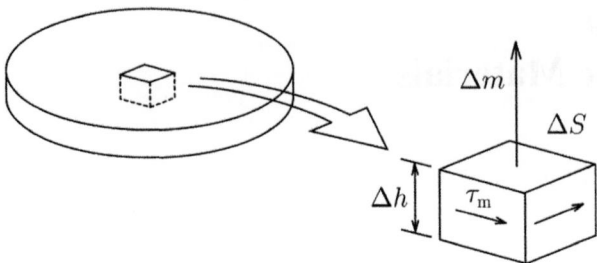

Fig. 9.1 Thin magnetic slab magnetized in the normal direction with an enlarged small part

where the dimensionless proportional constant χ_m is called the magnetic susceptibility.

Suppose a thin magnetic slab of thickness Δh that is magnetized uniformly in the normal direction. The magnetization of a small region in this slab is expressed by magnetizing current flowing around it (see Fig. 9.1). We denote the area of the top surface and the surface density of the magnetizing current by ΔS and τ_m, respectively. The magnetizing current flowing around the small region is $\tau_m \Delta h$, and the magnetic moment is $\Delta m = \tau_m \Delta h \Delta S = \tau_m \Delta V$ with $\Delta V = \Delta S \Delta h$ denoting the volume of this small region. Hence, the magnetization is given by

$$M = \frac{\Delta m}{\Delta V} = \tau_m. \tag{9.2}$$

Namely, the magnetization is equal to the surface density of the magnetizing current. When the magnetization is uniform, the magnetizing current flows only on the periphery of the magnetic material.

We assume that the magnetization is not uniform in region V occupied by a magnetic material. The magnetic moment of a small region of volume dV' at point A positioned at r' is $m(r') = M(r')dV'$. Using Eq. (6.25), the vector potential produced by this magnetic moment at observation point P at r is

$$dA = \frac{\mu_0}{4\pi} \cdot \frac{M(r') \times (r - r')}{|r - r'|^3} dV'. \tag{9.3}$$

Hence, the total vector potential is given by

$$A(r) = \frac{\mu_0}{4\pi} \int_V \frac{M(r') \times (r - r')}{|r - r'|^3} dV', \tag{9.4}$$

which transforms to

9.1 Magnetization

$$A(r) = \frac{\mu_0}{4\pi} \int_V \frac{\nabla' \times M(r')}{|r - r'|} dV', \quad (9.5)$$

where the operation, $\nabla' \times$, is curl with respect to r'. On comparing this and Eq. (6.20), we understand that this vector potential is produced by the magnetizing current density,

$$\nabla \times M = i_m. \quad (9.6)$$

In addition, $\mu_0 M$ represents the magnetic flux density produced by the magnetizing current. Thus, the following equation holds:

$$\nabla \cdot M = 0. \quad (9.7)$$

Suppose that a magnetic sphere of radius a is in a uniform magnetic flux density B_0, as shown in Fig. 9.2a. Here, we determine the magnetizing current density on the surface of the magnetic material. The magnitude of the magnetization, M, is denoted by M. We define spherical coordinates with the origin at the center of the sphere and the axis along the direction of the applied magnetic flux density. We denote by θ the zenithal angle measured from this axis, as shown in Fig. 9.2b. We slice the spherical magnetic material into thin plates and denote the surface magnetizing current density on the edge of a plate by τ. Equation (9.2) gives $\tau = M$. If we denote the thickness of the thin plate and the corresponding zenithal angle interval by dh and $d\theta$, respectively, we have $dh = a \sin\theta d\theta$. The magnetizing current that flows on the edge of this plate is $\tau dh = \tau_m a d\theta$. Thus, the surface magnetizing current density is given by

$$\tau_m = \tau \sin\theta = M \sin\theta. \quad (9.8)$$

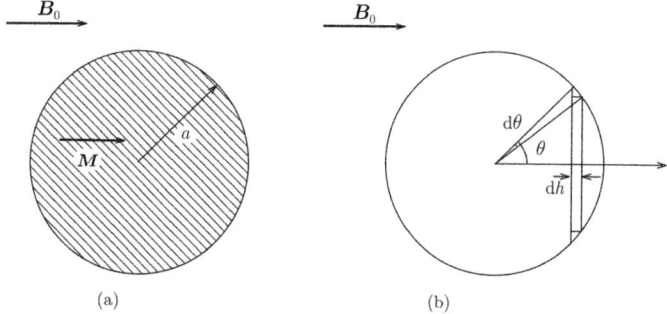

Fig. 9.2 **a** Magnetization of a magnetic sphere in a magnetic flux density and **b** thin plate of magnetic material

9.2 Magnetic Field

When a current of density i and a magnetizing current of density i_m coexist, the vector potential is given by

$$A(r) = \frac{\mu_0}{4\pi} \int_V \frac{i(r') + i_m(r')}{|r - r'|} dV'. \tag{9.9}$$

The curl of the magnetic flux density is

$$\nabla \times B = \mu_0 (i + i_m). \tag{9.10}$$

Here, we define a new physical quantity

$$H = \frac{1}{\mu_0} B - M. \tag{9.11}$$

Then, we have

$$\nabla \times H = i. \tag{9.12}$$

That is, H is a variable that corresponds only to the current and is called the magnetic field or magnetic field strength. The unit of the magnetic field is [A/m] and is the same as for the magnetization. Equation (5.8) is also satisfied in this case, and hence, this is the magnetic field produced by a steady current.

Since the magnetic field H is the variable associated only with a current, the Biot-Savart law and Ampere's law are the laws for H. Namely, the general form of the Biot-Savart law is given by

$$H(r) = \frac{1}{4\pi} \int_V \frac{i(r') \times (r - r')}{|r - r'|^3} dV \tag{9.13}$$

instead of Eq. (6.5). The general form of Ampere's law is now written as

$$\oint_C H \cdot ds = \int_S i \cdot dS \tag{9.14}$$

instead of Eq. (6.15). Equation (9.12) is the general differential form of Ampere's law.

Usually, the definition of the magnetization is given by

$$M = \chi_m H. \tag{9.15}$$

Since the measurement of a magnetic material is done in vacuum with $H = B/\mu_0$, there is no contradiction between Eqs. (9.1) and (9.15). The magnetic field strength is defined to be

$$H = \frac{1}{\mu} B, \tag{9.16}$$

where μ is the magnetic permeability given by

$$\mu = \mu_0 (1 + \chi_m). \tag{9.17}$$

Substituting Eqs. (9.12) and (9.16) into Eq. (6.18), we have

$$\nabla \times (\nabla \times A) = \mu i. \tag{9.18}$$

Using the Coulomb gauge, Eq. (6.19), for static magnetism, the above equation gives Poisson's equation:

$$\Delta A = -\mu i. \tag{9.19}$$

In a region in which $i = 0$, Eq. (9.19) leads to Laplace's equation.

9.3 Boundary Conditions

Here, we describe the boundary conditions to be fulfilled for the magnetic flux density and magnetic field at an interface between two different magnetic materials with magnetic permeabilities μ_1 and μ_2. Assume that a planar current density, τ, exists on the interface.

The boundary condition on the magnetic flux density is given by

$$n \cdot (B_1 - B_2) = 0, \tag{9.20}$$

where n is the normal unit vector on the boundary directed from magnetic material 2 to 1. The boundary condition on the magnetic field strength is given by

$$n \times (H_1 - H_2) = \tau. \tag{9.21}$$

Using the vector potential, the boundary conditions (9.20) and (9.21) are, respectively, written as

$$\boldsymbol{n} \cdot (\nabla \times \boldsymbol{A}_1 - \nabla \times \boldsymbol{A}_2) = 0, \tag{9.22}$$

and

$$\boldsymbol{n} \times \left(\frac{1}{\mu_1} \nabla \times \boldsymbol{A}_1 - \frac{1}{\mu_2} \nabla \times \boldsymbol{A}_2\right) = \boldsymbol{\tau}. \tag{9.23}$$

If we denote the unit vector along the planar current on the interface by $\boldsymbol{a}$, we have

$$\nabla \times \boldsymbol{A} = -\frac{\partial A_a}{\partial n} \boldsymbol{t} + \frac{\partial A_a}{\partial t} \boldsymbol{n},$$

where A_a is the component of the vector potential along $\boldsymbol{a}$, $\boldsymbol{t} = \boldsymbol{a} \times \boldsymbol{n}$ is a unit vector on the interface, and $\partial/\partial n$ and $\partial/\partial t$ are the derivatives along the directions of $\boldsymbol{n}$ and $\boldsymbol{t}$. Hence, Eq. (9.22) gives

$$\frac{\partial A_{a1}}{\partial t} = \frac{\partial A_{a2}}{\partial t}.$$

Integrating this along the direction of $\boldsymbol{t}$, we have

$$A_{a1} = A_{a2}. \tag{9.24}$$

Equation (9.23) leads to

$$\frac{1}{\mu_1} \cdot \frac{\partial A_{a1}}{\partial n} - \frac{1}{\mu_2} \cdot \frac{\partial A_{a1}}{\partial n} = -\tau. \tag{9.25}$$

Here, we discuss refraction of magnetic flux lines at a boundary using the boundary conditions. Suppose an interface between two magnetic materials with magnetic permeabilities μ_1 and μ_2. Assume that magnetic flux density B_1 is applied to magnetic material 1 in the direction of angle θ_1 measured from the normal direction to the interface, as shown in Fig. 9.3. The strength and angle of the magnetic flux density in magnetic material 2 are denoted by B_2 and θ_2. The continuity of the normal component of the magnetic flux density gives

$$B_1 \cos\theta_1 = B_2 \cos\theta_2. \tag{9.26}$$

Since planar current does not usually exist on the interface, the continuity of the parallel component of the magnetic field is written as

9.3 Boundary Conditions

Fig. 9.3 Refraction of magnetic flux lines at interface

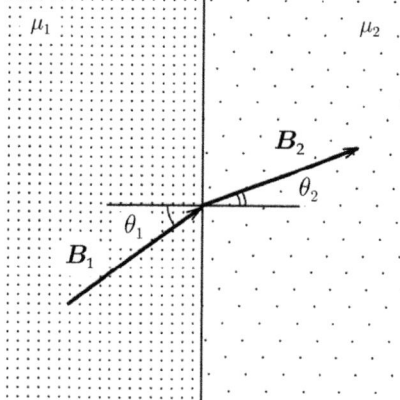

$$\frac{B_1}{\mu_1}\sin\theta_1 = \frac{B_2}{\mu_2}\sin\theta_2. \tag{9.27}$$

These equations give

$$\frac{\tan\theta_1}{\tan\theta_2} = \frac{\mu_1}{\mu_2}. \tag{9.28}$$

This is the law of refraction. We obtain B_2 and θ_2 as

$$B_2 = B_1 \left[\left(\frac{\mu_2}{\mu_1}\right)^2 \sin^2\theta_1 + \cos^2\theta_1\right]^{1/2}, \tag{9.29}$$

$$\theta_2 = \tan^{-1}\left(\frac{\mu_2}{\mu_1}\tan\theta_1\right). \tag{9.30}$$

Example 9.1 Magnetic materials 1 and 2 with magnetic permeabilities μ_1 and μ_2 each occupy half of the space between two long parallel superconducting plates, as shown in Fig. 9.4. We apply current I to each superconducting plate in opposite directions. Determine the magnetic flux density and magnetic field in each magnetic material and the self-inductance in a unit length of superconducting plates. Assume that the width of the plate, w, is sufficiently long in comparison with the distance, d, between the two plates.

Fig. 9.4 Two superconducting plates with magnetic materials with different magnetic permeabilities

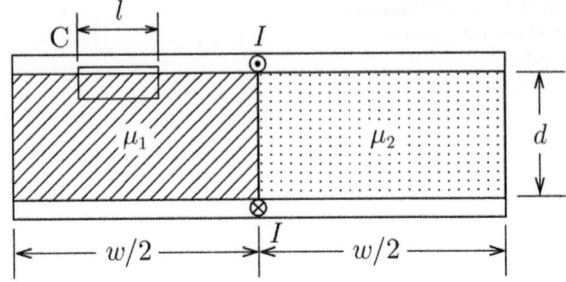

Solution 9.1 The magnetic flux density and magnetic field are directed parallel to the superconducting plates in the space between them. We denote these values in magnetic materials 1 and 2 by B_1, H_1, B_2, and H_2. From Eq. (9.20) we have

$$B_1 = B_2,$$

which gives

$$H_2 = \frac{\mu_1}{\mu_2} H_1.$$

We apply Eq. (9.14) to rectangle C in Fig. 9.4. If the surface current density on the superconducting plate in this region is τ_1, we have $H_1 l = \tau_1 l$. This reduces to

$$H_1 = \tau_1.$$

The surface current density on the superconducting plate in contact with magnetic material 2 is

$$\tau_2 = H_2 = \frac{\mu_1}{\mu_2} \tau_1.$$

Since the total current I is equal to $(w/2)(\tau_1 + \tau_2)$, we obtain

$$H_1 = \frac{2\mu_2 I}{w(\mu_1 + \mu_2)}, \quad H_2 = \frac{2\mu_1 I}{w(\mu_1 + \mu_2)},$$

and

$$B_1 = B_2 = \frac{2\mu_1 \mu_2 I}{w(\mu_1 + \mu_2)}.$$

9.3 Boundary Conditions

The magnetic flux that penetrates the superconducting plates in a unit length is

$$\Phi' = \frac{2\mu_1\mu_2 dI}{w(\mu_1 + \mu_2)},$$

and we obtain the self-inductance in a unit length as

$$L' = \frac{2\mu_1\mu_2 d}{w(\mu_1 + \mu_2)}.$$

◆

Example 9.2 We apply current I to two long parallel superconducting plates with magnetic materials that have magnetic permeabilities μ_1 and μ_2 between them, as shown in Fig. 9.5. Assume that the width of the plate, w, is sufficiently larger than the distance, d, between the plates. Determine the magnetic flux density and magnetic field in each magnetic material and the self-inductance in a unit length.

Fig. 9.5 Two parallel superconducting plates with two magnetic materials that have different magnetic permeabilities

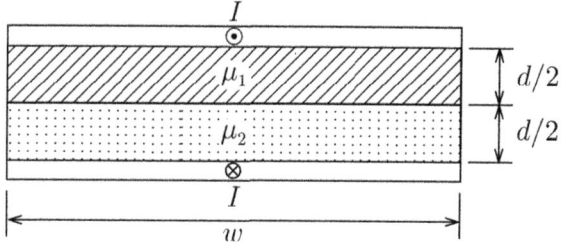

Solution 9.2 The magnetic flux density and magnetic field are parallel to the plates. We denote these values in magnetic materials 1 and 2 by B_1, H_1, B_2, and H_2, respectively. Ampere's law derives $H_1 = H_2 = I/w$, and these satisfy the continuity of the parallel component of the magnetic field on the boundary. These yield $B_1 = \mu_1 I/w$ and $B_2 = \mu_2 I/w$. The magnetic flux in a unit length is $\Phi' = d(B_1 + B_2)/2$ and the self-inductance in a unit length is

$$L' = \frac{(\mu_1 + \mu_2)d}{2w}.$$

◆

Example 9.3 Determine the self-inductance of a unit length for the superconducting coaxial transmission line with two magnetic materials of different magnetic permeabilities in Fig. 9.6.

Fig. 9.6 Coaxial superconducting transmission line with two magnetic materials that have different magnetic permeabilities

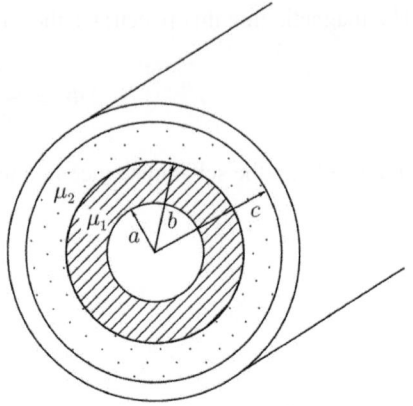

Solution 9.3 We denote the distance from the central axis by R. When we apply current I to the transmission line, the magnetic field is $H(R) = I/(2\pi R)$ in the region $a < R < c$ and zero in other regions. Hence, the magnetic flux densities in magnetic materials 1 and 2 are $B_1 = \mu_1 I/(2\pi R)$ and $B_2 = \mu_2 I/(2\pi R)$, respectively. The magnetic flux in a unit length is

$$\Phi' = \int_a^b \frac{\mu_1 I}{2\pi R} dR + \int_b^c \frac{\mu_2 I}{2\pi R} dR = \frac{I}{2\pi}\left(\mu_1 \log\frac{b}{a} + \mu_2 \log\frac{c}{b}\right).$$

The self-inductance in a unit length is

$$L' = \frac{1}{2\pi}\left(\mu_1 \log\frac{b}{a} + \mu_2 \log\frac{c}{b}\right).$$

The magnetic energy densities in magnetic materials 1 and 2 are $\mu_1(I/2\pi R)^2/2$ and $\mu_2(I/2\pi R)^2/2$, respectively. The magnetic energy in a unit length is

$$U'_m = \int_a^b \frac{\mu_1}{2}\left(\frac{I}{2\pi R}\right)^2 2\pi R dR + \int_b^c \frac{\mu_2}{2}\left(\frac{I}{2\pi R}\right)^2 2\pi R dR$$

$$= \frac{I^2}{4\pi}\left(\mu_1 \log\frac{b}{a} + \mu_2 \log\frac{c}{b}\right).$$

So, the same self-inductance is obtained from the magnetic energy. ◆

9.3 Boundary Conditions

Example 9.4 A magnetic sphere with radius a is in a uniform magnetic flux density of B_0, as shown in Fig. 9.2a. Determine the magnetic flux density, magnetic field, and surface magnetizing current density.

Solution 9.4 We define spherical coordinates as shown in Fig. 9.2b. We can assume that the magnetization is uniform in the magnetic material. Hence, the magnetic flux density outside the sphere is given by the sum of the applied magnetic flux density and the contribution of the magnetic moment placed at the origin after removal of the sphere. We expect a uniform magnetic flux density inside the sphere due to the uniform magnetization. We denote the magnetic moment directed to the z-axis by m. Equation (6.29) gives the radial and zenithal components of the magnetic flux density outside the sphere due to the magnetic moment:

$$B_r = \frac{\mu_0 m \cos\theta}{2\pi r^3}, \quad B_\theta = \frac{\mu_0 m \sin\theta}{4\pi r^3}.$$

We use B to denote the internal magnetic flux density along the z-axis. Then, the continuities of the normal component of the magnetic flux density and the parallel component of the magnetic field on the surface ($r = a$) are given by

$$B_0 \cos\theta + \frac{\mu_0 m \cos\theta}{2\pi a^3} = B \cos\theta,$$

$$\frac{1}{\mu_0}\left(-B_0 \sin\theta + \frac{\mu_0 m \sin\theta}{4\pi a^3}\right) = -\frac{1}{\mu}B \sin\theta.$$

From these equations we have

$$m = \frac{\mu - \mu_0}{\mu + 2\mu_0} \cdot \frac{4\pi a^3 B_0}{\mu_0}, \quad B = \frac{3\mu}{\mu + 2\mu_0} B_0.$$

Using these results, the magnetic flux density outside the sphere ($r > a$) is

$$B_r = \mu_0 H_r = \left(1 + \frac{\mu - \mu_0}{\mu + 2\mu_0} \cdot \frac{2a^3}{r^3}\right) B_0 \cos\theta,$$

$$B_\theta = \mu_0 H_\theta = -\left(1 - \frac{\mu - \mu_0}{\mu + 2\mu_0} \cdot \frac{a^3}{r^3}\right) B_0 \sin\theta,$$

and that inside the sphere ($r < a$) is

$$B_r = \mu_0 H_r = \frac{3\mu}{\mu + 2\mu_0} B_0 \cos\theta,$$

$$B_\theta = \mu_0 H_\theta = -\frac{3\mu}{\mu + 2\mu_0} B_0 \sin\theta.$$

We determine the magnetization to be

$$M = \left(\frac{1}{\mu_0} - \frac{1}{\mu}\right) B = \frac{3(\mu - \mu_0)}{\mu_0(\mu + 2\mu_0)} B_0.$$

Here, we suppose small closed loop ΔC that contains a part of the interface on a plane including the z-axis with an arbitrary azimuthal angle (see Fig. 9.7). We apply the integrated form of Eq. (9.10),

$$\oint_{\Delta C} \mathbf{B} \cdot d\mathbf{s} = \mu_0 \int_{\Delta S} (\mathbf{i} + \mathbf{i}_m) \cdot d\mathbf{S},$$

to this region, where ΔS is the surface surrounded by ΔC. Since there is no true current, the difference in the parallel component of the magnetic flux density divided by μ_0 is equal to the surface magnetizing current density. Thus, we have

$$\tau_m(\theta) = \frac{3(\mu - \mu_0)}{\mu_0(\mu + 2\mu_0)} B_0 \sin\theta = M \sin\theta.$$

This agrees with Eq. (9.8).

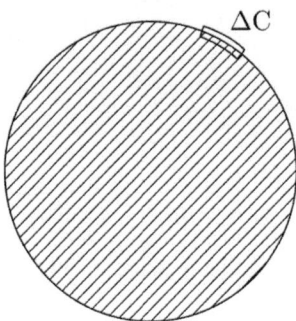

Fig. 9.7 Small closed loop ΔC on a plane that contains a small part of the surface of a magnetic sphere

◆

***

Example 9.5 Solve the problem of Example 9.4 using the vector potential.

Solution 9.5 It is assumed that the magnetic flux density is applied along the z-axis. Hence, its vector potential has an azimuthal component and is given by

$$A_{\mathrm{f}\varphi} = \frac{B_0 r}{2}\sin\theta.$$

We put a magnetic moment m directed along the applied magnetic flux density placed on the origin after the magnetic material is virtually removed. Its vector potential is given by

$$A_{\mathrm{d}\varphi} = \frac{\mu_0 m}{4\pi r^2}\sin\theta.$$

Thus, the vector potential outside the magnetic material is

$$A_{1\varphi} = A_{\mathrm{f}\varphi} + A_{\mathrm{d}\varphi} = \left(\frac{B_0 r}{2} + \frac{\mu_0 m}{4\pi r^2}\right)\sin\theta.$$

The magnetic flux density inside the spherical magnetic material is uniform along the z-axis, and its value is denoted by B. Thus, the vector potential inside is

$$A_{2\varphi} = \frac{Br}{2}\sin\theta.$$

Equation (9.24) leads to

$$B_0 + \frac{\mu_0 m}{2\pi a^3} = B.$$

In Eq. (9.25), $\partial A_\varphi/\partial n$ means $-(1/r)\partial(rA_\varphi)/\partial r$, and we have

$$\frac{1}{\mu_0}\left(B_0 - \frac{\mu_0 m}{4\pi a^3}\right) = \frac{1}{\mu}B.$$

Thus, the same results as in Example 9.4 are obtained.

◆

Example 9.6 We apply current I to a thin straight line at distance a above the flat surface of a magnetic material with magnetic permeability μ, as shown in Fig. 9.8. Determine the vector potential.

Fig. 9.8 Straight current and flat surface of magnetic material

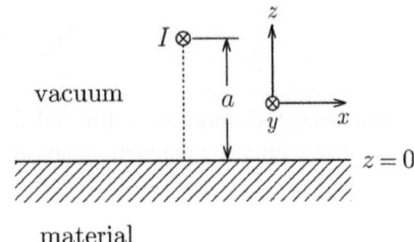

Solution 9.6 We use the method of images. We define the y-axis along the direction of the current and the x-y plane ($z = 0$) on the surface of the magnetic material with $x = 0$ as the position of the current.

To determine the vector potential in the vacuum region ($z > 0$), we virtually assume that all the space is vacuum and the vector potential is given by the sum of the component caused by I and that caused by the image current I' placed at the mirror position with respect to the surface of the magnetic material, as shown in Fig. 9.9a. Hence, the vector potential has only the y-component, A_y, and depends only on x and z. The vector potential at point (x, z) in the vacuum is given by

$$A_{vy}(x, z) = \frac{\mu_0}{2\pi} \left\{ I \log \frac{R_0}{\left[x^2 + (z-a)^2\right]^{1/2}} + I' \log \frac{R_0}{\left[x^2 + (z+a)^2\right]^{1/2}} \right\},$$

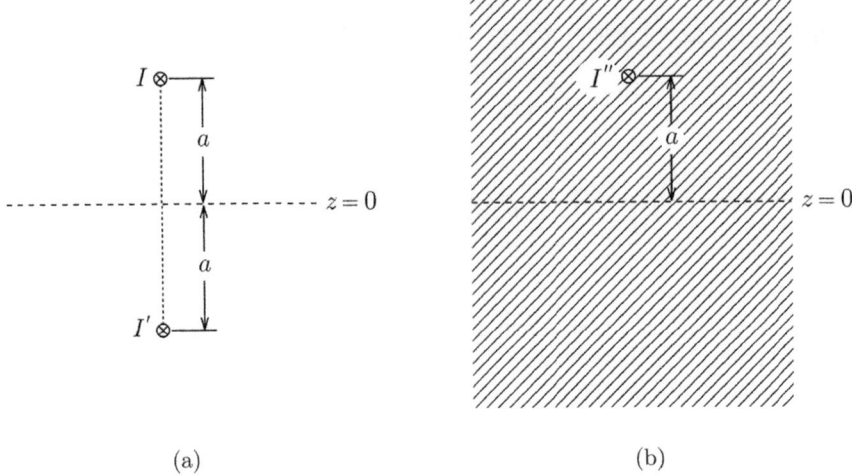

Fig. 9.9 Solution using method of images: assumed conditions for **a** vacuum and **b** magnetic material

where R_0 is the distance to the reference point of the vector potential. On the other hand, to determine the vector potential in the magnetic material ($z < 0$), we virtually assume that all the space is occupied by the magnetic material and the vector potential is produced by the image current I'' placed at the original position, as shown in Fig. 9.9b. The vector potential at point (x, z) in the magnetic material is

$$A_{my}(x, z) = \frac{\mu_0}{2\pi} I'' \log \frac{R_0}{[x^2 + (z-a)^2]^{1/2}}.$$

The continuity condition of the normal component of the magnetic flux density on the surface is given by Eq. (9.24), i.e., $A_{vy}(z = 0) = A_{my}(z = 0)$. This condition gives

$$\mu_0(I + I') = \mu I''.$$

The continuity condition of the parallel component of the magnetic field on the surface is given by Eq. (9.25) in which the derivative along the direction of n is that with respect to z. This condition gives

$$I - I' = I''.$$

Thus, we have

$$I' = \frac{\mu - \mu_0}{\mu + \mu_0} I, \quad I'' = \frac{2\mu_0}{\mu + \mu_0} I.$$

We obtain the vector potential as

$$A_{vy}(x, z) = \frac{\mu_0 I}{2\pi} \left\{ \log \frac{R_0}{[x^2 + (z-a)^2]^{1/2}} + \frac{\mu - \mu_0}{\mu + \mu_0} \log \frac{R_0}{[x^2 + (z+a)^2]^{1/2}} \right\}; z > 0,$$

$$= \frac{\mu_0 \mu I}{\pi(\mu + \mu_0)} \log \frac{R_0}{[x^2 + (z-a)^2]^{1/2}}; \quad z < 0.$$

◆

***

Example 9.7 A long magnetic cylinder with radius a and magnetic permeability μ is in a uniform normal magnetic flux density B_0. Determine the magnetic flux density, magnetic field, magnetization, and surface magnetizing current density.

Solution 9.7 We define cylindrical coordinates with the z-axis at the central axis of the magnetic cylinder and the azimuthal angle φ measured from the direction of the applied magnetic flux density. We assume that the magnetic flux density outside the magnetic material ($R > a$) due to its magnetization is given by a magnetic dipole line of moment $\hat{m}$ in a unit length placed at the central axis after virtually removing the magnetic material. The magnetic flux density B inside the magnetic material ($R < a$) is assumed to be constant. The directions of the moment of the magnetic dipole line and the inner magnetic flux density are parallel to that of the applied magnetic flux density. The continuities of the normal (radial) component of the magnetic flux density and the parallel (azimuthal) component of the magnetic field at the surface ($R = a$) give

$$\hat{m} = \frac{\mu - \mu_0}{\mu + \mu_0} \cdot \frac{2\pi a^2 B_0}{\mu_0}, \quad B = \frac{2\mu}{\mu + \mu_0} B_0.$$

Using these results, the magnetic flux density outside the magnetic material ($R > a$) is

$$B_R = \mu_0 H_R = \left(1 + \frac{\mu - \mu_0}{\mu + \mu_0} \cdot \frac{a^2}{R^2}\right) B_0 \cos\varphi,$$

$$B_\varphi = \mu_0 H_\varphi = -\left(1 - \frac{\mu - \mu_0}{\mu + \mu_0} \cdot \frac{a^2}{R^2}\right) B_0 \sin\varphi,$$

and that inside the magnetic material ($R < a$) is

$$B_R = \mu_0 H_R = \frac{2\mu}{\mu + \mu_0} B_0 \cos\varphi, \quad B_\varphi = \mu_0 H_\varphi = -\frac{2\mu}{\mu + \mu_0} B_0 \sin\varphi.$$

The magnetization of the magnetic material is

$$M = \left(\frac{1}{\mu_0} - \frac{1}{\mu}\right) B = \frac{2(\mu - \mu_0)}{\mu_0(\mu + \mu_0)} B_0.$$

Here, we apply the integral form of Eq. (9.10) to a small rectangle on a plane normal to the central axis that includes the surface of the magnetic material. Since there is no true current on the surface, the surface magnetizing current density is given by the difference in the parallel component of the magnetic flux density on the surface divided by μ_0:

$$\tau_m(\varphi) = \frac{2(\mu - \mu_0)}{\mu_0(\mu + \mu_0)} B_0 \sin\varphi = M \sin\varphi.$$

◆

9.4 Magnetic Energy in Magnetic Materials

The magnetic energy in a system made of magnetic materials is formally the same as that in a current system, due to the change from Eq. (6.16) to Eq. (9.12). That is, the magnetic energy density in magnetic materials is

$$u_m = \frac{1}{2\mu}B^2 = \frac{1}{2}\boldsymbol{B}\cdot\boldsymbol{H} = \frac{1}{2}\mu H^2, \quad (9.31)$$

and the magnetic energy is given by its volume integral:

$$U_m = \int_V \frac{1}{2\mu}B^2 dV = \int_V \frac{1}{2}\boldsymbol{B}\cdot\boldsymbol{H} dV = \int_V \frac{1}{2}\mu H^2 dV. \quad (9.32)$$

Example 9.8 The space between the superconductors in a superconducting coaxial transmission line is occupied by two magnetic materials with magnetic permeabilities μ_1 and μ_2, as shown in Fig. 9.10. Determine the magnetic energy in a unit length of the transmission line, when current I is applied.

Fig. 9.10 Coaxial superconducting transmission line with two magnetic materials that have different magnetic permeabilities

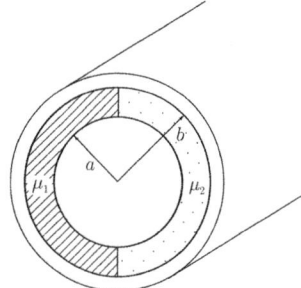

Solution 9.8 It is assumed that the surface currents with densities τ_1 and τ_2 flow on the interface of the inner superconductor surface ($r = a$) in contact with magnetic materials 1 and 2, respectively. Then, the magnetic field strength at radius R in the space between the two superconductors is

$$H_1(R) = \frac{a\tau_1}{R}, \quad H_2(R) = \frac{a\tau_2}{R},$$

and the corresponding magnetic flux density is

$$B_1(R) = \frac{\mu_1 a\tau_1}{R}, \quad B_2(R) = \frac{\mu_2 a\tau_2}{R}.$$

These quantities must be equal from the continuity of the magnetic flux density and we have $\mu_1 \tau_1 = \mu_2 \tau_2$. From the condition of the total current, $I = \pi a(\tau_1 + \tau_2)$, the surface current densities are obtained, and the magnetic flux density is determined:

$$B(R) = \frac{\mu_1 \mu_2 I}{\pi(\mu_1 + \mu_2) R}.$$

The magnetic flux in a unit length is

$$\Phi' = \frac{\mu_1 \mu_2 I}{\pi(\mu_1 + \mu_2)} \int_a^b \frac{dR}{R} = \frac{\mu_1 \mu_2 I}{\pi(\mu_1 + \mu_2)} \log \frac{b}{a},$$

and we determine the magnetic energy in a unit length to be

$$U'_m = \frac{1}{2} \Phi' I = \frac{\mu_1 \mu_2 I^2}{2\pi(\mu_1 + \mu_2)} \log \frac{b}{a}.$$

◆

9.5 Exercises

9.1. A magnetic cylinder with diameter a is placed in a uniform transverse magnetic flux density B_0. Determine the magnetizing current density on the surface of the magnetic material. The magnitude of magnetization, M, is M.

9.2. Magnetic materials 1 and 2 with magnetic permeabilities μ_1 and μ_2 each occupy half of the space between two parallel superconducting plates, as shown in Fig. E9.1. Determine the inductance of this parallel-plate transmission line.

9.3. In the superconducting parallel-plate transmission line shown in Fig. E9.2, the magnetic permeability varies along the width as $\mu'(x) = \mu(1 + kx)$. Determine the inductance in a unit length of the transmission line.

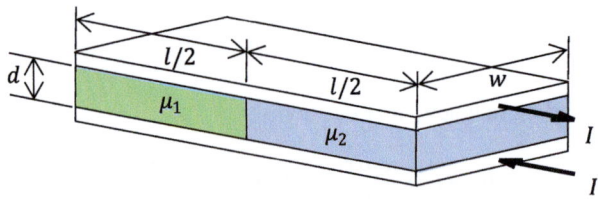

Fig. E9.1 Superconducting parallel-plate transmission line with two kinds of magnetic material between the plates

9.5 Exercises

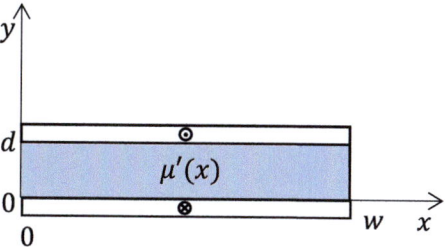

Fig. E9.2 Superconducting parallel-plate transmission line containing a magnetic material with magnetic permeability that varies along the width

9.4. In the superconducting parallel-plate transmission line shown in Fig. E9.3, the magnetic permeability varies with the distance y from the lower plate as $\mu'(y) = \mu(1 + ky)$. Determine the inductance in a unit length of the transmission line.

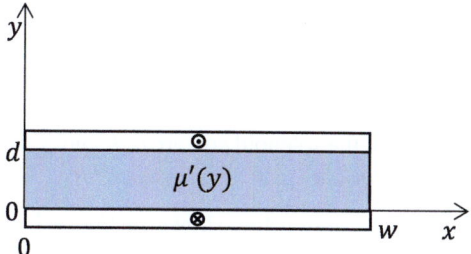

Fig. E9.3 Superconducting parallel-plate transmission line containing a magnetic material with magnetic permeability that varies with the distance from the lower plate

9.5. In the superconducting coaxial transmission line shown in Fig. E9.4, the magnetic permeability varies along the radius as $\mu'(R) = \mu(1 + kR)$. Determine the inductance in a unit length.

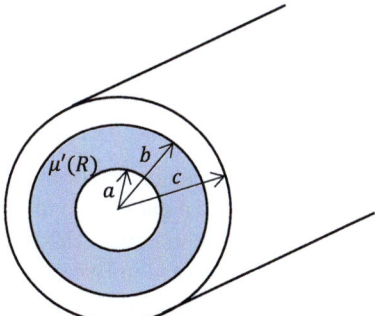

Fig. E9.4 Superconducting coaxial transmission line containing a magnetic material with magnetic permeability that varies radially

9.6. When a magnetic rod of magnetic permeability μ is inserted into a long solenoid coil, as shown in Fig. E9.5, determine the self-inductance of the coil. The radius,

Fig. E9.5 Long solenoid coil with a magnetic rod having magnetic permeability μ

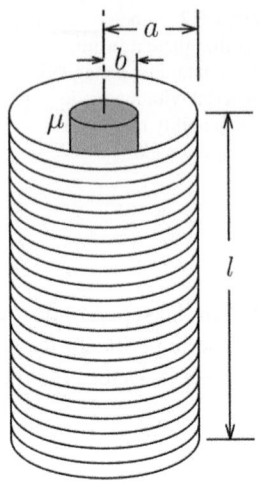

length, and number of turns of the solenoid coil are a, l, and N, respectively, and the radius of the magnetic rod is $b(<a)$.

9.7. In the parallel-plate transmission line shown in Fig. 9.4 in Example 9.2, the magnetic flux produced in each magnetic material is additive. So, the inductance of this transmission line can also be regarded as additive and behave like two inductances connected in series. Prove that this concept is correct.

9.8. In the parallel-plate transmission line shown in Fig. 9.3 in Example 9.1, the current is shared to produce the same magnetic flux in each magnetic material. So, this situation is similar to the case of two resistors connected in parallel. This suggests that the inductance is given by that of two inductances connected in parallel. Prove that this concept is correct.

9.9. The superconducting parallel-plate transmission line treated in Exercise 9.4 can be regarded as an assembly of narrow transmission lines with constant permeabilities. Prove that the inductance can be obtained as a summation of each of these inductances of narrow regions.

9.10. It is shown in Exercise 9.8 that the inductance of the transmission line treated in Example 9.1 can be obtained as that of a parallel connection of the transmission lines, each composed of a single magnetic material. This suggest that the superconducting parallel-plate transmission line treated in Exercise 9.3 can also be obtained as that of a parallel connection of narrow transmission lines with constant magnetic permeabilities. Prove that this speculation is valid.

9.11. Determine the magnetizing current density on the interface between different magnetic materials for the refraction discussed in Sect. 9.3 (see Fig. 9.3).

9.12. A cylindrical superconductor with radius a carrying current I is surrounded by a magnetic material with magnetic permeability μ. Determine the magnetic

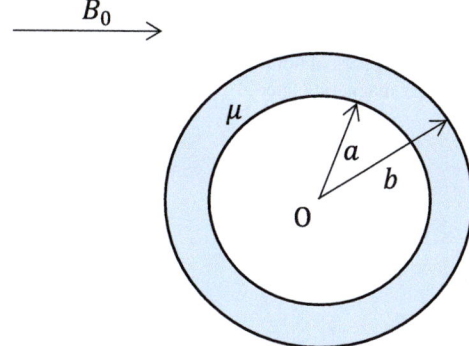

Fig. E9.6 Hollow magnetic sphere in a uniform magnetic flux density

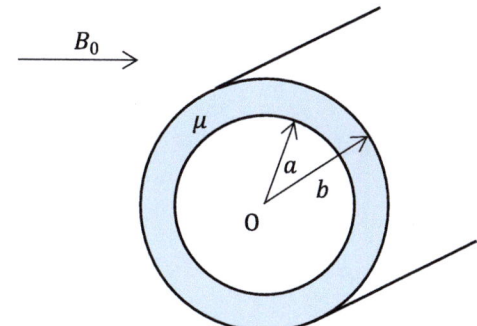

Fig. E9.7 Hollow magnetic cylinder in a uniform perpendicular magnetic flux density

flux density, magnetic field, and magnetization in the magnetic material, and the magnetizing current density on the superconductor surface.

9.13. The magnetic condition inside and outside of a magnetic cylinder in a uniform perpendicular magnetic flux density is discussed in Example 9.7. Determine the vector potential.

9.14. A hollow magnetic sphere is placed in a uniform magnetic flux density B_0, as shown in Fig. E9.6. Determine the magnetic flux density in each region.

9.15. A long, hollow magnetic cylinder is placed in a uniform perpendicular magnetic flux density B_0, as shown in Fig. E9.7. Determine the magnetic flux density in each region.

9.16. Suppose that a magnetic moment m is placed at the center of a spherical hollow space with radius a in a magnetic material, as shown in Fig. E9.8. Determine the magnetic flux density in each region and the magnetizing current density on the inner surface of the magnetic material. Assume that a magnetic moment m' is placed at the center after replacing vacuum by the magnetic material to determine the magnetic flux density in the magnetic material.

Fig. E9.8 Magnetic moment m placed at the center of a spherical hollow space in a magnetic material

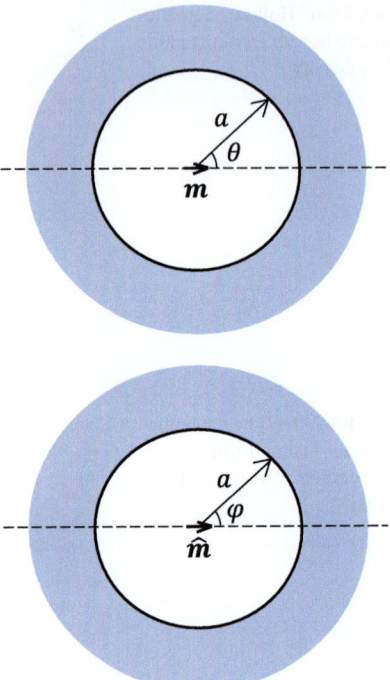

Fig. E9.9 Magnetic moment $\hat{m}$ placed at the central axis of a cylindrical hollow space in a magnetic material

9.17. Suppose that a magnetic dipole line of moment $\hat{m}$ in a unit length is placed at the central axis of a cylindrical hollow space with radius a in a long magnetic material, as shown in Fig. E9.9. Determine the magnetic flux density and the magnetizing current density on the inner surface of the magnetic material. Assume that a magnetic dipole line of moment $\hat{m}'$ in a unit length is placed at the central axis after replacing vacuum by the magnetic material to determine the magnetic flux density in the magnetic material.

9.18. The vector potential is determined inside and outside a magnetic sphere in a uniform magnetic flux density in Example 9.5. Prove that the obtained vector potential satisfies Laplace's equation.

9.19. The vector potential is determined inside and outside a magnetic cylinder in a uniform perpendicular magnetic flux density in Exercise 9.13. Prove that the obtained vector potential satisfies Laplace's equation.

9.20. The magnetic condition is discussed for a hollow magnetic sphere placed in a uniform magnetic flux density B_0 in Exercise 9.14. Determine the vector potential in each region and prove that Laplace's equation holds.

9.21. The magnetic condition is discussed for a hollow magnetic cylinder placed in a uniform perpendicular magnetic flux density B_0 in Exercise 9.15. Determine the vector potential in each region and prove that Laplace's equation holds.

9.5 Exercises

9.22. The magnetic condition is treated for a magnetic moment m placed at the center of a spherical hollow interior with radius a in a magnetic material in Exercise 9.16. Prove that the vector potential satisfies Laplace's equation both in the hollow space and in the magnetic material.

9.23. The magnetic condition is treated for a linear magnetic moment $\hat{m}$ in a unit length placed at the central axis of a cylindrical hollow interior with radius a in a long magnetic material in Exercise 9.17. Prove that the vector potential satisfies Laplace's equation both in the hollow space and in the magnetic material.

9.24. Determine the equivector potential surface for the magnetic cylinder in a transverse magnetic flux density discussed in Exercise 9.13.

9.25. Determine the equivector potential surface in vacuum and in a magnetic material, when a moment of magnetic dipole line $\hat{m}$ in a unit length is placed on the central axis of the cylindrical hollow interior in a long magnetic material that is discussed in Exercise 9.23.

9.26. The magnetic condition is discussed for a magnetic cylinder placed in a uniform perpendicular magnetic flux density B_0 in Example 9.7. Solve the same problem using the magnetic potential.

9.27. Determine the magnetizing current density on the surface of a magnetic material for the case when a straight current, I, is placed at a distance a from the surface, as treated in Example 9.6.

9.28. Suppose that a straight current, I, is placed at a distance a from the surface of a magnetic material, as treated in Example 9.6. Determine the Lorentz force between the given current and the magnetizing current induced on the surface of the magnetic material.

9.29. When current I is applied to the superconducting parallel-plate transmission line treated in Exercise 9.3, determine the magnetic energy in a unit length of the transmission line.

9.30. When current I is applied to the superconducting parallel-plate transmission line treated in Exercise 9.4, determine the magnetic energy in a unit length of the transmission line.

9.31. Suppose that currents $\pm I$ are applied to the superconducting transmission line discussed in Example 9.8. Determine the magnetic energy in a unit length of the transmission line using Eq. (9.32).

9.32. When current I is applied to the cylindrical superconducting transmission line treated in Exercise 9.5, determine the magnetic energy in a unit length of the transmission line.

9.33. Solve the problem of Example 9.3 using the vector potential.

9.34. Solve the problem of Exercise 9.31 using the vector potential.

9.35. Solve the problem of Exercise 9.32 using the vector potential.

Answer to Exercises

9.1. We define cylindrical coordinates with the z-axis on the central axis of the magnetic cylinder and the azimuthal angle φ measured from the direction of the applied magnetic flux density. We slice the magnetic material into thin plates that are perpendicular to the applied magnetic flux density as shown Fig. B9.1, and denote the surface magnetizing current density on the edge of the plate by τ. If we denote the thickness of the thin plate and the corresponding azimuthal angle interval by dh and $d\varphi$, respectively, we have $dh = a \sin\varphi \, d\varphi$. The magnetizing current that flows on the edge of this plate is $\tau dh = \tau_m a d\varphi$. Thus, the surface magnetizing current density is given by

$$\tau_m = \tau \sin\varphi = M \sin\varphi.$$

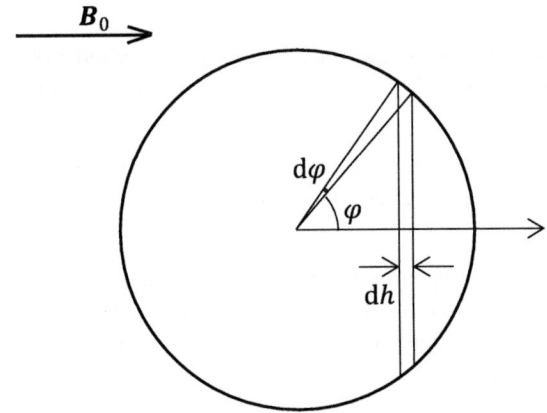

Fig. B9.1 Thin plate of magnetic material and definition of angle φ

9.2. When we apply current I to the transmission line, the magnetic flux density in each magnetic material is

$$B_1 = \frac{\mu_1 I}{w}, \quad B_2 = \frac{\mu_2 I}{w}.$$

Hence, the magnetic flux in the transmission line is

$$\Phi = \frac{dl(\mu_1 + \mu_2)}{2w} I$$

and the inductance is determined to be

$$L = \frac{dl(\mu_1 + \mu_2)}{2w}.$$

9.5 Exercises

9.3. We apply current I to the transmission line. The current density on the superconductor surface is denoted by $\tau(x)$. The magnetic flux density, which is constant, is denoted by B_0. Then, the current density is given by $\tau(x) = B_0/\mu'(x)$. The total current is

$$I = \int_0^w \tau(x)dx = \frac{B_0}{\mu}\int_0^w \frac{dx}{1+kx} = \frac{B_0}{\mu k}\log(1+kw).$$

The total magnetic flux in a unit length is $\Phi' = B_0 d$. Hence, the inductance in a unit length is

$$L' = \frac{\mu d k}{\log(1+kw)}.$$

9.4. We apply current I to the transmission line. The magnetic field is $H(y) = I/w$ and the magnetic flux density is $B(y) = \mu'(y)I/w$. So, the magnetic flux in a unit length is

$$\Phi' = \frac{\mu I}{w}\int_0^d (1+ky)dy = \frac{\mu d I}{w}\left(1+\frac{kd}{2}\right)$$

and the inductance in a unit length is determined to be

$$L' = \frac{\mu d}{w}\left(1+\frac{kd}{2}\right).$$

9.5. We apply current I. The magnetic flux density at radius R is $B(R) = \mu'(R)I/2\pi R$, and the magnetic flux in a unit length is

$$\Phi' = \frac{\mu I}{2\pi}\int_a^b \frac{1+kR}{R}dR = \frac{\mu I}{2\pi}\left[\log\frac{b}{a} + k(b-a)\right].$$

The inductance in a unit length is determined to be

$$L' = \frac{\mu}{2\pi}\left[\log\frac{b}{a} + k(b-a)\right].$$

9.6. When we apply current I to the coil, the magnetic field strength inside the coil is obtained using Eq. (9.14) to be $H = NI/l$. So, the magnetic flux density is $B_1 = \mu_0 NI/l$ and $B_2 = \mu NI/l$ in vacuum and the magnetic material, respectively.

The magnetic flux that penetrates one turn of the coil is

$$\phi = \frac{\pi[\mu_0(a^2 - b^2) + \mu b^2]NI}{l},$$

and the magnetic flux that penetrates the coil is

$$\Phi = N\phi = \frac{\pi[\mu_0(a^2 - b^2) + \mu b^2]N^2 I}{l}.$$

Thus, the inductance is determined to be

$$L = \frac{\pi[\mu_0(a^2 - b^2) + \mu b^2]N^2}{l}.$$

9.7. The inductance in a unit length of the upper part is

$$L'_1 = \frac{\mu_1 d}{2w}$$

and that of the lower part is similarly obtained as

$$L'_2 = \frac{\mu_2 d}{2w}.$$

Thus, we have

$$L' = L'_1 + L'_2 = \frac{(\mu_1 + \mu_2)d}{2w}$$

and the assumed concept is proved to be correct.

9.8. The inductance in a unit length of the left part with magnetic permeability μ_1 is

$$L'_1 = \frac{2\mu_1 d}{w}$$

and that of the right part with magnetic permeability μ_2 is similarly obtained as

$$L'_2 = \frac{2\mu_2 d}{w}.$$

Thus, we have

$$L' = \frac{L'_1 L'_2}{L'_1 + L'_2} = \frac{2d\mu_1\mu_2}{w(\mu_1 + \mu_2)}$$

and the assumed concept is proved to be correct.

9.5 Exercises

9.9. The magnetic flux in a region from y to $y+dy$ is given by $d\Phi(y) = [\mu'(y)I/w]dy$, and the corresponding inductance is $dL = d\Phi(y)/I = [\mu'(y)/w]dy$. So, the total inductance is given by

$$L = \int dL = \int_0^d \frac{\mu(1+ky)}{w} dy = \frac{\mu d}{w}\left(1 + \frac{kd}{2}\right).$$

This result agrees with the inductance obtained directly.

9.10. The inductance in a unit length in the region from x to $x+dx$ is given by

$$dL' = \frac{B_0 d}{\tau(x)dx},$$

where $\tau(x) = B_0/\mu'(x)$ is the surface current density and B_0 is a constant magnetic flux density produced by current I and is given by $B_0 = \mu k I / \log(1 + kw)$. If we denote the reciprocal of dL' by dK', we have

$$dK' = \frac{dx}{\mu'(x)d}.$$

Hence, the reciprocal of the inductance is determined to be

$$\frac{1}{L'} = K' = \frac{1}{\mu d}\int_0^w \frac{dx}{1+kx} = \frac{1}{\mu dk}\log(1+kw).$$

Thus, we obtain

$$L' = \frac{\mu dk}{\log(1+kw)},$$

which agrees with the result in Exercise 9.3.

9.11. The magnetizing current density is equal to the difference between the parallel components of the magnetization:

$$\tau_m = M_1 \sin\theta_1 - M_2 \sin\theta_2 = \left(\frac{1}{\mu_0} - \frac{1}{\mu_1}\right)B_1 \sin\theta_1 - \left(\frac{1}{\mu_0} - \frac{1}{\mu_2}\right)B_2 \sin\theta_2.$$

Using Eq. (9.27), this leads to

$$\tau_m = \frac{1}{\mu_0}\left(1 - \frac{\mu_2}{\mu_1}\right)B_1 \sin\theta_1.$$

This result can also be directly derived from $(1/\mu_0)(B_1 \sin\theta_1 - B_2 \sin\theta_2)$.

9.12. The magnetic field and magnetic flux density are

$$H(R) = \frac{I}{2\pi R}, \quad B(R) = \frac{\mu I}{2\pi R},$$

for $R > a$. So, the magnetization is given by

$$M(R) = \frac{1}{\mu_0}B(R) - H(R) = \frac{(\mu - \mu_0)I}{2\pi \mu_0 R}.$$

Since the magnetization is parallel to the surface, the magnetizing current density on the surface is

$$\tau_m = M(a) = \frac{(\mu - \mu_0)I}{2\pi \mu_0 a}.$$

9.13. The vector potential outside the cylindrical magnetic material is given by

$$A_z(R, \varphi) = \left(B_0 R + \frac{\mu_0 \hat{m}}{2\pi R}\right)\sin\varphi,$$

where $\hat{m}$ is the moment of the magnetic dipole line in a unit length on the central axis. If the uniform magnetic flux density in the magnetic material is denoted by B, the vector potential inside the cylindrical magnetic material is given by

$$A_z(R, \varphi) = BR \sin\varphi.$$

From the condition given by Eq. (9.24) at $R = a$, we have

$$B_0 + \frac{\mu_0 \hat{m}}{2\pi a^2} = B. \tag{A9.1}$$

In Eq. (9.25), the derivative with respect to n means that with respect to R, and the continuity condition at $R = a$ leads to

$$\frac{1}{\mu_0}\left(B_0 - \frac{\mu_0 \hat{m}}{2\pi a^2}\right) = \frac{1}{\mu}B. \tag{A9.2}$$

From Eqs. (A9.1) and (A9.2), we have

$$\hat{m} = \frac{\mu - \mu_0}{\mu + \mu_0} \cdot \frac{2\pi a^2}{\mu_0}B_0, \quad B = \frac{2\mu}{\mu + \mu_0}B_0.$$

9.5 Exercises

These results are the same as in Example 9.7. Thus, the vector potential is given by

$$A_z(R, \varphi) = \frac{2\mu B_0}{\mu + \mu_0} R \sin \varphi; \qquad 0 \le R < a,$$

$$= B_0 \left(R + \frac{\mu - \mu_0}{\mu + \mu_0} \cdot \frac{a^2}{R} \right) \sin \varphi; \quad R > a.$$

9.14. Magnetizing current is induced on the inner and outer surfaces by application of the external magnetic flux density. The magnetizing current on each surface produces a uniform magnetic flux density inside it and a dipole magnetic flux density outside it. The uniform magnetic flux density in the hollow space is denoted by B_1. The magnetic moments equivalent to the magnetizing currents on the inner and outer surfaces are denoted by m and m', respectively. We denote by B_2 a uniform magnetic flux density in the magnetic material produced by the magnetizing current on the outer surface. The magnetic flux density distribution is:

$$\boldsymbol{B} = \boldsymbol{B}_1; \qquad 0 \le r < a,$$

$$= \boldsymbol{B}_2 + \frac{\mu_0}{4\pi} \nabla \times \frac{\boldsymbol{m} \times \boldsymbol{r}}{r^3}; \qquad a < r < b,$$

$$= \boldsymbol{B}_0 + \frac{\mu_0}{4\pi} \nabla \times \frac{(\boldsymbol{m} + \boldsymbol{m}') \times \boldsymbol{r}}{r^3}; \quad r > b.$$

So, the boundary conditions at $r = a$ are

$$B_1 \cos \theta = B_2 \cos \theta + \frac{\mu_0 m \cos \theta}{2\pi a^3}, \quad -\frac{1}{\mu_0} B_1 \sin \theta = \frac{1}{\mu} \left(-B_2 \sin \theta + \frac{\mu_0 m \sin \theta}{4\pi a^3} \right),$$

and the boundary conditions at $r = b$ are

$$B_2 \cos \theta + \frac{\mu_0 m \cos \theta}{2\pi b^3} = B_0 \cos \theta + \frac{\mu_0 (m + m') \cos \theta}{2\pi b^3},$$

$$\frac{1}{\mu} \left(-B_2 \sin \theta + \frac{\mu_0 m \sin \theta}{4\pi b^3} \right) = \frac{1}{\mu_0} \left[-B_0 \sin \theta + \frac{\mu_0 (m + m') \sin \theta}{4\pi b^3} \right],$$

where θ is the zenithal angle. From these conditions, m, m', B_1, and B_2 are determined to be

$$m = -\frac{4\pi a^3}{3\alpha} \cdot \frac{\mu - \mu_0}{\mu_0^2} B_0, \quad m' = -\frac{2\pi b^3}{\mu_0} \left(1 - \frac{2\mu + \mu_0}{3\alpha \mu_0} \right) B_0,$$

$$B_1 = \frac{1}{\alpha} B_0, \quad B_2 = \frac{2\mu + \mu_0}{3\alpha \mu_0} B_0,$$

where α is a constant given by

$$\alpha = \frac{(\mu + 2\mu_0)(2\mu + \mu_0)}{9\mu_0\mu} - \frac{2a^3}{9b^3} \cdot \frac{(\mu - \mu_0)^2}{\mu_0\mu}.$$

9.15. Magnetizing current is induced on the inner and outer surfaces by application of the external magnetic flux density. The magnetizing current on each surface produces a uniform magnetic flux density inside it and a dipole magnetic flux density outside it. The uniform magnetic flux density in the hollow space is denoted by $\boldsymbol{B}_1$. The moments of magnetic dipole lines responsible for the magnetizing currents on the inner and outer surfaces are denoted by $\hat{m}$ and $\hat{m}'$, respectively. We denote by $\boldsymbol{B}_2$ a uniform magnetic flux density in the magnetic material produced by the magnetizing current on the outer surface. The magnetic flux density distribution is:

$$\boldsymbol{B} = \boldsymbol{B}_1; \qquad 0 \leq R < a,$$

$$= \boldsymbol{B}_2 + \frac{\mu_0}{2\pi}\nabla \times \frac{\hat{m} \times \boldsymbol{R}}{R^2}; \qquad a < R < b,$$

$$= \boldsymbol{B}_0 + \frac{\mu_0}{2\pi}\nabla \times \frac{(\hat{m} + \hat{m}') \times \boldsymbol{R}}{R^2}; \qquad R > b.$$

So, the boundary conditions at $R = a$ are

$$B_1 \cos\varphi = B_2 \cos\varphi + \frac{\mu_0 \hat{m}\cos\varphi}{2\pi a^2}, \quad -\frac{1}{\mu_0}B_1 \sin\varphi = \frac{1}{\mu}\left(-B_2 \sin\varphi + \frac{\mu_0 \hat{m}\sin\varphi}{2\pi a^2}\right),$$

and the boundary conditions at $R = b$ are

$$B_2 \cos\varphi + \frac{\mu_0 \hat{m}\cos\varphi}{2\pi b^2} = B_0 \cos\varphi + \frac{\mu_0(\hat{m}+\hat{m}')\cos\varphi}{2\pi b^2},$$

$$\frac{1}{\mu}\left(-B_2 \sin\varphi + \frac{\mu_0\hat{m}\sin\varphi}{2\pi b^2}\right) = \frac{1}{\mu_0}\left[-B_0 \sin\varphi + \frac{\mu_0(\hat{m}+\hat{m}')\sin\varphi}{2\pi b^2}\right],$$

where φ is the azimuthal angle. From these conditions, $\hat{m}$, $\hat{m}'$, B_1, and B_2 are determined to be

$$\hat{m} = -\frac{\pi a^2}{\beta}\cdot\frac{\mu - \mu_0}{\mu_0^2}B_0, \quad \hat{m}' = -\frac{2\pi b^2}{\mu_0}\left(1 - \frac{\mu+\mu_0}{2\beta\mu_0}\right)B_0,$$

$$B_1 = \frac{1}{\beta}B_0, \quad B_2 = \frac{\mu+\mu_0}{2\beta\mu_0}B_0,$$

where β is a constant given by

9.5 Exercises

$$\beta = \frac{(\mu + \mu_0)^2}{4\mu_0 \mu} - \frac{a^2}{4b^2} \cdot \frac{(\mu - \mu_0)^2}{\mu_0 \mu}.$$

9.16. The magnetic flux density in vacuum produced by the magnetic moment m is

$$B_r = \frac{\mu_0 m}{2\pi r^3} \cos\theta, \quad B_\theta = \frac{\mu_0 m}{4\pi r^3} \sin\theta.$$

The uniform magnetic flux density in vacuum produced by the magnetizing current on the surface is denoted by B_0. Its radial and zenithal components are

$$B_0 \cos\theta, \quad -B_0 \sin\theta.$$

The magnetic flux density in the magnetic material is also influenced by the magnetizing current, but this can be reproduced by some magnetic moment placed at the center. So, the assumed magnetic moment m' includes this effect. After a virtual occupation of vacuum by the magnetic material, the magnetic flux density in the magnetic material produced by the magnetic moment m' is

$$B_r = \frac{\mu m'}{2\pi r^3} \cos\theta, \quad B_\theta = \frac{\mu m'}{4\pi r^3} \sin\theta.$$

So, the boundary conditions on the magnetic flux density and magnetic field at $r = a$ are

$$\left(B_0 + \frac{\mu_0 m}{2\pi a^3}\right) \cos\theta = \frac{\mu m'}{2\pi a^3} \cos\theta,$$

$$\left(-\frac{B_0}{\mu_0} + \frac{m}{4\pi a^3}\right) \sin\theta = \frac{m'}{4\pi a^3} \sin\theta.$$

These yield

$$m' = \frac{3\mu_0}{2\mu + \mu_0} m, \quad B_0 = \frac{\mu_0(\mu - \mu_0)}{2\pi(2\mu + \mu_0)a^3} m.$$

Thus, we have

$$B_r = \frac{\mu_0 m}{2\pi}\left[\frac{1}{r^3} + \frac{\mu - \mu_0}{(2\mu + \mu_0)a^3}\right]\cos\theta, \quad B_\theta = \frac{\mu_0 m}{4\pi}\left[\frac{1}{r^3} - \frac{2(\mu - \mu_0)}{(2\mu + \mu_0)a^3}\right]\sin\theta$$

for $0 < r < a$ and

$$B_r = \frac{3\mu\mu_0}{2\pi(2\mu + \mu_0)r^3} m \cos\theta, \quad B_\theta = \frac{3\mu\mu_0}{4\pi(2\mu + \mu_0)r^3} m \sin\theta$$

for $r > a$. The magnetizing current density is

$$\tau_m = -\frac{1}{\mu_0}[B_\theta(r = a_{+0}) - B_\theta(r = a_{-0})] = -\frac{3(\mu - \mu_0)}{2\pi(2\mu + \mu_0)a^3}m\sin\theta.$$

9.17. The magnetic flux density in vacuum produced by the magnetic moment $\hat{m}$ is

$$B_{mR} = \frac{\mu_0 \hat{m}}{2\pi R^2}\cos\varphi, \quad B_{m\varphi} = \frac{\mu_0 \hat{m}}{2\pi R^2}\sin\varphi.$$

The uniform magnetic flux density in vacuum produced by the magnetizing current on the surface is denoted by B_0. Its radial and azimuthal components are

$$B_0 \cos\varphi, \quad -B_0 \sin\varphi.$$

The magnetic flux density in the magnetic material is also influenced by the magnetizing current, but this can be reproduced by some magnetic moment placed at the central axis. So, the assumed linear magnetic moment $\hat{m}'$ includes this effect. After a virtual occupation of vacuum by the magnetic material, the magnetic flux density in the magnetic material produced by the linear magnetic moment $\hat{m}'$ is

$$B_R = \frac{\mu \hat{m}'}{2\pi R^2}\cos\varphi, \quad B_\varphi = \frac{\mu \hat{m}'}{2\pi R^2}\sin\varphi.$$

So, the boundary conditions on the magnetic flux density and magnetic field at $R = a$ are

$$\left(B_0 + \frac{\mu_0 \hat{m}}{2\pi a^2}\right)\cos\varphi = \frac{\mu \hat{m}'}{2\pi a^2}\cos\varphi,$$

$$\left(-\frac{B_0}{\mu_0} + \frac{\hat{m}}{2\pi a^2}\right)\sin\varphi = \frac{\hat{m}'}{2\pi a^3}\sin\varphi.$$

These yield

$$\hat{m}' = \frac{2\mu_0}{\mu + \mu_0}\hat{m}, \quad B_0 = \frac{\mu_0(\mu - \mu)}{2\pi(\mu + \mu_0)a^2}\hat{m}.$$

Thus, we have

$$B_R = \frac{\mu_0 \hat{m}}{2\pi}\left[\frac{1}{R^2} + \frac{\mu - \mu_0}{(\mu + \mu_0)a^2}\right]\cos\varphi, \quad B_\varphi = \frac{\mu_0 \hat{m}}{2\pi}\left[\frac{1}{R^2} - \frac{\mu - \mu_0}{(\mu + \mu_0)a^2}\right]\sin\varphi$$

for $0 < R < a$ and

9.5 Exercises

$$B_R = \frac{\mu\mu_0 \hat{m}}{\pi(\mu+\mu_0)R^2}\cos\varphi, \quad B_\varphi = \frac{\mu\mu_0 \hat{m}}{\pi(\mu+\mu_0)R^2}\sin\varphi$$

for $R > a$. The magnetizing current density is

$$\tau_m = -\frac{1}{\mu_0}[B_\varphi(R=a_{+0}) - B_\varphi(R=a_{-0})] = -\frac{(\mu-\mu_0)\hat{m}}{\pi(\mu+\mu_0)a^2}\sin\varphi.$$

9.18. The vector potential inside the magnetic sphere has only an azimuthal component and is given by

$$A_\varphi = \frac{3\mu B_0}{2(\mu+2\mu_0)} r \sin\theta.$$

Using Eq. (A.18) in the Appendix, we have

$$\nabla \times \nabla \times A = -i_\varphi \frac{1}{r}\left\{\frac{\partial^2}{\partial r^2}(rA_\varphi) + \frac{\partial}{\partial\theta}\left[\frac{1}{r\sin\theta}\cdot\frac{\partial}{\partial\theta}(\sin\theta A_\varphi)\right]\right\}$$
$$= -i_\varphi\left(\frac{3\mu B_0}{\mu+2\mu_0}\cdot\frac{\sin\theta}{r} - \frac{3\mu B_0}{\mu+2\mu_0}\cdot\frac{\sin\theta}{r}\right) = 0.$$

The vector potential outside the magnetic material is given by

$$A_\varphi = \frac{B_0}{2}\left(r + \frac{\mu-\mu_0}{\mu+2\mu_0}\cdot\frac{2a^3}{r^2}\right)\sin\theta.$$

Then, we have

$$\nabla \times \nabla \times A = -i_\varphi\left[\frac{B_0}{r}\cdot\left(1 + \frac{\mu-\mu_0}{\mu+2\mu_0}\cdot\frac{2a^3}{r^3}\right) - \frac{B_0}{r}\cdot\left(1 + \frac{\mu-\mu_0}{\mu+2\mu_0}\cdot\frac{2a^3}{r^3}\right)\right] = 0.$$

Thus, Laplace's equation holds inside and outside the magnetic sphere.

9.19. The vector potential inside the magnetic cylinder has only a z component and is given by

$$A_z(R,\varphi) = \frac{2\mu B_0}{\mu+\mu_0} R \sin\varphi.$$

Using Eq. (A.14) in the Appendix, we have

$$\nabla \times \nabla \times \boldsymbol{A} = \frac{\boldsymbol{i}_z}{R}\left[-\frac{\partial}{\partial R}\left(R\frac{\partial A_z}{\partial R}\right) - \frac{\partial}{\partial \varphi}\left(\frac{1}{R}\cdot\frac{\partial A_z}{\partial \varphi}\right)\right]$$
$$= \frac{\boldsymbol{i}_z}{R}\left(-\frac{2\mu B_0}{\mu+\mu_0}\sin\varphi + \frac{2\mu B_0}{\mu+\mu_0}\sin\varphi\right) = 0.$$

The vector potential outside the magnetic cylinder is

$$A_z(R,\varphi) = B_0\left(R + \frac{\mu-\mu_0}{\mu+\mu_0}\cdot\frac{a^2}{R}\right)\sin\varphi,$$

and we have

$$\nabla \times \nabla \times \boldsymbol{A} = \frac{\boldsymbol{i}_z}{R}\left[-\left(1+\frac{\mu-\mu_0}{\mu+\mu_0}\cdot\frac{a^2}{R^2}\right)\sin\varphi + \left(1+\frac{\mu-\mu_0}{\mu+\mu_0}\cdot\frac{a^2}{R^2}\right)\sin\varphi\right] = 0.$$

Thus, Laplace's equation holds inside and outside the magnetic cylinder.

9.20. The vector potential has an azimuthal component and its value in each region is given by

$$A_\varphi = \frac{B_1 r}{2}\sin\theta; \qquad\qquad 0 \le r < a,$$
$$= \left(\frac{B_2 r}{2} + \frac{\mu_0 m}{4\pi r^2}\right)\sin\theta; \qquad a < r < b,$$
$$= \left[\frac{B_0 r}{2} + \frac{\mu_0(m+m')}{4\pi r^2}\right]\sin\theta; \quad r > c,$$

where B_1, B_1, m, and m' are given by

$$B_1 = \frac{1}{\alpha}B_0, \quad B_2 = \frac{2\mu+\mu_0}{3\alpha\mu_0}B_0,$$

$$m = -\frac{4\pi a^3}{3\alpha}\cdot\frac{\mu-\mu_0}{\mu_0^2}B_0, \quad m' = -\frac{2\pi b^3}{\mu_0}\left(1 - \frac{2\mu+\mu_0}{3\alpha\mu_0}\right)B_0$$

with

$$\alpha = \frac{(\mu+2\mu_0)(2\mu+\mu_0)}{9\mu_0\mu} - \frac{2a^3}{9b^3}\cdot\frac{(\mu-\mu_0)^2}{\mu_0\mu}.$$

Since the vector potential in each region has a similar form, we prove for the vector potential in the region $a < r < b$. So, we have

9.5 Exercises

$$\nabla \times A = i_r \frac{1}{r \sin\theta} \cdot \frac{\partial}{\partial \theta}(\sin\theta A_\varphi) - i_\theta \frac{1}{r} \cdot \frac{\partial}{\partial r}(rA_\varphi)$$

$$= i_r \left(B_2 + \frac{\mu_0 m}{2\pi r^3}\right) \cos\theta - i_\theta \left(B_2 - \frac{\mu_0 m}{4\pi r^3}\right) \sin\theta$$

and

$$\nabla \times \nabla \times A = -i_\varphi \frac{1}{r} \left[\frac{\partial}{\partial r}\left(B_2 - \frac{\mu_0 m}{4\pi r^3}\right) \sin\theta + \frac{\partial}{\partial \theta}\left(B_2 + \frac{\mu_0 m}{2\pi r^3}\right) \cos\theta\right] = 0.$$

Thus, Laplace's equation holds in the whole region.

9.21. The vector potential has an axial component and its value in each region is given by

$$A_z = B_1 R \sin\varphi; \qquad 0 \le R < a,$$

$$= \left(B_2 R + \frac{\mu_0 \hat{m}}{2\pi R}\right) \sin\varphi; \qquad a < R < b,$$

$$= \left[B_0 R + \frac{\mu_0(\hat{m} + \hat{m}')}{2\pi R}\right] \sin\varphi; \qquad R > c,$$

where B_1, B_1, $\hat{m}$, and $\hat{m}'$ are given by

$$B_1 = \frac{1}{\beta} B_0, \quad B_2 = \frac{\mu + \mu_0}{2\beta\mu_0} B_0,$$

$$\hat{m} = -\frac{\pi a^2}{\beta} \cdot \frac{\mu - \mu_0}{\mu_0^2} B_0, \quad \hat{m}' = -\frac{2\pi b^2}{\mu_0}\left(1 - \frac{\mu + \mu_0}{2\beta\mu_0}\right) B_0$$

with

$$\beta = \frac{(\mu + \mu_0)^2}{4\mu_0\mu} - \frac{a^2}{4b^2} \cdot \frac{(\mu - \mu_0)^2}{\mu_0\mu}.$$

Since the vector potential in each region has a similar form, we prove for the vector potential in the region $a < R < b$. So, we have

$$\Delta A_z = \frac{1}{R} \cdot \frac{\partial}{\partial R}\left(R \frac{\partial A_z}{\partial R}\right) + \frac{1}{R^2} \cdot \frac{\partial^2 A_z}{\partial R^2}$$

$$= \left(\frac{B_2}{R} + \frac{\mu_0 \hat{m}}{2\pi R^3}\right) \sin\varphi - \left(\frac{B_2}{R} + \frac{\mu_0 \hat{m}}{2\pi R^3}\right) \sin\varphi = 0.$$

Thus, Laplace's equation holds in the whole region.

9.22. The vector potential in the hollow space ($0 < r < a$) is

$$A_\varphi = \frac{\mu_0 m}{4\pi r^2} \sin\theta - \frac{B_0 r}{2} \sin\theta.$$

In this case,

$$\nabla \times A = i_r \left(\frac{\mu_0 m}{2\pi r^3} - B_0 \right) \cos\theta + i_\theta \left(\frac{\mu_0 m}{4\pi r^3} + B_0 \right) \sin\theta$$

and we have

$$\nabla \times \nabla \times A = i_\varphi \left[\frac{1}{r} \cdot \frac{\partial}{\partial r} \left(\frac{\mu_0 m}{4\pi r^2} + B_0 r \right) \sin\theta - \frac{1}{r} \cdot \frac{\partial}{\partial \theta} \left(\frac{\mu_0 m}{2\pi r^3} - B_0 \right) \cos\theta \right]$$

$$= i_\varphi \left[\left(-\frac{\mu_0 m}{2\pi r^4} + \frac{B_0}{r} \right) \sin\theta + \left(\frac{\mu_0 m}{2\pi r^4} - \frac{B_0}{r} \right) \sin\theta \right] = 0.$$

The vector potential in the magnetic material ($r > a$) is

$$A_\varphi = \frac{\mu m'}{4\pi r^2} \sin\theta.$$

In this case

$$\nabla \times A = i_r \frac{\mu m'}{2\pi r^2} \cos\theta + i_\theta \frac{\mu m'}{4\pi r^2} \sin\theta$$

and we have

$$\nabla \times \nabla \times A = i_\varphi \left[\frac{1}{r} \cdot \frac{\partial}{\partial r} \left(\frac{\mu m'}{4\pi r^2} \right) \sin\theta - \frac{1}{r} \cdot \frac{\partial}{\partial \theta} \left(\frac{\mu m'}{2\pi r^2} \right) \cos\theta \right]$$

$$= i_\varphi \left(-\frac{\mu m'}{4\pi r^2} \sin\theta + \frac{\mu m'}{4\pi r^2} \sin\theta \right) = 0.$$

Thus, Laplace's equation is satisfied for the vector potential in the hollow space and in the magnetic material.

9.23. The vector potential in the hollow space ($0 < R < a$) is

$$A_z = \frac{\mu_0 \hat{m}}{2\pi R} \sin\varphi + B_0 R \sin\varphi.$$

In this case,

$$\nabla \times A = i_R \left(\frac{\mu_0 \hat{m}}{2\pi R^2} + B_0 \right) \cos\varphi + i_\varphi \left(\frac{\mu_0 \hat{m}}{2\pi R^2} - B_0 \right) \sin\varphi$$

9.5 Exercises

and we have

$$\nabla \times \nabla \times \mathbf{A} = \mathbf{i}_z \left[\frac{1}{R} \cdot \frac{\partial}{\partial R} \left(\frac{\mu_0 \hat{m}}{2\pi R} - B_0 R \right) \sin \varphi - \frac{1}{R} \cdot \frac{\partial}{\partial \varphi} \left(\frac{\mu_0 \hat{m}}{2\pi R^2} - B_0 \right) \cos \varphi \right]$$

$$= \mathbf{i}_z \left[-\left(\frac{\mu_0 \hat{m}}{2\pi R^3} + \frac{B_0}{R} \right) \sin \varphi + \left(\frac{\mu_0 \hat{m}}{2\pi R^3} + \frac{B_0}{R} \right) \sin \varphi \right] = 0.$$

The vector potential in the magnetic material ($R > a$) is

$$A_z = \frac{\mu \hat{m}'}{2\pi R} \sin \varphi.$$

In this case

$$\nabla \times \mathbf{A} = \mathbf{i}_R \frac{\mu \hat{m}'}{2\pi R^2} \cos \varphi + \mathbf{i}_\varphi \frac{\mu \hat{m}'}{2\pi R^2} \sin \varphi$$

and we have

$$\nabla \times \nabla \times \mathbf{A} = \mathbf{i}_z \left[\frac{1}{R} \cdot \frac{\partial}{\partial R} \left(\frac{\mu \hat{m}'}{2\pi R} \right) \sin \varphi - \frac{1}{R} \cdot \frac{\partial}{\partial \varphi} \left(\frac{\mu \hat{m}'}{2\pi R^2} \right) \cos \varphi \right]$$

$$= \mathbf{i}_z \left(-\frac{\mu_0 \hat{m}'}{2\pi R^3} \sin \varphi + \frac{\mu_0 \hat{m}'}{2\pi R^3} \sin \varphi \right) = 0.$$

Thus, Laplace's equation is satisfied for the vector potential in the hollow space and in the magnetic material.

9.24. We define the x-axis in the direction of the applied magnetic flux density. The equivector potential surface outside the magnetic material ($R > a$) is given by

$$B_0 \left(R - \frac{a'^2}{R} \right) \sin \varphi = K,$$

where $a' = [(\mu - \mu_0)/(\mu + \mu_0)]^{1/2} a$ and K is a constant. Using the relationships $R^2 = x^2 + y^2$ and $R \sin \varphi = y$, the above relation leads to

$$x = \pm a' \left(1 + \frac{K}{B_0 y - K} - \frac{y^2}{a'^2} \right)^{1/2}$$

in the x-y plane.

Inside the magnetic material ($R < a$), the equivector potential surface is given by

$$BR \sin \varphi = K',$$

where $B = 2\mu B_0/(\mu + \mu_0)$ and K' is a constant. This leads the equivector potential surface:

$$y = \frac{K'}{B}.$$

9.25. We define the x-y plane normal to the central axis of the hollow space with the x-axis in the direction of the moment of the magnetic dipole line. The equivector potential surface in the hollow space ($0 < R < a$) is given by

$$\left(\frac{b}{R} + B_0 R\right) \sin\varphi = K,$$

where $b = \mu_0 \hat{m}/2\pi$ and K is a constant. Using the relationships $R^2 = x^2 + y^2$ and $R\sin\varphi = y$, the above relation leads to

$$x = \pm\left(\frac{by}{K - B_0 y} - y^2\right)^{1/2}$$

in the x-y plane.

In the magnetic material ($R > a$), the equivector potential surface is given by

$$\frac{b'}{R} \sin\varphi = K',$$

where $b' = \mu \hat{m}'/2\pi$ and K' is a constant. This leads the equivector potential surface:

$$x^2 + \left(y - \frac{b'}{2K'}\right)^2 = \left(\frac{b'}{2K'}\right)^2.$$

9.26. The effect of the magnetizing current on the surface is represented by the moment of magnetic dipole line $\hat{m}$. The resultant uniform magnetic flux density in the magnetic material is denoted by B. Then, the magnetic potential is

$$\phi_m = \left(-B_0 R + \frac{\mu_0 \hat{m}}{2\pi R}\right) \cos\varphi, \quad R > a,$$
$$= -BR\cos\varphi, \quad 0 \leq R < a.$$

From the continuities of the normal component of the magnetic flux density and the parallel component of the magnetic field on the surface ($R = a$), we have

$$B_0 + \frac{\mu_0 \hat{m}}{2\pi a^2} = B, \quad \frac{1}{\mu_0}\left(B_0 - \frac{\mu_0 \hat{m}}{2\pi a^2}\right) = \frac{B}{\mu}.$$

9.5 Exercises

These conditions yield

$$\hat{m} = \frac{\mu - \mu_0}{\mu + \mu_0} \cdot \frac{2\pi a^2}{\mu_0} B_0, \quad B = \frac{2\mu}{\mu + \mu_0} B_0.$$

These agree with the results derived in Example 9.7.

9.27. The magnetic flux density in the vacuum side on the magnetic material surface is

$$B_v(0) = -\frac{\mu_0^2 aI}{\pi(\mu + \mu_0)(x^2 + a^2)},$$

and that in the magnetic material side on its surface is

$$B_m(0) = -\frac{\mu\mu_0 aI}{\pi(\mu + \mu_0)(x^2 + a^2)}.$$

Hence, the magnetizing current density is determined to be

$$\tau_m = \frac{1}{\mu_0}[B_v(0) - B_m(0)] = \frac{(\mu - \mu_0)aI}{\pi(\mu + \mu_0)(x^2 + a^2)}.$$

The total magnetizing current is

$$I_m = \int \tau_m dx = \frac{2(\mu - \mu_0)aI}{\pi(\mu + \mu_0)} \int_0^\infty \frac{dx}{(x^2 + a^2)} = \frac{\mu - \mu_0}{\mu + \mu_0} I,$$

which is equal to the image currents I' obtained in Example 9.6

9.28. A perpendicular plane is drawn from the straight current and the y-axis ($x = 0$) is defined at the foot of the perpendicular plane on the surface ($z = 0$). The magnetic flux density at the straight current produced by the magnetizing current in a narrow region from x to $x + dx$ on the surface is

$$dB = \frac{\mu_0 \tau_m(x) dx}{2\pi(x^2 + a^2)^{1/2}}.$$

Only the horizontal component, $dBa/(x^2 + a^2)^{1/2}$, remains from symmetry. Using the result of $\tau_m(x)$ obtained in Exercise 9.27, the total magnetic flux density at the straight current is determined to be

$$B = \frac{\mu_0(\mu - \mu_0)Ia^2}{\pi^2(\mu + \mu_0)} \int_0^\infty \frac{dx}{(x^2 + a^2)^2} = \frac{\mu_0(\mu - \mu_0)I}{\pi^2(\mu + \mu_0)a} \int_0^{\pi/2} \cos^2\theta \, d\theta$$

$$= \frac{\mu_0(\mu - \mu_0)I}{4\pi(\mu + \mu_0)a}.$$

Thus, the Lorentz force in a unit length is

$$F' = \frac{\mu_0 II'}{4\pi a},$$

which is equal to the force between the given and image currents separated by distance $2a$.

9.29. The magnetic flux density in the magnetic material is uniform and is given by

$$B = \frac{\mu k I}{\log(1 + kw)}.$$

So, the magnetic energy in a unit length is

$$U'_m = d \int_0^w \frac{B^2 dx}{2\mu} = \frac{\mu k^2 d I^2}{2[\log(1 + kw)]^2} \int_0^w \frac{dx}{1 + kx} = \frac{\mu k d I^2}{2 \log(1 + kw)} = \frac{1}{2} L' I^2,$$

where L' is the inductance in a unit length determined in Exercise 9.3.

9.30. The magnetic field in the magnetic material is uniform and is given by $H = I/w$. So, the magnetic energy in a unit length is

$$U'_m = w \int_0^d \frac{1}{2} \mu(1 + ky) H^2 dy = \frac{\mu d I^2}{2w}\left(1 + \frac{kd}{2}\right) = \frac{1}{2} L' I^2,$$

where L' is the inductance in a unit length determined in Exercise 9.4.

9.31. The magnetic energy density in each magnetic material at distance R from the central axis is $B^2(R)/2\mu_1$ and $B^2(R)/2\mu_2$ with

$$B(R) = \frac{\mu_1 \mu_2 I}{\pi(\mu_1 + \mu_2)R},$$

and the magnetic flux density is zero in other places. Hence, the magnetic energy of the transmission in a unit length is given by

9.5 Exercises

$$U'_m = \int_a^b \left(\frac{B^2}{2\mu_1} + \frac{B^2}{2\mu_2}\right)\pi R dR = \frac{\mu_1\mu_2 I^2}{2\pi(\mu_1+\mu_2)}\log\frac{b}{a}.$$

9.32. The magnetic field at radius R is $H(R) = I/2\pi R$. Hence, the magnetic energy in a unit length is

$$U'_m = \int_0^d \frac{1}{2}\mu' H^2 2\pi R dR = \frac{\mu I^2}{4\pi}\int_0^d \left(\frac{1}{R}+k\right)dR$$
$$= \frac{\mu I^2}{4\pi}\left[\log\frac{b}{a}+k(b-a)\right] = \frac{1}{2}L'I^2,$$

where L' is the inductance in a unit length determined in Exercise 9.5.

9.33. The vector potential in the magnetic materials 1 and 2 are

$$A_{z2}(R) = \frac{\mu_2 I}{2\pi}\log\frac{c}{R}; \qquad b < R < c,$$
$$A_{z1}(R) = \frac{\mu_1 I}{2\pi}\log\frac{b}{R} + \frac{\mu_2 I}{2\pi}\log\frac{c}{b}; \quad a < R < b.$$

Hence, the vector potential at the inner superconductor with current I is $A_{z1}(a)$ and that at the outer superconductor $A_{z2}(c)$ is zero. So, the magnetic energy of the transmission line in a unit length is given by

$$U'_m = \frac{1}{2}A_{z1}(a)I = \frac{I^2}{4\pi}\left(\mu_1\log\frac{b}{a} + \mu_2\log\frac{c}{b}\right).$$

9.34. The vector potential in the magnetic material ($a \leq R \leq b$) is

$$A_z(R) = \frac{\mu_1\mu_2 I}{\pi(\mu_1+\mu_2)}\log\frac{b}{R}.$$

Hence, it is $A_z(a)$ at the inner superconductor with current I and is zero at the outer superconductor ($R = b$). Hence, the magnetic energy of the transmission line in a unit length is given by

$$U'_m = \frac{1}{2}A_z(a)I = \frac{\mu_1\mu_2 I^2}{2\pi(\mu_1+\mu_2)}\log\frac{b}{a}.$$

9.35. The magnetic flux density has an azimuthal component given by

$$B_\varphi(R) = \frac{\mu(1+kR)I}{2\pi R}$$

for $a \leq R \leq b$. The vector potential in this region is

$$A_z(R) = -\frac{\mu I}{2\pi} \int_b^R \left(\frac{1}{R} + k\right) dR = \frac{\mu I}{2\pi}\left[\log\frac{b}{R} + k(b-R)\right].$$

Hence, the magnetic energy in a unit length is determined to be

$$U'_m = \frac{1}{2}[A_z(a)I - A_z(b)I] = \frac{\mu I^2}{4\pi}\left[\log\frac{b}{a} + k(b-a)\right].$$

Chapter 10
Electromagnetic Induction

10.1 Induction Law

Suppose that there are two coils 1 and 2. When the current applied to coil 1 changes, a current is induced in coil 2. A similar thing happens when coil 2 is moved, even when the current does not change in coil 1. This phenomenon is called electromagnetic induction. The electromotive force to induce current is given by

$$V_{\text{em}} = -\frac{d\Phi}{dt}, \qquad (10.1)$$

where Φ is the magnetic flux that penetrates coil 1. This is called Faraday's law or the magnetic flux law. In this case, Φ and V_{em} follow the right-hand rule. That is, the directions of Φ and V_{em} correspond to the directions of the motions of a screw and a screw driver. Equation (10.1) is transformed to

$$\int_S \nabla \times \mathbf{E} \cdot d\mathbf{S} = -\int_S \frac{\partial \mathbf{B}}{\partial t} \cdot d\mathbf{S}, \qquad (10.2)$$

where S is a surface. Since this holds for arbitrary S, we have

$$\nabla \times \mathbf{E} = -\frac{\partial \mathbf{B}}{\partial t}. \qquad (10.3)$$

Next, we consider the second case where coil 2 in the form of closed loop C moves with velocity $\mathbf{v}$ in a magnetic flux density $\mathbf{B}$ that does not change with time. The magnetic flux that enters the loop during a period Δt is

© The Editor(s) (if applicable) and The Author(s), under exclusive license to Springer Nature Switzerland AG 2025
T. Matsushita, *Exercises in Electricity and Magnetism*,
https://doi.org/10.1007/978-3-031-67940-7_10

$$\Delta \Phi = \Delta t \oint_C (\boldsymbol{B} \times \boldsymbol{v}) \cdot d\boldsymbol{s}. \tag{10.4}$$

It should be noted that the directions of $d\boldsymbol{s}$ and $\boldsymbol{B}$ follow the right-hand rule. In the limit $\Delta t \to 0$, we have

$$\frac{\Delta \Phi}{\Delta t} \to \frac{d\Phi}{dt} = \oint_C (\boldsymbol{B} \times \boldsymbol{v}) \cdot d\boldsymbol{s}. \tag{10.5}$$

Since the electromotive force is written as

$$V_{\mathrm{em}} = \oint_C \boldsymbol{E} \cdot d\boldsymbol{s}, \tag{10.6}$$

we obtain the relationship describing the induced electric field:

$$\boldsymbol{E} = \boldsymbol{v} \times \boldsymbol{B}. \tag{10.7}$$

This is called the motional law.

Example 10.1 AC current $I(t) = I_{\mathrm{m}} \sin \omega t$ flows along a straight line. Calculate the electromotive force induced along ABCD in a rectangular coil separated by distance d from the current (see Fig. 10.1).

Fig. 10.1 Straight line carrying AC current and rectangular coil

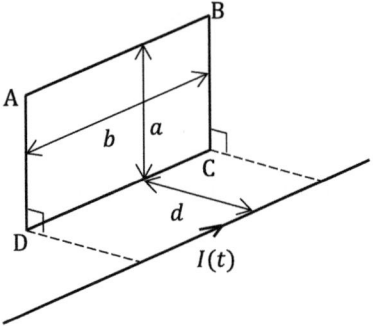

Solution 10.1 The direction of the magnetic flux produced by the current is opposite to that when the current flows along ABCD. So, the magnetic flux penetrating the coil is

10.1 Induction Law

$$\Phi = -\frac{\mu_0 I(t) b}{2\pi} \int_d^{(a^2+d^2)^{1/2}} \frac{dR}{R} = -\frac{\mu_0 I(t) b}{2\pi} \log \frac{(a^2+d^2)^{1/2}}{d}.$$

The induced electromotive force is

$$V_{em} = -\frac{d\Phi}{dt} = \frac{\mu_0 b}{2\pi} \log \frac{(a^2+d^2)^{1/2}}{d} \cdot \frac{dI(t)}{dt}$$

$$= \frac{\mu_0 I_m b \omega}{2\pi} \log \frac{(a^2+d^2)^{1/2}}{d} \cos \omega t.$$

◆

Example 10.2 A triangular closed circuit and a straight current, I, are placed on a common plane. The closed circuit is moving away with velocity v from the current, as shown in Fig. 10.2. The distance between the closed circuit and the current is $r = r_0$ in the initial condition. Determine the electromotive force induced in the closed circuit with the magnetic flux law. We define the electromotive force to be positive along the direction of ABC.

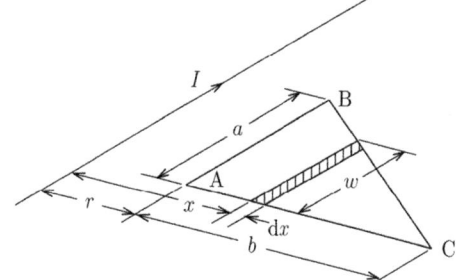

Fig. 10.2 Straight current and triangular closed circuit moving away with constant velocity

Solution 10.2 At time t, the distance between the circuit and current is $r(t) = r_0 + vt$. The width of the triangle at distance x ($r \leq x \leq r+b$) from the current is

$$w(x) = a - \frac{a(x-r)}{b}.$$

The direction of the magnetic flux produced by current I inside the circuit is the same as that of the magnetic flux produced by the current flowing along ABC. Hence, the magnetic flux produced by current I is positive. The magnetic flux density at distance x from the current is $B(x) = \mu_0 I/(2\pi x)$. The magnetic flux penetrating the narrow region x to $x + dx$ in the circuit is

$$d\Phi = B(x)w(x)dx = \frac{\mu_0 aI}{2\pi b}\left(\frac{r+b}{x} - 1\right)dx.$$

Thus, the total magnetic flux penetrating the circuit is

$$\Phi = \frac{\mu_0 aI}{2\pi b}\int_r^{r+b}\left(\frac{r+b}{x} - 1\right)dx = \frac{\mu_0 aI}{2\pi b}\left[(r+b)\log\frac{r+b}{r} - b\right].$$

The electromotive force induced in the circuit is

$$V_{em} = -\frac{d\Phi}{dt} = -\frac{\partial\Phi}{\partial r}\cdot\frac{dr}{dt} = \frac{\mu_0 Iav}{2\pi b}\left[\frac{b}{r_0 + vt} - \log\left(1 + \frac{b}{r_0 + vt}\right)\right].$$

◆

Example 10.3 Determine the electromotive force in Example 10.2 with the motional law.

Solution 10.3 Figure 10.3a shows the direction of the induced electric field on each side. We determine the electromotive force induced on each side. On side AB, the magnetic flux density is $B = \mu_0 I/(2\pi r)$, and $v \times B$ is directed from A to B and its magnitude is $\mu_0 Iv/(2\pi r)$. Hence, the contribution from this side to the electromotive force is

$$\int_A^B (v \times B)\cdot ds = \frac{\mu_0 Iv}{2\pi r}a.$$

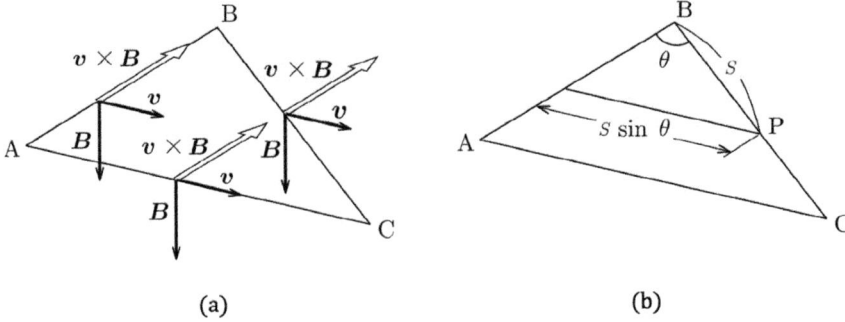

(a) (b)

Fig. 10.3 **a** Direction of induced electric field on each side and **b** a point on side BC

10.1 Induction Law

Next, the magnetic flux density at point P at distance s from B is $\mu_0 I/[2\pi(r+s\sin\theta)]$ with θ denoting the angle of B (see Fig. 10.3b). The induced electric field has magnitude $\mu_0 I v/[2\pi(r+s\sin\theta)]$ and its direction is tilted by $\pi-\theta$ from the direction of integration, ds. Thus, we have

$$(v \times B) \cdot ds = \frac{\mu_0 I v \cos(\pi-\theta)}{2\pi(r+s\sin\theta)} ds = -\frac{\mu_0 I v \cos\theta}{2\pi(r+s\sin\theta)} ds,$$

and the contribution from side BC is

$$\int_B^C (v \times B) \cdot ds = -\frac{\mu_0 I v \cos\theta}{2\pi} \int_0^{\overline{BC}} \frac{ds}{r+s\sin\theta} = -\frac{\mu_0 I a v}{2\pi b} \log \frac{r+b}{r}.$$

Finally, the induced electric field is perpendicular to the direction of integration on side CA. Hence, there is no contribution from this side. As a result, the induced electromotive force is

$$V_{em} = \oint (v \times B) \cdot ds = \frac{\mu_0 I a v}{2\pi b} \left[\frac{b}{r_0 + vt} - \log\left(1 + \frac{b}{r_0 + vt}\right) \right],$$

which agrees with the result obtained in Example 10.2.

◆

Example 10.4 A rectangular closed circuit starts to fall down from the distance d_0 at $t = 0$ below a straight line carrying current I, as shown in Fig. 10.4. Determine the electromotive force induced in the direction of ABCD of the rectangular circuit. The gravity acceleration is g.

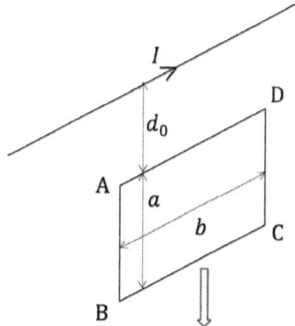

Fig. 10.4 Rectangular circuit that starts to fall down below the straight current

Solution 10.4 The magnetic flux law is used to determine the electromotive force first. The magnetic flux that the straight current produces in the circuit in the direction of ABCD is negative. This magnetic flux is

$$\Phi = -\frac{\mu_0 I b}{2\pi} \int_d^{d+a} \frac{dx}{x} = -\frac{\mu_0 I b}{2\pi} \log \frac{d+a}{d},$$

where $d = d_0 + gt^2/2$ is the distance between the straight current and side AD. Hence, the electromotive force is

$$V_{\text{em}} = -\frac{d\Phi}{dt} = -\frac{\partial \Phi}{\partial d} \cdot \frac{dd}{dt} = -\frac{\mu_0 I a b g t}{2\pi \left(d_0 + gt^2/2\right)\left(a + d_0 + gt^2/2\right)}.$$

Next, the electromotive force is determined using the motional law. On side AB, the induced electric field is normal to the direction of integration, and there is no contribution from this side. On side BC, the magnetic flux density is $\mu_0 I/2\pi(a+d)$ and the induced electric field is $\mu_0 I v/2\pi(a+d)$ directed from B to C. Hence, the contribution from this side is

$$V_{\text{BC}} = \frac{\mu_0 I b v}{2\pi (a+d)} = \frac{\mu_0 I b g t}{2\pi \left(a + d_0 + gt^2/2\right)}.$$

On side CD, the induced electric field is normal to the direction of integration, and there is no contribution from this side. On side DA, the magnetic flux density is $\mu_0 I/2\pi d$ and the induced electric field is $\mu_0 I v/2\pi d$ directed from A to D. Hence, the contribution from this side is

$$V_{\text{DA}} = -\frac{\mu_0 I b v}{2\pi d} = -\frac{\mu_0 I b g t}{2\pi \left(d_0 + gt^2/2\right)}.$$

As a result, the total electromotive force is

$$V_{\text{em}} = V_{\text{BC}} + V_{\text{DA}} = -\frac{\mu_0 I a b g t}{2\pi \left(d_0 + gt^2/2\right)\left(a + d_0 + gt^2/2\right)}.$$

♦

Suppose that a conducting circular plate of radius a is rotated with angular frequency ω around its axis in a uniform magnetic flux density B, as shown in Fig. 10.5. We determine the electromotive force induced between the center O of the plate and point P on the edge. Since the magnetic flux penetrating the closed loop composed of the straight line connecting O and P and the line C outside the plate does not change with time, it seems that no electromotive force is induced in it, although an electromotive force is induced in reality. This is called unipolar induction.

According to the motional law, since the magnetic flux crosses line OP, the electromotive force is induced there. In the arrangement in Fig. 10.5, the induced electric field is directed from O to P. At a point at a distance R from the central axis, the velocity of rotation, v, is equal to $R\omega$. Hence, the induced electric field is $E = R\omega B$. Integrating this from O to P, the electromotive force is

$$V_{em} = \int_0^a R\omega B dR = \frac{1}{2}\omega B a^2. \tag{10.8}$$

We assume that the external magnetic field is increasing with time. In this case the magnetic flux density inside the material also increases because of the penetrating magnetic flux. Thus, we can define the velocity of the magnetic flux lines and denote it as V. If a coil stays stationary but the magnetic flux lines move with velocity $V = -v$, the same amount of magnetic flux penetrates into the coil as in the case in which the coil moves with velocity v in the stationary magnetic flux density. Hence, repeating a similar argument up to Eq. (10.5), the time-variation in the magnetic flux Φ that penetrates the coil is given by

$$\frac{d\Phi}{dt} = -\oint_C (\boldsymbol{B} \times \boldsymbol{V}) \cdot d\boldsymbol{s}. \tag{10.9}$$

Thus, we obtain the local relationship:

$$\nabla \times (\boldsymbol{B} \times \boldsymbol{V}) = -\frac{\partial \boldsymbol{B}}{\partial t}. \tag{10.10}$$

This is called the continuity equation of magnetic flux and is frequently used to determine the velocity of quantized magnetic flux lines in superconductors. Comparing

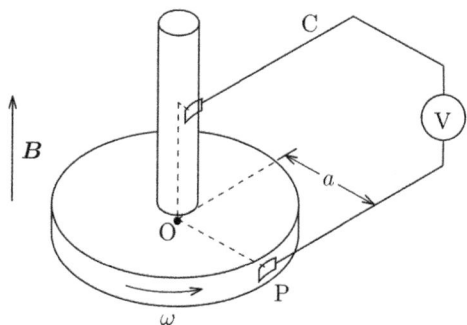

Fig. 10.5 Unipolar induction system

this equation with Eq. (10.3), we have

$$E = B \times V. \qquad (10.11)$$

Example 10.5 Suppose that we increase by ΔB the magnetic flux density B_0 applied parallel to a long cylindrical conductor of radius a during a short period, Δt. Determine with Eq. (10.11) the electromotive force measured with potential leads with the different arrangements shown in Fig. 10.6a, b.

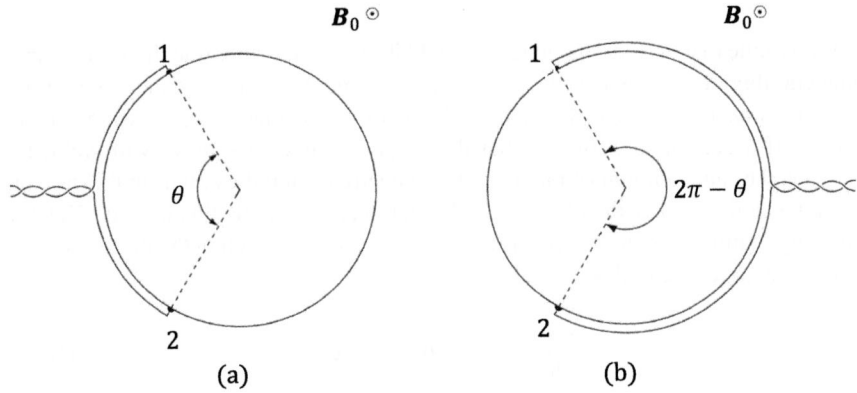

Fig. 10.6 Different arrangements of potential leads for measuring electromotive force induced in cylindrical conductor

Solution 10.5 We denote the velocity of the magnetic flux as

$$V = -i_R V,$$

where i_R is the radial unit vector. Then, the continuity equation of magnetic flux shown earlier reduces to

$$-\frac{1}{R} \cdot \frac{\partial}{\partial R}(RB_0 V) = -\frac{\Delta B}{\Delta t},$$

and we have

$$E = B_0 V = \frac{1}{2} \cdot \frac{\Delta B}{\Delta t} a$$

on the surface ($R = a$). The induced electric field is directed counterclockwise. One can easily show that the induced electric field integrated along the circumference in this direction, $2\pi a E$, is equal to the total electromotive force, $-\Delta \Phi / \Delta t = \pi a^2 \Delta B / \Delta t$.

Here, we consider the case shown in Fig. 10.6a. The azimuthal angle between potential terminals 1 and 2 is θ, and we place the potential leads on the surface of the conductor and twist to eliminate the electromotive force outside the conductor. The measured electromotive force is

$$V_{em} = a\theta B_0 V = \frac{1}{2} \cdot \frac{\Delta B}{\Delta t} a^2 \theta,$$

where we set a reference point on terminal 1.

Second, we determine the electromotive force for the arrangement shown in Fig. 10.6b. We assume an integral path of the induced electric field on the right conductor surface. In this case, we can neglect the magnetic flux penetrating the closed loop composed of this path and the potential leads, and similarly determine the electromotive force only by integrating Eq. (10.11) along the path. Thus, we have

$$V'_{em} = -(2\pi - \theta) B_0 V = -\frac{1}{2} \cdot \frac{\Delta B}{\Delta t} a^2 (2\pi - \theta)$$

for the same reference point.

On the other hand, it is possible to choose the left conductor surface for the integral path. The integral gives $a^2\theta(\Delta B/\Delta t)/2$. In this case the electromotive force due to the magnetic flux penetrating the integral path should be taken into account. This additional component is $-\pi a^2(\Delta B/\Delta t)$, and we obtain the same result by adding it.

Thus, there is a freedom in choosing the integral path. For example, it is also possible to choose the path shown by the dotted line in Fig. 10.6b. In this case the induced electric field is perpendicular to the path, resulting in a zero line integral. From the magnetic flux penetrating the area surrounded by the path and the potential leads (denoted by the azimuthal angle $2\pi - \theta$), we directly obtain the same result. ◆

10.2 Potential

In general, the electric field $\boldsymbol{E}$ contains not only the electrostatic field but also the induced electric field. Hence, such a general electric field cannot be described only by the electric potential. Here, we note that the right side of Eq. (10.3) is written in terms of the vector potential $\boldsymbol{A}$ as

$$-\frac{\partial \boldsymbol{B}}{\partial t} = -\nabla \times \left(\frac{\partial \boldsymbol{A}}{\partial t}\right), \tag{10.12}$$

where we have changed the order of the time differentiation and spatial differentiation. Comparing this with the left side of Eq. (10.3), it is obvious that the induced electric field is given by $-\partial \boldsymbol{A}/\partial t$. Hence, with the electrostatic field, the general electric field is given by

$$E = -\nabla \phi - \frac{\partial A}{\partial t}. \tag{10.13}$$

This satisfies Eq. (10.3) and reduces to Eq. (1.18) in the static condition.

Now, the vector potential depends on time due to a variation in the magnetic condition. Even in this case, Poisson's equation holds:

$$\Delta A = -\mu i. \tag{10.14}$$

Example 10.6 Suppose that alternating current $I(t) = I_0 \sin \omega t$ flows in a straight wire, as shown in Fig. 10.7. Determine the electromotive force induced in the rectangular circuit with two parallel sides using the vector potential. The straight current and rectangular circuit are on the common plane. We define the electromotive force to be positive in the direction of ABCD.

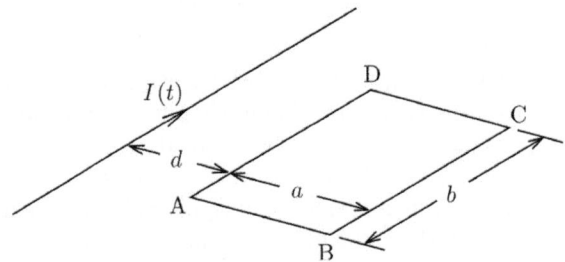

Fig. 10.7 Straight wire with alternating current and rectangular circuit

---

Solution 10.6 Suppose that the AC current flows in the direction of the z-axis. The magnetic flux density produced by the current is an azimuthal component given by

$$B_\varphi = \frac{\mu_0 I}{2\pi R} \sin \omega t,$$

and the vector potential, A_z, is

$$A_z = -\int B_\varphi dR = -\frac{\mu_0 I}{2\pi} \sin \omega t \log \frac{R}{R_0},$$

where R_0 is the distance to the reference point. Using Eq. (10.13), the induced electric field on side BC is $(\mu_0 I_0 \omega / 2\pi) \cos \omega t \log(a+d)/R_0$ in the direction from B to C. So, the contribution from side BC to the electromotive force is

$$V_{BC} = \frac{\mu_0 I_0 b \omega}{2\pi} \cos \omega t \log \frac{(a+d)}{R_0}.$$

The induced electric field on side DA is $(\mu_0 I_0 \omega/2\pi) \cos \omega t \log d/R_0$ in the direction from A to D. So, the contribution from side DA to the electromotive force is

$$V_{DA} = -\frac{\mu_0 I_0 b \omega}{2\pi} \cos \omega t \log \frac{a}{R_0}.$$

On sides AB and CD, the electric field is normal to these sides, and there is no contribution from these sides. Thus, the electromotive force is determined to be

$$V_{em} = V_{BC} + V_{DA} = \frac{\mu_0 I_0 b \omega}{2\pi} \cos \omega t \log \frac{a+d}{a}.$$

♦

10.3 Magnetic Energy

If we try to estimate the magnetic energy in a current system from the mechanical work to carry current from infinity until the final current condition is attained, similarly to the estimation of the electrostatic energy, we will surely fail because of the electromagnetic induction. The reason why we can estimate the magnetic energy from the work done by the Lorentz force in Chap. 8 is that the perfect diamagnetism of the superconductor keeps the magnetic flux constant. So, it is usual to estimate the magnetic energy as an equivalent electric energy due to the electromagnetic induction. Here, we follow this approach, since an explanation from a different viewpoint is helpful for a deep understanding of the phenomenon.

Suppose a coil of self-inductance L. When we apply current I' to this coil, the magnetic flux penetrating this coil is

$$\Phi' = LI'. \tag{10.15}$$

When the current is increased by a small amount dI' in a short period of time dt, the induced electromotive force is

$$V_{em} = -L\frac{dI'}{dt}. \tag{10.16}$$

This acts to restrict the increase in the current. This phenomenon is called self-induction. Thus, the electric power source must work against this electromotive force to increase the current, and the electric power in this period is

$$P = -V_{em}I' = LI'\frac{dI'}{dt}. \tag{10.17}$$

The energy stored in the coil when we apply the current I to the coil is equal to the energy supplied by the electric power source until the current increases from 0 to I, and is given by

$$U_m = W = \int LI' \frac{dI'}{dt} = L \int_0^I I' dI' = \frac{1}{2} LI^2. \tag{10.18}$$

Using the magnetic flux $\Phi = LI$, this is also written as

$$U_m = \frac{1}{2} LI^2 = \frac{1}{2} \Phi I = \frac{1}{2L} \Phi^2. \tag{10.19}$$

This result agrees with Eq. (8.17), and we can understand that this energy is exactly the magnetic energy.

Example 10.7 Prove Eq. (8.18) for a system composed of two coils and prove the reciprocity, $L_{12} = L_{21}$.

Solution 10.7 Currents I_1 and I_2 flow in coils 1 and 2, and the resultant penetrating magnetic fluxes in these coils are Φ_1 and Φ_2, respectively. Assume that this final situation is reached after we apply the currents to coil 1 and then to coil 2. We suppose an intermediate situation where coil 2 has no current and coil 1 has current I_1'. With the inductance coefficient the magnetic flux that penetrates coil 1 is $L_{11} I_1'$. Hence, the work done to apply current I_1 to coil 1 is

$$W_1 = \int_0^{I_1} L_{11} I_1' dI_1' = \frac{1}{2} L_{11} I_1^2.$$

Next, we consider the situation where coils 1 and 2 have currents I_1 and I_2', respectively. The magnetic flux penetrating coil 2 is $L_{21} I_1 + L_{22} I_2'$. Hence, the work done to apply current I_2 to coil 2 is

$$W_2 = \int_0^{I_2} (L_{21} I_1 + L_{22} I_2') dI_2' = L_{21} I_1 I_2 + \frac{1}{2} L_{22} I_2^2.$$

Thus, the magnetic energy of this system is

$$U_m = W_1 + W_2 = \frac{1}{2} L_{11} I_1^2 + L_{21} I_1 I_2 + \frac{1}{2} L_{22} I_2^2.$$

If we apply the currents in reversed order, the magnetic energy is

$$U_\mathrm{m} = \frac{1}{2}L_{11}I_1^2 + L_{12}I_1I_2 + \frac{1}{2}L_{22}I_2^2.$$

Since the two results must be the same, we can prove the reciprocity,

$$L_{12} = L_{21}.$$

Thus, if we write $L_{21} = (L_{12} + L_{21})/2$ in the first result, the magnetic energy is

$$U_\mathrm{m} = \frac{1}{2}I_1(L_{11}I_1 + L_{12}I_2) + \frac{1}{2}I_2(L_{21}I_1 + L_{22}I_2) = \frac{1}{2}(\Phi_1 I_1 + \Phi_2 I_2).$$

Thus, Eq. (8.18) holds for $n = 2$.

◆

10.4 Skin Effect

Here, we learn about the skin effect for an example of dynamic phenomena. Suppose that we apply an AC magnetic flux density of amplitude B_0 and angular frequency ω along the z-axis parallel to the surface of a semi-infinite conductor that occupies $x \geq 0$ (see Fig. 10.8). The equations that we use in this analysis are Eqs. (5.14), (9.12), (9.16), and (10.3). Using Eqs. (5.14) and (9.16), Eq. (9.12) is written as

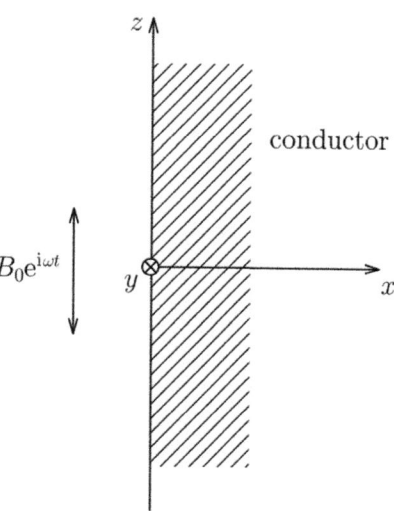

Fig. 10.8 Magnetic flux density applied parallel to the surface of a semi-infinite conductor

$$\nabla \times \boldsymbol{B} = \mu \sigma_c \boldsymbol{E}. \tag{10.20}$$

We can solve this equation with Eq. (10.3). We can assume that no physical variable changes along the y- and z-axes:

$$\frac{\partial}{\partial y}, \frac{\partial}{\partial y} \to 0. \tag{10.21}$$

The variation with time can be expressed with the factor $e^{i\omega t}$, and we replace the time differentiation with $i\omega$:

$$\frac{\partial}{\partial t} \to i\omega. \tag{10.22}$$

In addition, since the external magnetic flux density is directed along the z-axis, we can assume that the internal magnetic flux density has only a z-component. Hence, the left-hand side of Eq. (10.20) leads to

$$\nabla \times \boldsymbol{B} = -\boldsymbol{i}_y \frac{dB_z}{dx}. \tag{10.23}$$

Thus, we find that the electric field has only a y-component, E_y, and Eq. (10.20) reduces to

$$\frac{dB_z}{dx} = -\mu \sigma_c E_y. \tag{10.24}$$

Since the left side of Eq. (10.3) is

$$\nabla \times \boldsymbol{E} = \boldsymbol{i}_z \frac{dE_y}{dx}, \tag{10.25}$$

Equation (10.3) leads to

$$\frac{dE_y}{dx} = -i\omega B_z. \tag{10.26}$$

Eliminating E_y in Eqs. (10.24) and (10.26), we have

$$\frac{d^2 B_z}{dx^2} - i\omega \mu \sigma_c B_z = 0. \tag{10.27}$$

10.4 Skin Effect

We can derive this equation generally. Taking a curl of Eq. (10.20) and substituting Eq. (10.3), we have

$$\nabla \times (\nabla \times \mathbf{B}) = -\mu \sigma_c \frac{\partial \mathbf{B}}{\partial t}. \tag{10.28}$$

Using Eqs. (A.9) in the Appendix and (6.13), this equation becomes

$$\Delta \mathbf{B} - \mu \sigma_c \frac{\partial \mathbf{B}}{\partial t} = 0. \tag{10.29}$$

This is a differential equation of the second-order called a diffusion equation.

From the condition that the magnetic flux density must be finite in the limit $x \to \infty$ and the boundary condition, $B_z(x=0) = B_0$, we obtain the solution of the magnetic flux density as

$$B_z(x,t) = B_0 e^{-x/\delta} \exp\left[i\left(\omega t - \frac{x}{\delta}\right)\right] \to B_0 e^{-x/\delta} \cos\left(\omega t - \frac{x}{\delta}\right), \tag{10.30}$$

where δ is a quantity with the dimension of length called the skin depth and is given by

$$\delta = \left(\frac{2}{\omega \mu \sigma_c}\right)^{1/2}. \tag{10.31}$$

Substituting this solution into Eq. (10.24), the electric field is given by

$$E_y(x,t) = B_0 \left(\frac{\omega}{\mu \sigma_c}\right)^{1/2} e^{-x/\delta} \exp\left[i\left(\omega t - \frac{x}{\delta} + \frac{\pi}{4}\right)\right]$$

$$\to B_0 \left(\frac{\omega}{\mu \sigma_c}\right)^{1/2} e^{-x/\delta} \cos\left(\omega t - \frac{x}{\delta} + \frac{\pi}{4}\right). \tag{10.32}$$

The magnetic field and current density are obtained from $H_z(x,t) = B_z(x,t)/\mu$ and $i_y(x,t) = \sigma_c E_y(x,t)$ with the above results.

Example 10.8 Suppose that we apply an AC electric field of amplitude E_0 and angular frequency ω along the z-axis parallel to the surface of a semi-infinite conductor of electric conductivity σ_c that occupies $x \geq 0$. Determine the electric field and magnetic flux density in the conductor.

Solution 10.8 We can assume that the derivatives with respect to y and z are zero from spatial symmetry and replace the time derivative by $i\omega$. We can also assume that the inner electric field has only a z-component, E_z. Equation (10.3) leads to $dE_z/dx = i\omega B_y$, showing that the magnetic flux density has only a y-component. Thus, Eq. (10.20) leads to $dB_y/dx = \mu\sigma_c E_z$. The above two equations yield

$$\frac{d^2 E_z}{dx^2} - i\omega\mu\sigma_c E_z = 0.$$

We can easily solve this equation under the boundary condition $E_z(x=0) = E_0$. Taking the real part, we have

$$E_z(x,t) = E_0 e^{-x/\delta} \exp\left[i\left(\omega t - \frac{x}{\delta}\right)\right] \rightarrow E_0 e^{-x/\delta} \cos\left(\omega t - \frac{x}{\delta}\right).$$

Substituting the complex solution into the first equation yields

$$B_y(x,t) = E_0 \left(\frac{\mu\sigma_c}{\omega}\right)^{1/2} e^{-x/\delta} \exp\left[i\left(\omega t - \frac{x}{\delta} + \frac{3\pi}{4}\right)\right]$$
$$\rightarrow E_0 \left(\frac{\mu\sigma_c}{\omega}\right)^{1/2} e^{-x/\delta} \cos\left(\omega t - \frac{x}{\delta} + \frac{3\pi}{4}\right).$$

♦

10.5 Exercises

10.1. We apply current $I(t) = I_0 \sin\omega t$ to a straight line. Determine the electromotive force induced in a circular coil, as shown in Fig. E10.1, in the direction shown by the arrow with the magnetic flux law.

10.2. Suppose that a rectangular circuit is moving away with velocity v from a straight current I, as shown in Fig. E10.2. Determine the electromotive force induced in the

Fig. E10.1 AC current $I(t)$ and a circular coil placed on a common plane

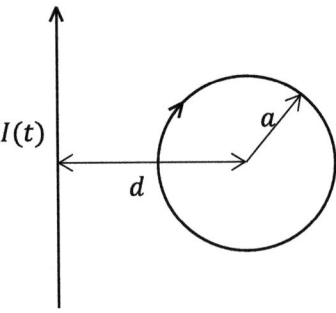

10.5 Exercises

Fig. E10.2 Straight current I and rectangular circuit moving away from the current

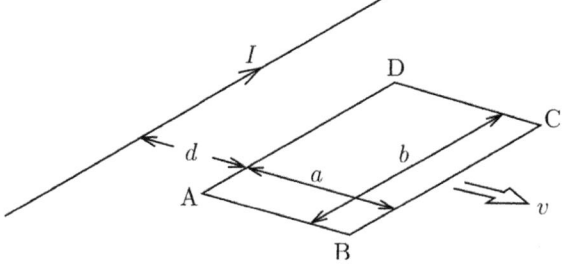

Fig. E10.3 Closed circuit in the form of an equilateral triangle moving away from the current

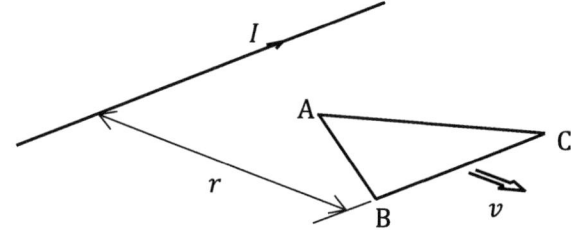

circuit with the magnetic flux law. The straight current and the rectangular circuit are placed on a common plane, and the electromotive force is defined to be positive in the direction of ABCD.

10.3. Solve the problem in Exercise 10.2 with the motional law.

10.4. A closed circuit composed of an equilateral triangle with side length a is moving away with velocity v from the current I, as shown in Fig. E10.3. The circuit and the current are on a common plane, and the distance is $r(t) = r_0 + vt$. Determine the induced electromotive force in the direction of ABC with the magnetic flux law.

10.5. Solve the problem of Exercise 10.4 with the motional law.

10.6. Solve the problem of Example 10.1 using the electric field determined with the vector potential.

10.7. Solve the problem in 10.2 with the vector potential.

10.8. Solve the problem in Example 10.2 with the vector potential.

10.9. Solve the problem of Example 10.4 with the vector potential.

10.10. Solve the problem of Example 10.6 using Eq. (10.3).

10.11. Solve the problem in Exercise 10.2 with Eq. (10.3).

10.12. Solve the problem in Example 10.2 with Eq. (10.3).

10.13. Solve the problem in Example 10.4 with Eq. (10.3).

10.14. Suppose that the current changes as $I(t) = I_m \sin \omega t$ and a rectangular circuit is moving away with velocity v from the current, as shown in Fig. E10.4. Determine the electromotive force induced in the circuit along the direction ABCD with the magnetic flux law.

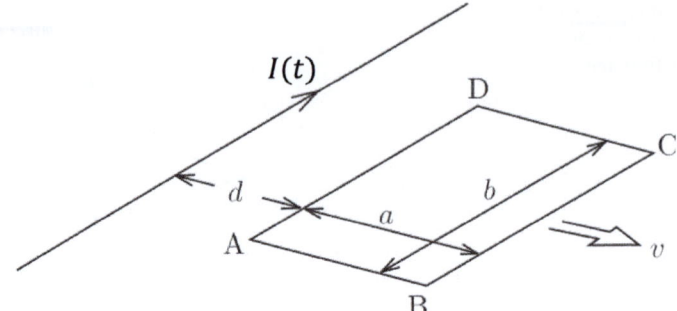

Fig. E10.4 Straight current $I(t)$ and rectangular circuit moving away from the current

10.15. Solve the problem of Exercise 10.14 with the vector potential.

10.16. Solve the problem of Exercise 10.14 with Eq. (10.3).

10.17. Derive the equation for the electric field when an AC electric field of amplitude E_0 and angular frequency ω is applied parallel to a long cylindrical material of radius a.

10.18. Derive the equation for the electric field when an AC magnetic flux density of amplitude B_0 and angular frequency ω is applied parallel to a long cylinder of radius a.

10.19. Suppose two concentric coils, as shown in Fig. E10.5. We apply voltage $V_B = V_0 \sin \omega t$ to winding B of radius a_B and number of turns N_B. Determine the voltage induced in winding A of radius a_A and number of turns N_A. Winding A is composed of a magnetic material of magnetic permeability μ, and the length of the two coils is l.

10.20. In Example 10.5, the electromotive force is calculated for the integral path on the surface of the cylindrical conductor. When the dotted line is chosen for the integral path shown in Fig. 10.6a, what is the electromotive force?

Fig. E10.5 Two magnetically coupled concentric coils

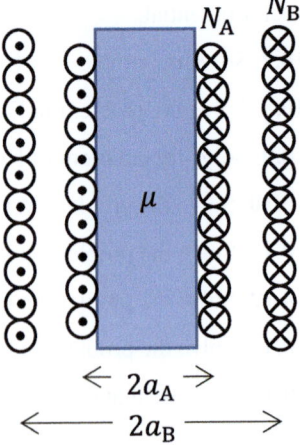

10.5 Exercises

Fig. E10.6 Unipolar induction system with two plates that have different radii

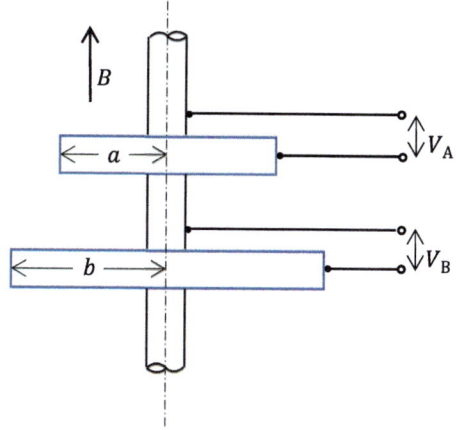

10.21. Suppose a unipolar induction system with two rotating plates, as shown in Fig. E10.6. We apply DC voltage V_A to terminal A connected to the edge of radius a in magnetic flux density B to rotate the machine with angular frequency ω. Determine the voltage induced in terminal B.

10.22. Determine the voltage on the terminals of the circular one-turn coil shown in Fig. E10.7 with Eq. (10.13), when current is increased to I. The resistance of the coil is R.

10.23. When a magnetic moment placed on the origin changes with time as $m(t) = m_0 \sin \omega t$, determine the electromagnetic fields.

10.24. We apply current I to a superconducting parallel-plate transmission line and insert a magnetic material of magnetic permeability μ into the space between the two superconducting plates by a distance x from the edge, as shown in Fig. E10.8. Determine the magnetic energy and the force on the magnetic material.

10.25. Determine the force between the wide superconducting surface and the superconducting cylinder with an applied current treated in Example 7.5 and Exercise 8.3.

Fig. E10.7 Circular one-turn coil

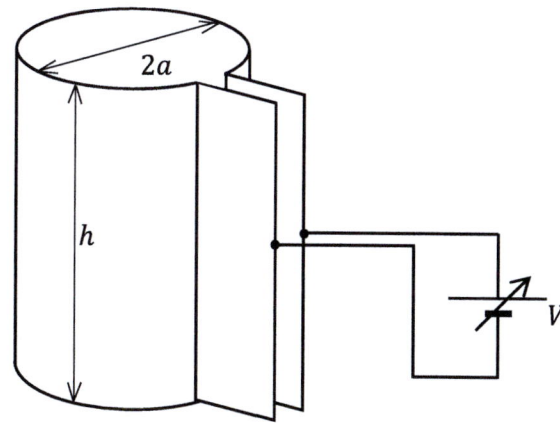

Fig. E10.8 Superconducting parallel-plate transmission line with inserted magnetic material

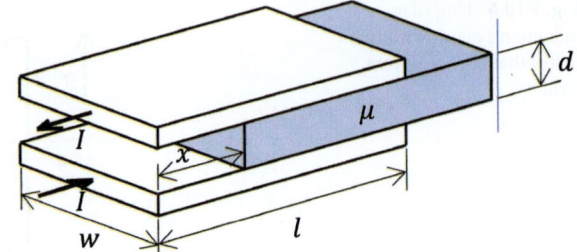

10.26. A superconducting parallel-plate transmission line is immersed in a liquid of magnetic permeability μ and specific weight ρ, and a balanced situation is attained in the condition of applied current I, as shown in Fig. E10.9. Derive the equation to determine the height h of the liquid level between the superconducting plates. The length of the transmission line is l, and the gravitational acceleration is denoted by g. Neglect the effect of disorder near the edges of the transmission line.

10.27. The skin effect in an AC electric field is discussed in Example 10.8. Derive the vector potential and magnetic flux density, and prove that the Coulomb gauge is satisfied.

10.28. The skin effect in an AC magnetic flux density is discussed in Sect. 10.5 and the internal magnetic flux density is given by Eq. (10.30). Derive Eq. (10.32) for the electric field with the vector potential, and prove that the Coulomb gauge is satisfied.

10.29. The skin effect is treated in Sect. 10.5, and Eqs. (10.30) and (10.32) give the magnetic flux density and electric field, respectively. Prove that the corresponding vector potential satisfies Poisson's equation.

10.30. The skin effect in an AC electric field is treated in Example 10.8 and Exercise 10.26. Prove that the vector potential satisfies Poisson's equation.

Fig. E10.9 Superconducting transmission line immersed in a magnetic liquid

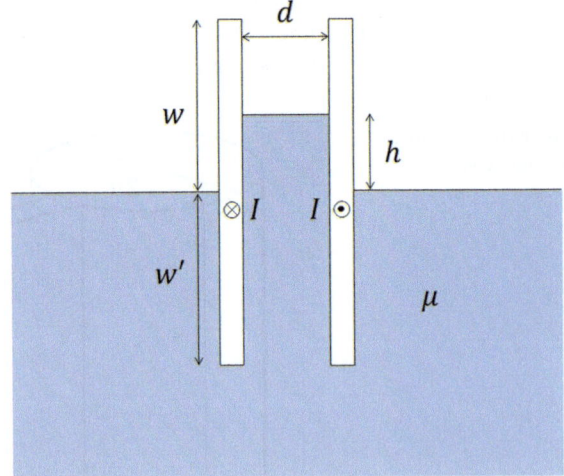

10.31. Prove that the vector potential follows the diffusion equation with the same form as Eq. (10.29) when Poisson's equation holds and the electrostatic field does not exist.

Answers to Exercises

10.1. The magnetic flux produced by the current is positive with respect to the direction shown by the arrow of the circle. We denote the distance and angle measured from the center of the circle by R and φ, respectively, as shown in Fig. B10.1. Then, the magnetic density at point (R, φ) is

$$B = \frac{\mu_0 I}{2\pi (d + R\cos\varphi)}.$$

So, the magnetic flux penetrating the circle is

$$\Phi = \frac{\mu_0 I}{2\pi} \int_0^a \int_0^{2\pi} \frac{R dR d\varphi}{d + R\cos\varphi}.$$

Using Eq. (A.23) in the Appendix, the integral with respect to φ is carried out and we have

$$\Phi = \mu_0 I \int_0^a \frac{R dR}{(d^2 - R^2)^{1/2}} = \mu_0 I \left[d - (d^2 - a^2)^{1/2} \right].$$

Hence, the electromotive force is determined to be

$$V_{em} = -\frac{\partial \Phi}{\partial t} = -\mu_0 I_0 \omega \left[d - (d^2 - a^2)^{1/2} \right] \cos \omega t.$$

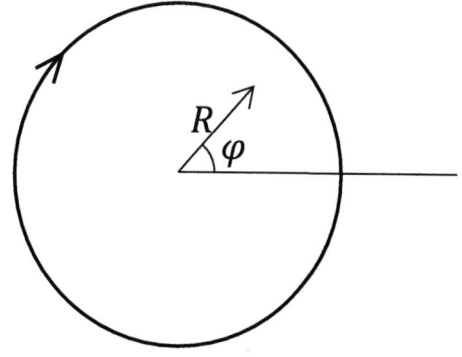

Fig. B10.1 Distance R and angle φ from the center of the circle

10.2. The direction of the magnetic flux due to the current I that penetrates the circuit is opposite to that produced by the current flowing in the direction ABCD. The magnetic flux density at distance R from the current is $B(R) = \mu_0 I/2\pi R$, and the magnetic flux that penetrates the circuit is

$$\Phi = -\frac{\mu_0 I b}{2\pi} \int_d^{a+d} \frac{dR}{R} = -\frac{\mu_0 I b}{2\pi} \log \frac{a+d}{d}.$$

The distance d changes as $d(t) = d_0 + vt$. Thus, the induced electromotive force is determined to be

$$V_{em} = -\frac{d\Phi}{dt} = \frac{\mu_0 I b}{2\pi} \cdot \frac{d}{dt} \log \frac{a+d}{d} = -\frac{\mu_0 I a b v}{2\pi (a + d_0 + vt)(d_0 + vt)}.$$

10.3. The electric field induced on sides AB and CD is perpendicular to each side, and hence, there is no contribution from these sides to the electromotive force. The induced electric field is $\mu_0 I v/[2\pi(a + d_0 + vt)]$ and is in the same direction to that of integration for side BC. So, the contribution to the electromotive force from this side is $\mu_0 I b v/[2\pi(a + d_0 + vt)]$. On side DA, the electric field is $\mu_0 I v/[2\pi(d_0 + vt)]$ and is directed opposite to integration. The contribution from this side is $-\mu_0 I b v/[2\pi(d_0 + vt)]$. As a result, the electromotive force induced in the circuit is determined to be

$$V_{em} = \frac{\mu_0 I b v}{2\pi (a + d_0 + vt)} - \frac{\mu_0 I b v}{2\pi (d_0 + vt)} = -\frac{\mu_0 I a b v}{2\pi (a + d_0 + vt)(d_0 + vt)}.$$

10.4. The magnetic flux produced by current I is in the opposite direction as that due to a current along ABC. The width of the circuit at a distance x between point A and the current is

$$w(x) = \frac{2}{\sqrt{3}}\left(x - r + \frac{\sqrt{3}}{2}a\right); \quad r - \frac{\sqrt{3}}{2}a < x < r.$$

Hence, the magnetic flux penetrating the circuit is

$$\Phi = -\int_{r-(\sqrt{3}a/2)}^{r} w(x) \frac{\mu_0 I}{2\pi x} dx = -\frac{\mu_0 I}{2\pi}\left[a - \left(\frac{2}{\sqrt{3}}r - a\right)\log \frac{r}{r - (\sqrt{3}a/2)}\right].$$

Thus, the induced electromotive force is

10.5 Exercises

$$V_{em} = -\frac{d\Phi}{dt} = -v\frac{\partial \Phi}{\partial r} = -\frac{\mu_0 I}{2\pi} v \left[\frac{2}{\sqrt{3}} \log \frac{r}{r - (\sqrt{3}a/2)} - \frac{a}{r} \right].$$

10.5. The induced electric field is parallel to the current and directed from B to C. Hence, the induced electric field at distance x is $\mu_0 I v/(2\pi x)$ and the angle of the induced electric field from side AB is $2\pi/3$. Here, we denote the distance of a point on side AB from point A by s, as shown in Fig. B10.2. Then, we have $x = (\sqrt{3}/2)(s-a) + r$. Hence, the contribution to the electromotive force from this side is

$$V_{AB} = \frac{\mu_0 I v}{2\pi} \cos\frac{2\pi}{3} \int_0^a \frac{1}{x} ds = -\frac{\mu_0 I v}{2\sqrt{3}\pi} \log \frac{r}{r - (\sqrt{3}a/2)}.$$

On side BC, the induced electric field is $\mu_0 I v/(2\pi r)$, and its angle from the side is 0. Hence, the contribution to the electromotive force from this side is

$$V_{BC} = \frac{\mu_0 I v}{2\pi r} a.$$

The contribution from side CA is the same as that from side AB:

$$V_{CA} = -\frac{\mu_0 I v}{2\sqrt{3}\pi} \log \frac{r}{r - (\sqrt{3}a/2)}.$$

The electromotive force is determined to be

$$V_{em} = V_{AB} + V_{BC} + V_{CA} = -\frac{\mu_0 I}{2\pi} v \left[\frac{2}{\sqrt{3}} \log \frac{r}{r - (\sqrt{3}a/2)} - \frac{a}{r} \right].$$

Fig. B10.2 Distance s of a point on side AB from point A

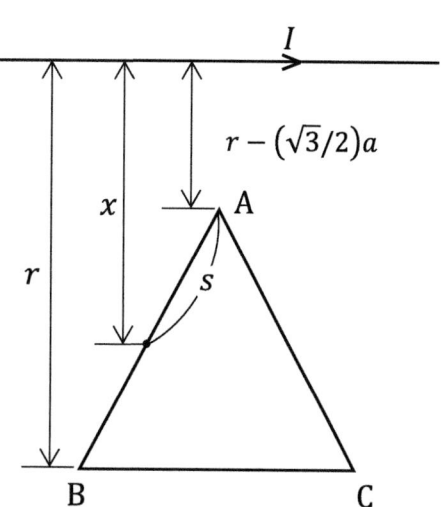

10.6. Using the same method as in Example 10.6, the vector potential at a distance R from the current is given by

$$A_z = -\frac{\mu_0 I_m}{2\pi} \sin \omega t \log \frac{R}{R_0},$$

where R_0 is the distance to the reference point of the vector potential. So, the electric field is in the direction of the current and its value on side AB is

$$E_{z1} = -\frac{\partial A_z}{\partial t}\bigg|_{R=(a^2+d^2)^{1/2}} = \frac{\mu_0 I_m \omega}{2\pi} \cos \omega t \log(a^2+d^2)^{1/2},$$

and that on side CD is

$$E_{z2} = -\frac{\partial A_z}{\partial t}\bigg|_{R=d} = \frac{\mu_0 I_m \omega}{2\pi} \cos \omega t \log d.$$

The electric fields on other two sides are perpendicular to the sides and do not contribute to the electromotive force. Hence, the induced electromotive force is determined to be

$$V_{em} = b(E_{z1} - E_{z2}) = \frac{\mu_0 I_m b \omega}{2\pi} \log \frac{(a^2+d^2)^{1/2}}{d} \cos \omega t.$$

10.7. The magnetic flux density produced by the current is

$$B_\varphi(R) = \frac{\mu_0 I}{2\pi R}$$

and the vector potential, which is parallel to the current, is

$$A_z = -\int B_\varphi(R) dR = -\frac{\mu_0 I}{2\pi} \log \frac{R}{R_0}.$$

So, the electric field on side DA is directed from A to D and its value is

$$E_{z1} = -\frac{\partial A_z}{\partial t}\bigg|_{R=d(t)} = \frac{\mu_0 I v}{2\pi(d_0 + vt)},$$

and that on side BC is directed from B to C and its value is

$$E_{z2} = -\frac{\partial A_z}{\partial t}\bigg|_{R=a+d(t)} = \frac{\mu_0 I v}{2\pi(a+d_0+vt)}.$$

The electric field is perpendicular to sides AB and CD, and there is no contribution to the electromotive force from these sides. Hence, the electromotive force is determined to be

10.5 Exercises

$$V_{em} = b(-E_{z1} + E_{z2}) = -\frac{\mu_0 I a b v}{2\pi (a + d_0 + vt)(d_0 + vt)}.$$

10.8. The distance from the current is denoted by R. The magnetic flux density produced by the current is $B_\varphi = \mu_0 I / 2\pi R$, and the vector potential is

$$A_z = -\int B_\varphi dR = -\frac{\mu_0 I}{2\pi} \log \frac{R}{R_0}.$$

The electric field is

$$E_z(R) = -\frac{\partial A_z}{\partial t} = \frac{\mu_0 I}{2\pi R} \cdot \frac{\partial R}{\partial t} = \frac{\mu_0 I v}{2\pi R}.$$

On side AB on which $R = r_0 + vt$, the electric field is parallel to the side, and the contribution to the electromotive force is

$$V_{AB} = aE_z(r_0 + vt) = \frac{\mu_0 I a v}{2\pi (r_0 + vt)}.$$

On side BC, the electric field is tilted by angle $\pi - \theta$ with $\theta = \tan^{-1}(b/a)$ from the side. We denote by s the distance of point P on this side from vertex B. The electric field at this point is

$$E_z(P) = \frac{\mu_0 I v}{2\pi (r_0 + vt + s \sin \theta)}.$$

Hence, the contribution to the electromotive force from this side is

$$V_{BC} = \int_0^{b/\sin\theta} \frac{\mu_0 I v \cos(\pi - \theta)}{2\pi (r_0 + vt + s \sin \theta)} ds$$
$$= -\frac{\mu_0 I v \cot \theta}{2\pi} [\log(r_0 + vt + s \sin \theta)]_0^{b/\sin\theta}$$
$$= -\frac{\mu_0 I a v}{2\pi b} \log \left(1 + \frac{b}{r_0 + vt}\right).$$

On side CA, the electric field is perpendicular to the side, and there is no contribution to the electromotive force from this side. The electromotive force is given by

$$V_{em} = V_{AB} + V_{BC} = \frac{\mu_0 I a v}{2\pi b} \left[\frac{b}{(r_0 + vt)} - \log\left(1 + \frac{b}{r_0 + vt}\right)\right].$$

10.9. The magnetic flux density at a distance R from the current is $B_\varphi = \mu_0 I / 2\pi R$. Hence, the vector potential is

$$A_z = -\int B_\varphi dR = \frac{\mu_0 I}{2\pi} \log \frac{R_0}{R},$$

where R_0 is the distance to the reference point. Then, the electric field is

$$E_z = -\frac{\partial A_z}{\partial t} = \frac{\mu_0 I g t}{2\pi R}.$$

The electric field is in the same direction as the current. So, there is no contribution to the electromotive force from sides AB and CD. The contribution from side BC ($R = a + d_0 + gt^2/2$) is

$$V_{\text{BC}} = \frac{\mu_0 I b g t}{2\pi (a + d_0 + gt^2/2)}$$

and the contribution from side DA ($R = d_0 + gt^2/2$) is

$$V_{\text{DA}} = -\frac{\mu_0 I b g t}{2\pi (d_0 + gt^2/2)}.$$

So, the electromotive force is determined to be

$$V_{\text{em}} = V_{\text{BC}} + V_{\text{DA}} = -\frac{\mu_0 I a b g t}{2\pi (d_0 + gt^2/2)(a + d_0 + gt^2/2)}.$$

10.10. In the given condition, the magnetic flux density has an azimuthal component, B_φ, and the electric field has a z-component, E_z. These quantities depend only on the radius, R. Thus, Eq. (10.3) leads to

$$\frac{\partial E_z}{\partial R} = \frac{\partial B_\varphi}{\partial t}.$$

The right-side is

$$\frac{\partial B_\varphi}{\partial t} = \frac{\mu_0 I_0 \omega}{2\pi R} \cos \omega t,$$

and from the above equation, we have

$$E_z(R) = \frac{\mu_0 I_0 \omega}{2\pi} \cos \omega t \log \frac{R}{R_a},$$

where R_a is an integral constant. Hence, the induced electric field on side BC is $(\mu_0 I_0 \omega / 2\pi) \cos \omega t \log(a+d)/R_a$ in the direction from B to C. So, the contribution from side BC to the electromotive force is

$$V_{BC} = \frac{\mu_0 I_0 b \omega}{2\pi} \cos \omega t \log \frac{(a+d)}{R_a}.$$

The induced electric field on side DA is $(\mu_0 I_0 \omega / 2\pi) \cos \omega t \log d/R_a$ in the direction from A to D. So, the contribution from side DA to the electromotive force is

$$V_{DA} = -\frac{\mu_0 I_0 b \omega}{2\pi} \cos \omega t \log \frac{a}{R_a}.$$

On sides AB and CD, the electric field is normal to these sides, and there is no contribution from these sides. Thus, the electromotive force is determined to be

$$V_{em} = V_{BC} + V_{DA} = \frac{\mu_0 I_0 b \omega}{2\pi} \cos \omega t \log \frac{a+d}{a}.$$

10.11. In the given condition, the magnetic flux density has an azimuthal component, B_φ, and the electric field has a z-component, E_z. These quantities depend only on the distance, $R = d$. Thus, Eq. (10.3) leads to

$$\frac{\partial E_z}{\partial R} = \frac{\partial B_\varphi}{\partial t}.$$

Noting that the distance changes with time as $R = \text{const.} + vt$, the right-side is

$$\frac{\partial B_\varphi}{\partial t} = -\frac{\mu_0 I v}{2\pi R^2}.$$

From the above equation, we have

$$E_z(R) = \frac{\mu_0 I v}{2\pi R}.$$

Hence, the contribution from side DA ($R = d_0 + vt$) to the electromotive force is

$$V_{DA} = -\frac{\mu_0 I b v}{2\pi (d_0 + vt)},$$

and the contribution from side BC ($R = d_0 + a + vt$) is

$$V_{BC} = \frac{\mu_0 I b v}{2\pi (d_0 + a + vt)}.$$

Hence, the electromotive force is determined to be

$$V_{em} = V_{DA} + V_{BC} = -\frac{\mu_0 I a b v}{2\pi (a + d_0 + vt)(d_0 + vt)}.$$

10.12. In the given condition, the magnetic flux density has an azimuthal component, B_φ, and the electric field has a z-component, E_z. These quantities depend only on the distance, R. Thus, Eq. (10.3) leads to

$$\frac{\partial E_z}{\partial R} = \frac{\partial B_\varphi}{\partial t}.$$

Noting that the distance changes with time as $R = \text{const.} + vt$, the right-side is

$$\frac{\partial B_\varphi}{\partial t} = -\frac{\mu_0 I v}{2\pi R^2}.$$

From the above equation, we have

$$E_z(R) = \frac{\mu_0 I v}{2\pi R}.$$

On side AB ($R = r = r_0 + vt$), the electric field is parallel to the side and is directed from A to B. Hence, the contribution from this side to the electromotive force is

$$V_{AB} = E_z(r)a = \frac{\mu_0 I a v}{2\pi (r_0 + vt)}.$$

On side BC, the electric field is given by

$$E_z = \frac{\mu_0 I v}{2\pi (r + s \sin \theta)},$$

as defined in Fig. 10.3b in Example 10.3, and is tilted by angle $\pi - \theta$ from the direction of integration, where $\theta = \tan^{-1}(b/a)$. Thus, the contribution from this side is

$$V_{BC} = -\frac{\mu_0 I v}{2\pi} \int_0^{\overline{BC}} \frac{\cos(\pi - \theta)}{r + s \sin \theta} ds = -\frac{\mu_0 I a v}{2\pi b} \log\left(1 + \frac{b}{r_0 + vt}\right).$$

On side CA, the electric field is normal to the side, and there is no contribution from this side.

As a result, the electromotive force is determined to be

10.5 Exercises

$$V_{em} = V_{AB} + V_{BC} = \frac{\mu_0 I a v}{2\pi}\left[\frac{1}{r_0 + vt} - \frac{1}{b}\log\left(1 + \frac{b}{r_0 + vt}\right)\right].$$

10.13. In the given condition, the magnetic flux density has an azimuthal component, B_φ, and the electric field has a z-component, E_z. These quantities depend only on the distance, R. Thus, Eq. (10.3) leads to

$$\frac{\partial E_z}{\partial R} = \frac{\partial B_\varphi}{\partial t}.$$

Noting that the distance changes with time as $R = \text{const.} + gt^2/2$, the right-side is

$$\frac{\partial B_\varphi}{\partial t} = -\frac{\mu_0 I g t}{2\pi R^2}.$$

From the above equation, we have

$$E_z(R) = \frac{\mu_0 I g t}{2\pi R}.$$

Hence, the contribution from side DA ($R = d_0 + gt^2/2$) to the electromotive force is

$$V_{DA} = -\frac{\mu_0 I b g t}{2\pi\left(d_0 + gt^2/2\right)},$$

and the contribution from side BC ($R = d_0 + a + gt^2/2$) is

$$V_{BC} = \frac{\mu_0 I b g t}{2\pi\left(d_0 + a + gt^2/2\right)}.$$

Hence, the electromotive force is determined to be

$$V_{em} = V_{DA} + V_{BC} = -\frac{\mu_0 I a b g t}{2\pi\left(a + d_0 + gt^2/2\right)\left(d_0 + gt^2/2\right)}.$$

10.14. The direction of the magnetic flux due to current $I(t)$ that penetrates the circuit is opposite to that produced by the current flowing in the direction ABCD. The magnetic flux density at a distance R from the current is $B(R) = \mu_0 I(t)/2\pi R$, and the magnetic flux that penetrates the circuit is

$$\Phi = -\frac{\mu_0 I(t) b}{2\pi}\int_d^{a+d}\frac{dR}{R} = -\frac{\mu_0 I(t) b}{2\pi}\log\frac{a+d(t)}{d(t)},$$

where the distance d changes as $d(t) = d_0 + vt$. Thus, the induced electromotive force is determined to be

$$V_{em} = -\frac{d\Phi}{dt} = \frac{\mu_0 b}{2\pi} \cdot \frac{dI(t)}{dt} \log \frac{a+d(t)}{d(t)} + \frac{\mu_0 I(t) b}{2\pi} \cdot \frac{\partial}{\partial d} \log \frac{a+d}{d} \cdot \frac{dd}{dt}$$

$$= \frac{\mu_0 I_m b}{2\pi} \left[\omega \cos \omega t \log \left(1 + \frac{a}{d_0 + vt}\right) - \frac{av \sin \omega t}{(a + d_0 + vt)(d_0 + vt)} \right].$$

10.15. The magnetic flux density produced by the current is

$$B_\varphi(R) = \frac{\mu_0 I(t)}{2\pi R}$$

and the vector potential, which is parallel to the current, is

$$A_z = -\int B_\varphi(R) dR = -\frac{\mu_0 I(t)}{2\pi} \log \frac{R}{R_0}.$$

So, the electric field on side DA is directed from A to D and its value is

$$E_{z1} = -\left.\frac{\partial A_z}{\partial t}\right|_{R=d(t)} = \frac{\mu_0 I_m}{2\pi} \left(\omega \cos \omega t \log \frac{d_0 + vt}{R_0} + \frac{v \sin \omega t}{d_0 + vt} \right),$$

and that on side BC is directed from B to C and its value is

$$E_{z2} = -\left.\frac{\partial A_z}{\partial t}\right|_{R=a+d(t)} = \frac{\mu_0 I_m}{2\pi} \left(\omega \cos \omega t \log \frac{a + d_0 + vt}{R_0} + \frac{v \sin \omega t}{a + d_0 + vt} \right).$$

The electric field is perpendicular to sides AB and CD, and there is no contribution to the electromotive force from these sides. Hence, the electromotive force is determined to be

$$V_{em} = b(-E_{z1} + E_{z2})$$

$$= \frac{\mu_0 I_m b}{2\pi} \left[\omega \cos \omega t \log \left(1 + \frac{a}{d_0 + vt}\right) - \frac{av \sin \omega t}{(a + d_0 + vt)(d_0 + vt)} \right].$$

10.16. In the given condition, the magnetic flux density has an azimuthal component, B_φ, which is given by

$$B_\varphi = \frac{\mu_0 I(t)}{2\pi R},$$

and the electric field has a z-component, E_z. Thus, Eq. (10.3) leads to

10.5 Exercises

$$\frac{\partial E_z}{\partial R} = \frac{\partial B_\varphi}{\partial t} = \frac{\mu_0 I_m \omega}{2\pi R} \cos \omega t - \frac{\mu_0 I_m v}{2\pi R^2} \sin \omega t.$$

So, the electric field on side DA $[R = d(t)]$ is directed from A to D and its value is

$$E_{z1} = \frac{\mu_0 I_m}{2\pi} \left(\omega \cos \omega t \log \frac{d_0 + vt}{R_0} + \frac{v \sin \omega t}{d_0 + vt} \right),$$

and that on side BC $[R = a + d(t)]$ is directed from B to C and its value is

$$E_{z2} = \frac{\mu_0 I_m}{2\pi} \left(\omega \cos \omega t \log \frac{a + d_0 + vt}{R_0} + \frac{v \sin \omega t}{a + d_0 + vt} \right),$$

where R_0 is the distance to the reference point for the vector potential. The electric field is normal to the integral path for other two sides. Hence, the electromotive force is determined to be

$$V_{em} = b(-E_{z1} + E_{z2})$$
$$= \frac{\mu_0 I_m b}{2\pi} \left[\omega \cos \omega t \log \left(1 + \frac{a}{d_0 + vt} \right) - \frac{av \sin \omega t}{(a + d_0 + vt)(d_0 + vt)} \right].$$

10.17. We can assume that the electric field and magnetic flux density have the axial and azimuthal components, respectively. The external electric field is expressed as $E_0 e^{i\omega t}$. Then, Eqs. (10.3) and (6.16) are, respectively, reduced to

$$\frac{\partial E_z}{\partial R} = i\omega B_\varphi, \quad -\frac{\partial B_z}{\partial R} = \mu \sigma_c E_z.$$

Eliminating B_φ from these equations, we have

$$\frac{\partial^2 E_z}{\partial R^2} + \frac{1}{R} \cdot \frac{\partial E_z}{\partial R} - i\omega \mu \sigma_c E_z = 0.$$

Here, we put as $\xi = (-i\omega\mu\sigma_c)^{1/2} R = (1-i)(\omega\mu\sigma_c/2)^{1/2} R$. Then, the above equation is reduced to

$$\frac{\partial^2 E_z}{\partial \xi^2} + \frac{1}{\xi} \cdot \frac{\partial E_z}{\partial \xi} + E_z = 0.$$

The solution is given by

$$E_z = A J_0(\xi),$$

where $J_0(\xi)$ is the Bessel function of zeroth order and A is determined with the boundary condition:

$$AJ_0(\xi_a) = E_0$$

with $\xi_a = (-i\omega\mu\sigma_c)^{1/2}a$.

10.18. We can assume that the magnetic flux density and electric field have the axial and azimuthal components, respectively. The external magnetic flux density is expressed as $B_0 e^{i\omega t}$. Then, Eqs. (10.3) and (6.16) are, respectively, reduced to

$$\frac{1}{R} \cdot \frac{\partial(RE_\varphi)}{\partial R} = -i\omega B_z, \quad -\frac{\partial B_z}{\partial R} = \mu\sigma_c E_\varphi.$$

Eliminating B_z from these equations, we have

$$\frac{\partial^2 E_\varphi}{\partial R^2} + \frac{1}{R} \cdot \frac{\partial E_\varphi}{\partial R} - \frac{E_\varphi}{R^2} - i\omega\mu\sigma_c E_\varphi = 0.$$

Here, we put as $\xi = (-i\omega\mu\sigma_c)^{1/2}R = (1-i)(\omega\mu\sigma_c/2)^{1/2}R$. Then, the above equation is reduced to

$$\frac{\partial^2 E_\varphi}{\partial \xi^2} + \frac{1}{\xi} \cdot \frac{\partial E_\varphi}{\partial \xi} + \left(1 - \frac{1}{\xi^2}\right) E_\varphi = 0.$$

The solution is given by

$$E_\varphi = AJ_1(\xi),$$

where $J_1(\xi)$ is the Bessel function of first order and A is a constant. Using the formula, $\partial J_0(\xi)/\partial \xi = -J_1(\xi)$ with $J_0(\xi)$ denoting the Bessel function of zeroth order, the magnetic flux density is given by

$$B_z = \frac{B_0}{J_0(\xi_a)} J_0(\xi)$$

with $\xi_a = (-i\omega\mu\sigma_c)^{1/2}a$. Hence, the electric field is given by

$$E_\varphi = (1+i)\left(\frac{\omega}{2\mu\sigma_c}\right)^{1/2} \frac{B_0}{J_0(\xi_a)} J_1(\xi).$$

10.19. We denote the current in winding B by $I_B = I_0 \cos \omega t$. Then, the magnetic field in winding B is $(N_B I_0/l) \cos \omega t$. The magnetic flux induced in winding B is

$$\Phi_B = \pi \left[\mu_0(a_B^2 - a_A^2) + \mu a_A^2\right] \frac{N_B I_0}{l} \cos \omega t.$$

10.5 Exercises

Then, the induced voltage in winding B is

$$V_B = -N_B \frac{d\Phi_B}{dt} = \pi\omega\left[\mu_0\left(a_B^2 - a_A^2\right) + \mu a_A^2\right]\frac{N_B^2 I_0}{l} \sin\omega t.$$

The magnetic flux induced in winding A is

$$\Phi_A = \pi\mu a_A^2 \frac{N_B I_0}{l} \cos\omega t$$

and the voltage induced in winding A is

$$V_A = -N_A \frac{d\Phi_A}{dt} = \pi\omega\mu a_A^2 \frac{N_A N_B I_0}{l}\sin\omega t = \frac{N_A}{N_B} \cdot \frac{\mu a_A^2}{\mu_0\left(a_B^2 - a_A^2\right) + \mu a_A^2} V_B.$$

10.20. The integral of $E = B \times V$ is zero on the dotted line, since V is parallel to and E is normal to the dotted line. On the other hand, the magnetic flux comes into the closed region composed of the surface and the dotted line, and the resultant induced electromotive force must be taken into account. This is given by the variation rate of the magnetic flux density $\Delta B/\Delta t$ multiplied by the area $a^2\theta/2$. Hence, the electromotive for is given by

$$V_{em} = \frac{a^2\theta}{2} \cdot \frac{\Delta B}{\Delta t},$$

which is the same result as in Example 10.5.

10.21. Under the given condition, the angular frequency is given by

$$\omega = \frac{2V_A}{Ba^2}.$$

Then, the DC voltage

$$V_B = \frac{1}{2}\omega B b^2$$

appears on the other side with different radius. Thus, this unipolar induction system is useful for DC voltage transformation:

$$V_B = \left(\frac{b}{a}\right)^2 V_A.$$

10.22. The electromotive force necessary to apply the current is from Eq. (10.13) given by

$$V = \oint \boldsymbol{E} \cdot d\boldsymbol{s} = \oint \left(-\nabla \phi - \frac{\partial \boldsymbol{A}}{\partial t} \right) \cdot d\boldsymbol{s}.$$

The first term gives the potential difference due to the resistance, RI, and the second term is written as

$$-\frac{\partial}{\partial t} \oint \boldsymbol{A} \cdot d\boldsymbol{s} = -\frac{\partial \Phi}{\partial t},$$

where Φ is the magnetic flux in the coil and we used Eq. (6.21). Thus, we have

$$V = RI - \frac{\partial \Phi}{\partial t},$$

which is well known in electric circuits.

10.23. Spherical coordinates are defined with the zenithal angle θ measured from the direction of the magnetic moment. The vector potential has an azimuthal component given by

$$A_\varphi(t) = \frac{\mu_0 m_0}{4\pi r^2} \sin\theta \sin\omega t,$$

which satisfies $\nabla \cdot \boldsymbol{A} = 0$. Then, the magnetic flux density is

$$B_r = \frac{\mu_0 m_0}{2\pi r^3} \cos\theta \sin\omega t, \quad B_\theta = \frac{\mu_0 m_0}{4\pi r^3} \sin\theta \sin\omega t.$$

which leads to

$$\nabla \cdot \boldsymbol{B} = \frac{1}{r^2} \cdot \frac{\partial}{\partial r}(r^2 B_r) + \frac{1}{r\sin\theta} \cdot \frac{\partial}{\partial \theta}(\sin\theta B_\theta)$$

$$= \frac{\mu_0 m_0}{2\pi} \sin\omega t \left(-\frac{1}{r^4}\cos\theta + \frac{1}{r^4}\cos\theta \right) = 0.$$

Thus, the magnetic flux density satisfies $\nabla \cdot \boldsymbol{B} = 0$. The electric field has an azimuthal component given by

$$E_\varphi(t) = -\frac{\partial A_\varphi(t)}{\partial t} = -\frac{\mu_0 m_0 \omega}{4\pi r^2} \sin\theta \cos\omega t,$$

and satisfies $\nabla \cdot \boldsymbol{E} = 0$.

10.24. We apply current I. Then, the magnetic flux densities in vacuum and the magnetic material are

$$B_0 = \frac{\mu_0 I}{w}, \quad B = \frac{\mu I}{w}.$$

10.5 Exercises

Then, the magnetic energy is given by

$$U_m = dw \left[\frac{B_0^2}{2\mu_0} x + \frac{B^2}{2\mu}(l-x) \right] = \frac{dI^2}{2w}[\mu l - (\mu - \mu_0)x].$$

It should be noted that the electromagnetic induction occurs when the magnetic material moves, since the magnetic flux density changes. The magnetic flux is

$$\Phi = \frac{dI}{w}[\mu l - (\mu - \mu_0)x].$$

The electromotive force that appears in the circuit is $-d\Phi/dt$. Thus, the corresponding energy is

$$W = -\int I \frac{d\Phi}{dt} dt = -I\Phi = -\frac{dI^2}{w}[\mu l - (\mu - \mu_0)x].$$

Hence, the force on the magnetic material is

$$F = -\frac{\partial}{\partial x}(U_m + W) = -\frac{(\mu - \mu_0)dI^2}{2w}.$$

Since F is negative ($\mu > \mu_0$), it is directed towards decreasing x, i.e., attractive.

10.25. The aimed force cannot be directly obtained from the magnetic energy due to the electromagnetic induction, as pointed out in Exercise 8.33. Using the result of Exercise 8.3, the magnetic flux between the two superconductors in a unit length is given by

$$\Phi' = \frac{\mu_0 I}{2\pi} \log \frac{l + \sqrt{l^2 - a^2}}{a},$$

where I is the current applied to the cylindrical superconductor and l is the distance between the center of the cylindrical superconductor and the superconductor surface. Thus, the magnetic energy in a unit length is

$$U'_m = \frac{1}{2}\Phi' I = \frac{\mu_0 I^2}{4\pi} \log \frac{l + \sqrt{l^2 - a^2}}{a}.$$

When l changes, the electromotive force induced in a unit length is given by

$$V'_{em} = -\frac{d\Phi'}{dt} = -\frac{\mu_0 I}{2\pi\sqrt{l^2 - a^2}} \cdot \frac{dl}{dt},$$

and the work in a unit length done by the power source is

$$W' = \int V'_{em} I \, dt = -\frac{\mu_0 I^2}{2\pi} \int \frac{dl}{\sqrt{l^2 - a^2}} = -\frac{\mu_0 I^2}{2\pi} \log \frac{l + \sqrt{l^2 - a^2}}{a}.$$

The integral constant was determined by the condition $W'(l = a) = 0$. Thus, the force on the cylindrical superconductor in a unit length is

$$F' = -\frac{\partial}{\partial l}(U'_m + W') = \frac{\mu_0 I^2}{4\pi \sqrt{l^2 - a^2}}.$$

This is the force between two image currents (see (Exercise 8.33)).

10.26. We suppose that currents of surface densities τ_0 and τ flow in the regions of the superconductor facing the vacuum and the magnetic liquid, respectively. The magnetic field in the gap region is parallel to the superconductors, and its strength is $H_0 = \tau_0$ and $H = \tau$ in the vacuum and the magnetic liquid, respectively, and the corresponding magnetic flux density is $B_0 = \mu_0 \tau_0$ and $B = \mu \tau$. The boundary condition yields $\mu_0 \tau_0 = \mu \tau$. Since the total current is $\tau_0(w - h) + \tau(w' + h) = I$, we obtain the surface current densities as

$$\tau_0 = \frac{\mu I}{\mu w + \mu_0 w' - (\mu - \mu_0)h},$$

$$\tau = \frac{\mu_0 I}{\mu w + \mu_0 w' - (\mu - \mu_0)h}.$$

The magnetic flux density is

$$B_0 = B = \frac{\mu \mu_0 I}{\mu w + \mu_0 w' - (\mu - \mu_0)h}.$$

Thus, we calculate the magnetic energy as

$$U_m = dl \left[\frac{B_0^2}{2\mu_0}(w - h) + \frac{B^2}{2\mu}(w' + h) \right] = \frac{\mu \mu_0 dl I^2}{2[\mu w + \mu_0 w' - (\mu - \mu_0)h]}.$$

It should be noted that the electromagnetic induction occurs when h changes, since the magnetic flux density changes. The magnetic flux is given by $\Phi = dlB$, and the electromotive force that appears in the circuit is $-\partial \Phi / \partial t$. Thus, the corresponding energy is

$$W = -\int I \frac{d\Phi}{dt} dt = -I\Phi = -\frac{\mu \mu_0 dl I^2}{\mu w + \mu_0 w' - (\mu - \mu_0)h}.$$

10.5 Exercises

Hence, the force on the magnetic material is

$$F = -\frac{\partial}{\partial h}(U_m + W) = \frac{\mu\mu_0(\mu - \mu_0)dlI^2}{2[\mu w + \mu_0 w' - (\mu - \mu_0)h]^2},$$

which is directed towards increasing h. From the force balance condition, $F = \rho d l h g$, the height h is given by the solution of the equation:

$$h = \frac{\mu\mu_0(\mu - \mu_0)I^2}{2\rho g[\mu w + \mu_0 w' - (\mu - \mu_0)h]^2}.$$

This cubic equation on h has a single real solution with a positive number.

10.27. The electric field is

$$E_z = E_0 e^{i\omega t} \exp\left[-\frac{(1+i)}{\delta}x\right].$$

Then, the vector potential is obtained as

$$A_z = -\int E_z dt = \frac{i}{\omega} E_0 e^{i\omega t} \exp\left[-\frac{(1+i)}{\delta}x\right],$$

and the magnetic flux density is determined to be

$$B_y = -\frac{\partial A_z}{\partial x} = -\frac{1-i}{\omega\delta} E_0 e^{i\omega t} \exp\left[-\frac{(1+i)}{\delta}x\right]$$

$$= E_0\left(\frac{\mu\sigma_c}{\omega}\right)^{1/2} \exp\left[i\left(\omega t + \frac{3\pi}{4}\right)\right] \exp\left[-\frac{(1+i)}{\delta}x\right],$$

which agrees with the result of Example 10.8.

Using the above vector potential, we have

$$\nabla \cdot A = \frac{\partial A_z}{\partial z} = 0.$$

Thus, the Coulomb gauge is satisfied.

10.28. The magnetic flux density is

$$B_z = B_0 e^{i\omega t} \exp\left[-\frac{(1+i)}{\delta}x\right].$$

Then, the vector potential is obtained as

$$A_y = \int B_z dx = -\frac{(1-i)\delta}{2} B_0 e^{i\omega t} \exp\left[-\frac{(1+i)}{\delta}x\right].$$

Using Eq. (10.13), the electric field is determined to be

$$E_y = -\frac{\partial A_y}{\partial t} = \frac{(1+i)\omega\delta}{2} B_0 e^{i\omega t} \exp\left[-\frac{(1+i)}{\delta}x\right]$$

$$= B_0 \left(\frac{\omega}{\mu\sigma_c}\right)^{1/2} \exp\left[i\left(\omega t + \frac{\pi}{4}\right)\right] \exp\left[-\frac{(1+i)}{\delta}x\right],$$

which agrees with Eq. (10.32). Using the above vector potential, we have

$$\nabla \cdot \mathbf{A} = \frac{\partial A_y}{\partial y} = 0.$$

Thus, the Coulomb gauge is satisfied.

10.29. The magnetic flux density given by Eq. (10.30) originates from the vector potential:

$$A_y = -\frac{(1-i)}{2}\delta B_0 \exp\left(i\omega t - \frac{1+i}{\delta}x\right).$$

Using the electric field given by Eq. (10.32), the current density is

$$i_y = \sigma_c E_y = \frac{1+i}{\delta} \cdot \frac{B_0}{\mu} \exp\left(i\omega t - \frac{1+i}{\delta}x\right).$$

Hence, we have

$$\Delta A_y = \frac{\partial^2 A_y}{\partial x^2} = -\frac{1+i}{\delta} B_0 \exp\left(i\omega t - \frac{1+i}{\delta}x\right) = -\mu i_y.$$

Thus, Poisson's equation is satisfied.

10.30. The electric field is

$$E_z = E_0 \exp\left(i\omega t - \frac{1+i}{\delta}x\right).$$

Thus, the current density and the vector potential are given by

10.5 Exercises

$$i_z = \sigma_c E_0 \exp\left(i\omega t - \frac{1+i}{\delta}x\right),$$

$$A_z = -\int E_z dt = \frac{i}{\omega} E_0 \exp\left(i\omega t - \frac{1+i}{\delta}x\right).$$

Hence, we have

$$\Delta A_z = \frac{\partial^2 A_z}{\partial x^2} = -\mu \sigma_c E_0 \exp\left(i\omega t - \frac{1+i}{\delta}x\right) = -\mu i_z.$$

Thus, Poisson's equation is satisfied.

10.31. Poisson's equation is given by

$$\Delta A = -\mu i.$$

The left-hand side is equal to $-\mu\sigma_c E$. Using Eq. (10.13) with $\nabla \phi = 0$, the above equation leads to

$$\Delta A - \mu \sigma_c \frac{\partial A}{\partial t} = 0.$$

Coffee Break: Is Magnetic Flux a Magnetic Potential?

There is a well-known analogy between the electric energy and magnetic energy: The electric energy is given by the product of the electric source, i.e., the electric charge Q and the resultant electric potential ϕ divided by 2. On the other hand, the magnetic energy is given by the product of the magnetic source, i.e., the current I and the resultant magnetic flux Φ divided by 2. Does this analogy mean that the magnetic flux is a magnetic potential, i.e., the vector potential? It is clear that the magnetic flux is not the vector potential. How can we explain such a disagreement between electricity and magnetism?

The answer is that this difference is caused by the difference in the dimension of the sources. Originally, the electric charge density corresponds to the current density, and the electric and magnetic energies are, respectively, given by

$$U_e = \frac{1}{2} \int_V \phi \rho \, dV,$$

$$U_m = \frac{1}{2} \int_V \mathbf{A} \cdot \mathbf{i} \, dV,$$

where ρ is the electric charge density, i is the current density, and A is the vector potential. For a conductor system, the electric potential is constant in each conductor. Then, the electric energy reduces to

$$U_e = \frac{1}{2}\phi \int_V \rho dV = \frac{1}{2}\phi Q.$$

On the other hand, we obtain the current by integrating the current density in the cross-sectional area. Then, the magnetic energy reduces to

$$U_m = \frac{1}{2} \int_C A \cdot ds \int_S i \cdot dS = \frac{1}{2}\Phi I.$$

In the above the volume integral was divided into the cross-sectional integral and the integral along the current path as $dV = dS \cdot ds$. That is, the magnetic flux is not the vector potential but is the vector potential integrated along a closed loop. This may be simply understood from dimensions. In this sense the magnetic flux is a kind of magnetic potential.

Chapter 11
Displacement Current and Maxwell's Equations

11.1 Displacement Current

In the steady state, Ampere's law, Eq. (9.14), holds for a closed line, C, with different surfaces on it, S_1 and S_2, as shown in Fig. 11.1a, b. If the magnetic field H is integrated on C in opposite directions as drawn in Fig. 11.1a, b, the sum of the two integrations is naturally zero. At the same time, the sum of the surface integrals of the current density i on S_1 and S_2 is also zero. This sum becomes the surface integral on closed surface S_{12} composed of S_1 and S_2 (see Fig. 11.1c). Thus, we have

$$\int_{S_{12}} i \cdot dS = 0. \tag{11.1}$$

This agrees with Eq. (5.5) in the steady state. It means, however, that Ampere's law contradicts Eq. (5.5) in a non-steady state. In such a general case, the law of conservation of electric charge, Eq. (5.5), must be satisfied. Substituting Eq. (4.9) into this equation, we have

$$\int_{S_{12}} \left(i + \frac{\partial D}{\partial t} \right) \cdot dS = 0. \tag{11.2}$$

This strongly suggests that, for generalizing to a non-steady state, we can assume

$$\oint_C H \cdot ds = \int_S \left(i + \frac{\partial D}{\partial t} \right) \cdot dS, \tag{11.3}$$

© The Editor(s) (if applicable) and The Author(s), under exclusive license to Springer Nature Switzerland AG 2025
T. Matsushita, *Exercises in Electricity and Magnetism*,
https://doi.org/10.1007/978-3-031-67940-7_11

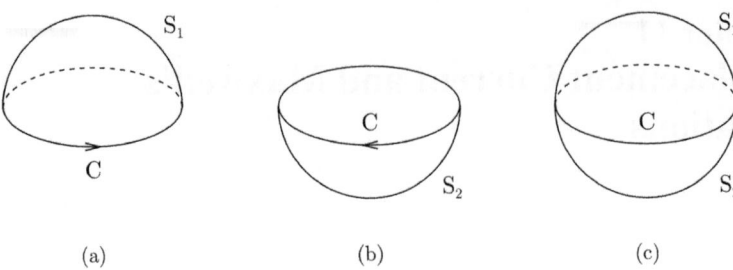

Fig. 11.1 **a** Surface S_1 and **b** surface S_2 surrounded by closed line C, and (**c**) closed surface composed of S_1 and S_2

instead of Ampere's law. In the above, S is the surface surrounded by C. The second term on the right side, $\partial \boldsymbol{D}/\partial t$, is called the displacement current, and has the same unit as electric current density [A/m²]. In the steady state, Eq. (11.3) reduces to the usual form of Ampere's law, Eq. (9.14), and there is no problem. Equation (11.3) is called the generalized form of Ampere's law. The corresponding generalized differential form of Ampere's law is

$$\nabla \times \boldsymbol{H} = \boldsymbol{i} + \frac{\partial \boldsymbol{D}}{\partial t}. \tag{11.4}$$

Here, we show the validity of the displacement current. Suppose that a capacitor is energized using an electric power source. When we apply current I to the capacitor, as shown in Fig. 11.2a, the electric charge Q in the electrode changes. We assume a closed line, C, around a wire through which the current flows and a surface, S_1, as in the figure. We apply Eq. (11.3) to C and S_1. The displacement current is zero on S_1 and the right side is equal to the current I applied to the capacitor. Next, we assume another surface S_2 that does not contain the wire, as shown in Fig. 11.2b. In this case, while the left side does not change, $\boldsymbol{i}$ is zero on S_2, and the right side is

Fig. 11.2 Closed line C around a current-carrying wire and surfaces surrounded by C: **a** surface that includes the wire and **b** surface that does not include the wire

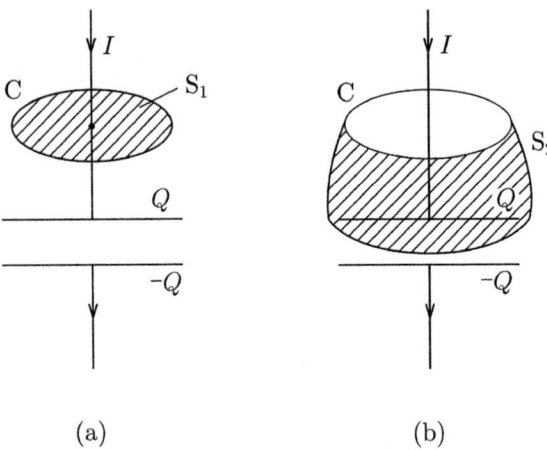

11.1 Displacement Current

$$\frac{\partial}{\partial t}\int_{S_2} \mathbf{D}\cdot d\mathbf{S} = \frac{dQ}{dt}. \tag{11.5}$$

In the above, we changed the order of the time differential and surface integral, since S_2 does not change with time. Hence, we have

$$I = \frac{dQ}{dt}, \tag{11.6}$$

and no contradiction results from Eq. (11.3).

***

Example 11.1. We apply alternating current $I(t) = I_0 \sin\omega t$ to a capacitor with circular electrodes of radius a separated by d, as shown in Fig. 11.3. Determine the displacement current and magnetic field in the space between the electrodes.

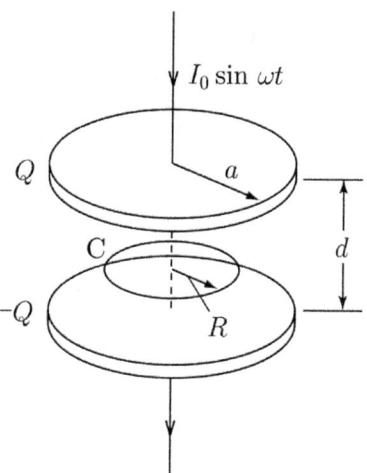

Fig. 11.3 AC flowing through circular parallel-plate capacitor

***

Solution 11.1 The electric charge on the electrodes is $Q(t) = -(I_0/\omega)\cos\omega t$ and the electric flux density is directed downward with magnitude

$$D(t) = \frac{Q(t)}{\pi a^2} = -\frac{I_0}{\pi a^2 \omega}\cos\omega t.$$

Hence, the displacement current is

$$\frac{\partial D(t)}{\partial t} = \frac{I_0}{\pi a^2}\sin\omega t.$$

This is similar to a virtual situation in which a current of the same density flows uniformly in the space between the electrodes, suggesting the continuity of current.

In fact, this situation is realized if the space is occupied by a material with resistivity sufficiently higher than that of the electrodes.

The magnetic field produced by the displacement current is concentric around the central axis of the capacitor. If the magnetic field on a circle of radius R from the axis is $H(R)$, the left side of Eq. (11.3) is $2\pi R H(R)$. The right side is $\pi R^2 \partial D(t)/\partial t = (R^2 I_0/a^2) \sin \omega t$. Thus, we have

$$H(t) = \frac{I_0 R}{2\pi a^2} \sin \omega t.$$

◆

11.2 Maxwell's Equations

All of the equations that describe electromagnetic phenomena have been introduced. They are summarized as follows:

$$\nabla \times \boldsymbol{E} = -\frac{\partial \boldsymbol{B}}{\partial t}, \tag{11.7}$$

$$\nabla \times \boldsymbol{H} = \boldsymbol{i} + \frac{\partial \boldsymbol{D}}{\partial t}, \tag{11.8}$$

$$\nabla \cdot \boldsymbol{D} = \rho, \tag{11.9}$$

$$\nabla \cdot \boldsymbol{B} = 0. \tag{11.10}$$

These are Maxwell's equations. Equation (11.7) describes the law of electromagnetic induction, which directly connects the first fundamental variables in electricity and magnetism. Equation (11.8) is the generalized differential form of Ampere's law that directly connects the second fundamental variables in electricity and magnetism. Equations (11.9) and (11.10) are Gauss's laws on electric flux and magnetic flux, respectively, and represent the conditions of divergence.

These equations are solved for electromagnetic fields with material relationships:

$$\boldsymbol{D} = \epsilon \boldsymbol{E}, \tag{11.11}$$

$$\boldsymbol{B} = \mu \boldsymbol{H}, \tag{11.12}$$

$$\boldsymbol{i} = \sigma_c \boldsymbol{E}. \tag{11.13}$$

11.2 Maxwell's Equations

The variables to be obtained are the electric field E, magnetic flux density B, electric flux density D, magnetic field H, and current density i. These five variables are obtained with five equations, (11.7), (11.8), and (11.11)–(11.13). Equations (11.9) and (11.10) provide supplementary conditions.

The integral equations corresponding to Eqs. (11.7)–(11.10) are

$$\oint_C E \cdot ds = -\frac{d}{dt} \int_S B \cdot dS, \tag{11.14}$$

$$\oint_C H \cdot ds = \int_S \left(i + \frac{\partial D}{\partial t}\right) \cdot dS, \tag{11.15}$$

$$\int_S H \cdot dS = \int_V \rho dV, \tag{11.16}$$

$$\int_S B \cdot dS = 0. \tag{11.17}$$

In the above, C is the closed line that surrounds the surface S in Eqs. (11.14) and (11.15), S is the closed surface and V is its internal region in Eqs. (11.16) and (11.17).

Here, we derive a characteristic equation when there is no electric charge. In terms of only the electric field and magnetic flux density, Eq. (11.8) is rewritten as

$$\frac{1}{\mu} \nabla \times B = \sigma_c E + \epsilon \frac{\partial E}{\partial t}. \tag{11.18}$$

Substituting Eq. (11.18) into a curl of Eq. (11.7) gives

$$\nabla \times (\nabla \times E) = -\frac{\partial}{\partial t}(\nabla \times B) = -\mu\epsilon \frac{\partial^2 E}{\partial t^2} - \mu\sigma_c \frac{\partial E}{\partial t}. \tag{11.19}$$

When there is no electric charge ($\rho = 0$) in a material, the left side of Eq. (11.19) is equal to $-\Delta E$, and we have

$$\Delta E - \mu\epsilon \frac{\partial^2 E}{\partial t^2} - \mu\sigma_c \frac{\partial E}{\partial t} = 0. \tag{11.20}$$

This equation is the telegraphic equation. The second and third terms originate from the displacement current and true current, respectively. When the effect of displacement current is negligible as in a conductive material, Eq. (11.20) is reduced to the diffusion equation. Equation (10.25) is of this type in an AC condition. When the effect of true current is negligible, as in a dielectric material or in vacuum, Eq. (11.20) is reduced to

$$\Delta E - \mu\epsilon \frac{\partial^2 E}{\partial t^2} = 0. \qquad (11.21)$$

This is called the wave equation.

11.3 Electromagnetic Potential

Here, we summarize the potentials that describe the electromagnetic fields. The electric field E is given by Eq. (10.13) with the electric potential (scalar potential) ϕ and the vector potential A:

$$E = -\nabla\phi - \frac{\partial A}{\partial t}. \qquad (11.22)$$

On the other hand, the magnetic flux density B always satisfies Eq. (11.10) and is given by Eq. (6.18) with the vector potential A:

$$B = \nabla \times A. \qquad (11.23)$$

Thus, the electric field and magnetic flux density are given by ϕ and A, and the set of these potentials is called the electromagnetic potential.

When an electric charge of density ρ and a current of density i coexist in space, it is common to use the condition given by

$$\epsilon\mu \frac{\partial \phi}{\partial t} + \nabla \cdot A = 0 \qquad (11.24)$$

to determine the vector potential. This condition is the Lorentz gauge. Hence, Eq. (11.9) leads to

$$\Delta\phi - \epsilon\mu \frac{\partial^2 \phi}{\partial t^2} = -\frac{\rho}{\epsilon}. \qquad (11.25)$$

Equation (11.8) leads similarly to

$$\Delta A - \epsilon\mu \frac{\partial^2 A}{\partial t^2} = -\mu i. \qquad (11.26)$$

When there is neither electric charge nor current, all these equations reduce to the wave equation.

11.4 The Poynting Vector

The total energy density of electromagnetic fields varying with time is given by

$$u = \frac{1}{2}\epsilon E^2 + \frac{1}{2\mu}B^2 + \int \boldsymbol{i} \cdot \boldsymbol{E} \, dt. \tag{11.27}$$

The first, second, and third terms are the electric energy, magnetic energy, and mechanical energy of charged particles, respectively. Hence, the variation in the total energy in space V with time is

$$\frac{\partial U}{\partial t} = \frac{\partial}{\partial t}\int_V u \, dV = \int_V \left(\epsilon \boldsymbol{E} \cdot \frac{\partial \boldsymbol{E}}{\partial t} + \frac{\boldsymbol{B}}{\mu} \cdot \frac{\partial \boldsymbol{B}}{\partial t} + \boldsymbol{i} \cdot \boldsymbol{E}\right) dV. \tag{11.28}$$

Using Eqs. (11.7) and (11.8), Eq. (11.28) is rewritten as

$$\frac{\partial}{\partial t}\int_V u \, dV + \int_S \boldsymbol{S}_\mathrm{P} \cdot d\boldsymbol{S} = 0, \tag{11.29}$$

where S is the surface of V, and the vector

$$\boldsymbol{S}_\mathrm{P} = \boldsymbol{E} \times \boldsymbol{H} \tag{11.30}$$

is called the Poynting vector. This equation says that the variation in the energy in space V with time is equal to the Poynting vector that enters V through the surface S. Hence, we understand that the Poynting vector represents the flow of electromagnetic energy, and Eq. (11.29) represents the law of conservation of energy. When this equation is written in differential form, we have

$$\frac{\partial u}{\partial t} + \nabla \cdot \boldsymbol{S}_\mathrm{P} = 0. \tag{11.31}$$

This gives the continuity equation of energy.

Example 11.2 We apply magnetic flux density B parallel to the thin cylinder of radius a and height h shown in Fig. 11.4. Determine the magnetic energy stored in the cylinder using the Poynting vector.

Fig. 11.4 Thin cylinder

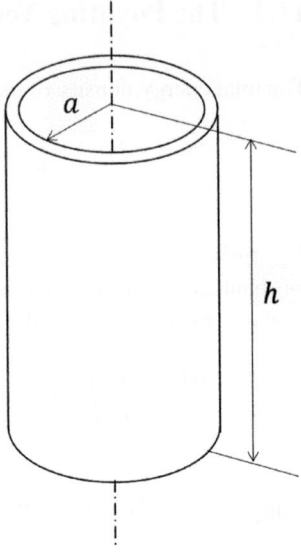

Solution 11.2 We denote the magnetic flux density applied to the cylinder by b. The electromotive force induced on the cylinder surface is

$$V_{em} = -\pi a^2 \frac{db}{dt}.$$

The electric field is

$$E = \frac{V_{em}}{2\pi a} = -\frac{a}{2} \cdot \frac{db}{dt}$$

and is directed counterclockwise (in the direction of increasing azimuthal angle) from the top view. The magnetic field is $H = b/\mu_0$ and is directed upward. Thus, the Poynting vector directed inward into the cylinder is

$$S_P = -EH = \frac{a}{2\mu_0} b \frac{db}{dt}.$$

The magnetic energy penetrating the cylinder until b reaches B is determined to be

$$U_m = \int 2\pi a h S_P dt = \frac{\pi a^2 h}{\mu_0} \int_0^B b \, db = \frac{\pi a^2 h}{2\mu_0} B^2.$$

This is equal to the product of the magnetic energy density, $B^2/2\mu_0$, and the volume of the cylinder, $\pi a^2 h$.   ◆

11.4 The Poynting Vector

Example 11.3 We apply current I to a wide parallel-plate transmission line using an electric power source of output voltage V, as shown in Fig. 11.5a. Determine the Poynting vector and discuss the flow of energy. Neglect the electric resistance of the conductor. The width of the plate is w and the plate separation is d. When we cannot neglect the electric resistance, how does the energy flow?

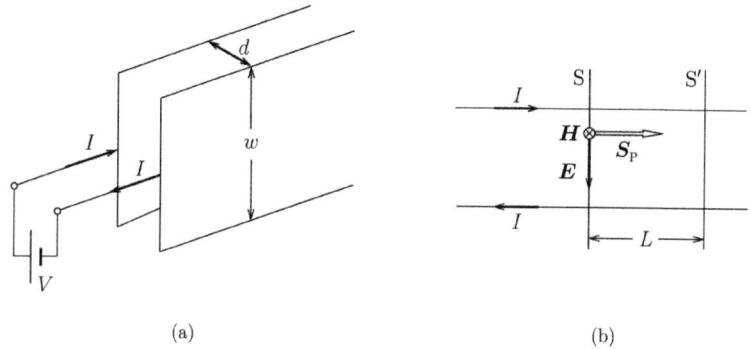

Fig. 11.5 **a** Parallel-plate transmission line and **b** the Poynting vector in the transmission line

Solution 11.3 First, we consider the case where the electric resistance can be neglected. We determine the Poynting vector on a surface, S, at some distance from the electric power source (see Fig. 11.5b). The electric field of strength $E = V/d$ is directed downward and the magnetic field of strength $H = I/w$ is directed backward. Hence, the magnitude of the Poynting vector is $S_\mathrm{P} = VI/(wd)$ and the vector is directed from the electric power source to the terminal of the transmission line. Hence, the energy that flows from the electric power source to the transmission line in unit time is

$$S_\mathrm{P} wd = VI$$

and is equal to the electric power as well known. This value is constant and independent of the position of surface S.

Second, we consider the case where the electric resistance cannot be neglected. We use R'_r and V' to denote the electric resistance of the conducting plate in a unit length and the potential difference between the conductors on surface S, respectively. The potential difference on surface S' at distance L from S is then $V' - 2LR'_\mathrm{r}I$. The electric power that enters through surface S is $V'I$ and the electric power that exits through S' is $(V' - 2LR'_\mathrm{r}I)I = V'I - 2LR'_\mathrm{r}I^2$. Hence, the difference, $2LR'_\mathrm{r}I^2$, is the power consumed as the Joule heat in the region between S and S'. Exactly speaking, the conductors are not equipotential and hence, the electric field is not perpendicular to the conductor surfaces (see Fig. 11.6). Thus, the Poynting vector is not parallel to the surface and the dissipated energy enters the conductor.

Fig. 11.6 Equipotential surface (*solid line*), electric field, and the Poynting vector

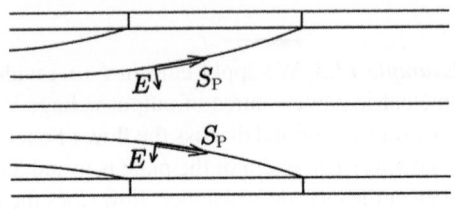

Example 11.4 We apply current to a cylindrical one-turn coil made of a thin conducting plate shown in Fig. 11.7. Determine the Poynting vector as the current increases from zero to I and discuss the flow of the energy. The height h is sufficiently greater and the width of the gap δ is sufficiently smaller than the radius a. Neglect the electric resistance of the conducting plate.

Fig. 11.7 Cylindrical one-turn coil

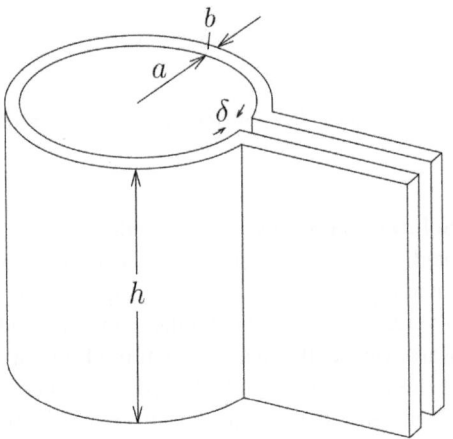

Solution 11.4 The magnetic flux density produced in the coil when the current I' flows is $B' = \mu_0 I'/h$ and the electric field induced in the conducting plate is $E'_i = -(\mu_0 a/2h) dI'/dt$. The electric field provided by the electric power source to keep the current constant is $E'_s = -E'_i$. Thus, the electric field inside the conductor is $E' = E'_i + E'_s = 0$. Hence, the Poynting vector on the conductor surface is zero, and there is no energy flow into the conductor. On the other hand, since the voltage between the gap at the terminal is $V = 2\pi a E'_s$, the electric field there is

$$E' = \frac{V}{\delta} = \frac{\mu_0 \pi a^2}{h\delta} \cdot \frac{dI'}{dt}.$$

Hence, the Poynting vector at the terminal is directed inside the coil and the magnitude of the vector is

11.4 The Poynting Vector

$$S_P = \frac{B'E'}{\mu_0} = \frac{\mu_0 \pi a^2}{h^2 \delta} I' \frac{dI'}{dt}.$$

The energy supplied to the coil until the current reaches I is

$$U_m = h\delta \int_0^I \frac{\mu_0 \pi a^2}{h^2 \delta} I' \frac{dI'}{dt} = \frac{\mu_0 \pi a^2}{2h} I^2 = \pi a^2 h \frac{B^2}{2\mu_0},$$

where $B = \mu_0 I/h$ is the magnetic flux density in the final state and $\pi a^2 h$ is the volume of the space in which the magnetic flux is stored. Hence, we can see that all the energy fed by the energy source is stored in the coil as the magnetic energy. ◆

Example 11.5 We apply a current to a thin superconducting parallel-plate transmission line, as shown in Fig. 11.8. Determine the energy stored in the transmission line when the applied current is I with the Poynting vector.

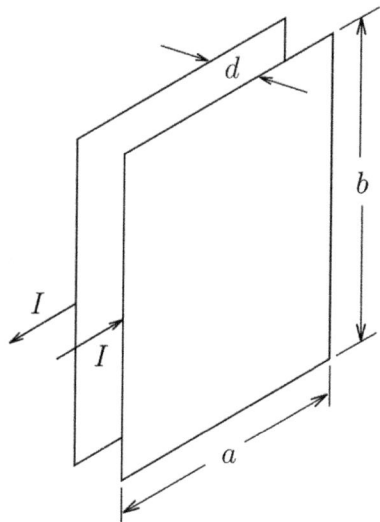

Fig. 11.8 Parallel-plate transmission line

Solution 11.5 We define Cartesian coordinates as shown in Fig. 11.9. When the applied current is I', the magnetic flux density in the transmission line is

$$B_z = \frac{\mu_0 I'}{b}.$$

Fig. 11.9 Arrangement of parallel-plate transmission line

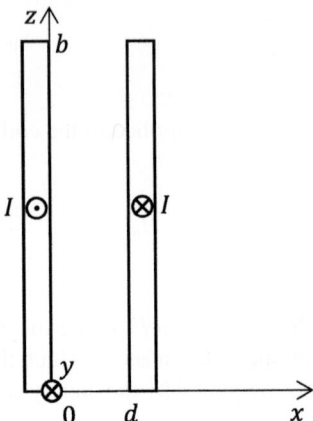

In this case, the magnetic flux does not go through the superconducting plates but penetrates from the both ends (along the y-axis). That is, the electric field is directed along the x-axis and is given by

$$E_x(y) = \frac{\mu_0}{b} \cdot \frac{\partial I'}{\partial t}\left(y - \frac{a}{2}\right).$$

Then, the Poynting vector at the edge at $y = 0$ is

$$S_P = -\frac{E_x(y=0)B_z}{\mu_0} = \frac{\mu_0 a}{2b^2} I' \frac{\partial I'}{\partial t}$$

and is directed inward. There is also the same contribution from the other side at $y = a$. Hence, the power that enters the transmission line is

$$P = 2bdS_P = \frac{\mu_0 ad}{b} I' \frac{\partial I'}{\partial t}.$$

The energy stored in the transmission line is

$$U_m = \frac{\mu_0 ad}{b} \int I' \frac{\partial I'}{\partial t} dt = \frac{\mu_0 ad}{b} \int_0^I I' dI' = \frac{\mu_0 ad}{2b} I^2 = \frac{1}{2} L I^2,$$

where $L = \mu_0 ad/b$ is the inductance of the parallel-plate transmission line. This is the same result as in the case in which the transmission line is made of a conductor, although the direction of the penetration of the power is different (see Exercise 11.13). ◆

Example 11.6 We assume that current I is applied to the resistive material shown in Fig. 5.5 in Sect. 5.4 and that the voltage between the top and bottom is V. Prove that the dissipated power is given by $P = IV$ using the Poynting vector.

11.4 The Poynting Vector

Solution 11.6 We define the ξ-axis on the side of the truncated cone that connects the points on the top and bottom of the same azimuthal angle, as shown in Fig. 11.10. The radius of the cone at distance ξ from the bottom is denoted by $R(\xi)$. The magnetic field at this position is

$$H(\xi) = \frac{I}{2\pi R(\xi)}.$$

The electric field denoted as $E(\xi)$ is directed along the ξ-axis and is normal to the magnetic field. Thus, the Poynting vector is $E(\xi)H(\xi)$ and is directed inward into the cone. The elementary surface vector is $|dS| = ds\, d\xi$ with ds denoting the line element on the perimeter. Since $E(\xi)$ is constant on the perimeter, the dissipated power is

$$P = \int E(\xi)d\xi \oint H(\xi)ds.$$

Using Ampere's law, we have

$$\oint H(\xi)ds = I$$

and the dissipated power is determined to be

$$P = I \int E(\xi)d\xi = IV.$$

Fig. 11.10 Electric and magnetic fields on the surface of a truncated cone

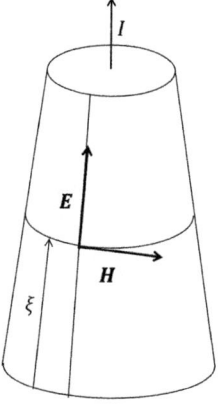

Example 11.7 The magnetic charging process of a superconducting parallel-plate transmission line is treated in Example 11.5. Determine the electromagnetic potential and prove that the Lorentz gauge is satisfied.

Solution 11.7 We define Cartesian coordinates as shown in Fig. 11.9. When the applied current is I, the magnetic flux density in the transmission line is

$$B_z = \frac{\mu_0 I}{b}$$

for $0 < x < d$. In this case, the vector potential has an x-component and is obtained as

$$A_x = -\int B_z dy = \frac{\mu_0 I}{b}\left(\frac{a}{2} - y\right).$$

Then, the electric field is

$$E_x = -\frac{\partial A_x}{\partial t} = \frac{\mu_0}{b} \cdot \frac{\partial I}{\partial t}\left(y - \frac{a}{2}\right),$$

which is reasonable since the electric field does not appear along the current in the superconductor. In reality, the electric field occurs between the superconducting plates due to the electromagnetic induction caused by the penetration of the magnetic flux along the y-axis. So, the electric potential is zero:

$$\phi = 0.$$

In addition, we have

$$\nabla \cdot \mathbf{A} = \frac{\partial A_x}{\partial x} = 0.$$

Thus, the Lorentz gauge is satisfied. Note that the vector potential is normal to the direction of the current, and Eq. (6.20) does not hold in this case. ◆

11.5 Exercises

11.1. Equation (11.3) cannot be proved theoretically. Explain the reason.

11.5 Exercises

11.2. Point charge q is moving with velocity v along the x-axis. Determine the displacement current.

11.3. AC current $I(t) = I_0 \sin \omega t$ is applied to a straight wire along the z-axis with a gap of length d, as shown in Fig. E11.1. Determine the displacement current and magnetic flux density. Electric charges $\pm Q(t)$ are concentrated at the ends of the gap.

Fig. E11.1 Terminated current at a gap in a wire

11.4. Determine the electric field, displacement current, and magnetic flux density when the electric dipole moment at the origin changes with time. Neglect the electromagnetic induction.

11.5. Determine the vector potential using the Lorentz gauge for the electromagnetic fields produced by the time-varying electric dipole treated in Exercise 11.4.

11.6. We are applying a current to a circular parallel-plate capacitor in Example 11.1. Determine the electromagnetic potential during this process and prove that the Lorentz gauge is satisfied.

11.7. It is assumed that the parallel-plate transmission line treated in Example 11.7 is made of a conductor. Determine the electromagnetic potential when a current is applied, and prove that the Lorentz gauge is satisfied.

11.8. Suppose that we apply current to the coaxial conducting transmission line shown in Fig. E11.2. Determine the electromagnetic potential and prove that the Lorentz gauge is satisfied. Neglect the thickness of the conductors.

11.9. Suppose that the coaxial transmission line treated in Exercise 11.8 is made of a superconductor. Determine the electromagnetic potential and prove that the Lorentz gauge is satisfied. The length of the transmission line is l.

11.10. We treat the electric charging process by applying a current to a cylindrical capacitor, as shown in Fig. E11.3. Determine the electromagnetic potential and prove that the Lorentz gauge is satisfied.

Fig. E11.2 Thin coaxial conducting transmission line

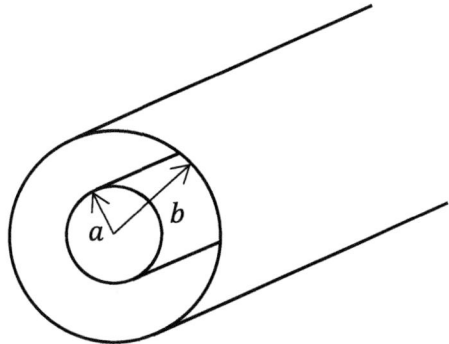

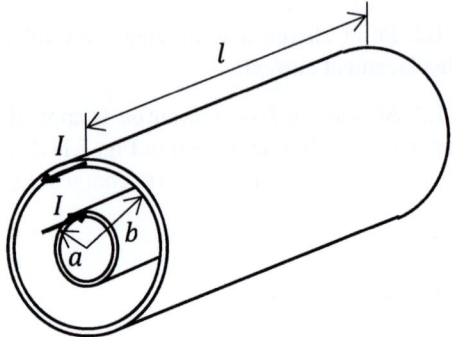

Fig. E11.3 Current flow during the process of charging a cylindrical capacitor

11.11. We apply a steady current I between the inner and outer electrodes in the cylindrical coaxial resistor shown in Fig. E11.4. Determine the electromagnetic potential and prove that the Lorentz gauge is satisfied. The resistivity of the resistive material is ρ_r.

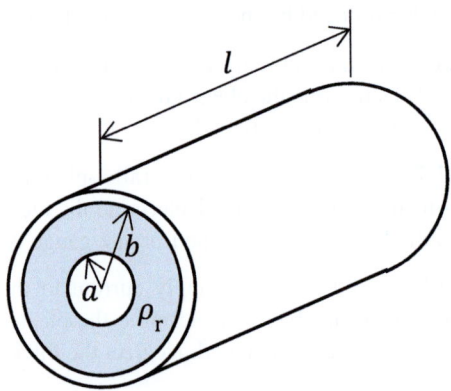

Fig. E11.4 Cylindrical resistor

11.12. Show that the third term in Eq. (11.27) for the Poynting vector is the mechanical work done by the electromagnetic field.

11.13. Assume that the parallel-plate transmission line in Example 11.5 is made of a conductor. Determine the energy stored in the transmission line with the Poynting vector when the applied current is I. Neglect the thickness of the plates.

11.14. Determine the energy stored in a unit length of the conducting transmission line discussed in Exercise 11.8 using the Poynting vector when current I is applied.

11.15. Determine the energy stored in a unit length of the superconducting transmission line discussed in Exercise 11.9 using the Poynting vector when current I is applied.

11.16. Determine the energy stored in the capacitor treated in Exercise 11.10 while applying current I for T seconds with the Poynting vector. Neglect the resistance of the electrodes and the thicknesses of the inner and outer electrodes. It is assumed that the length, l, is sufficiently long, and the effect of both edges can be neglected.

11.5 Exercises

11.17. We apply a steady current I to the cylindrical coaxial resistor treated in Exercise 11.11. Discuss the flow of energy in the resistor.

11.18. When voltage V is applied to the material shown in Fig. 5.4 in Sect. 5.4, current I flows. Prove that the loss power is given by $P = VI$ with the Poynting vector.

11.19. When voltage V is applied to the material shown in Fig. 5.6 in Example 5.1, current I flows. Prove that the loss power is given by $P = VI$ with the Poynting vector.

11.20. Determine the dissipated power in the resistor shown in Fig. 5.11 in Example 5.5 with the Poynting vector, when current I is applied to the resistor.

11.21. Determine the dissipated power in the resistor shown in Fig. 5.12 in Example 5.6 with the Poynting vector, when voltage V is applied to the resistor. Disregard the disorder of the magnetic field at the interfaces between the two resistive materials.

11.22. Suppose that we apply current I to the capacitor with tilted electrodes treated in Exercise 3.10 until the voltage between the electrodes reaches V, as shown in Fig. E11.5. Determine the energy stored in this capacitor. Assume that a is much longer than b.

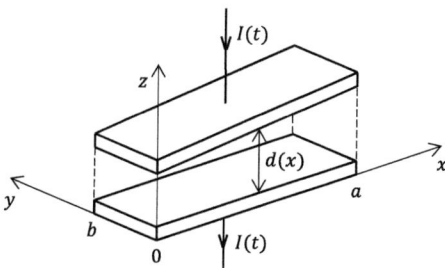

Fig. E11.5 Capacitor with a tilted electrode

11.23. We apply current I for T seconds to a cylindrical capacitor with two dielectric materials with dielectric constants ϵ_1 and ϵ_2, as shown in Fig. 4.8 in Example 4.8. The length of the capacitor is l. Determine the stored energy using the Poynting vector.

11.24. We apply current I for T seconds to a cylindrical capacitor with two dielectric materials with dielectric constants ϵ_1 and ϵ_2, as shown in Fig. E4.1 in Exercise 4.2. The length of the capacitor is l. Determine the stored energy using the Poynting vector.

11.25. We apply current I for T seconds to a cylindrical capacitor with dielectric constant that varies radially as $\epsilon' = \epsilon(1 + kR)$, as shown in Fig. E4.4 in Exercise 4.8. The length of the capacitor is l. Determine the stored energy using the Poynting vector.

11.26. Suppose that we apply current I to a conducting transmission line with a tilted plate, as shown in Fig. E11.6. The distance between the plates changes as $d(y) = d_0(1 + ky)$ along the length. Determine the energy stored in this transmission line.

Fig. E11.6 Conducting parallel-plate transmission line with a tilted plate

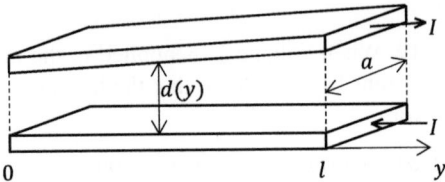

11.27. If the transmission line with a tilted plate treated in Exercise 11.26 is made of a superconductor, how is the energy stored in the transmission line when current I is applied?

11.28. We apply currents $\pm I$ to a cylindrical superconducting transmission line with two magnetic materials with magnetic permeabilities μ_1 and μ_2, as shown in Fig. 9.6 in Example 9.3. The length of the transmission line is l. Determine the stored magnetic energy in the transmission line using the Poynting vector.

11.29. We apply currents $\pm I$ to a cylindrical superconducting transmission line with two magnetic materials with magnetic permeabilities μ_1 and μ_2, as shown in Fig. 9.10 in Example 9.8. The length of the transmission line is l. Determine the stored magnetic energy in the transmission line using the Poynting vector.

11.30. We apply current I to a cylindrical superconducting transmission line with magnetic permeability that varies radially as $\mu' = \mu(1 + kR)$, as shown in Fig. E9.4 in Exercise 9.5. The length of the transmission line is l. Determine the stored magnetic energy in the transmission line using the Poynting vector.

11.31. We apply current I_0 to the outer coil of the superconducting coaxial coils shown in Fig. E11.7 to produce magnetic flux density B_0. Then, we apply current to the inner coil. Discuss the flow of energy until the current of the inner coil reaches I, at which point, the magnetic flux density in the inner bore is $B_0 + B$. The radii and number of windings of the inner and outer coils are a and b, and N and N_0, respectively. The height of these coils is h.

11.32. The problem of Exercise 11.31 is discussed from the viewpoint of the electric circuit shown in Fig. E11.8. The energy supplied by the power source in the right circuit for the inner coil is

$$U_1 = \int (-V') I' dt = \frac{\pi \mu_0 a^2 N^2}{2h} I^2 = \frac{\pi \mu_0 a^2}{2\mu_0} B^2.$$

Discuss the difference from the solution to Exercise 11.31.

11.5 Exercises

Fig. E11.7 Two coaxial superconducting solenoid coils

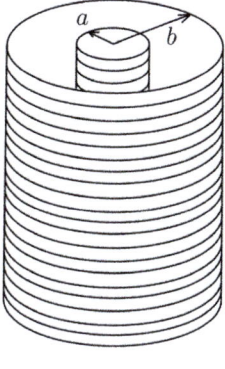

Fig. E11.8 Electric circuit for the coaxial coils

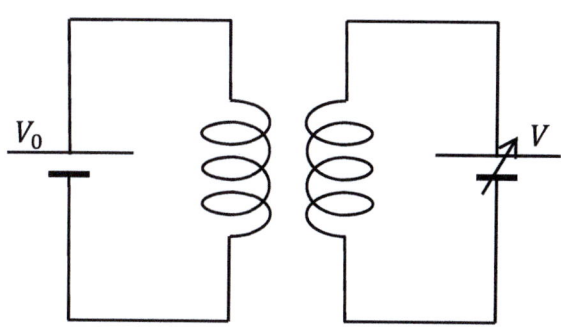

11.33. Suppose that the magnetic flux density of the inner solenoid coil treated in Exercises 11.31 and 11.32 is B. Then, we apply current I_0 to the outer solenoid coil so that the magnetic flux density reaches B_0 in its bore. Discuss the flow of energy during the magnetic charging process for the outer solenoid coil.

11.34. We apply currents I_1 and I_2 to two long parallel-plate transmission lines connected to the power sources of voltages V_1 and V_2, as shown in Fig. E11.9. Discuss the flow of energy. Neglect the electric resistance of the conductors. The width of the plates normal to the sheet is w.

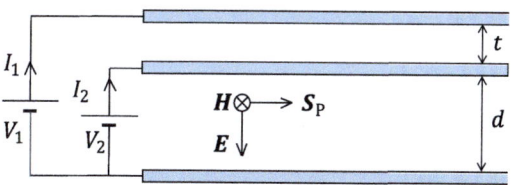

Fig. E11.9 Two long parallel-plate transmission lines operated independently

Answers to Exercises

11.1. When we try to derive Eq. (11.3) from Eq. (11.2), the same result is obtained even if we add a term given by a curl of an arbitrary vector function K to the right side of Eq. (11.4), since the equation,

$$\int_S (\nabla \times \boldsymbol{K}) \cdot d\boldsymbol{S} = \int_V \nabla \cdot (\nabla \times \boldsymbol{K}) dV = 0,$$

holds. In the above, Gauss's theorem and Eq. (A.7) in the Appendix are used. This is the same as the reason why the Poynting vector does not always give the correct flow of electromagnetic energy. So, the validity of Eq. (11.3) can only be checked by experiments.

11.2. The position of the point charge is denoted as $(vt, 0, 0)$. the electric field at observation point (x, y, z) is

$$\boldsymbol{E}(x, y, z) = \frac{q}{4\pi \epsilon_0 [(x-vt)^2 + y^2 + z^2]^{3/2}} \big[(x-vt)\boldsymbol{i}_x + y\boldsymbol{i}_y + z\boldsymbol{i}_z\big].$$

The displacement current is

$$\boldsymbol{i}_d = \frac{qv\{[2(x-vt)^2 - y^2 - z^2]\boldsymbol{i}_x + 3(x-vt)y\boldsymbol{i}_y + 3(x-vt)z\boldsymbol{i}_z\}}{4\pi [(x-vt)^2 + y^2 + z^2]^{5/2}}.$$

11.3. The electric charge at the left end is $Q(t) = -(I_0/\omega)\cos\omega t$. The electric field produced by the electric charges at the observation point P (R, φ, z) is

$$\boldsymbol{E}(R, \varphi, z) = (E_R, 0, E_z),$$

$$E_R = \frac{Q(t)}{4\pi \epsilon_0}\left\{\frac{R}{[R^2 + (z+d/2)^2]^{3/2}} - \frac{R}{[R^2 + (z-d/2)^2]^{3/2}}\right\},$$

$$E_z = \frac{Q(t)}{4\pi \epsilon_0}\left\{\frac{z+d/2}{[R^2 + (z+d/2)^2]^{3/2}} - \frac{z-d/2}{[R^2 + (z-d/2)^2]^{3/2}}\right\}.$$

The displacement current is

$$i_{dR} = \frac{\partial \epsilon_0 E_R}{\partial t} = \frac{I_0 \sin\omega t}{4\pi}\left\{\frac{R}{[R^2 + (z+d/2)^2]^{3/2}} - \frac{R}{[R^2 + (z-d/2)^2]^{3/2}}\right\},$$

$$i_{dz} = \frac{\partial \epsilon_0 E_z}{\partial t} = \frac{I_0 \sin\omega t}{4\pi}\left\{\frac{z+d/2}{[R^2 + (z+d/2)^2]^{3/2}} - \frac{z-d/2}{[R^2 + (z-d/2)^2]^{3/2}}\right\}.$$

The magnetic flux density is an azimuthal component, B_φ. Equation (11.3) for the displacement current derives the magnetic flux density:

11.5 Exercises

$$B_\varphi(R,z) = \frac{\mu_0}{2\pi R} \int_0^R 2\pi R i_{dz} dR$$

$$= \frac{\mu_0 I_0 \sin \omega t}{2\pi R} \left\{ 1 - \frac{z+d/2}{2[R^2 + (z+d/2)^2]^{1/2}} + \frac{z-d/2}{2[R^2 + (z-d/2)^2]^{1/2}} \right\}$$

for $-d/2 < z < d/2$. In other regions, the contribution of the current flowing in the wire must also be taken into consideration. The result is the same also for other regions, however. Check it by yourself.

11.4. When the electric dipole moment $p(t)$ at the origin changes with time, the electric potential in spherical coordinates is given by

$$\phi(t) = \frac{p(t)}{4\pi \epsilon_0 r^2} \cos\theta$$

and the components of the electric field are

$$E_r = \frac{p(t)}{2\pi \epsilon_0 r^3} \cos\theta, \quad E_\theta = \frac{p(t)}{4\pi \epsilon_0 r^3} \sin\theta.$$

Then, the components of the displacement current are

$$\frac{\partial D_r}{\partial t} = \frac{\dot{p}(t)}{2\pi r^3} \cos\theta, \quad \frac{\partial D_\theta}{\partial t} = \frac{\dot{p}(t)}{4\pi r^3} \sin\theta.$$

The magnetic flux density produced by the displacement current has only the azimuthal component, B_φ, which satisfies

$$\frac{\dot{p}(t)}{2\pi r^3} \cos\theta = \frac{1}{\mu_0 r \sin\theta} \cdot \frac{\partial}{\partial \theta}(\sin\theta \, B_\varphi),$$

$$\frac{\dot{p}(t)}{4\pi r^3} \sin\theta = -\frac{1}{\mu_0 r} \cdot \frac{\partial}{\partial r}(r B_\varphi).$$

Then, from the second equation we have

$$B_\varphi = \frac{\mu_0 \dot{p}(t)}{4\pi r^2} \sin\theta,$$

which satisfies also the first equation. So, the above magnetic flux density is the solution.

Exactly speaking, the variation in the magnetic flux density induces the electromotive force, and the electric field is influenced.

11.5. The radial and zenithal components of the vector potential are denoted by A_r and A_θ, respectively. Then, the azimuthal magnetic flux density is

$$\frac{1}{r}\left[\frac{\partial}{\partial r}(rA_\theta) - \frac{\partial A_r}{\partial \theta}\right] = \frac{\mu_0 \dot{p}(t)}{4\pi r^2} \sin\theta. \tag{B11.1}$$

The Lorentz gauge is written as

$$\frac{1}{r^2} \cdot \frac{\partial}{\partial r}(r^2 A_r) + \frac{1}{r\sin\theta} \cdot \frac{\partial}{\partial \theta}(\sin\theta\, A_\theta) = -\frac{\mu_0 \dot{p}(t)}{4\pi r^2}\cos\theta. \tag{B11.2}$$

Since A_θ is proportional to $(1/r)$, the first term on the left side in Eq. (B11.1) is zero. Thus, we have

$$A_r = \frac{\mu_0 \dot{p}(t)}{4\pi r}\cos\theta.$$

Substituting this into Eq. (B11.2), the azimuthal component is determined to be

$$A_\theta = \frac{\mu_0 \dot{p}(t)}{4\pi r}\sin\theta.$$

11.6. When the applied current is I, the electric charge on the electrode is

$$Q(t) = \int I\,dt.$$

The displacement current in the space between the electrodes is

$$i_d = \frac{1}{\pi a^2} \cdot \frac{dQ(t)}{dt} = \frac{I}{\pi a^2}.$$

This produces the magnetic flux density:

$$B_\varphi = \frac{1}{R}\int \mu_0 i_d R\,dR = \frac{\mu_0 I}{2\pi a^2}R.$$

The vector potential is given by

$$A_R = \int B_\varphi\,dz = \frac{\mu_0 I}{2\pi a^2}Rz,$$

where the z-axis is normal to the electrodes. The contribution to the electric field by the vector potential is $E_R = -\partial A_R/\partial t = 0$. Since the electric field is

11.5 Exercises

$$E_z = -\frac{Q(t)}{\pi\epsilon_0 a^2} = \frac{1}{\pi\epsilon_0 a^2}\int I\,dt,$$

the electric potential is given by

$$\phi = -\frac{z}{\pi\epsilon_0 a^2}\int I\,dt.$$

So, we have

$$\epsilon_0\mu_0\frac{\partial\phi}{\partial t} = -\frac{\mu_0 I}{\pi a^2}z$$

and

$$\nabla\cdot\mathbf{A} = \frac{1}{R}\cdot\frac{\partial(RA_R)}{\partial R} = \frac{\mu_0 I}{\pi a^2}z.$$

Thus, the relationship holds:

$$\epsilon_0\mu_0\frac{\partial\phi}{\partial t} + \nabla\cdot\mathbf{A} = 0.$$

11.7. We define Cartesian coordinates as in Fig. 11.9. When the applied current is I, the magnetic flux density in the transmission line is

$$B_z = \frac{\mu_0 I}{b}.$$

The vector potential that satisfies the symmetry with respect to the plane at $x = d/2$ is

$$A_y = \frac{\mu_0 I}{b}\left(x - \frac{d}{2}\right).$$

The electric field is

$$E_y = -\int\frac{\partial B_z}{\partial t}dx = -\frac{\mu_0}{b}\cdot\frac{\partial I}{\partial t}\left(x - \frac{d}{2}\right).$$

Since the electric field comes from the electromagnetic induction, it is given by the vector potential, $E_y = -\partial A_y/\partial t$. So, the electric potential is zero:

$$\phi = 0.$$

In addition, we have

$$\nabla \cdot \mathbf{A} = \frac{\partial A_y}{\partial y} = 0.$$

Thus, the Lorentz gauge is satisfied. In this case, Eq. (6.20) holds for the vector potential. This is different from the superconducting parallel-plate transmission line discussed in Example 11.7. If we denote as $\mathbf{A}_1$ the vector potential in the conducting transmission line, the vector potential in the superconducting transmission line is

$$\mathbf{A} = \mathbf{A}_1 + \nabla \alpha$$

with

$$\alpha = \frac{\mu_0 I}{b}\left(x - \frac{d}{2}\right)\left(\frac{a}{2} - y\right).$$

11.8. When the applied current is I, the azimuthal magnetic flux density in the transmission line is

$$B_\varphi = \frac{\mu_0 I}{2\pi R},$$

and the vector potential in the space ($a < R < b$) is

$$A_z(R) = \frac{\mu_0 I}{2\pi} \log \frac{b}{R},$$

which satisfies $A_z(b) = 0$. This leads to the electric field:

$$E_z(R) = -\frac{\partial A_z}{\partial t} = -\frac{\mu_0}{2\pi} \log \frac{b}{R} \cdot \frac{\partial I}{\partial t},$$

Since the electric field is produced by the electromagnetic induction, the electric potential is zero:

$$\phi = 0.$$

In addition, we have

$$\nabla \cdot \mathbf{A} = \frac{\partial A_z}{\partial z} = 0.$$

Thus, the Lorentz gauge is satisfied.

11.9. When the applied current is I, the azimuthal magnetic flux density in the transmission line is

$$B_\varphi = \frac{\mu_0 I}{2\pi R}.$$

In this case, the vector potential has a radial component and is obtained as

$$A_R = \int B_\varphi \mathrm{d}z = \frac{\mu_0 I}{2\pi R}\left(z - \frac{l}{2}\right),$$

so that it satisfies an antisymmetric condition with respect to $z = l/2$. This leads to the electric field:

$$E_R = -\frac{\partial A_R}{\partial t} = -\frac{\mu_0}{2\pi R}\left(z - \frac{l}{2}\right) \cdot \frac{\partial I}{\partial t}.$$

Note that the electric field does not appear along the length of the superconductor, which is the difference from the case of conducting transmission line. This electric field is caused by the penetration of the magnetic flux along the length of the transmission line. The same result will be shown in Exercise 11.15. Since the electric field is produced by the electromagnetic induction, the electric potential is zero:

$$\phi = 0.$$

In addition, we have

$$\nabla \cdot \mathbf{A} = \frac{1}{R} \cdot \frac{\partial (RA_R)}{\partial R} = 0.$$

Thus, the Lorentz gauge is satisfied.

Note that the vector potential is normal to the direction of the current, and Eq. (6.20) does not hold in this case. This is similar to the superconducting parallel-plate transmission line discussed in Example 11.7. If we denote as $\mathbf{A}_1$ the vector potential in the conducting coaxial transmission line treated in Exercise 11.8, the vector potential in the superconducting coaxial transmission line is

$$\mathbf{A} = \mathbf{A}_1 + \nabla \alpha$$

with

$$\alpha = \frac{\mu_0 I}{2\pi}\left(z - \frac{l}{2}\right)\log\frac{R}{b}.$$

11.10. The current changes along the length, i.e., the z-axis, to distribute the electric charge on the electrode surface as

$$I'(z) = I\left(1 - \frac{z}{l}\right).$$

(see Exercise 5.1) Then, the magnetic flux density has a φ-component given by

$$B_\varphi(z) = \frac{\mu_0 I}{2\pi R}\left(1 - \frac{z}{l}\right).$$

The vector potential has a z-component:

$$A_z = -\int B_\varphi dR = \frac{\mu_0 I}{2\pi}\left(1 - \frac{z}{l}\right)\log\frac{b}{R},$$

which satisfies $A_z(b) = 0$. Noting that the total electric charge is $Q = It$, the electric field is

$$E_R = \frac{It}{2\pi\epsilon_0 lR},$$

and hence, the electric potential is

$$\phi = \frac{It}{2\pi\epsilon_0 l}\log\frac{b}{R},$$

which satisfies $\phi(b) = 0$. From the above results, we have

$$\epsilon_0\mu_0\frac{\partial\phi}{\partial t} = \frac{\mu_0 I}{2\pi l}\log\frac{b}{R}, \quad \nabla\cdot\mathbf{A} = \frac{\partial A_z}{\partial z} = -\frac{\mu_0 I}{2\pi l}\log\frac{b}{R}.$$

Thus, the relationship holds:

$$\epsilon_0\mu_0\frac{\partial\phi}{\partial t} + \nabla\cdot\mathbf{A} = 0.$$

11.11. We define the z-axis along the length and the positions of the two edges are denoted by $z = \pm l/2$. The radial current produces the radial electric field,

$$E_R = \frac{\rho_r I}{2\pi lR},$$

and the electric potential is given by

$$\phi(R) = -\frac{\rho_r I}{2\pi l}\log\frac{R}{a},$$

11.5 Exercises

which satisfies $\phi(a) = 0$. The magnetic flux density has an azimuthal component,

$$B_\varphi = -\frac{\mu_0 Iz}{2\pi lR},$$

and the vector potential is given by

$$A_R = -\frac{\mu_0 Iz^2}{4\pi lR}.$$

From the above results, we have $\partial \phi/\partial t = 0$ and $\nabla \cdot \boldsymbol{A} = 0$. Thus, the Lorentz gauge is satisfied.

11.12. If we denote the electric charge, number density, and velocity of charged particles by q, n, and $\boldsymbol{v}$, respectively, from Eq. (5.3), the current density is expressed as

$$\boldsymbol{i} = qn\boldsymbol{v}.$$

Hence, $\boldsymbol{i} \cdot \boldsymbol{E}$ is given by

$$\boldsymbol{i} \cdot \boldsymbol{E} = n(q\boldsymbol{E}) \cdot \boldsymbol{v}. \tag{B11.3}$$

In the above, $n(q\boldsymbol{E})$ is the Coulomb force on electric charges in a unit volume, and hence, Eq. (B11.3) gives the density of the mechanical work done by the Coulomb force in unit time. On the other hand, the Lorentz force does not do any mechanical work on electric charges. Thus, the third term in Eq. (11.27) is the kinetic energy density, i.e., the mechanical work done on charged particles by the electromagnetic fields.

11.13. We use Fig. 11.9 in Example 11.5. When the applied current is I', the magnetic flux density in the transmission line is $B_z = \mu_0 I'/b$. The magnetic flux penetrates through the conductor, and the electric field is induced along the y-axis:

$$E_y = -\int \frac{\mu_0}{b} \cdot \frac{\partial I'}{\partial t} dx = \frac{\mu_0}{b} \cdot \frac{\partial I'}{\partial t}\left(\frac{d}{2} - x\right),$$

which satisfies an antisymmetric condition with respect to $x = d/2$. Thus, the Poynting vector on the left surface at $x = 0$ is

$$S_P = \frac{\mu_0 d}{2b^2} I' \frac{\partial I'}{\partial t}$$

and directed inward. There is also a contribution from the right surface. Hence, the power that enters into the transmission line is

$$P = 2abS_P = \frac{\mu_0 ad}{b} I' \frac{\partial I'}{\partial t}.$$

The energy stored in the transmission line is

$$U_m = \frac{\mu_0 ad}{b} \int I' \frac{\partial I'}{\partial t} dt = \frac{\mu_0 ad}{b} \int_0^I I' dI' = \frac{\mu_0 ad}{2b} I^2 = \frac{1}{2} LI^2,$$

where $L = \mu_0 ad/b$ is the inductance of the parallel-plate transmission line. This is the same result as in the case in which the transmission line is made of a superconductor.

11.14. When the applied current is I', the azimuthal magnetic flux density in the transmission line is

$$B_\varphi(R) = \frac{\mu_0 I'}{2\pi R}$$

for $a < R < b$, and the electric field induced by the penetrating magnetic flux from the outer conductor is

$$E_z(R) = \int_a^R \frac{\mu_0}{2\pi R} \cdot \frac{\partial I'}{\partial t} dR = \frac{\mu_0}{2\pi} \log \frac{R}{a} \cdot \frac{\partial I'}{\partial t}.$$

So, the Poynting vector on the inner surface ($R = b$) of the outer conductor is

$$S_P = \frac{\mu_0}{4\pi^2 b} \log \frac{b}{a} I' \frac{\partial I'}{\partial t}$$

and is directed inward. Hence, the power in a unit length that enters from the outside is

$$P' = 2\pi b S_P = \frac{\mu_0}{2\pi} \log \frac{b}{a} I' \frac{\partial I'}{\partial t}.$$

The energy stored in the transmission line in a unit length is

$$U'_m = \frac{\mu_0}{2\pi} \log \frac{b}{a} \int I' \frac{\partial I'}{\partial t} dt = \frac{\mu_0}{2\pi} \log \frac{b}{a} \int_0^I I' dI' = \frac{\mu_0}{4\pi} \log \frac{b}{a} I^2 = \frac{1}{2} L' I^2,$$

where $L' = (\mu_0/2\pi) \log(b/a)$ is the inductance of the coaxial transmission line in a unit length.

11.5 Exercises

11.15. When the applied current is I', the azimuthal magnetic flux density in the transmission line is

$$B_\varphi(R) = \frac{\mu_0 I'}{2\pi R}$$

for $a < R < b$. In this case, the magnetic flux does not go through the outer superconductor but penetrates from the both ends (along the z-axis). So, the induced electric field is different from the case of conducting transmission line. That is, the electric field is radial because of the penetration of magnetic flux from both ends, and is given by

$$E_R(z) = -\int \frac{\mu_0}{2\pi R} \cdot \frac{\partial I'}{\partial t} dz = -\frac{\mu_0}{2\pi R} \cdot \frac{\partial I'}{\partial t}\left(z - \frac{l}{2}\right),$$

which satisfies an antisymmetric condition with respect to $z = l/2$. Thus, the Poynting vector on the surface at $z = 0$ is

$$S_P = \frac{\mu_0 l}{8\pi^2 R^2} I' \frac{\partial I'}{\partial t}$$

and is directed inward. There is also a contribution from the surface at $z = l$. Hence, the power that enters into the transmission line is

$$P = 2\int_a^b \frac{\mu_0 l}{8\pi^2 R^2} I' \frac{\partial I'}{\partial t} 2\pi R dR = \frac{\mu_0 l}{2\pi} \log\frac{b}{a} I' \frac{\partial I'}{\partial t}.$$

The energy stored in the transmission line in a unit length is

$$U'_m = \frac{\mu_0}{2\pi}\log\frac{b}{a}\int I'\frac{\partial I'}{\partial t}dt = \frac{\mu_0}{2\pi}\log\frac{b}{a}\int_0^I I'dI' = \frac{\mu_0}{4\pi}\log\frac{b}{a}I^2 = \frac{1}{2}L'I^2,$$

where $L' = (\mu_0/2\pi)\log(b/a)$ is the inductance of the coaxial transmission line in a unit length.

11.16. When current I is applied for t seconds, the electric charges stored in the capacitor are $\pm Q' = \pm It$. Hence, the electric charge density on the surface of the inner electrode is $\sigma = Q'/2\pi al$, and the electric field directed radially from the inner electrode is $E(R) = \sigma a/\epsilon_0 R = (I/2\pi\epsilon_0 lR)t$ at radius R. The current is I at the edge at which the current is applied but decreases to zero at the other edge. Then, the resultant magnetic field is $H(R) = I/2\pi R$ and the Poynting vector is $S_P(R) = EH = I^2 t/4\pi^2 \epsilon_0 lR^2$ directed inward at the edge at which the current is applied. Hence, the power that enters into the capacitor is

$$P = \int_a^b S_{\mathrm{P}}(R) 2\pi R \mathrm{d}R = \frac{I^2 t}{2\pi \epsilon_0 l} \log \frac{b}{a}.$$

Integrating this until $t = T$, the final energy put into the capacitor is

$$U_{\mathrm{e}} = \int_0^T \frac{I^2 t}{2\pi \epsilon_0 l} \log \frac{b}{a} \mathrm{d}t = \frac{Q^2}{4\pi \epsilon_0 l} \log \frac{b}{a} = \frac{Q^2}{2C},$$

where $Q = IT$ is the electric charge in the final state and $C = 2\pi \epsilon_0 l / \log(b/a)$ is the capacitance of the cylindrical concentric capacitor.

11.17. We define the z-axis along the length and the positions of the two edges are denoted by $z = \pm l/2$. The radial current produces the radial electric field:

$$E_R = \frac{\rho_{\mathrm{r}} I}{2\pi l R}.$$

It also produces the azimuthal magnetic flux density:

$$B_\varphi = -\frac{\mu_0 I z}{2\pi l R}.$$

Hence, the Poynting vector is directed inward. The power from the two edges is

$$P = 2 \int_a^b \frac{E_R B_\varphi(z = -l/2)}{\mu_0} 2\pi R \mathrm{d}R = \frac{\rho_{\mathrm{r}} I^2}{2\pi l} \log \frac{b}{a} = R_{\mathrm{r}} I^2,$$

where $R_{\mathrm{r}} = \rho_{\mathrm{r}} \log(b/a)/2\pi l$ is the resistance.

11.18. The equipotential surface is a plane extending from the origin, as shown in Fig. B5.1 in Exercise 5.27. We consider a narrow region between the two equipotential surfaces of angles θ and $\theta + \mathrm{d}\theta$. Then, the potential difference between these two equipotential surfaces is $2V \mathrm{d}\theta / \pi$. Using the Poynting vector, the power that enters from the narrow surface into this region is

$$\mathrm{d}P = \int_{\mathrm{d}S} HE \mathrm{d}S = \int E \mathrm{d}l \int H \mathrm{d}s = \frac{2V \mathrm{d}\theta}{\pi} \int H \mathrm{d}s = \frac{2VI \mathrm{d}\theta}{\pi},$$

where we have used Ampere's law, although we do not know the magnetic field strength on the surface. Integrating this with respect to angle θ, we have

$$P = \int_0^{\pi/2} \frac{2VI}{\pi} d\theta = VI.$$

11.19. The equipotential surface is a plane extending from the origin, as shown in Fig. B11.1. We consider a narrow region between the two equipotential surfaces at angles θ and $\theta + d\theta$. Then, the potential difference between these two equipotential surfaces is $2V d\theta/\pi$. Using the Poynting vector, the power that enters from the narrow surface into this region is

$$dP = \int_{dS} HE dS = \int E dl \int H ds = \frac{2V d\theta}{\pi} \int H ds = \frac{2VI d\theta}{\pi},$$

where we have used Ampere's law, although we do not know the magnetic field strength on the surface. Integrating this with respect to angle θ, we have

$$P = \int_0^{\pi/2} \frac{2VI}{\pi} d\theta = VI.$$

Fig. B11.1 Equipotential surfaces at angles θ and $\theta + d\theta$

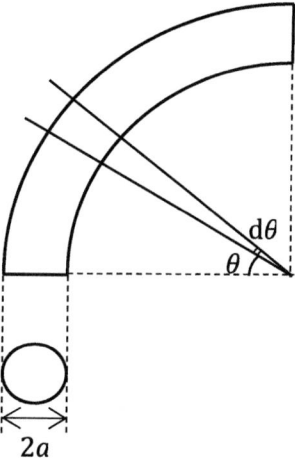

11.20. This problem can be solved using the same method as in Exercise 11.17. The current density is $i = I/2\pi lR$ at a point R from the central axis. So, the electric field is

$$E(R) = \frac{\rho_{r1} I}{2\pi lR}; \quad a < R < b,$$
$$= \frac{\rho_{r2} I}{2\pi lR}; \quad b < R < c.$$

The magnetic flux density is azimuthal and is given by $B(R, z) = -\mu_0 Iz/2\pi lR$ $(-l/2 < z < l/2)$. Then, the power that enters into the resistor from the edge at $z = -l/2$ is the same as that at $z = l/2$. Hence, the total power into the resistor is

$$P = \frac{2}{\mu_0} \int_S EB(z = -l/2) 2\pi R\, dR = \frac{I^2}{2\pi l}\left(\rho_{r1} \int_a^b \frac{dR}{R} + \rho_{r2} \int_b^c \frac{dR}{R}\right)$$

$$= \frac{I^2}{2\pi l}\left(\rho_{r1} \log \frac{b}{a} + \rho_{r2} \log \frac{c}{b}\right).$$

This is equal to $R_r I^2$ with the resistance R_r determined in Example 5.5.

11.21. We denote the electric field by $E = k/R$ as a function of radius R, where k is a constant. Then, the voltage is $V = k \log(b/a)$. So, we have $k = V/\log(b/a)$, and the current density is

$$i(R) = \frac{V}{\rho_{r1} \log(b/a) R}; \quad \text{in resistive material 1,}$$

$$= \frac{V}{\rho_{r2} \log(b/a) R}; \quad \text{in resistive material 2.}$$

The magnetic flux density is azimuthal and is given by

$$B(R, z) = -\frac{\mu_0 Vz}{\rho_{r1} \log(b/a) R}; \quad \text{in resistive material 1,}$$

$$= -\frac{\mu_0 Vz}{\rho_{r2} \log(b/a) R}; \quad \text{in resistive material 2}$$

for $-l/2 < z < l/2$. Then, the power that enters into the resistor from the edge at $z = -l/2$ is the same as that at $z = l/2$. Hence, the total power into the resistor is

$$P = \frac{2}{\mu_0} \int_S EB(z = -l/2) 2\pi R\, dR = \frac{\pi l V^2}{[\log(b/a)]^2}\left(\frac{1}{\rho_{r1}} + \frac{1}{\rho_{r2}}\right) \int_a^b \frac{dR}{R}$$

$$= \frac{\pi l(\rho_{r1} + \rho_{r2}) V^2}{\rho_{r1}\rho_{r2} \log(b/a)}.$$

This is equal to V^2/R_r with the resistance R_r determined in Example 5.6.

11.22. The voltage between the electrodes is denoted as $V'(t)$. Then, the electric field between the electrodes has a z-component and is given by is $E_z(t) = -V'(t)/d(x)$ and the displacement current has also a z-component and is given by

11.5 Exercises

$$\frac{\partial D(t)}{\partial t} = -\frac{\epsilon_0}{d(x)} \cdot \frac{\partial V'(t)}{\partial t}.$$

The resultant magnetic field has an x-component, and from symmetry, it is given by

$$H_x(t) = \frac{\epsilon_0}{d(x)} \cdot \frac{\partial V'(t)}{\partial t}\left(y - \frac{b}{2}\right).$$

We consider the power that enters into the space between the electrodes from the surfaces at $y = 0$ and $y = b$. The Poynting vector is directed inward on these surfaces. For example, the Poynting vector on the surface at $y = 0$ is

$$S_P(y = 0) = -E_z(t)H_x(t) = \frac{\epsilon_0 b}{2d^2(x)} V'(t) \frac{\partial V'(t)}{\partial t}.$$

Taking account that the integration along the y-axis is identical with multiplication by $d(x)$, the input power is

$$P = 2 \iint S_P(y = 0) dx dy = \epsilon_0 b V'(t) \frac{\partial V'(t)}{\partial t} \int_0^a \frac{dx}{d_0 + kx}$$

$$= \frac{\epsilon_0 b}{k} \log\left(1 + \frac{ka}{d_0}\right) V'(t) \frac{\partial V'(t)}{\partial t}.$$

The energy stored in the capacitor until the voltage reaches V is determined to be

$$U_e = \int P dt = \frac{\epsilon_0 b}{k} \log\left(1 + \frac{ka}{d_0}\right) \int_0^V V' dV' = \frac{\epsilon_0 b V^2}{2k} \log\left(1 + \frac{ka}{d_0}\right).$$

This is equal to $CV^2/2$ with the capacitance C obtained in Exercise 3.10.

11.23. When current I is applied for t seconds, the electric charges stored in the capacitor are $\pm Q' = \pm It$. Hence, the electric charge density on the surface of the inner electrode ($R = a$) is $\sigma = Q'/2\pi a l$ and the electric field directed radially from the inner electrode is

$$E(R) = \frac{\sigma a}{\epsilon_1 R}; \quad a < R < b,$$

$$= \frac{\sigma a}{\epsilon_2 R}; \quad b < R < c.$$

The current is I at the edge at which the current is applied but decreases to zero at the other edge. Then, the resultant magnetic field is $H(R) = I/2\pi R$ and the Poynting vector is given by

$$S_P(R) = \frac{I^2 t}{4\pi^2 \epsilon_1 l R^2}; \quad a < R < b,$$

$$= \frac{I^2 t}{4\pi^2 \epsilon_2 l R^2}; \quad b < R < c$$

and is directed inward at the edge at which the current is applied. Hence, the power that enters into the capacitor is

$$P = \int_a^c S_P(R) 2\pi R dR = \frac{I^2 t}{2\pi l} \left(\frac{1}{\epsilon_1} \log \frac{b}{a} + \frac{1}{\epsilon_2} \log \frac{c}{b} \right).$$

Integrating this until $t = T$, the final energy put into the capacitor is

$$U_e = \int_0^T P dt = \frac{Q^2}{4\pi l} \left(\frac{1}{\epsilon_1} \log \frac{b}{a} + \frac{1}{\epsilon_2} \log \frac{c}{b} \right),$$

where $Q = IT$ is the electric charge in the final state. This agrees with the energy obtained in Example 4.8.

11.24. When current I is applied for t seconds, the electric charges stored in the capacitor are $\pm Q' = \pm It$. We denote by σ_1 and σ_2 the electric charge densities on the surface of the inner electrode ($R = a$) facing dielectric materials 1 and 2. Then, the electric field at radius R is $E(R) = \sigma_1 a / \epsilon_1 R$ in dielectric material 1 and $E(R) = \sigma_2 a / \epsilon_2 R$ in dielectric material 2. Hence, the condition,

$$\frac{\sigma_1}{\epsilon_1} = \frac{\sigma_2}{\epsilon_2},$$

must be satisfied. Since the total electric charge is $Q' = \pi a l (\sigma_1 + \sigma_2)$, we have

$$\sigma_1 = \frac{\epsilon_1 Q'}{\pi a l (\epsilon_1 + \epsilon_2)}, \quad \sigma_2 = \frac{\epsilon_2 Q'}{\pi a l (\epsilon_1 + \epsilon_2)}.$$

Then, the electric field is

$$E = \frac{Q'}{\pi l (\epsilon_1 + \epsilon_2) R}.$$

The current is I at the edge at which the current is applied but decreases to zero at the other edge. Then, the resultant magnetic field is $H(R) = I/2\pi R$ and the Poynting vector is $S_P(R) = EH = I^2 t / 2\pi^2 l (\epsilon_1 + \epsilon_2) R^2$ directed inward at the edge at which the current is applied. Hence, the power that enters into the capacitor is

11.5 Exercises

$$P = \int_a^b S_P(R) 2\pi R dR = \frac{I^2 t}{\pi l(\epsilon_1 + \epsilon_2)} \log \frac{b}{a}.$$

Integrating this until $t = T$, the final energy put into the capacitor is

$$U_e = \int_0^T P dt = \frac{Q^2}{2\pi l(\epsilon_1 + \epsilon_2)} \log \frac{b}{a},$$

where $Q = IT$ is the electric charge in the final state. This energy is given by $Q^2/2C'l$ with the capacitance in a unit length C' obtained in Exercise 4.2.

11.25. When current I is applied for t seconds, the electric charges stored in the capacitor are $\pm Q' = \pm It$. Hence, the electric charge density on the surface of the inner electrode ($R = a$) is $\sigma = Q'/2\pi a l$ and the electric field directed radially from the inner electrode is $E(R) = \sigma a/\epsilon(1 + kR)R = It/2\pi\epsilon l(1 + kR)R$ at radius R. The current is I at the edge at which the current is applied but decreases to zero at the other edge. Then, the resultant magnetic field is $H(R) = I/2\pi R$ and the Poynting vector is $S_P(R) = EH = I^2t/4\pi^2\epsilon l(1 + kR)R^2$ directed inward at the edge at which the current is applied. Hence, the power that enters into the capacitor is

$$P = \int_a^b S_P(R) 2\pi R dR = \frac{I^2 t}{2\pi\epsilon l} \log \frac{b(1 + ka)}{a(1 + kb)}.$$

Integrating this until $t = T$, the final energy put into the capacitor is

$$U_e = \int_0^T P dt = \frac{Q^2}{4\pi\epsilon l} \log \frac{b(1 + ka)}{a(1 + kb)},$$

where $Q = IT$ is the electric charge in the final state. This energy is given by $Q^2/2C'l$ with the capacitance in a unit length C' obtained in Exercise 4.8.

11.26. Assume that the applied current is I'. Then, the magnetic field is uniform and is $H = I'/a$. The induced electric field due to the penetration of the magnetic flux from both plates is directed along the current. Since the induced electromotive force is

$$V_{em} = -\int_0^l \mu_0 \frac{dH}{dt} d(y) dy = -\mu_0 d_0 l \left(1 + \frac{kl}{2}\right) \frac{dH}{dt},$$

the electric field strength on the upper plate is

$$E = -\frac{V_{em}}{2l} = \frac{\mu_0 d_0}{2}\left(1 + \frac{kl}{2}\right)\frac{dH}{dt}$$

along the y-axis. The Poynting vector is directed inward the space from the two plates, and the power into the space is

$$P = 2alEH = \frac{\mu_0 d_0 l}{a}\left(1 + \frac{kl}{2}\right)I'\frac{dI'}{dt}.$$

Hence, the stored energy when the current reaches I is determined to be

$$U_m = \int P dt = \frac{\mu_0 d_0 l}{a}\left(1 + \frac{kl}{2}\right)\int_0^I I' dI' = \frac{\mu_0 d_0 l}{2a}\left(1 + \frac{kl}{2}\right)I^2.$$

11.27. We define the x- and z-axes, as shown in Fig. B11.2. Assume that the applied current is I'. The magnetic field H in the space between the plates is the same as the case of conducting plates and is given by $H = I'/a$. The magnetic flux can penetrate into the space only from the two edges, and hence, the induced electric field is different between the two edges. The variation rate of the magnetic flux along the z-axis in the space between the two plates is

$$\frac{d\Phi}{dt} = \mu_0 d_0 l\left(1 + \frac{kl}{2}\right)\frac{dH}{dt}.$$

This induces the electric field at the two edges. The electric fields at $y = 0$ and at $y = l$ are in the negative and positive directions along the x-axis, respectively, and we have

$$-\int E(y=0)dx + \int E(y=l)dx \equiv \oint \mathbf{E} \cdot d\mathbf{s} = \mu_0 d_0 l\left(1 + \frac{kl}{2}\right)\frac{dH}{dt}.$$

The power that enters into the space is given by

Fig. B11.2 Cross-section of superconducting transmission line with a tilted plate

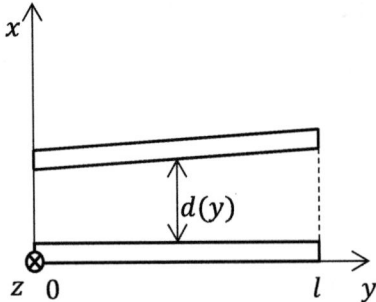

$$P = aH \oint \mathbf{E} \cdot d\mathbf{s} = \frac{\mu_0 d_0 l}{a}\left(1 + \frac{kl}{2}\right) I' \frac{dI'}{dt}.$$

Hence, the stored energy when the current reaches I is determined to be

$$U_m = \int P dt = \frac{\mu_0 d_0 l}{a}\left(1 + \frac{kl}{2}\right) \int_0^I I' dI' = \frac{\mu_0 d_0 l}{2a}\left(1 + \frac{kl}{2}\right) I^2,$$

which is the same as in the case of transmission line made of conductors.

11.28. When the applied current is I', the azimuthal magnetic flux density in the transmission line is

$$B_\varphi(R) = \frac{\mu_1 I'}{2\pi R}; \quad a < R < b,$$
$$= \frac{\mu_2 I'}{2\pi R}; \quad b < R < c.$$

In this case, the magnetic flux does not go through the outer superconductor but penetrates from both ends (along the z-axis). So, the electric field is different from the case of normal conducting transmission line. That is, the electric field is radial because of the penetration of the magnetic flux from the two ends, and is given by

$$E_R(z) = -\int \frac{\partial B_\varphi}{\partial t} dz = -\frac{\mu_1}{2\pi R} \cdot \frac{\partial I'}{\partial t}\left(z - \frac{l}{2}\right); \quad a < R < b,$$
$$= -\frac{\mu_2}{2\pi R} \cdot \frac{\partial I'}{\partial t}\left(z - \frac{l}{2}\right); \quad b < R < c.$$

Thus, the Poynting vector on the surface at $z = 0$ is

$$S_P = \frac{\mu_1 l}{8\pi^2 R^2} I' \frac{\partial I'}{\partial t}; \quad a < R < b,$$
$$= \frac{\mu_2 l}{8\pi^2 R^2} I' \frac{\partial I'}{\partial t}; \quad b < R < c,$$

and directed inward. There is also a contribution from the surface at $z = l$. Hence, the power that enters into the transmission line is

$$P = 2\left(\int_a^b \frac{\mu_1 l}{8\pi^2 R^2} 2\pi R dR + \int_b^c \frac{\mu_2 l}{8\pi^2 R^2} 2\pi R dR\right) I' \frac{\partial I'}{\partial t}$$
$$= \frac{l}{2\pi}\left(\mu_1 \log \frac{b}{a} + \mu_2 \log \frac{c}{b}\right) I' \frac{\partial I'}{\partial t}.$$

The energy stored in the transmission line is

$$U_m = \frac{l}{2\pi}\left(\mu_1 \log\frac{b}{a} + \mu_2 \log\frac{c}{b}\right)\int I'\frac{\partial I'}{\partial t}dt = \frac{l}{4\pi}\left(\mu_1 \log\frac{b}{a} + \mu_2 \log\frac{c}{b}\right)I'^2,$$

which agrees with result obtained in Example 9.3.

11.29. We apply currents $\pm I'$. We denote by τ_1 and τ_2 the current densities on the surface of the inner superconductor in the regions facing magnetic materials 1 and 2, respectively. Then, the azimuthal magnetic flux densities in the magnetic materials 1 and 2 are

$$B_{\varphi 1} = \frac{\mu_1 a \tau_1}{R}, \quad B_{\varphi 2} = \frac{\mu_2 a \tau_2}{R}.$$

So, we have $\mu_1 \tau_1 = \mu_2 \tau_2$. Since the total current is $I' = \pi a(\tau_1 + \tau_2)$, the magnetic flux density is determined to be

$$B_\varphi = \frac{\mu_1 \mu_2 I'}{\pi(\mu_1 + \mu_2)R}.$$

The magnetic flux does not go through the outer superconductor but penetrates from both ends (along the z-axis). So, the electric field is different from the case of normal conducting transmission line. That is, the electric field is radial because of the penetration of the magnetic flux from the two ends, and is given by

$$E_R(z) = -\int \frac{\partial B_\varphi}{\partial t}dz = -\frac{\mu_1 \mu_2}{\pi(\mu_1 + \mu_2)R} \cdot \frac{\partial I'}{\partial t}\left(z - \frac{l}{2}\right),$$

which satisfies an antisymmetric condition with respect to $z = l/2$. Thus, the Poynting vector in magnetic materials 1 and 2 on the surface at $z = 0$ are

$$S_{P1} = \frac{\mu_1 \mu_2^2 l}{2\pi^2(\mu_1 + \mu_2)^2 R^2}I'\frac{\partial I'}{\partial t}, \quad S_{P2} = \frac{\mu_1^2 \mu_2 l}{2\pi^2(\mu_1 + \mu_2)^2 R^2}I'\frac{\partial I'}{\partial t}$$

and is directed inward. There is also the same contribution from the surface at $z = l$. Hence, the power that enters into the transmission line is

$$P = 2\int_a^b (S_{P1} + S_{P2})\pi R dR = \frac{\mu_1 \mu_2 l}{\pi(\mu_1 + \mu_2)}\log\frac{b}{a}I'\frac{\partial I'}{\partial t}.$$

The energy stored in the transmission line is

$$U_m = \frac{\mu_1 \mu_2 l}{\pi(\mu_1 + \mu_2)}\log\frac{b}{a}\int I'\frac{\partial I'}{\partial t}dt = \frac{\mu_1 \mu_2 l I^2}{2\pi(\mu_1 + \mu_2)}\log\frac{b}{a},$$

11.5 Exercises

which agrees with the result obtained in Example 9.8.

11.30. When the applied current is I', the azimuthal magnetic flux density in the transmission line is

$$B_\varphi(R) = \frac{\mu(1+kR)I'}{2\pi R}$$

for $a < R < b$. In this case, the magnetic flux does not go through the outer superconductor but penetrates from both ends (along the z-axis). So, the electric field is different from the case of normal conducting transmission line. That is, the electric field is radial because of the magnetic flux penetration from the two ends, and is given by

$$E_R(z) = -\int \frac{\mu(1+kR)}{2\pi R} \cdot \frac{\partial I'}{\partial t} dz = -\frac{\mu(1+kR)}{2\pi R} \cdot \frac{\partial I'}{\partial t}\left(z - \frac{l}{2}\right),$$

which satisfies an antisymmetric condition with respect to $z = l/2$. Thus, the Poynting vector on the surface at $z = 0$ is

$$S_P = \frac{\mu l(1+kR)}{8\pi^2 R^2} I' \frac{\partial I'}{\partial t}$$

and directed inward. There is also a contribution from the surface at $z = l$. Hence, the power that enters into the transmission line is

$$P = 2\int_a^b \frac{\mu l(1+kR)}{8\pi^2 R^2} I' \frac{\partial I'}{\partial t} 2\pi R dR = \frac{\mu l}{2\pi}\left[\log\frac{b}{a} + k(b-a)\right] I' \frac{\partial I'}{\partial t}.$$

The energy stored in the transmission line is

$$U_m = \frac{\mu l}{2\pi}\left[\log\frac{b}{a} + k(b-a)\right]\int I' \frac{\partial I'}{\partial t} dt = \frac{\mu l}{4\pi}\left[\log\frac{b}{a} + k(b-a)\right] I^2,$$

which is written as $lL'I^2/2$ with the inductance in a unit length L' obtained in Exercise 9.5.

11.31. The energy can be simply regarded to flow through the winding wall of the inner coil of radius a, although we have to calculate as in Example 11.4 in a strict sense. We denote the voltage of the inner coil by V' when the current is I'. At this moment, the magnetic flux density in the inner coil is $B_0 + \mu_0 NI'/h$ and the electric field is $E' = V'/2\pi aN = -(\mu_0 aN/2h)(\partial I'/\partial t)$. Hence, the Poynting vector directed inward is

$$S_P = \frac{aN}{2h}\left(B_0 + \frac{\mu_0 N}{h}I'\right)\frac{\partial I'}{\partial t}.$$

The energy that penetrates into the inner coil until the current I' reaches I is

$$U_m = 2\pi ah \int_0^I S_P dt = \pi a^2 N \int_0^I \left(B_0 + \frac{\mu_0 N}{h} I'\right) dI' = \frac{\pi a^2 h}{\mu_0}\left(B_0 B + \frac{1}{2}B^2\right).$$

Since the magnetic energy inside the inner coil before and after the application of current to the inner coil is $\pi a^2 h B_0^2/2\mu_0$ and $\pi a^2 h (B_0 + B)^2/2\mu_0$, respectively. So, the energy penetrated to the inner coil agrees with the variation in the magnetic energy.

11.32. Application of the current to the inner coil induces an electromotive force in the left circuit for the outer coil to reduce the current of the outer coil, which is given by

$$-N_0 \frac{\partial}{\partial t}\left(\pi a^2 B'\right).$$

So, the power source for the outer coil supplies additional voltage to keep the current, and the resultant additional power is

$$P = I_0 N_0 \frac{\partial}{\partial t}\left(\pi a^2 B'\right),$$

where $I_0 = (h/\mu_0 N_0) B_0$ is the current of the outer coil. So, the energy supplied by this power source is

$$U_{m0} = \int P dt = \frac{\pi a^2 h}{\mu_0} B_0 \int_0^B dB' = \frac{\pi a^2 h}{\mu_0} B_0 B.$$

Thus, the sum of the energies supplied by the two power sources is equal to the increase in the magnetic energy, estimated in Exercise 11.31.

11.33. In the initial condition, the magnetic energy of the inner solenoid coil is

$$U_{mi}(0) = \frac{\pi a^2 h}{2\mu_0} B^2.$$

First, we determine the energy flow into the outer coil. When the value of the current in the outer coil is I_0', the magnetic flux density in the outer coil is $B_0' = \mu_0 N_0 I_0'/h$, and the induced electric field on the coil surface is $E_0' = -[\mu_0 N_0 (b^2 - a^2)/2bh](\partial I_0'/\partial t)$. The Poynting vector is directed inward and is given by

$$S_P = \frac{\mu_0 N_0^2 (b^2 - a^2) I_0'}{2bh^2} \cdot \frac{\partial I_0'}{\partial t}.$$

11.5 Exercises

Hence, the energy supplied to the outer coil is

$$U_{mo} = 2\pi bh \int S_p dt = \frac{\pi \mu_0 N_0^2 (b^2 - a^2)}{h} \int_0^{I_0} I_0' dI_0' = \frac{\pi \mu_0 N_0^2 (b^2 - a^2)}{2h} I_0^2$$

$$= \frac{\pi h}{2\mu_0}(b^2 - a^2) B_0^2.$$

Second, we determine the energy flow into the inner coil. When the value of the current in the outer coil is I_0', the magnetic flux density in the inner coil is $B' = B + \mu_0 N_0 I_0'/h$, and the induced electric field on the coil surface is $E_0' = -(\mu_0 N_0 a/2h)(\partial I_0'/\partial t)$. Hence, the energy supplied to the inner coil is

$$U_{mi} = \pi a^2 N_0 \int_0^{I_0} \left(B + \frac{\mu_0 N_0 I_0'}{h} \right) dI_0' = \frac{\pi h}{2\mu_0} a^2 (2BB_0 + B_0^2).$$

Thus, the final energy in the magnet system is

$$U_m = U_{mi}(0) + U_{mo} + U_{mi} = \frac{\pi h}{2\mu_0}\left[(b^2 - a^2)B_0^2 + a^2(B_0 + B)^2\right].$$

This agrees with the energy given by Eq. (8.20).

11.34. The electric field and magnetic field are $E = (V_1 - V_2)/t$ and $H = I_1/w$ in the upper space and are $E = V_2/d$ and $H = (I_1 + I_2)/w$ in the lower space, respectively. Hence, the magnitude of the Poynting vector into the transmission line is $S_{P1} = EH = (V_1 - V_2)I_1/tw$ and $S_{P2} = V_2(I_1 + I_2)/dw$ in the upper and lower spaces, respectively. Thus, the expected total power from the two power sources is

$$P = S_{P1} tw + S_{P2} dw = V_1 I_1 + V_2 I_2.$$

Thus, the Poynting vector gives the energy flow correctly.

Chapter 12
Electromagnetic Waves

12.1 Planar Electromagnetic Waves

Electromagnetic fields in a dielectric material follow the wave equation, Eq. (11.21), since there is no electric charge in the material. This chapter covers the properties of electromagnetic fields described by this equation. For simplicity, we focus only on the electric field and assume that it has only a y-component varying only along the x-axis. Thus, Eq. (11.21) reduces to

$$\frac{\partial^2 E_y}{\partial x^2} - \mu\epsilon \frac{\partial^2 E_y}{\partial t^2} = 0. \tag{12.1}$$

We assume that E_y varies with time as $e^{i\omega t}$, with ω denoting the angular frequency. Equation (12.1) then leads to

$$\frac{\partial^2 E_y}{\partial x^2} + \mu\epsilon\omega^2 E_y = 0. \tag{12.2}$$

This equation is easily solved as

$$E_y = E_1 e^{i(\omega t + kx)} + E_2 e^{i(\omega t - kx)}, \tag{12.3}$$

where E_1 and E_2 are constants determined by initial and boundary conditions, and

$$k = (\mu\epsilon)^{1/2}\omega \equiv \frac{\omega}{c} \tag{12.4}$$

is the wave number. As will be shown later,

$$c = \left(\frac{1}{\mu\epsilon}\right)^{1/2} \tag{12.5}$$

is the speed of electromagnetic waves, or the light speed, in the dielectric material. The wavelength is given by

$$\lambda = \frac{2\pi}{k}. \tag{12.6}$$

In the first term of Eq. (12.3),

$$\omega t + kx = \text{const.} \tag{12.7}$$

gives the position at which the phase of the wave is constant. Hence,

$$\frac{dx}{dt} = -\frac{\omega}{k} = -c \tag{12.8}$$

shows that the first term in Eq. (12.3) represents a wave that propagates with the velocity c along the negative x-axis. Namely, this wave is an electromagnetic wave. Similarly, the second term in Eq. (12.3) gives the electromagnetic wave propagating along the positive x-axis. Such an electromagnetic wave, which has its phase constant over a plane perpendicular to the direction of wave travel, as in Eq. (12.3), is generally called a plane wave.

From Eq. (11.7), the magnetic flux density is given by

$$B_z = -B_1 e^{i(\omega t + kx)} + B_2 e^{i(\omega t - kx)} = -\frac{1}{c} E_1 e^{i(\omega t + kx)} + \frac{1}{c} E_2 e^{i(\omega t - kx)}. \tag{12.9}$$

We can also easily show that, if the electric field has only a z-component, the magnetic flux density has only a y-component, which is similar to the above case. Hence, the planar electromagnetic wave is a transverse wave in which the electric field and magnetic flux density are perpendicular to each other and directed perpendicularly to the propagation direction. The ratio of these amplitudes is

$$\frac{E_1}{B_1} = \frac{E_2}{B_2} = c. \tag{12.10}$$

The planar electromagnetic wave is generally expressed in the form

$$\exp[i(\omega t - \mathbf{k} \cdot \mathbf{r})], \tag{12.11}$$

where $\mathbf{k}$ is the wave number vector and $\mathbf{r}$ is the position vector. $\mathbf{i}_k = \mathbf{k}/|\mathbf{k}|$ is a unit vector along the propagation direction.

In vacuum the speed of the electromagnetic wave is

$$c_0 = \frac{1}{(\epsilon_0 \mu_0)^{1/2}} = 2.997925 \times 10^8 \mathrm{m/s}. \tag{12.12}$$

Equations (12.3) and (12.9) deal with the case where the electric field and magnetic flux density each remain in its own direction perpendicular to the other. Such a direction is called the direction of polarization. A polarization that has a fixed direction is linear polarization. In general, a superposition of components is possible. We assume two components of the electric field:

$$E_y = E_1 \cos(\omega t - kx), \quad E_z = E_2 \cos(\omega t - kx + \delta). \tag{12.13}$$

The corresponding components of the magnetic flux density are

$$B_y = -\frac{E_2}{c} \cos(\omega t - kx + \delta) = -\frac{E_z}{c}, \quad B_z = \frac{E_1}{c} \cos(\omega t - kx) = \frac{E_y}{c}. \tag{12.14}$$

Usually, the direction of polarization is designated as that of the electric field, E. Eliminating t in the above equations, we obtain the relationship between E_y and E_z as

$$\left(\frac{E_y}{E_1}\right)^2 - 2\cos\delta \frac{E_y E_z}{E_1 E_1} + \left(\frac{E_z}{E_2}\right)^2 = \sin^2 \delta. \tag{12.15}$$

Such a phenomenon where the directions of E and B in the electromagnetic wave are not uniform but biased is referred to as the polarization of a wave, and such a wave is called a polarized wave.

12.2 Reflection and Refraction of the Planar Electromagnetic Wave

Here, we treat the reflection and refraction of a planar electromagnetic wave. Suppose a planar interface $z = 0$ between media 1 and 2 with dielectric constants ϵ_1 and ϵ_2 and magnetic permeabilities μ_1 and μ_2, respectively, as shown in Fig. 12.1a, and a planar electromagnetic wave propagates from medium 1 to the interface. The plane formed by the propagation direction and the direction (z-axis) normal to the interface is called the plane of incidence. We define the x-axis as the line at which the plane of incidence and the interface meet and the y-axis on the interface in such a way that it is normal to both the x- and z-axes.

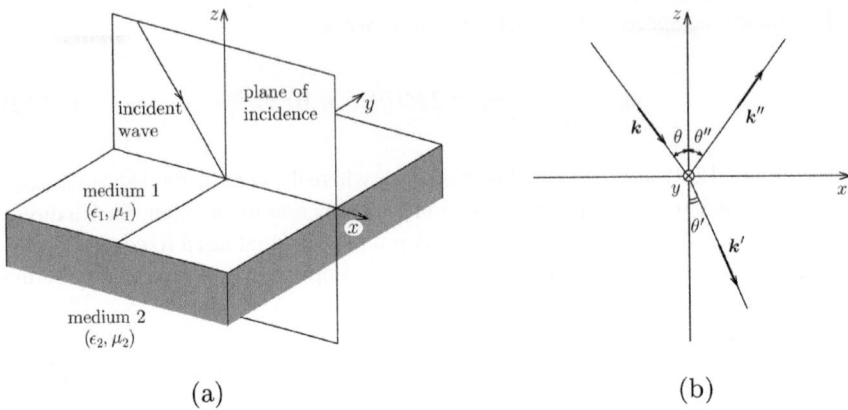

Fig. 12.1 Definitions of **a** axes on the interface and **b** angles of waves

In this case, the incident and reflected waves are in medium 1, and the transmitted wave is in medium 2. We denote by $\boldsymbol{k}$, $\boldsymbol{k}''$, and $\boldsymbol{k}'$ the wavenumber vectors of the incident, reflected, and transmitted waves, respectively. Each wave propagates in the direction of its wavenumber vector, and these wavenumber vectors lie on the plane of incident (x-z plane). The angles of the wavenumber vectors, $\boldsymbol{k}$, $\boldsymbol{k}''$, and $\boldsymbol{k}'$, from the z-axis are denoted by θ, θ'', and θ', respectively, as shown in Fig. 12.1b. The variation of each wave with time is the same, and the phase is given by ωt. The incident, reflected, and transmitted waves are expressed as

$$\exp[i(\omega t - \boldsymbol{k} \cdot \boldsymbol{r})], \quad \exp[i(\omega t - \boldsymbol{k}'' \cdot \boldsymbol{r})], \quad \exp[i(\omega t - \boldsymbol{k}' \cdot \boldsymbol{r})]. \tag{12.16}$$

The electric field strength $\boldsymbol{E}$ and the magnetic flux density $\boldsymbol{B}$ are perpendicular to each other, and both lie on a plane normal to the propagation direction. Here, we treat the boundary conditions as satisfied for $\boldsymbol{E}$ and $\boldsymbol{B}$. These are given by Eqs. (4.12), (4.13), (9.20), and (9.21). When there is no electric charge and no current on the interface, the parallel components of the electric field strength and magnetic field strength are continuous, and the normal components of the magnetic flux density and electric flux density are continuous across the interface. That is,

$$\boldsymbol{n} \times (\boldsymbol{E}_1 - \boldsymbol{E}_2) = 0, \tag{12.17}$$

$$\boldsymbol{n} \cdot (\epsilon_1 \boldsymbol{E}_1 - \epsilon_2 \boldsymbol{E}_2) = 0, \tag{12.18}$$

$$\boldsymbol{n} \times \left(\frac{\boldsymbol{B}_1}{\mu_1} - \frac{\boldsymbol{B}_2}{\mu_2} \right) = 0, \tag{12.19}$$

$$\boldsymbol{n} \cdot (\boldsymbol{B}_1 - \boldsymbol{B}_2) = 0, \tag{12.20}$$

12.2 Reflection and Refraction of the Planar Electromagnetic Wave

where $\boldsymbol{n}$ is the unit vector normal to the interface, and the subscripts 1 and 2 represent the quantities in media 1 and 2, respectively.

Considering the orthogonality between the electric field strength and the magnetic flux density, the incident wave is given by

$$\boldsymbol{E} = \boldsymbol{E}_0 \exp[i(\omega t - \boldsymbol{k} \cdot \boldsymbol{r})], \tag{12.21a}$$

$$\boldsymbol{B} = \frac{\boldsymbol{k}}{k} \times \frac{\boldsymbol{E}_0}{c_1} \exp[i(\omega t - \boldsymbol{k} \cdot \boldsymbol{r})], \tag{12.21b}$$

where $k = |\boldsymbol{k}|$ and c_1 is the light speed in medium 1. The magnetic flux density $\boldsymbol{B}$ is generally normal to both $\boldsymbol{E}$ and $\boldsymbol{k}$, and its magnitude is equal to the magnitude of $\boldsymbol{E}$ divided by the corresponding light speed. The reflected wave is similarly given by

$$\boldsymbol{E}'' = \boldsymbol{E}_0'' \exp[i(\omega t - \boldsymbol{k}'' \cdot \boldsymbol{r})], \tag{12.22a}$$

$$\boldsymbol{B}'' = \frac{\boldsymbol{k}''}{k''} \times \frac{\boldsymbol{E}_0''}{c_1} \exp[i(\omega t - \boldsymbol{k}'' \cdot \boldsymbol{r})], \tag{12.22b}$$

and the transmitted wave is given by

$$\boldsymbol{E}' = \boldsymbol{E}_0' \exp[i(\omega t - \boldsymbol{k}' \cdot \boldsymbol{r})], \tag{12.23a}$$

$$\boldsymbol{B}' = \frac{\boldsymbol{k}'}{k'} \times \frac{\boldsymbol{E}_0'}{c_2} \exp[i(\omega t - \boldsymbol{k}' \cdot \boldsymbol{r})]. \tag{12.23b}$$

In the above, $k'' = |\boldsymbol{k}''|$, $k' = |\boldsymbol{k}'|$, and c_2 is the light speed in medium 2. Thus, the electric field strength and magnetic flux density in medium 1 are

$$\boldsymbol{E}_1 = \boldsymbol{E} + \boldsymbol{E}'', \quad \boldsymbol{B}_1 = \boldsymbol{B} + \boldsymbol{B}'', \tag{12.24}$$

and those in medium 2 are

$$\boldsymbol{E}_2 = \boldsymbol{E}', \quad \boldsymbol{B}_2 = \boldsymbol{B}'. \tag{12.25}$$

To satisfy all the boundary conditions (12.17)–(12.20) at the interface ($z = 0$) at any time, the phase must be the same for the three waves. This condition is given by

$$\boldsymbol{k} \cdot \boldsymbol{r}|_{z=0} = \boldsymbol{k}'' \cdot \boldsymbol{r}|_{z=0} = \boldsymbol{k}' \cdot \boldsymbol{r}|_{z=0}. \tag{12.26}$$

Equation (12.26) is expressed as

$$\mathbf{k} \cdot \mathbf{r}_0 = \mathbf{k}'' \cdot \mathbf{r}_0 = \mathbf{k}' \cdot \mathbf{r}_0 \qquad (12.27)$$

in terms of an arbitrary position vector $\mathbf{r}_0$ on the interface. If $\mathbf{r}_0$ is given by

$$\mathbf{r}_0 = x\mathbf{i}_x + y\mathbf{i}_y, \qquad (12.28)$$

we have

$$k \sin \theta = k'' \sin \theta'' = k' \sin \theta', \qquad (12.29)$$

since the wavenumber vectors are normal to the y-axis. The speed is the same for the incident and reflected waves in the same medium. Thus, the wavenumbers of these waves are the same ($k = k''$), and we have

$$\theta = \theta''. \qquad (12.30)$$

That is, the incident and reflection angles are the same. This is the law of reflection. We also have the relationship between the incident and transmission angles as

$$\frac{\sin \theta}{\sin \theta'} = \frac{k'}{k} = \frac{c_1}{c_2} = \left(\frac{\epsilon_2 \mu_2}{\epsilon_1 \mu_1} \right)^{1/2}. \qquad (12.31)$$

This is called Snell's law for refraction.

Using the above results, Eqs. (12.17)–(12.20) are rewritten as

$$\mathbf{n} \times (\mathbf{E}_0 + \mathbf{E}_0'' - \mathbf{E}_0') = 0, \qquad (12.32)$$

$$\mathbf{n} \cdot [\epsilon_1 (\mathbf{E}_0 + \mathbf{E}_0'') - \epsilon_2 \mathbf{E}_0'] = 0, \qquad (12.33)$$

$$\mathbf{n} \times \left[\frac{1}{\mu_1} \left(\frac{\mathbf{k} \times \mathbf{E}_0}{kc_1} + \frac{\mathbf{k}'' \times \mathbf{E}_0''}{k''c_1} \right) - \frac{1}{\mu_2} \cdot \frac{\mathbf{k}' \times \mathbf{E}_0'}{k'c_2} \right] = 0, \qquad (12.34)$$

$$\mathbf{n} \cdot \left(\frac{\mathbf{k} \times \mathbf{E}_0}{kc_1} + \frac{\mathbf{k}'' \times \mathbf{E}_0''}{k''c_1} - \frac{\mathbf{k}' \times \mathbf{E}_0'}{k'c_2} \right) = 0. \qquad (12.35)$$

Although the electric field of the incident wave can be directed in various directions, we focus for simplicity on the case where the electric field is directed parallel to the y-axis (i.e., normal to the plane of incidence) as shown in Fig. 12.2. In this case, Eq. (12.32) reduces to

$$E_0 + E_0'' - E_0' = 0. \qquad (12.36)$$

12.2 Reflection and Refraction of the Planar Electromagnetic Wave

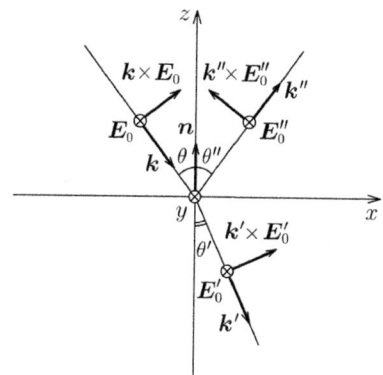

Fig. 12.2 Case where the electric field of the incident wave is normal to the plane of incidence, i.e., parallel to the y-axis

Since the electric field is perpendicular to the normal vector, n, Eq. (12.33) is already satisfied. Equation (12.34) becomes

$$\left(\frac{\epsilon_1}{\mu_1}\right)^{1/2}(E_0 - E_0'')\cos\theta - \left(\frac{\epsilon_2}{\mu_2}\right)^{1/2}E_0'\cos\theta' = 0. \qquad (12.37)$$

Equation (12.35) is written as

$$\frac{E_0}{c_1}\sin\theta + \frac{E_0''}{c_1}\sin\theta - \frac{E_0'}{c_2}\sin\theta\prime = 0, \qquad (12.38)$$

which is found to reduce to Eq. (12.36) using Eq. (12.31). Thus, from Eqs. (12.36) and (12.37), we obtain the amplitudes of the electric fields of the refracted and reflected waves as

$$E_0' = \frac{2(\epsilon_1/\mu_1)^{1/2}\cos\theta}{(\epsilon_1/\mu_1)^{1/2}\cos\theta + (\epsilon_2/\mu_2)^{1/2}\cos\theta'}E_0, \qquad (12.39a)$$

$$E_0'' = \frac{(\epsilon_1/\mu_1)^{1/2}\cos\theta - (\epsilon_2/\mu_2)^{1/2}\cos\theta'}{(\epsilon_1/\mu_1)^{1/2}\cos\theta + (\epsilon_2/\mu_2)^{1/2}\cos\theta'}E_0. \qquad (12.39b)$$

The amplitudes of the magnetic flux densities are

$$B_0' = \frac{E_0'}{c_2} = (\epsilon_2\mu_2)^{1/2}E_0', \qquad (12.40a)$$

$$B_0'' = \frac{E_0''}{c_1} = (\epsilon_1\mu_1)^{1/2}E_0''. \qquad (12.40b)$$

Example 12.1 Solve Eqs. (12.32)–(12.35) when the electric field of the incident wave is parallel to the plane of incidence, i.e., the magnetic flux density is parallel to the y-axis.

Solution 12.1 Under this condition, $(k \times E_0)$, $(k' \times E'_0)$, and $(k'' \times E''_0)$ are directed along the y-axis, as shown in Fig. 12.3. Hence, Eqs. (12.32) and (12.33) become

$$\left(E_0 - E''_0\right) \cos\theta - E'_0 \cos\theta' = 0,$$

$$\epsilon_1 \left(E_0 - E''_0\right) \sin\theta - \epsilon_2 E'_0 \sin\theta' = 0,$$

respectively. The latter equation is rewritten as

$$\left(\frac{\epsilon_1}{\mu_1}\right)^{1/2} (E_0 - E''_0) - \left(\frac{\epsilon_2}{\mu_2}\right)^{1/2} E'_0 = 0.$$

In Eq. (12.34), $n \times (k \times E)$ is parallel to the x-axis, and this equation agrees with the above result from Eq. (12.33). The condition given by Eq. (12.35) is satisfied. From the above two equations, we obtain the electric fields of the refracted and reflected waves as

$$E'_0 = \frac{2(\epsilon_1/\mu_1)^{1/2} \cos\theta}{(\epsilon_2/\mu_2)^{1/2} \cos\theta + (\epsilon_1/\mu_1)^{1/2} \cos\theta'} E_0,$$

$$E''_0 = \frac{(\epsilon_2/\mu_2)^{1/2} \cos\theta - (\epsilon_1/\mu_1)^{1/2} \cos\theta'}{(\epsilon_2/\mu_2)^{1/2} \cos\theta + (\epsilon_1/\mu_1)^{1/2} \cos\theta'} E_0.$$

The corresponding magnetic flux densities are derived by substituting these results into Eqs. (12.40a) and (12.40b).

Fig. 12.3 Case where the electric field of the incident wave is parallel to the plane of incidence, i.e., normal to the y-axis

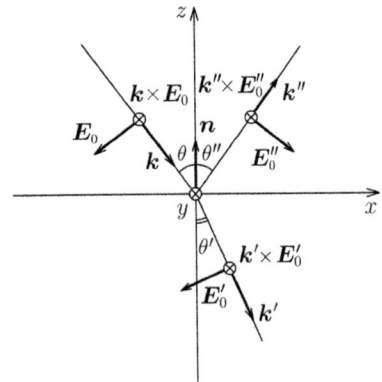

12.2 Reflection and Refraction of the Planar Electromagnetic Wave

Example 12.2 Discuss the reflection and refraction of a planar electromagnetic wave at the boundary when medium 1 is vacuum and medium 2 is a conductor. Assume that the electric field of the incident wave is normal to the plane of incidence.

Solution 12.2 Since the electric field is parallel to the conductor surface, electric charge does not appear on the surface. High frequency components of the magnetic flux density are completely shielded inside the conductor ($B_2 = 0$), and the current is consequently induced on the conductor surface. Hence, the fulfilled boundary conditions are only Eqs. (12.17) and (12.20). The corresponding Eqs. (12.32) and Eq. (12.35) are

$$\mathbf{n} \times (\mathbf{E}_0 + \mathbf{E}_0'') = 0, \quad \mathbf{n} \cdot \left(\frac{1}{k}\mathbf{k} \times \mathbf{E}_0 + \frac{1}{k''}\mathbf{k}'' \times \mathbf{E}_0''\right) = 0.$$

Using the definition in Fig. 12.2, the first equation reduces to

$$E_0 + E_0'' = 0.$$

The second equation gives also the same result. In this case, taking the real part, the electric field in the vacuum region is

$$E_y = E_0[\cos(\omega t - \mathbf{k} \cdot \mathbf{r}) - \cos(\omega t - \mathbf{k}'' \cdot \mathbf{r})]$$
$$= -2E_0 \sin(kz \cos\theta) \sin(\omega t - kx \sin\theta).$$

In the above $k = k''$ and we have used Eq. (12.30) and the following relations:

$$\mathbf{k} \cdot \mathbf{r} = kx \sin\theta - kz \cos\theta, \quad \mathbf{k}'' \cdot \mathbf{r} = kx \sin\theta + kz \cos\theta.$$

In this configuration the electric charge does not appear on the surface. The magnetic flux density is

$$B_x = \frac{E_0}{c_0} \cos(\omega t - \mathbf{k} \cdot \mathbf{r}) \cos\theta - \frac{E_0''}{c_0} \cos(\omega t - \mathbf{k}'' \cdot \mathbf{r}) \cos\theta$$
$$= \frac{2E_0}{c_0} \cos\theta \cos(kz \cos\theta) \cos(\omega t - kx \sin\theta),$$

$$B_z = \frac{E_0}{c_0} \cos(\omega t - \mathbf{k} \cdot \mathbf{r}) \sin\theta + \frac{E_0''}{c_0} \cos(\omega t - \mathbf{k}'' \cdot \mathbf{r}) \sin\theta$$
$$= -\frac{2E_0}{c_0} \sin\theta \sin(kz \cos\theta) \sin(\omega t - kx \sin\theta).$$

The surface current density is given by

$$\tau_y(x) = \frac{B_x(z=0)}{\mu_0} = 2\left(\frac{\epsilon_0}{\mu_0}\right)^{1/2} E_0 \cos\theta \cos(\omega t - kx\sin\theta).$$

So, we have $\nabla \cdot \boldsymbol{\tau} = 0$, and Eq. (5.7) is satisfied with $\sigma = 0$. ◆

12.3 Energy of the Electromagnetic Wave

Here, we discuss the energy of the planar electromagnetic wave described in Sect. 12.1. For simplicity, we treat the second terms in Eqs. (12.3) and (12.9). In this case, the electric field and magnetic flux density are given by

$$\boldsymbol{E} = E_2 \cos(\omega t - kx)\boldsymbol{i}_y, \tag{12.41}$$

$$\boldsymbol{B} = \frac{E_2}{c} \cos(\omega t - kx)\boldsymbol{i}_z. \tag{12.42}$$

Hence, the electric energy density and magnetic energy density are equal to each other and given by

$$\frac{1}{2}\epsilon E^2 = \frac{1}{2\mu}B^2 = \frac{1}{2}\epsilon E_2^2 \cos^2(\omega t - kx). \tag{12.43}$$

Since there is no current, from Eq. (11.27), the total energy density is

$$u = \epsilon E_2^2 \cos^2(\omega t - kx). \tag{12.44}$$

On the other hand, from Eq. (11.30), the Poynting vector is

$$\boldsymbol{S}_\mathrm{P} = \boldsymbol{E} \times \frac{\boldsymbol{B}}{\mu} = \left(\frac{\epsilon}{\mu}\right)^{1/2} E_2^2 \cos^2(\omega t - kx)\boldsymbol{i}_x = cu\boldsymbol{i}_x. \tag{12.45}$$

Thus, we find that the Poynting vector has a magnitude equal to the total energy density multiplied by the light speed and is directed along the x-axis, i.e., the propagation direction of the electromagnetic wave. This holds for all planar electromagnetic waves, including elliptically polarized waves. Hence, the Poynting vector expresses the energy that flows through a unit area in unit time, as defined in Sect. 11.4.

12.3 Energy of the Electromagnetic Wave

Example 12.3 Assume that the y- and z-components of the electric field of a polarized wave propagating along the positive x-axis are given by Eq. (12.13). Determine the energy density and the Poynting vector.

Solution 12.3 The electric energy density is

$$\frac{1}{2}\epsilon E^2 = \frac{1}{2}\epsilon \left(E_y^2 + E_z^2\right)$$
$$= \frac{1}{2}\epsilon \left[E_1^2 \cos^2(\omega t - kx) + E_2^2 \cos^2(\omega t - kx + \delta)\right].$$

The magnetic flux density is given by Eq. (12.14), and the magnetic energy density is

$$\frac{1}{2\mu}B^2 = \frac{1}{2\mu}\left(B_y^2 + B_z^2\right)$$
$$= \frac{1}{2}\epsilon\left[E_1^2 \cos^2(\omega t - kx) + E_2^2 \cos^2(\omega t - kx + \delta)\right].$$

Thus, the energy density of the electromagnetic wave is

$$u = \frac{1}{2}\epsilon E^2 + \frac{1}{2\mu}B^2 = \epsilon\left[E_1^2 \cos^2(\omega t - kx) + E_2^2 \cos^2(\omega - kx + \delta)\right].$$

The Poynting vector is

$$S_P = E \times \frac{B}{\mu} = \frac{1}{\mu}\left(E_y i_y + E_z i_z\right) \times \left(B_y i_y + B_z i_z\right) = \frac{1}{\mu}\left(E_y B_z - E_z B_y\right) i_x$$
$$= \frac{1}{\mu c}\left[E_1^2 \cos^2(\omega t - kx) + E_2^2 \cos^2(\omega t - kx + \delta)\right] i_x = cu i_x.$$

◆

Example 12.4 Discuss the energy flow using the Poynting vector for the reflection and refraction of a planar electromagnetic wave, as treated in Sect. 12.2. Assume that the electric field in the incident wave is normal to the place of incidence.

Solution 12.4 From Eqs. (12.21a, 12.21b) and (12.22a, 12.22b), we obtain the electric power flowing from medium 1 to medium 2 through a unit area as the incident and reflected waves as

$$-\frac{1}{\mu_1}[E(z=0) \times B(z=0)]_z = \frac{E_0^2}{\mu_1 c_1} \cos^2(\omega t - \mathbf{k} \cdot \mathbf{r}_0) \cos\theta,$$

$$-\frac{1}{\mu_1}[E''(z=0) \times B''(z=0)]_z = -\frac{E_0''^2}{\mu_1 c_1} \cos^2(\omega t - \mathbf{k}'' \cdot \mathbf{r}_0) \cos\theta'',$$

respectively. From Eqs. (12.23a, 12.23b) the electric power penetrating into medium 2 as the transmitted wave is

$$-\frac{1}{\mu_2}[E'(z=0) \times B'(z=0)]_z = \frac{E_0'^2}{\mu_2 c_2} \cos^2(\omega t - \mathbf{k}' \cdot \mathbf{r}_0) \cos\theta',$$

Because of Eq. (12.27), the factors dependent on time and space such as $\cos^2(\omega t - \mathbf{k} \cdot \mathbf{r}_0)$ are the same. Neglecting these factors, the rate of energy flow from medium 1 is

$$\frac{1}{\mu_1 c_1}\left(E_0^2 - E_0''^2\right)\cos\theta = \frac{4\alpha \cos^2\theta \cos\theta' E_0^2}{\mu_1 c_1 (\cos\theta + \alpha\cos\theta')^2},$$

where $\alpha = (\epsilon_2 \mu_1 / \epsilon_1 \mu_2)^{1/2}$ and we have used Eqs. (12.30) and (12.39b). On the other hand, Eq. (12.39a) yields the rate of energy penetration into medium 2:

$$\frac{1}{\mu_2 c_2} E_0'^2 \cos\theta' = \frac{4\cos^2\theta \cos\theta' E_0^2}{\mu_2 c_2 (\cos\theta + \alpha\cos\theta')^2}.$$

We can easily show that this is equal to the rate of energy flow from medium 1. ◆

12.4 Wave Guides

Hollow metal tubes called wave guides are used to transmit electromagnetic waves such as microwaves. The cross-section of a wave guide is usually rectangular or circular. Here, we treat a rectangular wave guide for simplicity. Assume that the wave guide is uniformly extended along the z-axis and that the internal vacuum region is $0 \leq x \leq a$ and $0 \leq y \leq b$, as shown in Fig. 12.4.

We assume that the factors for time variation and spatial variation are given by $e^{i\omega t}$ and $e^{-i\gamma z}$, respectively. Equations (11.7) and (11.8) then reduce to

$$\frac{\partial E_z}{\partial y} + i\gamma E_y = -i\omega B_x, \qquad (12.46a)$$

12.4 Wave Guides

Fig. 12.4 Rectangular wave guide

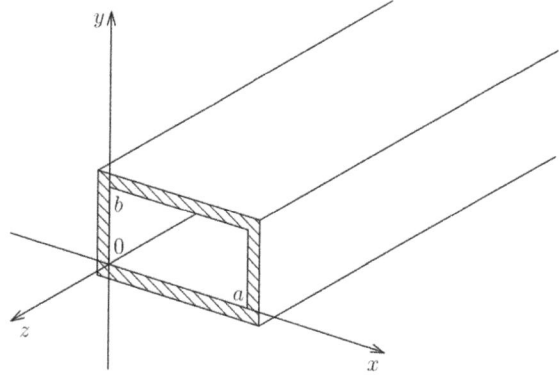

$$-i\gamma E_x - \frac{\partial E_z}{\partial x} = -i\omega B_y, \tag{12.46b}$$

$$\frac{\partial E_y}{\partial x} - \frac{\partial E_x}{\partial y} = -i\omega B_z, \tag{12.46c}$$

and

$$\frac{\partial B_z}{\partial y} + i\gamma B_y = i\frac{\omega}{c_0^2}E_x, \tag{12.47a}$$

$$-i\gamma B_x - \frac{\partial B_z}{\partial x} = i\frac{\omega}{c_0^2}E_y, \tag{12.47b}$$

$$\frac{\partial B_y}{\partial x} - \frac{\partial B_x}{\partial y} = i\frac{\omega}{c_0^2}E_z. \tag{12.47c}$$

Using these equations, the equations for the z-components, E_z and B_z, are obtained as

$$\frac{\partial^2 E_z}{\partial x^2} + \frac{\partial^2 E_z}{\partial y^2} + k^2 E_z = 0, \quad \frac{\partial^2 B_z}{\partial x^2} + \frac{\partial^2 B_z}{\partial y^2} + k^2 B_z = 0. \tag{12.48}$$

If these equations can be solved, we obtain other components from

$$E_x = -\frac{i}{k^2}\left(\gamma \frac{\partial E_z}{\partial x} + \omega \frac{\partial B_z}{\partial y}\right), \tag{12.49a}$$

$$E_y = -\frac{i}{k^2}\left(\gamma \frac{\partial E_z}{\partial y} - \omega \frac{\partial B_z}{\partial x}\right), \tag{12.49b}$$

$$B_x = \frac{i}{k^2}\left(\frac{\omega}{c_0^2}\cdot\frac{\partial E_z}{\partial y} - \gamma\frac{\partial B_z}{\partial x}\right), \tag{12.49c}$$

$$B_y = -\frac{i}{k^2}\left(\frac{\omega}{c_0^2}\cdot\frac{\partial E_z}{\partial x} + \gamma\frac{\partial B_z}{\partial y}\right), \tag{12.49d}$$

where

$$k^2 = \left(\frac{\omega}{c_0}\right)^2 - \gamma^2, \tag{12.50}$$

and $\nabla\cdot\mathbf{E}=0$ is used. Equations (12.49a)–(12.49d) hold for $k\neq 0$.

In the case of $k=0$, we have $\gamma=\pm\omega/c_0$, and we may consider that there is an electromagnetic wave propagating along the z-axis at light speed. For example, we assume an electromagnetic wave without z-components ($E_z = B_z = 0$) similar to a planar electromagnetic wave. This is called the transverse electromagnetic (TEM) wave. A TEM wave cannot exist however, in simple rectangular or circular wave guides. Such a field can exist only when the guide is composed of two or more conductors like those in Fig. 12.5, and a potential difference can appear between conductors with electric field lines extending from one conductor to another.

From the above discussion, we know that either the electric field $\mathbf{E}$ or the magnetic flux density $\mathbf{B}$ has at least one component in the propagation direction. An electromagnetic wave with a zero longitudinal component of the electric field is called a transverse electric (TE) wave, and one with a zero longitudinal component of the magnetic flux density is called a transverse magnetic (TM) wave. The general electromagnetic wave is given by a linear combination of these waves.

Here, we consider a TM wave. In this case $B_z = 0$. Since no electromagnetic wave with high frequency can penetrate the conductor, the electric field is perpendicular to and the magnetic flux density is parallel to the conductor surface. That is,

$$\begin{aligned}E_y = E_z = B_x = 0; \quad x = 0, a,\\ E_x = E_z = B_y = 0; \quad y = 0, b.\end{aligned} \tag{12.51}$$

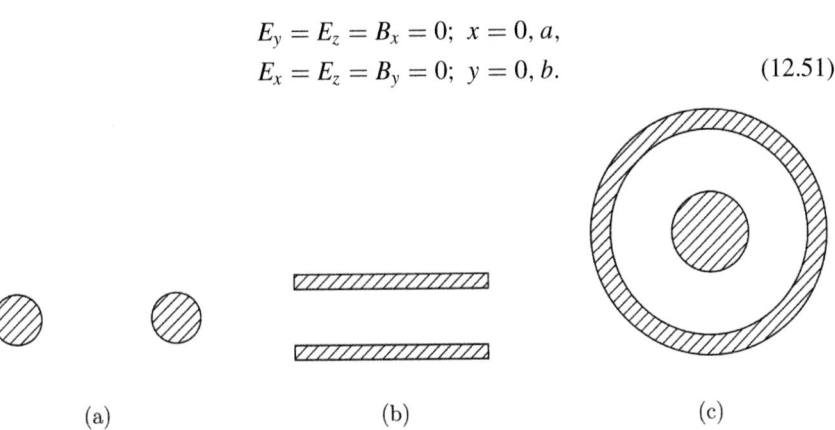

Fig. 12.5 Examples of the cross-section of a wave guide in which a TEM wave exists: **a** parallel cylindrical conductor, **b** parallel-plate conductor, and **c** coaxial conductor

12.4 Wave Guides

The general solution of Eq. (12.48) is given by

$$E_z(x, y, z, t) = K \exp[\pm i(k_x x + k_y y)] \exp[i(\omega t - \gamma z)] \tag{12.52}$$

with

$$k_x^2 + k_y^2 = k^2. \tag{12.53}$$

The dependence on x can be written as

$$E_z = K_1 e^{ik_x x} + K_2 e^{-ik_x x}. \tag{12.54}$$

From Eq. (12.51), the following conditions should be satisfied:

$$K_1 + K_2 = 0, \quad e^{i2k_x a} = 1. \tag{12.55}$$

The latter condition gives

$$k_x = \frac{m\pi}{a}; \quad m = 1, 2, \cdots. \tag{12.56}$$

The case of $m = 0$ also satisfies this condition. We have $E_z = 0$, however, which is meaningless. Thus, Eq. (12.54) reduces to

$$E_z = 2K_1' \sin\left(\frac{m\pi x}{a}\right) \tag{12.57}$$

with $K_1' = iK_1$. We similarly obtain the y-dependence with $k_y = n\pi/b (n = 1, 2, \cdots)$, and Eq. (12.52) is rewritten as

$$E_z(x, y, z, t) = A \sin\left(\frac{m\pi x}{a}\right) \sin\left(\frac{n\pi y}{b}\right) \exp[i(\omega t - \gamma z)], \tag{12.58}$$

where A is a constant, and there is a relationship between m and n written as

$$\left(\frac{m\pi}{a}\right)^2 + \left(\frac{n\pi}{b}\right)^2 = \left(\frac{\omega}{c_0}\right)^2 - \gamma^2. \tag{12.59}$$

The modes of the electromagnetic wave are different depending on the set of integers, (m, n), and each mode of the TM wave is expressed as TM_{mn}. The mode of the TE wave is represented similarly as TE_{mn}.

For the TM wave to propagate through the wave guide along the z-axis without damping, γ must be a real number and we have

$$\left(\frac{\omega}{c_0}\right)^2 - \left(\frac{m\pi}{a}\right)^2 - \left(\frac{n\pi}{b}\right)^2 \geq 0. \tag{12.60}$$

That is, the angular frequency ω should be larger than the cut-off frequency given by

$$\omega_0 = c_0 \left[\left(\frac{m\pi}{a}\right)^2 + \left(\frac{n\pi}{b}\right)^2\right]^{1/2}. \tag{12.61}$$

Example 12.5 Determine other components of electromagnetic fields of the above TM wave.

Solution 12.5 Substituting E_z in Eq. (12.58) and $B_z = 0$ into Eqs. (12.49a)–(12.49d), we have

$$E_x = -iA \frac{m\pi \gamma}{k^2 a} \cos\left(\frac{m\pi x}{a}\right) \sin\left(\frac{n\pi y}{b}\right),$$

$$E_y = -iA \frac{n\pi \gamma}{k^2 b} \sin\left(\frac{m\pi x}{a}\right) \cos\left(\frac{n\pi y}{b}\right),$$

$$B_x = iA \frac{n\pi \omega}{k^2 c_0^2 b} \sin\left(\frac{m\pi x}{a}\right) \cos\left(\frac{n\pi y}{b}\right),$$

$$B_y = -iA \frac{m\pi \omega}{k^2 c_0^2 a} \cos\left(\frac{m\pi x}{a}\right) \sin\left(\frac{n\pi y}{b}\right),$$

where the factor $\exp[i(\omega t - \gamma z)]$ is neglected. The real parts of these expressions give practical physical quantities. That is, the above factor is replaced by $\cos(\omega t - \gamma z)$ for E_z, and $i \exp[i(\omega t - \gamma z)]$ is replaced by $\sin(\omega t - \gamma z)$ for the above quantities. We easily find that $\mathbf{E} \cdot \mathbf{B} = 0$ is satisfied, indicating that the electric field and magnetic flux density are perpendicular to each other.

◆

Example 12.6 Determine the TEM wave that propagates along the length (z-axis) in a coaxial conductor, as shown in Fig. 12.5c, and discuss the relationship between the electric charge and the current. Assume $e^{i\omega t}$ and $e^{-i\gamma z}$ for the time and spatial variation, respectively, and that the vacuum space in the coaxial conductor is $a < R < b$.

12.4 Wave Guides

Solution 12.6 From the properties of electric and magnetic fields, we can assume that the electric field has only the radial component, E_R, and that the magnetic flux density has only the azimuthal component, B_φ. The differentials with respect to t and z can be replaced by $i\omega$ and $-i\gamma$, respectively. Thus, we have

$$\gamma E_R = \omega B_\varphi$$

and

$$\gamma B_\varphi = \omega \epsilon_0 \mu_0 E_R.$$

These conditions lead to

$$\frac{\omega}{\gamma} = \frac{1}{(\epsilon_0 \mu_0)^{1/2}} = c_0.$$

If the amplitude of the electric field at $R = a$ is denoted by E_0, we have

$$E_R = \frac{a}{R} E_0 \exp\bigl[i(\omega t - \gamma z)\bigr], \quad B_\varphi = \frac{a}{c_0 R} E_0 \exp\bigl[i(\omega t - \gamma z)\bigr].$$

The densities of the electric charge and the current flowing along the z-axis that appear on the surface $R = a$ are respectively given by

$$\sigma = \epsilon_0 E_R(R = a) = \epsilon_0 E_0 \exp\bigl[i(\omega t - \gamma z)\bigr],$$

$$\tau = \frac{B_\varphi(R = a)}{\mu_0} = \left(\frac{\epsilon_0}{\mu_0}\right)^{1/2} E_0 \exp\bigl[i(\omega t - \gamma z)\bigr].$$

The densities of the electric charge and the current on the surface $R = b$ are

$$\sigma = -\epsilon_0 E_R(R = b) = -\frac{a}{b} \epsilon_0 E_0 \exp\bigl[i(\omega t - \gamma z)\bigr],$$

$$\tau = -\frac{B_\varphi(R = b)}{\mu_0} = -\frac{a}{b}\left(\frac{\epsilon_0}{\mu_0}\right)^{1/2} E_0 \exp\bigl[i(\omega t - \gamma z)\bigr].$$

It can be easily shown that the continuity equation of the current given by Eq. (5.7),

$$\frac{\partial \tau}{\partial z} + \frac{\partial \sigma}{\partial t} = 0,$$

holds on the two surfaces.

♦

Example 12.7 Determine the electromagnetic fields and the electric power of the TE wave in the rectangular wave guide in Fig. 12.4.

Solution 12.7 The boundary conditions on E_x, E_y, B_x, and B_y are given by Eqs. (12.51). The boundary conditions on B_z are: $\partial B_z/\partial x = 0$ at $x = 0$ and a from Eqs. (12.49b) and (12.49c), and $\partial B_z/\partial y = 0$ at $y = 0$ and b from Eqs. (12.49a) and (12.49d). Using these conditions, the general solution of Eq. (12.48) is given by

$$B_z(x, y, z, t) = A' \cos\left(\frac{m\pi x}{a}\right) \cos\left(\frac{n\pi y}{b}\right).$$

Substituting this with $E_z = 0$ into Eqs. (12.49a)–(12.49d) yields

$$E_x = iA' \frac{n\pi \omega}{k^2 b} \cos\left(\frac{m\pi x}{a}\right) \sin\left(\frac{n\pi y}{b}\right),$$

$$E_y = -iA' \frac{m\pi \omega}{k^2 a} \sin\left(\frac{m\pi x}{a}\right) \cos\left(\frac{n\pi y}{b}\right),$$

$$B_x = iA' \frac{m\pi \gamma}{k^2 a} \sin\left(\frac{m\pi x}{a}\right) \cos\left(\frac{n\pi y}{b}\right),$$

$$B_y = iA' \frac{n\pi \gamma}{k^2 b} \cos\left(\frac{m\pi x}{a}\right) \sin\left(\frac{n\pi y}{b}\right).$$

For simplicity, the factor $\exp[i(\omega t - \gamma z)]$ is omitted. The Poynting vector along the z-axis is

$$S_{Pz} = A'^2 \frac{\pi^2 \gamma \omega}{\mu_0 k^4} \left[\frac{n^2}{b^2} \cos^2\left(\frac{m\pi x}{a}\right) \sin^2\left(\frac{n\pi y}{b}\right)\right.$$
$$\left. + \frac{m^2}{a^2} \sin^2\left(\frac{m\pi x}{a}\right) \cos^2\left(\frac{n\pi y}{b}\right)\right] \sin^2(\omega t - \gamma z).$$

Integrating this in the x-y plane yields the electric power through a unit area along the z-axis:

$$p = A'^2 \frac{\pi^2 \gamma \omega (n^2 a^2 + m^2 b^2)}{4 \mu_0 k^4 a^2 b^2} \sin^2(\omega t - \gamma z) = A'^2 \frac{\gamma \omega}{4 \mu_0 k^2} \sin^2(\omega t - \gamma z).$$

◆

Example 12.8 Discuss the energy flow and its speed for the TM wave in the wave guide treated in Sect. 12.4 and Example 12.5.

12.4 Wave Guides

Solution 12.8 From the condition of Eq. (12.51), the Poynting vector is zero on the surfaces of the wave guide, $(x = 0, a)$ and $(y = 0, b)$. Hence, there is no energy flow through these surfaces. Taking the real parts of the electric field and magnetic flux density, the z-component of the Poynting vector is

$$S_{Pz} = A^2 \frac{\pi^2 \epsilon_0 \gamma \omega}{k^4} \left[\frac{m^2}{a^2} \cos^2\left(\frac{m\pi x}{a}\right) \sin^2\left(\frac{n\pi y}{b}\right) \right.$$
$$\left. + \frac{n^2}{b^2} \sin^2\left(\frac{m\pi x}{a}\right) \cos^2\left(\frac{n\pi y}{b}\right) \right] \sin^2(\omega t - \gamma z).$$

Integrating this in the x-y plane, the electric power averaged over one period through a unit area along the z-axis is

$$\langle p \rangle = A^2 \frac{\pi^2 \epsilon_0 \gamma \omega (n^2 a^2 + m^2 b^2)}{4 k^4 a^2 b^2} \langle \sin^2(\omega t - \gamma z) \rangle = A^2 \frac{\epsilon_0 \gamma \omega}{8 k^2}.$$

The components of electric and magnetic energy densities are given by

$$\frac{1}{2}\epsilon_0 E_x^2 = \frac{1}{2}\epsilon_0 A^2 \left(\frac{m\pi \gamma}{k^2 a}\right)^2 \cos^2\left(\frac{m\pi x}{a}\right) \sin^2\left(\frac{n\pi y}{b}\right) \sin^2(\omega t - \gamma z),$$

$$\frac{1}{2}\epsilon_0 E_y^2 = \frac{1}{2}\epsilon_0 A^2 \left(\frac{n\pi \gamma}{k^2 b}\right)^2 \sin^2\left(\frac{m\pi x}{a}\right) \cos^2\left(\frac{n\pi y}{b}\right) \sin^2(\omega t - \gamma z),$$

$$\frac{1}{2}\epsilon_0 E_z^2 = \frac{1}{2}\epsilon_0 A^2 \sin^2\left(\frac{m\pi x}{a}\right) \sin^2\left(\frac{n\pi y}{b}\right) \cos^2(\omega t - \gamma z),$$

$$\frac{1}{2\mu_0} B_x^2 = \frac{1}{2\mu_0} A^2 \left(\frac{n\pi \omega}{k^2 c_0^2 b}\right)^2 \sin^2\left(\frac{m\pi x}{a}\right) \cos^2\left(\frac{n\pi y}{b}\right) \sin^2(\omega t - \gamma z),$$

$$\frac{1}{2\mu_0} B_y^2 = \frac{1}{2\mu_0} A^2 \left(\frac{m\pi \omega}{k^2 c_0^2 a}\right)^2 \cos^2\left(\frac{m\pi x}{a}\right) \sin^2\left(\frac{n\pi y}{b}\right) \sin^2(\omega t - \gamma z).$$

Integrating these in the region of $0 \leq x \leq a$ and $0 \leq y \leq b$, and averaging over one period, the averaged electromagnetic energy density is determined to be

$$\langle u_{em} \rangle = \frac{\epsilon_0 A^2}{8 k^4} \left\{ \gamma^2 \left[\left(\frac{m\pi}{a}\right)^2 + \left(\frac{n\pi}{b}\right)^2 \right] + k^4 + \frac{\omega^2}{c_0^2} \left[\left(\frac{m\pi}{a}\right)^2 + \left(\frac{n\pi}{b}\right)^2 \right] \right\} = A^2 \frac{\epsilon_0 \omega^2}{8 k^2 c_0^2}.$$

From the relation $\langle p \rangle = \langle u_{em} \rangle v$, the speed of energy flow is given by

$$v = \frac{c_0^2 \gamma}{\omega}.$$

Since $\omega/c_0 > \gamma$ from Eq. (12.59), this speed is smaller than the phase velocity, $\omega/\gamma (> c_0)$, and smaller than the light speed in vacuum, c_0.

♦

Example 12.9 Determine the TEM wave that propagates along the length (z-axis) in the parallel-plate conductor in Fig. 12.6 and discuss the relationship between the electric charge and the current induced on the conductor surface. Assume $e^{i\omega t}$ and $e^{-i\gamma z}$ for the time and spatial variation, respectively, and that the vacuum space in the parallel-plate conductor is $0 < x < a$ and $0 < y < b$.

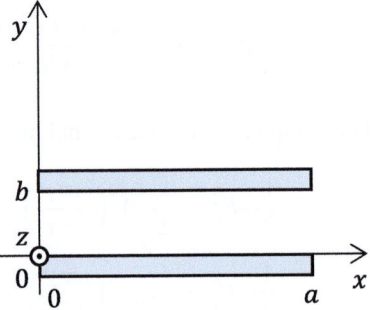

Fig. 12.6 Parallel-plate conductor

Solution 12.9 From the properties of electric and magnetic fields, we can assume that the electric field has only a y-component, E_y, and that the magnetic flux density has only a x-component, B_x. The differentials with respect to t and z can be replaced by $i\omega$ and $-i\gamma$, respectively. Thus, we have

$$\gamma E_y = \omega B_x, \quad \gamma B_x = \omega \epsilon_0 \mu_0 E_y.$$

These conditions lead to

$$\frac{\omega}{\gamma} = \frac{1}{(\epsilon_0 \mu_0)^{1/2}} = c_0.$$

If the amplitude of the electric field is denoted by E_0, we have

$$E_y = E_0 \exp[i(\omega t - \gamma z)], \quad B_x = \frac{E_0}{c_0} \exp[i(\omega t - \gamma z)].$$

12.4 Wave Guides

The densities of the electric charge and the current flowing along the z-axis that appear on the surface at $y = 0$ are

$$\sigma = \epsilon_0 E_y(y=0) = \epsilon_0 E_0 \exp[i(\omega t - \gamma z)],$$

$$\tau = \frac{B_x(y=0)}{\mu_0} = \left(\frac{\epsilon_0}{\mu_0}\right)^{1/2} E_0 \exp[i(\omega t - \gamma z)].$$

It can be easily shown that the continuity equation of current given by Eq. (5.7) holds:

$$\frac{\partial \tau}{\partial z} + \frac{\partial \sigma}{\partial t} = 0.$$

The densities of the electric charge and the current on the surface at $y = b$ are equal to the above quantities with the opposite sign.

◆

Example 12.10 Determine the electromagnetic fields of a TEM wave propagating along the length (z-axis) for the case of the two parallel cylindrical conductors of radius a and mean distance d in Fig. 12.7. Disregard the factor $e^{i(\omega t - \gamma z)}$ with $\gamma/\omega = 1/c_0$.

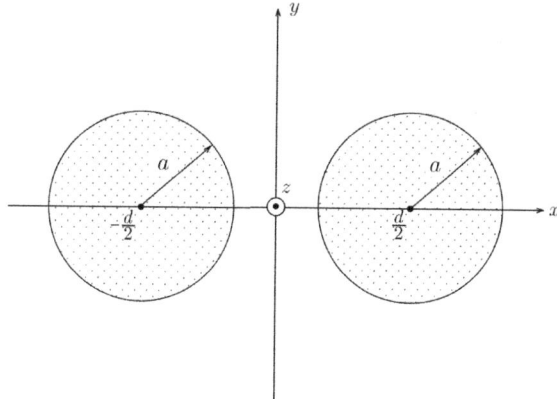

Fig. 12.7 Parallel cylindrical conductors

Solution 12.10 The electric field in the plane normal to the conductors is similar to that in the case where the line charges $\pm \lambda$ are given at the image axes ($\mp l, 0$) of the left and right conductors, respectively, where $l = [(d/2)^2 - a^2]^{1/2}$. The electric potential,

gives

$$E_x = \frac{\lambda}{2\pi\epsilon_0}\left[\frac{x+l}{(x+l)^2+y^2} - \frac{x-l}{(x-l)^2+y^2}\right],$$

$$E_y = \frac{\lambda}{2\pi\epsilon_0}\left[\frac{y}{(x+l)^2+y^2} - \frac{y}{(x-l)^2+y^2}\right].$$

$$\phi(x,y) = \frac{\lambda}{4\pi\epsilon_0}\log\frac{(x-l)^2+y^2}{(x+l)^2+y^2},$$

For the TEM wave, λ is an arbitrary parameter associated with the electric field strength. The magnetic flux density is

$$B_x = \frac{E_y}{c_0}, \quad B_y = -\frac{E_x}{c_0}.$$

Although a detailed calculation is not shown, the total electric charges that appear on the surface of each conductor of a unit length are equal to $\pm\lambda$, and the continuity equation of current holds with the surface charges (see Exercise 12.27).

◆

12.5 Exercises

12.1. We treat the plane electromagnetic wave given by the second terms of Eqs. (12.3) and (12.9). Determine the vector potential and prove that the Lorentz gauge is satisfied.

12.2. Derive Eq. (12.15) for the polarization of a wave.

12.3. Discuss the reflection and refraction of a planar electromagnetic wave at the boundary when medium 1 is vacuum and medium 2 is a superconductor. Assume that the electric field of the incident wave is parallel to the plane of incidence.

12.4. Discuss the reflection and refraction of a planar electromagnetic wave at the boundary when medium 1 is vacuum and medium 2 is a superconductor. Assume that the electric field of the incident wave is normal to the plane of incidence.

12.5. The reflection and refraction of a planar electromagnetic wave on a conductor surface is treated for the case in which the electric field of the incident wave is normal to the plane of incidence in Example 12.2. Discuss the energy flow with the Poynting vector.

12.5 Exercises

12.6. We treat the reflection and refraction of a planar electromagnetic wave on a conductor surface for the case in which the electric field of the incident wave is parallel to the plane of incidence. Determine the electromagnetic fields and discuss the energy flow with the Poynting vector.

12.7. The reflection and refraction of a planar electromagnetic wave is treated for the case in which the electric field of the incident wave is parallel to the plane of incidence in Example 12.1. Discuss the energy flow with the Poynting vector.

12.8. Determine the electromagnetic fields of a TEM wave propagating along the length dimension (z-axis) for the case of a cylindrical conductor of radius a and a wide conductor surface, as shown in Fig. E12.1. The distance between the central axis of the cylindrical conductor and the conductor surface is l.

Fig. E12.1 Cylindrical conductor and wide conductor surface

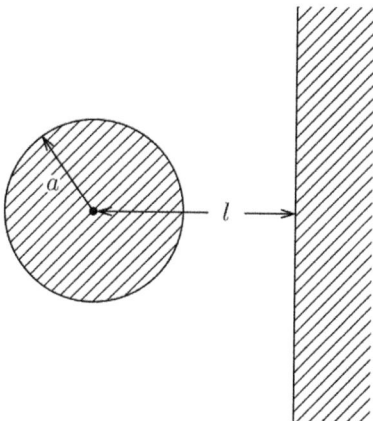

12.9. Determine the electromagnetic potential for the TEM wave in the parallel-plate conductor discussed in Example 12.9. Prove that the Lorentz gauge is satisfied for the electromagnetic potential.

12.10. The case of a TEM wave in a coaxial conductor is discussed in Example 12.6. Determine the electromagnetic potential in the space and prove that the Lorentz gauge is satisfied.

12.11. Determine the electromagnetic potential for the TEM wave around two cylindrical conductors discussed in Example 12.10 and prove that the Lorentz gauge is satisfied.

12.12. Determine the electromagnetic potential for the TE_{01} mode ($m = 0, n = 1$) of the TE wave in a rectangular wave guide discussed in Sect. 12.4. Prove that the Lorentz gauge is satisfied.

12.13. Determine the electromagnetic potential for the TE_{10} mode ($m = 1, n = 0$) of a TE wave in a rectangular wave guide. Prove that the Lorentz gauge is satisfied.

12.14. Determine the electromagnetic potential for the TM$_{11}$ mode ($m = 1, n = 1$) of a TM wave in a rectangular wave guide, as discussed in Sect. 12.4. Prove that the Lorentz gauge is satisfied.

12.15. Determine the electromagnetic potential for the TEM wave between a cylindrical conductor and a conductor surface that was treated in Exercise 12.8.

12.16. The electromagnetic potential is obtained in Exercise 12.10 for a TEM wave in a coaxial conductor. Prove that wave equation holds for the electromagnetic potential.

12.17. The vector potential is obtained for a TEM wave in the space between a cylindrical conductor and a conductor surface in Exercise 12.16. Prove that the vector potential satisfies the wave equation around the conductor surface.

12.18. The electric potential is zero and the vector potential only takes a nonzero value for a TEM wave in separated wave guides in Fig. 12.5a, b (Exercises 12.9 and 12.11) and for a TE wave in a rectangular wave guide (Exercises 12.12 and 12.13). On the other hand, the electric potential also has a nonzero value for a TEM wave in a coaxial wave guide (Exercise 12.10) and for a TM wave in a rectangular wave guide (Exercise 12.14). Discuss the reason for such a difference.

12.19. Discuss the flow of energy for the TEM wave in the parallel-plate conductor treated in Example 12.9.

12.20. Discuss the flow of energy for the TEM wave in the coaxial conductor treated in Example 12.6.

12.21. Discuss the flow of energy for the TEM wave around the parallel cylindrical conductors treated in Example 12.10.

12.22. Discuss the flow of energy for the TEM wave between a cylindrical conductor and a wide conductor surface treated in Exercise 12.8.

12.23. Determine the electric power through a unit area along a wave guide and the speed of energy flow for the TE$_{01}$ mode ($m = 0, n = 1$) of a TE wave.

12.24. Determine the electric power through a unit area along a wave guide and the speed of energy flow for the TE$_{10}$ mode ($m = 1, n = 0$) of a TE wave.

12.25. The energy flow in a wave guide is discussed for a TE wave in Example 12.7. Determine the speed of the energy flow.

12.26. Determine the electric power through a unit area along a wave guide and the speed of energy flow for the TM$_{11}$ mode ($m = 1, n = 1$) of a TM wave.

12.27. Discuss the relation between the electric charge and the current on the surfaces of the cylindrical conductors for the TEM wave treated in Example 12.10.

12.28. Discuss the relation between the electric charge and the current on the surfaces of the conductors for the TEM wave treated in Exercise 12.8.

12.29. Discuss the relation between the electric charge and the current on the surface of the wave guide for the TE wave treated in Example 12.7.

12.5 Exercises

12.30. Discuss the relation between the electric charge and the current on the surface of the wave guide for the TM wave treated in Example 12.5.

12.31. The relationship between the Poynting vector and the surface current is discussed for various cases. In the case of a TEM wave in a wave guide, both the Poynting vector and the current on the surface are directed along the direction of transmission of the wave. On the other hand, the Poynting vector and the current along the direction normal to the transmission direction exist on the surface of the wave guide for the TE (Exercise 12.29) and TM (Exercise 12.30) waves and on the conductor surface for the reflected planar waves (Example 12.2 and Exercises 12.5 and 12.6). Discuss the relation between the Poynting vector and the current on the surface in these cases.

Answers to Exercises

12.1. The real parts of the electric field and magnetic flux density are, respectively, given by

$$E_y = E_2 \cos(\omega t - kx), \quad B_z = \frac{E_2}{c} \cos(\omega t - kx).$$

From the magnetic flux density, the vector potential is determined to be

$$A_y = \int B_z dx = -\frac{E_2}{kc} \sin(\omega t - kx).$$

This produces the above electric field:

$$E_y = -\frac{\partial A_y}{\partial t} = E_2 \cos(\omega t - kx).$$

So, we can conclude that the electric potential ϕ is zero. This is reasonable, since the electric field comes from the electromagnetic induction. So, we can show that

$$\nabla \cdot A = \frac{\partial A_y}{\partial y} = 0.$$

Thus, the Lorentz gauge is satisfied.

12.2. Using Eq. (12.13), $(E_z/E_2)^2$ is given by

$$\left(\frac{E_z}{E_2}\right)^2 = [\cos(\omega t - kx) \cos\delta - \sin(\omega t - kx) \sin\delta]^2,$$

and $E_y E_z / E_1 E_2$ is given by

$$\frac{E_y E_z}{E_1 E_2} = \cos(\omega t - kx)[\cos(\omega t - kx)\cos\delta - \sin(\omega t - kx)\sin\delta].$$

Thus, we have

$$\left(\frac{E_z}{E_2}\right)^2 - 2\cos\delta \frac{E_y E_z}{E_1 E_2} = \sin^2\delta - \cos^2(\omega t - kx).$$

Since $(E_y/E_1)^2$ is equal to $\cos^2(\omega t - kx)$, we have Eq. (12.15).

12.3. The magnetic flux density and the electric field are completely shielded in the superconductor due to the diamagnetic current, which also does not allow a variation in the magnetic condition. So, we can assume $E_2 = 0$ and $B_2 = 0$ in the superconductor. Hence, the fulfilled boundary conditions are only Eqs. (12.17) and (12.20). In this case, it may be better to use B instead of E. Corresponding Eqs. (12.32) and (12.35) are

$$\boldsymbol{n} \times \left(\frac{1}{k}\boldsymbol{B}_0 \times \boldsymbol{k} + \frac{1}{k''}\boldsymbol{B}_0'' \times \boldsymbol{k}''\right) = 0, \quad \boldsymbol{n} \cdot \left(\boldsymbol{B}_0 + \boldsymbol{B}_0''\right) = 0.$$

The latter condition is automatically satisfied, since the magnetic flux density is parallel to the surface. The former condition leads to $(B_0 - B_0'')\cos\theta = 0$, where we used Eq. (12.30). Thus, we have

$$B_0 = B_0''.$$

Then, taking the real part, the magnetic flux density in the vacuum region is

$$B_y = B_0 \cos(\omega t - \boldsymbol{k} \cdot \boldsymbol{r}) + B_0'' \cos(\omega t - \boldsymbol{k}'' \cdot \boldsymbol{r})$$
$$= 2B_0 \cos(kz\cos\theta)\cos(\omega t - kx\sin\theta).$$

In the above, $k = k''$ and we have used the following relations:

$$\boldsymbol{k} \cdot \boldsymbol{r} = kx\sin\theta - kz\cos\theta, \quad \boldsymbol{k}'' \cdot \boldsymbol{r} = kx\sin\theta + kz\cos\theta.$$

The surface current density is

$$\tau_x(x) = -\frac{B_y(z=0)}{\mu_0} = \frac{2B_0}{\mu_0}\cos(\omega t - kx\sin\theta).$$

The electric field is

12.5 Exercises

$$E_x = -c_0 B_0 \cos(\omega t - \mathbf{k} \cdot \mathbf{r}) \cos\theta + c_0 B_0'' \cos(\omega t - \mathbf{k}'' \cdot \mathbf{r}) \cos\theta$$
$$= 2c_0 B_0 \cos\theta \sin(kz \cos\theta) \sin(\omega t - kx \sin\theta),$$

$$E_z = -c_0 B_0 \cos(\omega t - \mathbf{k} \cdot \mathbf{r}) \sin\theta - c_0 B_0'' \cos(\omega t - \mathbf{k}'' \cdot \mathbf{r}) \sin\theta$$
$$= -2c_0 B_0 \sin\theta \cos(kz \cos\theta) \cos(\omega t - kx \sin\theta).$$

The surface electric charge density is

$$\sigma(x) = \epsilon_0 E_z(z=0) = -2c_0 \epsilon_0 B_0 \sin\theta \cos(\omega t - kx \sin\theta).$$

Thus, the continuity equation of current holds:

$$\nabla \cdot \boldsymbol{\tau} + \frac{\partial \sigma}{\partial t} = 0.$$

Note that the condition is identical to that in the case of reflection by a conductor surface for the electric field parallel to the plane of incidence.

12.4. We can assume $\mathbf{E}_2 = 0$ and $\mathbf{B}_2 = 0$ in the superconductor similarly to the case treated in Exercise 12.3. We again use $\mathbf{B}$ instead of $\mathbf{E}$. Equations (12.32) and (12.35) are

$$\mathbf{n} \times \left(\frac{1}{k} \mathbf{B}_0 \times \mathbf{k} + \frac{1}{k''} \mathbf{B}_0'' \times \mathbf{k}'' \right) = 0, \quad \mathbf{n} \cdot (\mathbf{B}_0 + \mathbf{B}_0'') = 0.$$

The second equation leads to

$$B_0 + B_0'' = 0.$$

The first equation gives also the same result. In this case, taking the real part, the magnetic flux density in the vacuum region is

$$B_x = B_0 \cos(\omega t - \mathbf{k} \cdot \mathbf{r}) \cos\theta - B_0'' \cos(\omega t - \mathbf{k}'' \cdot \mathbf{r}) \cos\theta''$$
$$= 2B_0 \cos\theta \cos(kz \cos\theta) \cos(\omega t - kx \sin\theta),$$
$$B_z = B_0 \cos(\omega t - \mathbf{k} \cdot \mathbf{r}) \sin\theta + B_0'' \cos(\omega t - \mathbf{k}'' \cdot \mathbf{r}) \sin\theta''$$
$$= -2B_0 \sin\theta \sin(kz \cos\theta) \sin(\omega t - kx \sin\theta).$$

In the above, $k = k''$ and we have used Eq. (12.30) and the following relations:

$$\mathbf{k} \cdot \mathbf{r} = kx \sin\theta - kz \cos\theta, \quad \mathbf{k}'' \cdot \mathbf{r} = kx \sin\theta + kz \cos\theta.$$

Note that $B_z = 0$ on the surface ($z = 0$). The surface current density is given by

$$\tau_y(x) = \frac{B_x(z=0)}{\mu_0} = \frac{2B_0}{\mu_0}\cos\theta\cos(\omega t - kx\sin\theta).$$

The electric field is

$$E_y = c_0\left[B_0\cos(\omega t - \mathbf{k}\cdot\mathbf{r}) + B_0''\cos(\omega t - \mathbf{k}''\cdot\mathbf{r})\right]$$
$$= -2c_0 B_0 \sin(kz\cos\theta)\sin(\omega t - kx\sin\theta).$$

So, the electric field is zero on the surface ($z = 0$) and electric charge does not appear. In this case, $\nabla \cdot \boldsymbol{\tau} = 0$ and Eq. (5.7) is satisfied.

Note that the condition is identical to that in the case of reflection by a conductor surface for the electric field normal to the plane of incidence.

12.5. The electric field and magnetic flux density in vacuum are

$$E_y = -2E_0 \sin(kz\cos\theta)\sin(\omega t - kx\cos\theta),$$
$$B_x = \frac{2E_0}{c_0}\cos\theta\cos(kz\cos\theta)\cos(\omega t - kx\cos\theta),$$
$$B_z = -\frac{2E_0}{c_0}\sin\theta\sin(kz\cos\theta)\sin(\omega t - kx\sin\theta).$$

Thus, the electric power flowing from vacuum to the conductor through a unit area is obtained as

$$\frac{1}{\mu_0}\left(E_y B_x\right)_{z=0} = 0.$$

So, there is no energy flow from vacuum to the conductor. This is consistent with the zero electric field and magnetic flux density in the conductor.

In this case, the Poynting vector on the conductor surface along the x-axis is also zero:

$$\frac{1}{\mu_0}E_y(z=0)B_z(z=0) = 0.$$

The surface current flows along the y-axis, i.e., normal direction to the incident wave (see Example 12.2). This is different from the condition in the case of electric field normal to the incident wave. This is associated with the current flow on the surface (see Exercise 12.31).

12.6. The electric field and magnetic flux density in vacuum are

$$E_x = 2E_0 \cos\theta \sin(kz\cos\theta) \sin(\omega t - kx\cos\theta),$$
$$E_z = -2E_0 \sin\theta \cos(kz\cos\theta) \cos(\omega t - kx\cos\theta),$$
$$B_y = \frac{2E_0}{c_0} \cos(kz\cos\theta) \cos(\omega t - kx\sin\theta).$$

Thus, the electric power flowing from vacuum to the conductor through a unit area is obtained as

$$-\frac{1}{\mu_0}(E_x B_y)_{z=0} = 0.$$

So, there is no energy flow from vacuum to the conductor. This is consistent with the zero electric field and magnetic flux density in the conductor.

It should be noted, however, that the Poynting vector on the conductor surface is not zero:

$$S_P(z=0) = -E_z(z=0)B_y(z=0)i_x = \frac{2E_0^2}{c_0} \sin\theta \cos^2(\omega t - kx\sin\theta)i_x.$$

That is, there is an energy flow parallel to the conductor surface. This is associated with the current flow on the surface (see Exercise 12.31).

Since the normal component of the electric field is not zero on the surface, an electric charge appears on the surface and its density is

$$\sigma(x) = \epsilon_0 E_z(z=0) = -2\epsilon_0 E_0 \sin\theta \cos(\omega t - kx\cos\theta).$$

Since the parallel component of the magnetic flux density is not zero on the surface, a current flows on the surface and its density is

$$\tau_x(x) = -\frac{B_y(z=0)}{\mu_0} = -2\left(\frac{\epsilon_0}{\mu_0}\right)^{1/2} E_0 \cos(\omega t - kx\sin\theta).$$

In this case, the following continuity equation holds for the surface current:

$$\nabla \cdot \tau + \frac{\partial \sigma}{\partial t} = 0.$$

12.7. Taking the real components, from Eqs. (12.21a, 12.21b) and (12.22a, 12.22b), the electric powers flowing from medium 1 to medium 2 through a unit area as the incident and reflected waves are obtained as

$$-\frac{1}{\mu_1}[\boldsymbol{E}(z=0)\times\boldsymbol{B}(z=0)]_z = \frac{E_0^2}{c_1\mu_1}\cos^2(\omega t - \boldsymbol{k}\cdot\boldsymbol{r}_0)\cos\theta,$$

$$-\frac{1}{\mu_1}[\boldsymbol{E}''(z=0)\times\boldsymbol{B}''(z=0)]_z = -\frac{E_0''^2}{c_1\mu_1}\cos^2(\omega t - \boldsymbol{k}''\cdot\boldsymbol{r}_0)\cos\theta'',$$

respectively. From Eqs. (12.23a, 12.23b), the electric power penetrating into medium 2 as the transmitted wave is

$$-\frac{1}{\mu_2}[\boldsymbol{E}'(z=0)\times\boldsymbol{B}'(z=0)]_z = \frac{E_0'^2}{c_2\mu_2}\cos^2(\omega t - \boldsymbol{k}'\cdot\boldsymbol{r}_0)\cos\theta'.$$

Since the factors dependent on time and space such as $\cos^2(\omega t - \boldsymbol{k}\cdot\boldsymbol{r}_0)$ are the same because of Eq. (12.27), these factors are neglected. Using Eq. (12.30), the rate of energy flow from medium 1 is

$$\frac{1}{c_1\mu_1}\left(E_0^2 - E_0''^2\right)\cos\theta = \frac{4\alpha\cos^2\theta\cos\theta' E_0^2}{c_1\mu_1(\alpha\cos\theta + \cos\theta')^2},$$

where $\alpha = (\epsilon_2\mu_1/\epsilon_1\mu_2)^{1/2}$, and we have used the results in Example 12.1. On the other hand, the rate of energy penetration into medium 2 is

$$\frac{1}{c_2\mu_2}E_0'^2\cos\theta' = \frac{4\cos^2\theta\cos\theta' E_0^2}{c_2\mu_2(\alpha\cos\theta + \cos\theta')^2}.$$

It is easy to show that this is equal to the rate of energy from medium 1.

12.8. If we compare the given situation with the situation in Example 12.10, the vacuum region with the cylindrical conductor corresponds to the left half of the space in Example 12.10 with replacement of $d/2$ by l. The central axis of cylindrical coordinates is placed on the central axis of the left conductor. So, the electric field is given by

$$E_R = K\left[\frac{R - h\cos\varphi}{R^2 + h^2 - 2Rh\cos\varphi} - \frac{R - (2l-h)\cos\varphi}{R^2 + (2l-h)^2 - 2R(2l-h)\cos\varphi}\right],$$

$$E_\varphi = K\left[\frac{h\sin\varphi}{R^2 + h^2 - 2Rh\cos\varphi} - \frac{(2l-h)\sin\varphi}{R^2 + (2l-h)^2 - 2R(2l-h)\cos\varphi}\right],$$

$$E_z = 0,$$

where K is a parameter representing the electric field strength and $h = l - (l^2 - a^2)^{1/2}$, and the common factor, $\exp[i(\omega t - \gamma z)]$, is abbreviated. The magnetic flux density is obtained as

12.5 Exercises

$$B_R = -\frac{\gamma K}{\omega}\left[\frac{h\sin\varphi}{R^2+h^2-2Rh\cos\varphi} - \frac{(2l-h)\sin\varphi}{R^2+(2l-h)^2-2R(2l-h)\cos\varphi}\right],$$

$$B_\varphi = \frac{\gamma K}{\omega}\left[\frac{R-h\cos\varphi}{R^2+h^2-2Rh\cos\varphi} - \frac{R-(2l-h)\cos\varphi}{R^2+(2l-h)^2-2R(2l-h)\cos\varphi}\right],$$

$$B_z = 0.$$

It can be easily found that

$$E_R(R=a) = K\left[\frac{\sin\varphi}{2(l-a\cos\varphi)} - \frac{\sin\varphi}{2(l-a\cos\varphi)}\right] = 0,$$

$$B_\varphi(R=a) = \frac{\gamma K}{\omega}\left[\frac{\sin\varphi}{2(l-a\cos\varphi)} - \frac{\sin\varphi}{2(l-a\cos\varphi)}\right] = 0.$$

So, the boundary conditions on the cylindrical conductor are satisfied.

Since the situation around the flat conductor surface is not clear, we define Cartesian coordinates with $x = R\cos\varphi$ and $y = R\sin\varphi$. Then, we have

$$E_x = E_R\cos\varphi - E_\varphi\sin\varphi$$

$$= K\left[\frac{x-l-\sqrt{l^2-a^2}}{\left(x-l-\sqrt{l^2-a^2}\right)^2+y^2} - \frac{x-l+\sqrt{l^2-a^2}}{\left(x-l+\sqrt{l^2-a^2}\right)^2+y^2}\right],$$

$$E_y = E_R\sin\varphi + E_\varphi\cos\varphi$$

$$= K\left[\frac{y}{\left(x-l-\sqrt{l^2-a^2}\right)^2+y^2} - \frac{y}{\left(x-l+\sqrt{l^2-a^2}\right)^2+y^2}\right].$$

So, it is clear that the boundary condition, $E_y = 0$, is satisfied on the conductor surface ($x = l$). The magnetic flux density is similarly obtained:

$$B_x = -\frac{\gamma K}{\omega}\left[\frac{y}{\left(x-l-\sqrt{l^2-a^2}\right)^2+y^2} - \frac{y}{\left(x-l+\sqrt{l^2-a^2}\right)^2+y^2}\right],$$

$$B_y = \frac{\gamma K}{\omega}\left[\frac{x-l-\sqrt{l^2-a^2}}{\left(x-l-\sqrt{l^2-a^2}\right)^2+y^2} - \frac{x-l+\sqrt{l^2-a^2}}{\left(x-l+\sqrt{l^2-a^2}\right)^2+y^2}\right].$$

Thus, the boundary condition, $B_x = 0$, is satisfied on the conductor surface ($x = l$).

12.9. The magnetic flux density has a x-component, and we assume that the vector potential has a y-component, A_y, since the electric field has also a y-component:

$$A_y = -\int B_x dz = i\frac{E_0}{c_0\gamma}\exp[i(\omega t - \gamma z)].$$

This yields

$$E_y = -\frac{\partial A_y}{\partial t} = \frac{E_0\omega}{c_0\gamma}\exp[i(\omega t - \gamma z)] = E_0\exp[i(\omega t - \gamma z)],$$

which is exactly the electric field in the space. So, the electric potential, ϕ, is zero. Then, we have

$$\nabla \cdot \mathbf{A} = \frac{\partial A_y}{\partial y} = 0.$$

Thus, the Lorentz gauge is satisfied.

12.10. The real part of the magnetic flux density is

$$B_\varphi = \frac{a}{c_0 R}E_0\cos(\omega t - \gamma z).$$

So, the vector potential is determined to be

$$A_z = -\frac{a}{c_0}E_0\cos(\omega t - \gamma z)\log\frac{R}{R_0},$$

where R_0 is the distance to the reference point. The real part of the electric field is

$$E_R = \frac{a}{R}E_0\cos(\omega t - \gamma z).$$

So, it is expected that the electric potential is given by

$$\phi = -aE_0\cos(\omega t - \gamma z)\log\frac{R}{R_0}.$$

In fact, we have

$$E_R = -\frac{\partial\phi}{\partial R} = \frac{a}{R}E_0\cos(\omega t - \gamma z),$$

$$E_z = -\frac{\partial\phi}{\partial z} - \frac{\partial A_z}{\partial t} = aE_0\left(\gamma - \frac{\omega}{c_0}\right)\sin(\omega t - \gamma z)\log\frac{R}{R_0} = 0,$$

12.5 Exercises

which satisfy the condition. From the above results we have

$$\epsilon_0\mu_0 \frac{\partial \phi}{\partial t} = \epsilon_0\mu_0 \omega a E_0 \sin(\omega t - \gamma z) \log \frac{R}{R_0},$$

$$\nabla \cdot \mathbf{A} = \frac{\partial A_z}{\partial z} = -\frac{a\gamma}{c_0} E_0 \sin(\omega t - \gamma z) \log R = -\epsilon_0\mu_0 \omega a E_0 \sin(\omega t - \gamma z) \log \frac{R}{R_0}.$$

Thus, the Lorentz gauge is satisfied:

$$\epsilon_0\mu_0 \frac{\partial \phi}{\partial t} + \nabla \cdot \mathbf{A} = 0.$$

Note that, if we chose a R-component for the vector potential, we have $\phi = 0$ and the Lorentz gauge is not satisfied, since $\nabla \cdot \mathbf{A} \neq 0$.

12.11. The vector potential is determined with the magnetic flux density as

$$A_R = \frac{iK}{\omega}\left[\frac{R - h\cos\varphi}{R^2 + h^2 - 2Rh\cos\varphi} - \frac{R - (d-h)\cos\varphi}{R^2 + (d-h)^2 - 2R(d-h)\cos\varphi}\right],$$

$$A_\varphi = \frac{iK}{\omega}\left[\frac{h\sin\varphi}{R^2 + h^2 - 2Rh\cos\varphi} - \frac{(d-h)\sin\varphi}{R^2 + (d-h)^2 - 2R(d-h)\cos\varphi}\right],$$

$$A_z = 0,$$

where the common factor, $\exp[i(\omega t - \gamma z)]$, is abbreviated. From $\mathbf{E} = -\partial \mathbf{A}/\partial t$, the electric field is given by

$$E_R = K\left[\frac{R - h\cos\varphi}{R^2 + h^2 - 2Rh\cos\varphi} - \frac{R - (d-h)\cos\varphi}{R^2 + (d-h)^2 - 2R(d-h)\cos\varphi}\right],$$

$$E_\varphi = K\left[\frac{h\sin\varphi}{R^2 + h^2 - 2Rh\cos\varphi} - \frac{(d-h)\sin\varphi}{R^2 + (d-h)^2 - 2R(d-h)\cos\varphi}\right],$$

$$E_z = 0.$$

This is the electric field determined in Example 12.10. So, we can conclude that the electric potential ϕ is zero. It can be shown that the following equation holds:

$$\nabla \cdot \mathbf{A} = \frac{1}{R}\left[\frac{\partial(RA_R)}{\partial R} + \frac{\partial A_\varphi}{\partial \varphi}\right] = 0.$$

Thus, the Lorentz gauge holds.

12.12. We assume the components of the vector potential as

$$\mathbf{A} = (A_x, 0, 0).$$

From the conditions of $B_y = \partial A_x / \partial z$, we obtain

$$A_x = -A' \frac{b}{\pi} \sin\left(\frac{\pi y}{b}\right) \exp[i(\omega t - \gamma z)].$$

In this case, the z-component of the magnetic flux density is

$$B_z = -\frac{\partial A_x}{\partial y} = A' \cos\left(\frac{\pi y}{b}\right) \exp[i(\omega t - \gamma z)].$$

Thus, the condition for the magnetic flux density is satisfied. The electric field is

$$E_x = -\frac{\partial A_x}{\partial t} = iA\frac{b\omega}{\pi}\sin\left(\frac{\pi y}{b}\right)\exp[i(\omega t - \gamma z)],$$

$$E_y = -\frac{\partial A_y}{\partial t} = 0,$$

$$E_z = -\frac{\partial A_z}{\partial t} = 0.$$

Thus, we have $\phi = 0$, and

$$\nabla \cdot \mathbf{A} = \frac{\partial A_x}{\partial x} = 0.$$

So, the Lorentz gauge is satisfied.

12.13. We assume the components of the vector potential as $\mathbf{A} = (0, A_y, 0)$. From the conditions of $B_x = -\partial A_y / \partial z$, we obtain

$$A_y = A' \frac{a}{\pi} \sin\left(\frac{\pi x}{a}\right) \exp[i(\omega t - \gamma z)].$$

In this case, the z-component of the magnetic flux density is

$$B_z = \frac{\partial A_y}{\partial x} = A' \cos\left(\frac{\pi x}{a}\right) \exp[i(\omega t - \gamma z)].$$

Thus, the condition for the magnetic flux density is satisfied. The electric field is

$$E_x = -\frac{\partial A_x}{\partial t} = 0,$$

$$E_y = -\frac{\partial A_y}{\partial t} = -iA'\frac{a\omega}{\pi}\sin\left(\frac{\pi x}{a}\right)\exp[i(\omega t - \gamma z)],$$

$$E_z = -\frac{\partial A_z}{\partial t} = 0.$$

12.5 Exercises

Then, we have $\phi = 0$ and

$$\nabla \cdot \mathbf{A} = \frac{\partial A_y}{\partial y} = 0.$$

Thus, the Lorentz gauge is satisfied.

12.14. From the x-, y-, and z-components of the magnetic flux density, the vector potential is determined to be

$$\mathbf{A} = (0, 0, A_z),$$

$$A_z = iA \frac{\omega}{k^2 c_0^2} \sin\left(\frac{\pi x}{a}\right) \sin\left(\frac{\pi y}{b}\right) \exp[i(\omega t - \gamma z)].$$

The electric field due to the variation in the vector potential is

$$E_{z1} = -\frac{\partial A_z}{\partial t} = A \frac{\omega^2}{k^2 c_0^2} \sin\left(\frac{\pi x}{a}\right) \sin\left(\frac{\pi y}{b}\right) \exp[i(\omega t - \gamma z)].$$

The electric potential is determined from the electric field components E_x and E_y to be

$$\phi = iA \frac{\gamma}{k^2} \sin\left(\frac{\pi x}{a}\right) \sin\left(\frac{\pi y}{b}\right) \exp[i(\omega t - \gamma z)].$$

The z-component of the electric field due to the electric potential is

$$E_{z2} = -\frac{\partial \phi}{\partial z} = -A \frac{\gamma^2}{k^2} \sin\left(\frac{\pi x}{a}\right) \sin\left(\frac{\pi y}{b}\right) \exp[i(\omega t - \gamma z)].$$

Thus, the total z-component of the electric field is

$$E_z = E_{z1} + E_{z2} = \frac{A}{k^2}\left[\left(\frac{\omega}{c_0}\right)^2 - \gamma^2\right] \sin\left(\frac{\pi x}{a}\right) \sin\left(\frac{\pi y}{b}\right) \exp[i(\omega t - \gamma z)]$$

$$= A \sin\left(\frac{\pi x}{a}\right) \sin\left(\frac{\pi y}{b}\right) \exp[i(\omega t - \gamma z)],$$

where Eq. (12.50) is used. Thus, Eq. (12.58) is satisfied for the case of $m = 1$ and $n = 1$, and the above electromagnetic potential is correct.

Then, the first and second terms of Eq. (11.24) are, respectively, given by

$$\epsilon_0 \mu_0 \frac{\partial \phi}{\partial t} = -A \frac{\gamma \omega}{k^2 c_0^2} \sin\left(\frac{\pi x}{a}\right) \sin\left(\frac{\pi y}{b}\right) \exp[i(\omega t - \gamma z)],$$

$$\nabla \cdot \boldsymbol{A} = \frac{\partial A_z}{\partial z} = A \frac{\gamma \omega}{k^2 c_0^2} \sin\left(\frac{\pi x}{a}\right) \sin\left(\frac{\pi y}{b}\right) \exp[i(\omega t - \gamma z)].$$

Thus, the Lorentz gauge is satisfied.

12.15. The condition in the vacuum region corresponds to that in the left half space in Exercise 12.11. But the condition around the wide conductor is not clear, since cylindrical coordinates are used. Here, the same problem is treated using Cartesian coordinates (see Exercise 12.8). The components of the vector potential are

$$A_x = \frac{iK}{\omega} \left[\frac{x - l - \sqrt{l^2 - a^2}}{\left(x - l - \sqrt{l^2 - a^2}\right)^2 + y^2} - \frac{x - l + \sqrt{l^2 - a^2}}{\left(x - l + \sqrt{l^2 - a^2}\right)^2 + y^2} \right],$$

$$A_y = \frac{iK}{\omega} \left[\frac{y}{\left(x - l - \sqrt{l^2 - a^2}\right)^2 + y^2} - \frac{y}{\left(x - l + \sqrt{l^2 - a^2}\right)^2 + y^2} \right],$$

$$A_z = 0,$$

where the common factor, $\exp[i(\omega t - \gamma z)]$, is abbreviated. From $\boldsymbol{E} = -\partial \boldsymbol{A}/\partial t$, the electric field is given by

$$E_x = K \left[\frac{x - l - \sqrt{l^2 - a^2}}{\left(x - l - \sqrt{l^2 - a^2}\right)^2 + y^2} - \frac{x - l + \sqrt{l^2 - a^2}}{\left(x - l + \sqrt{l^2 - a^2}\right)^2 + y^2} \right],$$

$$E_y = K \left[\frac{y}{\left(x - l - \sqrt{l^2 - a^2}\right)^2 + y^2} - \frac{y}{\left(x - l + \sqrt{l^2 - a^2}\right)^2 + y^2} \right],$$

$$E_z = 0.$$

These results agree with the electric field components determined in Exercise 12.8. So, it can be concluded that the electric potential, ϕ, is zero. It can be shown that the following equation holds and the Lorentz gauge holds:

$$\nabla \cdot \boldsymbol{A} = \frac{\partial A_x}{\partial x} + \frac{\partial A_y}{\partial y} = 0.$$

12.5 Exercises

12.16. The vector and scalar potentials are obtained as

$$A_z = -\frac{a}{c_0} E_0 \cos(\omega t - \gamma z) \log \frac{R}{R_0},$$

$$\phi = -aE_0 \cos(\omega t - \gamma z) \log \frac{R}{R_0},$$

where R_0 is the distance to the reference point. So, we have

$$\nabla^2 A_z = \frac{1}{R} \cdot \frac{\partial}{\partial R}\left(R \frac{\partial A_z}{\partial R}\right) + \frac{\partial^2 A_z}{\partial z^2} = \frac{a\gamma^2}{c_0} E_0 \cos(\omega t - \gamma z) \log \frac{R}{R_0},$$

$$\frac{\partial^2 A_z}{\partial t^2} = \frac{a\omega^2}{c_0} E_0 \cos(\omega t - \gamma z) \log \frac{R}{R_0}.$$

Thus, it is proved the following wave equation holds:

$$\nabla^2 A_z - \epsilon\mu \frac{\partial^2 A_z}{\partial t^2} = 0.$$

The same equation also holds for ϕ.

12.17. We can easily show that

$$\frac{\partial^2 A_x}{\partial x^2} + \frac{\partial^2 A_x}{\partial y^2} = 0,$$

$$\frac{\partial^2 A_x}{\partial z^2} = -\frac{iK\gamma^2}{\omega}\left[\frac{x - l - \sqrt{l^2 - a^2}}{\left(x - l - \sqrt{l^2 - a^2}\right)^2 + y^2} - \frac{x - l + \sqrt{l^2 - a^2}}{\left(x - l + \sqrt{l^2 - a^2}\right)^2 + y^2}\right],$$

$$\frac{\partial^2 A_x}{\partial t^2} = -iK\omega\left[\frac{x - l - \sqrt{l^2 - a^2}}{\left(x - l - \sqrt{l^2 - a^2}\right)^2 + y^2} - \frac{x - l + \sqrt{l^2 - a^2}}{\left(x - l + \sqrt{l^2 - a^2}\right)^2 + y^2}\right],$$

where we abbreviate the common factor, $\exp[i(\omega t - \gamma z)]$. So, we have

$$\Delta A_x - \mu\epsilon \frac{\partial^2 A_x}{\partial t^2} = 0.$$

We can similarly prove that A_y follows the wave equation.

12.18. In the former cases, the vector potential and the resultant electric field produced by the electromagnetic induction are parallel to each other, and the electric potential does not appear. On the other hand, the situation is different for the latter cases. In the case of a TEM wave in the coaxial wave guide, the electric field is radial, and such an electric field cannot be produced by the electromagnetic induction in a closed space. As a result, the electric potential is necessary to produce the radial electric field. The validity of the obtained electromagnetic potential is proved by the fact that the Lorentz gauge is satisfied. For a TM wave, a part of the axial electric field is produced by the electric potential: The axial electric field produced by the vector potential is not enough. This situation can also be understood from the viewpoint of the Lorentz gauge.

12.19. Taking the real part, the electric energy density is

$$\frac{1}{2}\epsilon_0 E_y^2 = \frac{1}{2}\epsilon_0 E_0^2 \cos^2(\omega t - \gamma z),$$

and the magnetic energy density is

$$\frac{1}{2\mu_0}B_x^2 = \frac{1}{2\mu_0}\left(\frac{E_0}{c_0}\right)^2 \cos^2(\omega t - \gamma z) = \frac{1}{2}\epsilon_0 E_0^2 \cos^2(\omega t - \gamma z).$$

Then, the electromagnetic energy density is

$$u_{em} = \epsilon_0 E_0^2 \cos^2(\omega t - \gamma z).$$

The Poynting vector is directed along the z-axis and is given by

$$S_P = \frac{E_y B_x}{\mu_0} = \frac{1}{\mu_0 c_0}E_0^2 \cos^2(\omega t - \gamma z) = c_0 u_{em}.$$

Thus, the electromagnetic energy flows along the z-axis with the light speed.

12.20. The Poynting vector is directed along the z-axis and its real part is given by

$$S_{Pz} = \frac{E_R B_\varphi}{\mu_0} = \left(\frac{\epsilon_0}{\mu_0}\right)^{1/2}\left(\frac{aE_0}{R}\right)^2 \cos^2(\omega t - \gamma z).$$

Note that the analysis with the factor $e^{i(\omega t - \gamma z)}$ cannot be used. Hence, the electric power that flows through the cross-section of the coaxial conductor is

$$P = \int_a^b S_{Pz} 2\pi R dR = 2\pi \left(\frac{\epsilon_0}{\mu_0}\right)^{1/2}(aE_0)^2 \log\frac{b}{a}\cos^2(\omega t - \gamma z).$$

12.5 Exercises

The electric energy in a unit length is

$$U'_e = \int_a^b \frac{\epsilon_0}{2} E_r^2 2\pi R dR = \pi \epsilon_0 (aE_0)^2 \log \frac{b}{a} \cos^2(\omega t - \gamma z)$$

and the magnetic energy in a unit length is

$$U'_m = \int_a^b \frac{1}{2\mu_0} B_\varphi^2 2\pi R dR = \pi \epsilon_0 (aE_0)^2 \log \frac{b}{a} \cos^2(\omega t - \gamma z).$$

Thus, the electromagnetic energy in a unit length is

$$U' = U'_e + U'_m = 2\pi \epsilon_0 (aE_0)^2 \log \frac{b}{a} \cos^2(\omega t - \gamma z).$$

So, we have

$$P = c_0 U'.$$

12.21. Taking the real part, the electromagnetic energy density at point (R, φ, z) is given by

$$u = \frac{1}{2}\epsilon_0 E^2 + \frac{1}{2\mu_0} B^2$$

$$= \epsilon_0 K^2 \left\{ \left[\frac{R - h\cos\varphi}{R^2 + h^2 - 2Rh\cos\varphi} - \frac{R - (d-h)\cos\varphi}{R^2 + (d-h)^2 - 2R(d-h)\cos\varphi} \right]^2 \right.$$

$$\left. + \left[\frac{h\sin\varphi}{R^2 + h^2 - 2Rh\cos\varphi} - \frac{(d-h)\sin\varphi}{R^2 + (d-h)^2 - 2R(d-h)\cos\varphi} \right]^2 \right\} \cos^2(\omega t - \gamma z).$$

The corresponding Poynting vector directed along the z-axis is

$$S_P = \frac{1}{\mu_0}(E_R B_\varphi - E_\varphi B_R)$$

$$= \frac{\gamma}{\mu_0 \omega} K^2 \left\{ \left[\frac{R - h\cos\varphi}{R^2 + h^2 - 2Rh\cos\varphi} - \frac{R - (d-h)\cos\varphi}{R^2 + (d-h)^2 - 2R(d-h)\cos\varphi} \right]^2 \right.$$

$$\left. + \left[\frac{h\sin\varphi}{R^2 + h^2 - 2Rh\cos\varphi} - \frac{(d-h)\sin\varphi}{R^2 + (d-h)^2 - 2R(d-h)\cos\varphi} \right]^2 \right\} \cos^2(\omega t - \gamma z)$$

$$= c_0 u.$$

Thus, the electromagnetic energy with the density u is flowing with the light speed c_0.

12.22. The condition in vacuum region corresponds to that in the left half space in Exercise 12.21. It is shown in Exercise 12.21 that the electromagnetic energy flows with the light speed in the space. However, the condition around the wide conductor is not clear, since cylindrical coordinates are used. Here, the same problem is treated using Cartesian coordinates (see Example 12.10). Taking the real part, the electromagnetic energy density is

$$u = \frac{1}{2}\epsilon_0 E^2 + \frac{1}{2\mu_0}B^2$$

$$= \epsilon_0 K^2 \left\{ \left[\frac{x-l-\sqrt{l^2-a^2}}{\left(x-l-\sqrt{l^2-a^2}\right)^2 + y^2} - \frac{x-l+\sqrt{l^2-a^2}}{\left(x-l+\sqrt{l^2-a^2}\right)^2 + y^2} \right]^2 \right.$$

$$\left. + \left[\frac{y}{\left(x-l-\sqrt{l^2-a^2}\right)^2 + y^2} - \frac{y}{\left(x-l+\sqrt{l^2-a^2}\right)^2 + y^2} \right]^2 \right\} \cos^2(\omega t - \gamma z).$$

The corresponding Poynting vector directed along the z-axis is

$$S_{\mathrm{P}} = \frac{1}{\mu_0}(E_x B_y - E_y B_x)$$

$$= \frac{\gamma}{\mu_0 \omega} K^2 \left\{ \left[\frac{x-l-\sqrt{l^2-a^2}}{\left(x-l-\sqrt{l^2-a^2}\right)^2 + y^2} - \frac{x-l+\sqrt{l^2-a^2}}{\left(x-l+\sqrt{l^2-a^2}\right)^2 + y^2} \right]^2 \right.$$

$$\left. + \left[\frac{y}{\left(x-l-\sqrt{l^2-a^2}\right)^2 + y^2} - \frac{y}{\left(x-l+\sqrt{l^2-a^2}\right)^2 + y^2} \right]^2 \right\} \cos^2(\omega t - \gamma z)$$

$$= c_0 u.$$

Thus, the electromagnetic energy with the density u is flowing with the light speed c_0.

12.23. The real part of the longitudinal magnetic flux density is given by

$$B_z = A' \cos\left(\frac{\pi y}{b}\right)\cos(\omega t - \gamma z).$$

Taking the real part, other components are given by

12.5 Exercises

$$E_x = -A'\frac{b\omega}{\pi}\sin\left(\frac{\pi y}{b}\right)\sin(\omega t - \gamma z),$$

$$E_y = 0,$$

$$B_x = 0,$$

$$B_y = -A'\frac{b\gamma}{\pi}\sin\left(\frac{\pi y}{b}\right)\sin(\omega t - \gamma z).$$

In the above, we have used $k = \pi/b$ and

$$\gamma^2 = \left(\frac{\omega}{c_0}\right)^2 - \left(\frac{\pi}{b}\right)^2.$$

The Poynting vector is directed to the z-axis and is given by

$$S_{P_z} = A'^2 \frac{\gamma\omega}{\mu_0}\left(\frac{b}{\pi}\right)^2 \sin^2\left(\frac{\pi y}{b}\right)\sin^2(\omega t - \gamma z).$$

Integrating this in the x-y plane and averaging over one period, the electric power through a unit area along the z-axis is given by

$$\langle p \rangle = A'^2 \frac{\gamma\omega}{4\mu_0}\left(\frac{b}{\pi}\right)^2.$$

Using the above results, the electric and magnetic energy densities are

$$\frac{1}{2}\epsilon_0 E_x^2 = \frac{1}{2}\epsilon_0 A'^2 \left(\frac{b\omega}{\pi}\right)^2 \sin^2\left(\frac{\pi y}{b}\right)\sin^2(\omega t - \gamma z),$$

$$\frac{1}{2\mu_0}B_y^2 = \frac{1}{2\mu_0}A'^2 \left(\frac{b\gamma}{\pi}\right)^2 \sin^2\left(\frac{\pi y}{b}\right)\sin^2(\omega t - \gamma z),$$

$$\frac{1}{2\mu_0}B_z^2 = \frac{1}{2\mu_0}A'^2 \cos^2\left(\frac{\pi y}{b}\right)\cos^2(\omega t - \gamma z).$$

The electromagnetic energy density averaged in the x-y plane and over one period is

$$\langle u_{\text{em}} \rangle = \frac{A'^2}{8}\left[\epsilon_0\left(\frac{b\omega}{\pi}\right)^2 + \frac{1}{\mu_0}\left(\frac{b\gamma}{\pi}\right)^2 + \frac{1}{\mu_0}\right] = A'^2 \frac{\epsilon_0 \omega^2}{4}\left(\frac{b}{\pi}\right)^2,$$

where $\langle \ \rangle$ represents the time average. Thus, the speed of the energy flow is given by

$$v = \frac{c_0^2 \gamma}{\omega}.$$

12.24. The real part of the longitudinal magnetic flux density is given by

$$B_z = A' \cos\left(\frac{\pi x}{a}\right) \cos(\omega t - \gamma z).$$

Taking the real part, other components are given by

$$E_x = 0,$$
$$E_y = A' \frac{a\omega}{\pi} \sin\left(\frac{\pi x}{a}\right) \sin(\omega t - \gamma z),$$
$$B_x = -A' \frac{a\gamma}{\pi} \sin\left(\frac{\pi x}{a}\right) \sin(\omega t - \gamma z),$$
$$B_y = 0.$$

In the above, we have used $k = \pi/a$ and

$$\gamma^2 = \left(\frac{\omega}{c_0}\right)^2 - \left(\frac{\pi}{a}\right)^2.$$

The Poynting vector is directed to the z-axis and is given by

$$S_{Pz} = A'^2 \frac{\gamma\omega}{\mu_0} \left(\frac{a}{\pi}\right)^2 \sin^2\left(\frac{\pi x}{a}\right) \sin^2(\omega t - \gamma z).$$

Integrating this in the x-y plane and averaging over one period, the electric power through a unit area along the z-axis is given by

$$\langle p \rangle = A'^2 \frac{\gamma\omega}{4\mu_0} \left(\frac{a}{\pi}\right)^2.$$

Using the above results, the electric and magnetic energy densities are

$$\frac{1}{2}\epsilon_0 E_y^2 = \frac{1}{2}\epsilon_0 A'^2 \left(\frac{a\omega}{\pi}\right)^2 \sin^2\left(\frac{\pi x}{a}\right) \sin^2(\omega t - \gamma z),$$

$$\frac{1}{2\mu_0} B_x^2 = \frac{1}{2\mu_0} A'^2 \left(\frac{a\gamma}{\pi}\right)^2 \sin^2\left(\frac{\pi x}{a}\right) \sin^2(\omega t - \gamma z),$$

$$\frac{1}{2\mu_0} B_z^2 = \frac{1}{2\mu_0} A'^2 \cos^2\left(\frac{\pi x}{a}\right) \sin^2(\omega t - \gamma z).$$

12.5 Exercises

The averaged electromagnetic energy density is

$$\langle u_{em} \rangle = \frac{A'^2}{8}\left[\epsilon_0\left(\frac{a\omega}{\pi}\right)^2 + \frac{1}{\mu_0}\left(\frac{a\gamma}{\pi}\right)^2 + \frac{1}{\mu_0}\right] = A'^2\frac{\epsilon_0\omega^2}{4}\left(\frac{a}{\pi}\right)^2.$$

Thus, the speed of the energy flow is given by

$$v = \frac{c_0^2\gamma}{\omega}.$$

12.25. The components of the electric and magnetic energy densities are given by

$$\frac{1}{2}\epsilon_0 E_x^2 = \frac{1}{2}\epsilon_0 A'^2 \left(\frac{n\pi\omega}{k^2 b}\right)^2 \cos^2\left(\frac{m\pi x}{a}\right)\sin^2\left(\frac{n\pi y}{b}\right)\sin^2(\omega t - \gamma z),$$

$$\frac{1}{2}\epsilon_0 E_y^2 = \frac{1}{2}\epsilon_0 A'^2 \left(\frac{m\pi\omega}{k^2 a}\right)^2 \sin^2\left(\frac{m\pi x}{a}\right)\cos^2\left(\frac{n\pi y}{b}\right)\sin^2(\omega t - \gamma z),$$

$$\frac{1}{2\mu_0} B_x^2 = \frac{1}{2\mu_0} A'^2 \left(\frac{m\pi\gamma}{k^2 a}\right)^2 \sin^2\left(\frac{m\pi x}{a}\right)\cos^2\left(\frac{n\pi y}{b}\right)\sin^2(\omega t - \gamma z),$$

$$\frac{1}{2\mu_0} B_y^2 = \frac{1}{2\mu_0} A'^2 \left(\frac{n\pi\gamma}{k^2 b}\right)^2 \cos^2\left(\frac{m\pi x}{a}\right)\sin^2\left(\frac{n\pi y}{b}\right)\sin^2(\omega t - \gamma z),$$

$$\frac{1}{2\mu_0} B_z^2 = \frac{1}{2\mu_0} A'^2 \cos^2\left(\frac{m\pi x}{a}\right)\cos^2\left(\frac{n\pi y}{b}\right)\cos^2(\omega t - \gamma z).$$

The electromagnetic energy density averaged in the x-y plane and over one period is

$$\langle u_{em} \rangle = \frac{\epsilon_0 A'^2}{16k^4}\left\{\omega^2\left[\left(\frac{m\pi}{a}\right)^2 + \left(\frac{n\pi}{b}\right)^2\right] + c_0^2\gamma^2\left[\left(\frac{m\pi}{a}\right)^2 + \left(\frac{n\pi}{b}\right)^2\right] + k^4 c_0^2\right\}$$

$$= A'^2 \frac{\epsilon_0 \omega^2}{8k^2}.$$

The averaged electric power over one period through a unit area along the wave guide is obtained in Example 12.7 as

$$\langle p \rangle = A'^2 \frac{\omega\gamma}{8\mu_0 k^2}.$$

Thus, the speed of the energy flow is given by

$$v = \frac{c_0^2 \gamma}{\omega}.$$

This speed is the same as that of TM wave (see Example 12.8).

12.26. The real parts of the electromagnetic fields are given by

$$E_z = A \sin\left(\frac{\pi x}{a}\right) \sin\left(\frac{\pi y}{b}\right) \cos(\omega t - \gamma z),$$

$$E_x = A \frac{\pi \gamma}{k^2 a} \cos\left(\frac{\pi x}{a}\right) \sin\left(\frac{\pi y}{b}\right) \sin(\omega t - \gamma z),$$

$$E_y = A \frac{\pi \gamma}{k^2 b} \sin\left(\frac{\pi x}{a}\right) \cos\left(\frac{\pi y}{b}\right) \sin(\omega t - \gamma z),$$

$$B_x = -A \frac{\pi \omega}{k^2 c_0^2 b} \sin\left(\frac{\pi x}{a}\right) \cos\left(\frac{\pi y}{b}\right) \sin(\omega t - \gamma z),$$

$$B_y = A \frac{\pi \omega}{k^2 c_0^2 a} \cos\left(\frac{\pi x}{a}\right) \sin\left(\frac{\pi y}{b}\right) \sin(\omega t - \gamma z),$$

and $B_z = 0$. The electromagnetic energy density averaged in the x-y plane and over one period is determined to be

$$\langle u_{em} \rangle = \frac{\epsilon_0 A^2}{8k^4} \left\{ \gamma^2 \left[\left(\frac{\pi}{a}\right)^2 + \left(\frac{\pi}{b}\right)^2 \right] + k^4 + \frac{\omega^2}{c_0^2} \left[\left(\frac{\pi}{a}\right)^2 + \left(\frac{\pi}{b}\right)^2 \right] \right\}$$

$$= A^2 \frac{\epsilon_0 \omega^2}{8k^2 c_0^2}.$$

The averaged electric power through a unit area along the wave guide obtained using the Poynting vector is

$$\langle p \rangle = A^2 \frac{\epsilon_0 \gamma \omega}{4k^2} \langle \sin^2(\omega t - \gamma z) \rangle = A^2 \frac{\epsilon_0 \gamma \omega}{8k^2}.$$

Thus, the speed of the energy flow is given by

$$v = \frac{\langle p \rangle}{\langle u_{em} \rangle} = \frac{c_0^2 \gamma}{\omega}.$$

12.27. The real part of the radial electric field on the surface of the left conductor ($R = a$) is

$$E_R(R = a) = K \Psi \cos(\omega t - \gamma z),$$

where

$$\Psi = \frac{a - h \cos\varphi}{a^2 + h^2 - 2ah \cos\varphi} - \frac{a - (d - h) \cos\varphi}{a^2 + (d - h)^2 - 2a(d - h) \cos\varphi}$$

and $h = (d/2) - \sqrt{(d/2)^2 - a^2}$. So, the electric charge density on the conductor surface is

12.5 Exercises

$$\sigma(\varphi) = \epsilon_0 E_R(R=a) = \epsilon_0 K \Psi \cos(\omega t - \gamma z).$$

The real part of the azimuthal magnetic flux density on the corresponding surface is

$$B_\varphi(R=a) = \frac{\gamma K}{\omega} \Psi \cos(\omega t - \gamma z).$$

The current density on the conductor surface is

$$\tau_z(\varphi) = \frac{B_\varphi(R=a)}{\mu_0} = \frac{\gamma K}{\mu_0 \omega} \Psi \cos(\omega t - \gamma z).$$

Thus, the following relation holds:

$$\frac{\partial \sigma}{\partial t} + \nabla \cdot \boldsymbol{\tau} = 0.$$

12.28. It is shown in Exercise 12.27 that the continuity equation of current holds on the cylindrical conductor surface. Here, we investigate the condition on the wide superconductor surface. Using the result of Exercise 12.8, the electric charge density on the wide conductor surface is

$$\sigma = -\epsilon_0 E_x(x=l) = \frac{2\epsilon_0 K \sqrt{l^2 - a^2}}{y^2 + l^2 - a^2} \exp[i(\omega t - \gamma z)].$$

The current density on the wide conductor surface is

$$\tau_z = -\frac{1}{\mu_0} B_y(x=l) = \frac{2\gamma K \sqrt{l^2 - a^2}}{\mu_0 \omega (y^2 + l^2 - a^2)} \exp[i(\omega t - \gamma z)].$$

So, the left-hand side of Eq. (5.7) leads to

$$\frac{\partial \tau_z}{\partial z} + \frac{\partial \sigma}{\partial t} = \frac{2iK\sqrt{l^2 - a^2}}{y^2 + l^2 - a^2} \left(\epsilon_0 \omega - \frac{\gamma^2}{\mu_0 \omega} \right) \exp[i(\omega t - \gamma z)] = 0.$$

Thus, the continuity equation also holds on the wide conductor surface.

12.29. Here, we treat the y-z plane at $x = 0$. for simplicity, we omit the common factor $\exp[i(\omega t - \gamma z)]$. The electric charge density on this plane is

$$\sigma = \epsilon_0 E_x = iA' \frac{n\pi \epsilon_0 \omega}{k^2 b} \sin\left(\frac{n\pi y}{b}\right).$$

Fig. B12.1 The y-z plane at $x = 0$

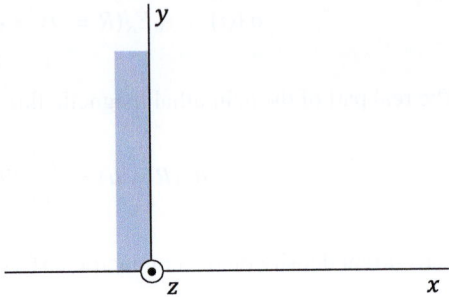

Noting the coordinates shown in Fig. B12.1, the surface current density is given by

$$\tau_y = -\frac{B_z}{\mu_0} = -\frac{A'}{\mu_0}\cos\left(\frac{n\pi y}{b}\right),$$

$$\tau_z = \frac{B_y}{\mu_0} = iA'\frac{n\pi \gamma}{k^2 b \mu_0}\sin\left(\frac{n\pi y}{b}\right).$$

From these results we have

$$\frac{\partial \sigma}{\partial t} = -A'\frac{n\pi \epsilon_0 \omega^2}{k^2 b}\sin\left(\frac{n\pi y}{b}\right)$$

and

$$\nabla \cdot \boldsymbol{\tau} = \frac{\partial \tau_y}{\partial y} + \frac{\partial \tau_z}{\partial z}$$

$$= A'\frac{n\pi}{\mu_0 b}\left(1 + \frac{\gamma^2}{k^2}\right)\sin\left(\frac{n\pi y}{b}\right) = A'\frac{n\pi \epsilon_0 \omega^2}{k^2 b}\sin\left(\frac{n\pi y}{b}\right).$$

In the above, we used Eq. (12.50). Thus, we obtain

$$\frac{\partial \sigma}{\partial t} + \nabla \cdot \boldsymbol{\tau} = 0.$$

The Poynting vector on the surface has two components. One is the z-component, which represents the main power flow, and the other is the y-component:

$$S_{Py}(x = 0) = -\frac{1}{\mu_0}E_x(x=0)B_y(x=0) = -iA'^2\frac{n\pi \omega}{\mu_0 k^2 b}\sin\left(\frac{n\pi y}{b}\right)\cos\left(\frac{n\pi y}{b}\right).$$

As shown above, the non-steady surface current of density τ_y flows along this direction.

The similar results are obtained for other planes.

12.5 Exercises

12.30. Here, we treat the y-z plane at $x = 0$. For simplicity, we omit the common factor $\exp[i(\omega t - \gamma z)]$. The electric charge density on this plane is

$$\sigma = \epsilon_0 E_x = -iA \frac{m\pi \epsilon_0 \gamma}{k^2 a} \sin\left(\frac{n\pi y}{b}\right),$$

and the surface current density is

$$\tau_z = \frac{B_y}{\mu_0} = -iA \frac{m\pi \epsilon_0 \omega}{k^2 a} \sin\left(\frac{n\pi y}{b}\right).$$

From the factor $\exp[i(\omega t - \gamma z)]$, the derivative with time $\partial/\partial t$ can be replaced by $i\omega$, and we have

$$\frac{\partial \sigma}{\partial t} = A \frac{m\pi \epsilon_0 \omega \gamma}{k^2 a} \sin\left(\frac{n\pi y}{b}\right).$$

Next, the divergence of the surface current density is given by

$$\nabla \cdot \boldsymbol{\tau} = \frac{\partial \tau_z}{\partial z} = -A \frac{m\pi \epsilon_0 \omega \gamma}{k^2 a} \sin\left(\frac{n\pi y}{b}\right),$$

where we replaced $\partial/\partial z$ by $-i\gamma$. Thus, it can be shown that the continuity equation of current holds:

$$\frac{\partial \sigma}{\partial t} + \nabla \cdot \boldsymbol{\tau} = 0.$$

In this case, the Poynting vector on the surface has only the z-component corresponding to the main power flow. And the surface current does not flow along the y-axis. These are different from TE wave.

The similar results are obtained for other planes.

12.31. In a TM wave, the Poynting vector along the direction of the wave transmission on the surface of the wave guide is zero, and there is no surface current. In a TE wave, there is a component of the Poynting vector perpendicular to the direction of the wave transmission on the surface, and a non-steady surface current also flows along this direction.

In the case of the reflection of a planar wave on the conductor surface, the situation is complicated: The Poynting vector parallel to the surface has a component parallel to the energy flow (transmission of incident and reflected waves) for an incident wave with the electric field parallel to the plane of incidence. In this case a non-steady surface current flows in this direction (see Exercise 12.6). On the other hand, this component of the Poynting vector is zero, and a steady surface current flows in the direction normal to the energy flow for an incident wave with the electric field normal to the plane of incidence (see Exercise 12.5).

Here, we discuss the difference between the last two cases. In both cases, the surface current flows. The properties of the surface current are different, however. In the former case, the current is a non-steady current associated with the induced electric charge, while it is a steady current without a change in the electric charge distribution in the latter case.

So, the key point is not the surface current but rather the induced surface electric charge. That is, the Poynting vector has a component in the direction of the surface current supplied by the induced surface electric charge. This principle explains the phenomenon for TE and TM waves in a wave guide. The zero Poynting vector in the TM wave can be explained by there being no change in the electric charge distribution.

Coffee Break: Breaking of Symmetry in Maxwell's Equations

When there is no electric charge or current as in the case of electromagnetic wave, Eq. (11.8) is given by

$$\nabla \times \boldsymbol{B} = \mu\epsilon \frac{\partial \boldsymbol{E}}{\partial t}. \tag{C12.1}$$

If we put $\boldsymbol{B} = (\mu\epsilon)^{1/2}\boldsymbol{B}'$ and $t = (\mu\epsilon)^{-1/2}t'$, Eq. (C12.1) is written as

$$\nabla \times \boldsymbol{B}' = \frac{\partial \boldsymbol{E}}{\partial t'}, \tag{C12.2}$$

and Eq. (11.7) is written as

$$\nabla \times \boldsymbol{E} = -\frac{\partial \boldsymbol{B}'}{\partial t'}. \tag{C12.3}$$

Equations (11.9) and (11.10) simply reduce to

$$\nabla \cdot \boldsymbol{E} = 0, \tag{C12.4}$$

$$\nabla \cdot \boldsymbol{B}' = 0. \tag{C12.5}$$

Thus, the symmetry of Maxwell's equations becomes clear between Eqs. (C12.2) and (C12.3) and between Eqs. (C12.4) and (C12.5) by such a normalization.

In mechanics, Hamilton's equations that describe the motion of mass show the clear symmetry:

12.5 Exercises

$$\frac{\partial q}{\partial t} = \frac{\partial \mathcal{H}}{\partial p},$$

$$\frac{\partial p}{\partial t} = -\frac{\partial \mathcal{H}}{\partial q},$$

where $\mathcal{H}$ is the Hamilton function, and q and p are coordinate and momentum of a mass, respectively.

When there are electric charges and currents, the corresponding equations change. In the case of electric charges, the quantity influenced by electric charges is E or D. In this case, electric charge is a scalar, and hence, it is given by a divergence of these quantities. Thus, Eq. (11.9) is modified. In the case of currents, the quantity influenced by currents is B or H. In this case, current is a vector, and hence, it is given by a curl of these quantities. Thus, Eq. (11.8) is modified. In such a way the symmetry of Maxwell's equations is broken.

Appendix

Differentiation of Products of Vectors

The following relations hold for various products:

$$\nabla(\phi\psi) = \phi\nabla\psi + \psi\nabla\phi, \tag{A.1}$$

$$\nabla(\boldsymbol{A}\cdot\boldsymbol{B}) = (\boldsymbol{A}\cdot\nabla)\boldsymbol{B} + (\boldsymbol{B}\cdot\nabla)\boldsymbol{A} + \boldsymbol{A}\times(\nabla\times\boldsymbol{B}) + \boldsymbol{B}\times(\nabla\times\boldsymbol{A}), \tag{A.2}$$

$$\nabla\cdot(\phi\boldsymbol{A}) = \phi\nabla\cdot\boldsymbol{A} + \nabla\phi\cdot\boldsymbol{A}, \tag{A.3}$$

$$\nabla\cdot(\boldsymbol{A}\times\boldsymbol{B}) = \boldsymbol{B}\cdot(\nabla\times\boldsymbol{A}) - \boldsymbol{A}\cdot(\nabla\times\boldsymbol{B}), \tag{A.4}$$

$$\nabla\times(\phi\boldsymbol{A}) = \phi\nabla\times\boldsymbol{A} - \boldsymbol{A}\times\nabla\phi, \tag{A.5}$$

$$\nabla\times(\boldsymbol{A}\times\boldsymbol{B}) = (\boldsymbol{B}\cdot\nabla)\boldsymbol{A} - (\boldsymbol{A}\cdot\nabla)\boldsymbol{B} + \boldsymbol{A}\nabla\cdot\boldsymbol{B} - \boldsymbol{B}\nabla\cdot\boldsymbol{A}. \tag{A.6}$$

Second Differentiation

$$\nabla\cdot(\nabla\times\boldsymbol{A}) = 0, \tag{A.7}$$

$$\nabla\times\nabla\phi = 0, \tag{A.8}$$

$$\nabla\times(\nabla\times\boldsymbol{A}) = \nabla(\nabla\cdot\boldsymbol{A}) - \nabla^2\boldsymbol{A}. \tag{A.9}$$

Appendix

Gauss's Theorem

$$\int_S A \cdot dS = \int_V \nabla \cdot A \, dV, \tag{A.10}$$

where V is the region surrounded by closed surface S.

Stokes' Theorem

$$\oint_C A \cdot ds = \int_S (\nabla \times A) \cdot dS, \tag{A.11}$$

where S is the surface surrounded by closed loop C.

Cylindrical Coordinates

We express the target position with the distance from this axis (R), the azimuthal angle (φ) on the plane normal to the z-axis and the position on this axis (z): (R, φ, z). When we use the common z-axis with Cartesian coordinates, as shown in Fig. A.1, the relationships between the two sets of coordinates are:

$$R = (x^2 + y^2)^{1/2}, \quad \varphi = \tan^{-1}\frac{y}{x}, \quad z = z.$$

The gradient, divergence and curl in cylindrical coordinates are

$$\nabla f = i_R \frac{\partial f}{\partial R} + i_\varphi \frac{1}{R} \cdot \frac{\partial f}{\partial \varphi} + i_z \frac{\partial f}{\partial z}, \tag{A.12}$$

Fig. A.1 Cylindrical coordinates

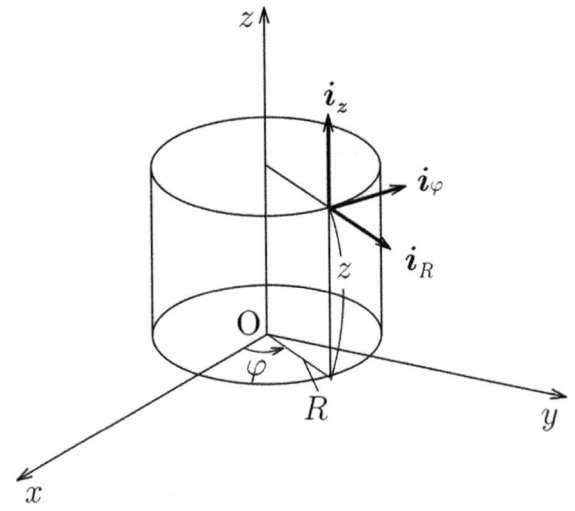

$$\nabla \cdot \mathbf{A} = \frac{1}{R} \cdot \frac{\partial (RA_R)}{\partial R} + \frac{1}{R} \cdot \frac{\partial A_\varphi}{\partial \varphi} + \frac{\partial A_z}{\partial z}, \tag{A.13}$$

$$\nabla \times \mathbf{A} = \mathbf{i}_R \left(\frac{1}{R} \cdot \frac{\partial A_z}{\partial \varphi} - \frac{\partial A_\varphi}{\partial z} \right) + \mathbf{i}_\varphi \left(\frac{\partial A_R}{\partial z} - \frac{\partial A_z}{\partial R} \right) + \mathbf{i}_z \frac{1}{R} \left[\frac{\partial (RA_\varphi)}{\partial R} - \frac{\partial A_R}{\partial \varphi} \right], \tag{A.14}$$

where $\mathbf{i}_R$, $\mathbf{i}_\varphi$, and $\mathbf{i}_z$ are the unit vectors along the radial, azimuthal, and z-axial directions, respectively. The second differentiation of a scalar function is given by

$$\nabla^2 f = \frac{1}{R} \cdot \frac{\partial}{\partial R} \left(R \frac{\partial f}{\partial R} \right) + \frac{1}{R^2} \cdot \frac{\partial^2 f}{\partial \varphi^2} + \frac{\partial^2 f}{\partial z^2}. \tag{A.15}$$

Spherical Coordinates

We first define the center with an axis that determines the two poles. Then, we express the target position with the distance from the center (r), the zenithal angle (θ) measured from the north pole, and the azimuthal angle (φ) on the plane normal to the axis: (r, θ, φ).

When we use the common center and z-axis with Cartesian coordinates, as shown in Fig. A.2, the relationships between the two sets of coordinates are:

$$r = \left(x^2 + y^2 + z^2\right)^{1/2}, \quad \theta = \tan^{-1} \frac{\left(x^2 + y^2\right)^{1/2}}{z}, \quad \varphi = \tan^{-1} \frac{y}{x}.$$

The gradient, divergence and curl in spherical coordinates are

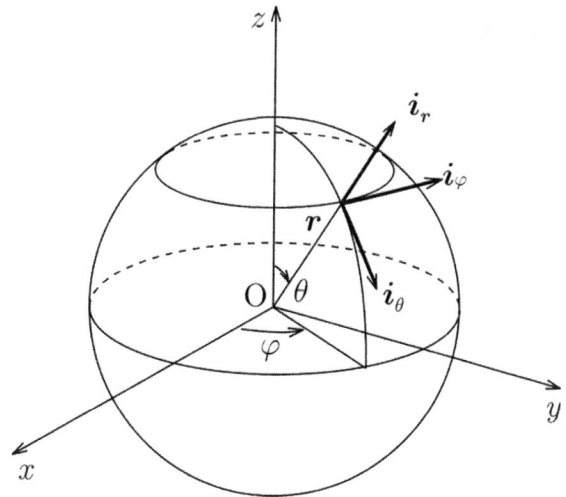

Fig. A.2 Spherical coordinates

$$\nabla f = i_r \frac{\partial f}{\partial r} + i_\theta \frac{1}{r} \cdot \frac{\partial f}{\partial \theta} + i_\varphi \frac{1}{r \sin\theta} \cdot \frac{\partial f}{\partial \varphi}, \tag{A.16}$$

$$\nabla \cdot A = \frac{1}{r^2} \cdot \frac{\partial (r^2 A_r)}{\partial r} + \frac{1}{r \sin\theta} \cdot \frac{\partial}{\partial \theta}(\sin\theta A_\theta) + \frac{1}{r \sin\theta} \cdot \frac{\partial A_\varphi}{\partial \varphi}, \tag{A.17}$$

$$\nabla \times A = i_r \frac{1}{r \sin\theta} \left[\frac{\partial}{\partial \theta}(\sin\theta A_\varphi) - \frac{\partial A_\theta}{\partial \varphi} \right] + i_\theta \frac{1}{r} \left[\frac{1}{\sin\theta} \cdot \frac{\partial A_r}{\partial \varphi} - \frac{\partial (rA_\varphi)}{\partial r} \right]$$
$$+ i_\varphi \frac{1}{r} \left[\frac{\partial (rA_\theta)}{\partial r} - \frac{\partial A_r}{\partial \theta} \right], \tag{A.18}$$

where i_r, i_θ, and i_φ are the unit vectors along the radial, zenithal, and azimuthal directions, respectively. The second differentiation of a scalar function is given by

$$\nabla^2 f = \frac{1}{r^2} \cdot \frac{\partial}{\partial r}\left(r^2 \frac{\partial f}{\partial r}\right) + \frac{1}{r^2 \sin\theta} \cdot \frac{\partial}{\partial \theta}\left(\sin\theta \frac{\partial f}{\partial \theta}\right) + \frac{1}{r^2 \sin\theta} \cdot \frac{\partial^2 f}{\partial \varphi^2}. \tag{A.19}$$

Formulae of Integration

$$\int \frac{dx}{x^2 + a^2} = \frac{1}{a} \tan^{-1} \frac{x}{a}, \tag{A.20}$$

$$\int \frac{dx}{(x^2 + a^2)^2} = \frac{x}{2a^2(x^2 + a^2)} + \frac{1}{2a^3} \tan^{-1} \frac{x}{a}, \tag{A.21}$$

$$\int \frac{dx}{(x^2 + a^2)^{1/2}} = \log\left(x + \sqrt{x^2 + a^2}\right), \tag{A.22}$$

$$\int_0^\pi \frac{d\theta}{a + b\cos\theta} = \frac{\pi}{\sqrt{a^2 - b^2}}; \quad a > |b|. \tag{A.23}$$

Table of Arrangement of Problems

There are many similar problems in each chapter. In this case, the tables shown here will assist the readers who want to search for good problems to understand the essential points of electromagnetic phenomena. For example, the table in Chap. 3 shows the arrangement of the problems on the capacitance and electrostatic energy of capacitors of various structures in given conditions. The bold shows an Example or Exercise in the textbook, *Electricity and Magnetism*. There is no overlapping in the exercises between this book and the textbook. This also clarifies the possibilities for the creation of new problems by yourself for empty columns in the tables. If the tables are compared between Chaps. 1 and 6, Chaps. 2 and 7, Chaps. 3 and 8, and Chaps. 4 and 9, your understanding of the *E-B* analogy will be deepened. It is hoped that these tables will be effectively used for lectures or for studying by oneself.

Chapter 1

Electric charge on a plane

	Electric field strength	Electric potential
Two point charges	1.11	1.11
Three point charges	1.5	
Triangle line charge	1.6	
Fan-shaped line charge	1.8	
Center of equilateral triangle line charge	1.13	1.13
Center of circular line charge		Example 1.8 Example 1.7
Central axis of equilateral triangle line charge	1.7	1.14
Central axis of square line charge	Example 1.3 **Exercise 1.3**	1.15
Central axis of circular line charge	Example 1.4 Example 1.3	1.12

Parallel line charges

	Electric field strength	Electric potential
2 lines	1.16	1.16
3 lines (equilateral triangle)	1.17	1.17
4 lines (square)	1.18	1.18

Electric dipole

	Electric field	Electric potential	Laplace Eq.	Equi-ϕ surface
3-dimensional	Section 1.6	Section 1.6	**Exercise 1.13**	
2-dimensional	Section 1.6 *Example 1.9*	Section 1.6 *Example 1.9*	1.29	**Exercise 1.12**
Quadrupole		1.30	1.32	
Linear quadrupole		1.31	1.33	

Bold shows an Example or Exercise used in the textbook

Chapter 2

	Solution with		Laplace Eq.	Equi-ϕ surface
	ϕ	Boundary condition on E		
Sphere and point charge Q	Section 2.2	2.33	2.20	
Cylinder and line charge λ	*Example 2.5* **Exercise 2.9**	2.34	2.21	2.18
Hollow sphere and inner Q	*Example 2.4* *Example 2.6*		2.22	
Hollow cylinder and inner λ	2.14		2.23	2.16
Sphere in E	Section 2.3	*Example 2.7* **Exercise 2.11**	2.26	
Cylinder in E	*Example 2.8* *Example 2.7*	2.35	2.27	2.25
p in spherical hollow	2.28		2.29	
$\hat{p}$ in cylindrical hollow	2.30		2.31	2.32

Table of Arrangement of Problems 493

Chapter 3

Capacitor, etc.

Style		Inductance	Energy		
			From $\epsilon_0 E^2/2$	From $\sum \phi Q/2$	From coefficients
Parallel plate capacitor		Section 3.2			
Parallel plate capacitor	Tilted	3.10	3.21(given V) 3.22(given Q)	3.23(given V)	
Concentric capacitor		Example 3.2 Example 3.3	Exercise 3.5	Exercise 3.5	Exercise 3.5
Coaxial capacitor		Example 3.6 Exercise 3.4	Example 3.6 Exercise 3.4	Example 3.6 Exercise 3.4	
Parallel wire		Example 3.3 Exercise 3.6		3.17(given Q)	
Concentric conductors	Given Q's			3.18	Example 3.4
Coaxial conductors	Given Q's		3.24 (λ_1, λ_2)	3.24 (λ_1, λ_2)	3.25($\pm Q$)
3 concentric conductors	Given Q's		3.19	3.20	Example 3.5 Exercise 3.7
3 coaxial conductors	Given Q's		3.26	3.27	3.28

Chapter 4

Capacitor with dielectric materials

Style	Space	Capacitance	Energy	
			From $\epsilon E^2/2$	From $\phi Q/2$
Parallel-plate	2 kinds parallel	Example 4.2		
	2 kinds series	Example 4.3		
	Variation in ϵ along width	4.6	4.28	
	Variation in ϵ along thickness	4.7	4.29	
Coaxial	2 kinds series	4.3	Example 4.8 Example 4.7	Example 4.8 Example 4.7
	2 kinds parallel	4.2	4.30	Exercise 4.11
	Radial variation in ϵ	4.8	4.32	4.33
Concentric	2 kinds parallel	Example 4.1 Exercise 4.1		
	2 kinds series	Example 4.2 Exercise 4.2		

(continued)

(continued)

Style	Space	Capacitance	Energy	
			From $\epsilon E^2/2$	From $\phi Q/2$
	Radial variation in ϵ	4.9	4.34	

Dielectric material

	Solution and σ_p		Laplace Eq.	Equi-ϕ surface
	Boundary conditions	ϕ		
Sphere in E	Example 4.3 Example 4.5	Example 4.6 Exercise 4.9	4.18	
Cylinder in E	Example 4.5 Exercise 4.8	4.13	4.19	4.22
Hollow sphere in E	4.14			
Hollow cylinder in E	4.15			
p in spherical hollow	4.16		4.20	
$\hat{p}$ in cylindrical hollow	4.17		4.21	4.23

Chapter 5

Shape	Structure		Resistance	Power loss
A quarter circle with square cross-section			Section 5.4	Example 5.7 **Exercise 5.5** 5.27
A quarter circle with circular cross-section			Example 5.1	5.14
A quarter circle with triangle cross-section			5.3	5.15
Truncated cone			Section 5.5	Example 5.8 Example 5.5
Rectangular frustum			Example 5.2 **Exercise 5.1**	5.16
Concentric sphere			Section 5.3, Eq. (5.20)	5.17
Coaxial cylinder			5.2	5.22
Parallel plate	Variation along length		5.10	5.23
	Variation along width		5.11	5.24
Concentric sphere	2 kinds in series		Example 5.3 Example 5.4	5.18
	Radial variation		5.13	5.26
	2 kinds in parallel		Example 5.4 **Exercise 5.2**	5.19

(continued)

Table of Arrangement of Problems 495

(continued)

Shape	Structure	Resistance	Power loss
Coaxial cylinder	2 kinds in series	Example 5.5 **Exercise 5.3**	5.20
	Radial variation	5.12	5.25
	2 kinds in parallel	Example 5.6 **Exercise 5.4**	5.21

Chapter 6

Current on a plane

	Magnetic flux density	Vector potential
Fan-shaped current	6.3	
Center of circular current	6.17	6.17
Center of equilateral triangle current	6.18	6.18
Center of square current	6.19	6.19
Central axis of square current	**Exercise 6.2**	
Central axis of circular current	Example 6.1 **Example 6.1**	6.20

Parallel straight currents

	Magnetic flux density	Force	Vector potential
2 lines	6.2	6.4, 6.5	6.14
3 lines (equilateral triangle)	6.1		
3 lines (equilateral triangle) 2	6.15		6.15
4 lines (square)	6.16		6.16

Two-dimensional current

	Magnetic flux density	Vector potential
Cylinder	Example 6.3 **Example 6.5**	Example 6.5 **Example 6.7**

Magnetic dipole

	Magnetic flux density	Vector potential	Laplace Eq.	Equi-A surface
3-dimensional	Section 6.6	Section 6.6	*Example 6.7* **Exercise 6.14**	
2-dimensional	Section 6.6 *Example 6.9*	Section 6.6 *Example 6.9*	6.28	*Example 6.8* **Exercise 6.12**
Transverse quadrupole		6.29	6.31	
Linear quadrupole		6.30		

Chapter 7

Image method

	Solution with			Current distribution	Laplace Eq.	Equi-A surface
	Vector potential	Boundary condition on B	Magnetic potential			
Cylinder and straight I	Section 7.2	7.27	7.33	Section 7.2	7.13	7.16
Hollow cylinder and inner I	*Example 7.6* **Exercise 7.10**	7.28		*Example 7.6* **Exercise 7.10**	7.14	7.17
Sphere in B	Section 7.3	**Exercise 7.11**	Section 7.3	Section 7.3	7.21	
Cylinder in B	*Example 7.7* *Example 7.6*	7.29	*Example 7.8* **Exercise 7.12**	*Example 7.7* *Example 7.6*	7.22	7.25
Cylinder with I and flat surface	*Example 7.5* **Exercise 7.9**	7.30		*Example 7.5* **Exercise 7.9**		
m in spherical hollow	7.19	7.31	7.34	7.19	7.23	
$\hat{m}$ in cylindrical hollow	7.20	7.32	7.35	7.20	7.24	7.26

Table of Arrangement of Problems

Chapter 8

Transmission line

Style	Material	Inductance	Energy	
			From $B^2/2\mu_0$	From $A \cdot i/2$
Parallel wire	Conductor	*Example 8.1* **Example 8.1**		8.13
	Superconductor	8.2(Exact solution)		8.14
Parallel plate	Conductor	*Example 8.9* † **Exercise 8.6** † (**Sect. 8.4***)	*Example 8.9* **Exercise 8.6**	8.17
	Superconductor	*Example 8.9* † **Exercise 8.6** †	*Example 8.9* **Exercise 8.6**	8.18
Coaxial	Conductor (thick inner)	*Example 8.6* † *Example 8.8** *Example 8.9* † *Example 8.10**	8.24	*Example 8.6* *Example 8.9*
	Conductor (thick outer)	8.25 †, 8.27*	8.25	8.26
	Conductor (both thick)	*Example 8.5* † **Exercise 8.3** †, 8.28 †, 8.29*	*Example 8.5* **Exercise 8.3**	8.28
	Superconductor	*Example 8.4* † *Example 8.7* † **Exercise 8.2**	*Example 8.4* *Example 8.7*	*Example 8.4* *Example 8.7*

* From mean magnetic flux
† From magnetic energy

Coil

Style	Inductance	Energy	
		From $B^2/2\mu_0$	From $A \cdot i/2$
Solenoid(thin)		*Example 8.8*	*Example 8.8*
Solenoid(thick)	8.30, 8.32*	8.30	8.31
Spherical (uniform B)	*Example 8.2* **Example 8.5**	*Example 8.3* **Example 8.6**	**Exercise 8.7**
Cylindrical (uniform B)	8.11	8.21	8.20

* From mean magnetic flux

Chapter 9

Transmission line with magnetic materials

Style	Space	Inductance	Energy		
			From $\Phi I/2$	From $B^2/2\mu_0$	From $A \cdot i/2$
Parallel-plate	2 kinds parallel	*Example 9.1* *Example 9.3*			
	2 kinds series	*Example 9.2* **Exercise 9.1**			
	Variation in μ along width	9.3		9.29	
	Variation in μ along thickness	9.4		9.30	
Coaxial	2 kinds parallel	*Example 9.8* *Example 9.8*	*Example 9.8* *Example 9.8*	9.31	9.33
	2 kinds series	*Example 9.3* **Exercise 9.2**		*Example 9.3* **Exercise 9.2**	9.34
	Radial variation in μ	9.5		9.32	9.35

Magnetic material

	Solution		Laplace Eq.	Equi-A surface	Magnetic potential
	Vector potential	Boundary condition			
Sphere in B	*Example 9.5* *Example 9.6*	*Example 9.4* *Example 9.5*	9.18		**Exercise 9.9**
Cylinder in B	9.13	*Example 9.7* **Exercise 9.8**	9.19	9.24	9.26
Hollow sphere in B	9.14		9.20		
Hollow cylinder in B	9.15		9.21		
m in spherical hollow	9.16		9.22		
$\hat{m}$ in cylindrical hollow	9.17		9.23	9.25	9.27

Table of Arrangement of Problems

Chapter 10

Determination of electromotive force

	Flux law	Motional law	Vector potential	Maxwell Eq.
AC current and square circuit	Example 10.1 Exercise 10.1		Example 10.6 Example 10.5	10.10
Square circuit moving with constant v	10.2	10.3	10.7	10.11
Right triangle moving with constant v	Example 10.2 Example 10.2	Example 10.3 Example 10.3	10.8	10.12
Equilateral triangle moving with constant v	10.4	10.5		
Falling square circuit	Exercise 10.5 Example 10.4	Exercise 10.5 Example 10.4	10.9	10.13
Square circuit moving away from AC current	10.14		10.15	10.16

Skin effect

	Solution	Coulomb gauge for A	Poisson's equation for A
AC magnetic field	Section 10.5	10.28	10.29
AC electric field	Example 10.8 Exercise 10.10	10.27	10.30

Chapter 11

	Material	Energy flow or loss	Electromagnetic potential
Parallel plate transmission line	conductor	11.13	11.7
	superconductor	Example 11.5 Example 11.10	Example 11.7 Example 11.5
Coaxial transmission line	conductor	11.14	11.8
	superconductor	11.15	11.9
Coaxial capacitor		11.16	11.10
Coaxial resistor		11.17	11.11

		Solutions on C, R_r, and L	Energy flow or loss with the Poynting vector
Coaxial resistor	2 kinds in series	Example 5.5 Exercise 5.4	11.20
Coaxial resistor	2 kinds in parallel	Example 5.6 Exercise 5.5	11.21

(continued)

(continued)

		Solutions on C, R_r, and L	Energy flow or loss with the Poynting vector
Parallel-plate capacitor	Tilted electrode	3.10	11.22
Coaxial capacitor	2 kinds in series	4.4	(*Example 4.8*) (**Example 4.7**) 11.23
	2 kinds in parallel	4.2	(**Exercise 4.11**), 11.24
	Axial variation	4.8	(4.31), 11.25
Parallel-plate transmission line	Tilted plate (conducting)		11.26
	Tilted plate (superconducting)	8.5	11.27
Coaxial superconducting transmission line	2 kinds in series	*Example 9.3* *Exercise 9.2*	11.28
	2 kinds in parallel		(*Example 9.8*) (**Example 9.8**) 11.29

(): Reference (the Poynting vector is not used)

Chapter 12

Reflection and refraction at interface

Direction of E	Solution	Energy flow
$\perp$ plane of incidence	Section 12.2	Example 12.4
$\parallel$ plane of incidence	Example 12.1	12.7

Reflection at the material surface

Material	Direction of E	Solution	Continuity of current	Energy flow
Conductor	$\perp$ plane of incidence	*Example 12.2*	*Example 12.2*	12.5
	$\parallel$ plane of incidence	12.6	12.6	12.6
Superconductor	$\perp$ plane of incidence	12.4	12.4	
	$\parallel$ plane of incidence	12.3	12.3	

Table of Arrangement of Problems 501

Wave guides

Wave	Shape	Solution	Potential, Lorentz gauge	Energy flow	Continuity
TEM	Parallel plate	*Example 12.9* **Exercise 12.4**	12.9	12.19	*Example 12.9* **Exercise 12.4**
	Coaxial	*Example 12.6* ***Example 12.5***	12.10	12.20	*Example 12.6* ***Example 12.5***
	Parallel cylinder	*Example 12.10* **Exercise 12.8**	12.11	12.21	12.27
	Cylinder and surface	12.8	12.15	12.22	12.28
TE	Wave guide	*Example 12.7* **Exercise 12.6**	***Example 12.6***	12.25	12.29
TM	Wave guide	*Example 12.5* ***Example 12.4***	**Exercise 12.5**	*Example 12.8* **Exercise 12.11**	12.30
TE_{01}	Wave guide		12.12	12.23	
TE_{10}	Wave guide		12.13	12.24	
TM_{11}	Wave guide		12.14	12.26	

Index

A
Ampere's law, 196, 316

B
Biot-Savart law, 191, 316

C
Capacitance, 83
Capacitance coefficient, 84
Capacitor, 85
Capacity, 83
Capacity coefficient, 84
Coefficient of electric potential, 83
Coil, 278
Conductance, 159
Conductor, 41
Continuity equation of current, 158
Coulomb's law, 2, 4
Coulomb force, 1
Coulomb gauge, 198
Current, 157
Current density, 157
Cut-off frequency, 452

D
Dielectric polarization, 117
Differential form of Ampere's law, 196, 316
Displacement current, 396

E
Electric charge, 1
Electric conductivity, 159
Electric dipole, 12
Electric dipole line, 14
Electric dipole moment, 13
Electric energy, 88
Electric energy density, 90
Electric field, 4, 363
Electric field line, 4
Electric flux density, 119
Electric polarization, 117
Electric potential, 10
Electric power, 169
Electric resistance, 158
Electromagnetic potential, 400
Electromagnetic wave, 438
Electrostatic energy, 88, 130
Electrostatic energy density, 90
Electrostatic force, 92
Electrostatic induction, 52
Equipotential surface, 10
Equivector potential surface, 199, 204

F
Faraday's law, 355

G
Gauss's divergence law, 8, 119
Gauss's divergence law for magnetic flux, 195
Gauss's law, 7, 119
Generalized differential form of Ampere's law, 396
Generalized form of Ampere's law, 396

I
Image charge, 43
Image current, 242
Inductance coefficient, 277

L
Laplace's equation, 10, 198, 317
Law of reflection, 442
Light speed, 442
Linear polarization, 439
Lorentz force, 195
Lorentz gauge, 400

M
Magnetic charge, 206
Magnetic dipole line, 202
Magnetic energy, 280, 329, 365
Magnetic energy density, 281
Magnetic field, 316
Magnetic flux, 195
Magnetic flux density, 191
Magnetic flux law, 357
Magnetic flux line, 195
Magnetic moment, 202
Magnetic potential, 207
Magnetization, 313
Magnetizing current, 313
Maxwell's equations, 398
Mean magnetic flux, 287
Meissner state, 237, 249
Method of images, 43, 125, 241, 326
Moment of electric dipole line, 14
Motional law, 356
Mutual inductance, 277

O
Ohm's law, 159

P
Parallel-plate capacitor, 85
Plane of incidence, 439
Plane wave, 438

Poisson's equation, 10, 120, 198, 317, 364
Polarization charge, 117
Polarization of wave, 439
Polarized wave, 439
Poynting vector, 401

R
Refraction of current, 161
Refraction of electric field line, 120
Refraction of magnetic flux line, 318
Resistivity, 159

S
Self-inductance, 277
Skin depth, 369
Skin effect, 367
Small closed current, 201
Snell's law, 442
Solenoid coil, 278
Spherical coil, 279
Steady current, 158
Superconductor, 237

T
Telegraphic equation, 399
Transverse electric (TE) wave, 450
Transverse electromagnetic (TEM) wave, 450
Transverse magnetic (TM) wave, 450

U
Unipolar induction, 361

V
Vector potential, 198

W
Wave equation, 400
Wave guide, 448

The manufacturer's authorised representative in the EU is Springer Nature Customer Service Centre GmbH, Europaplatz 3, 69115 Heidelberg, Germany. If you have any concerns regarding our products, please contact ProductSafety@springernature.com

Printed and bound by CPI Group (UK) Ltd, Croydon, CR0 4YY
03/06/2025
01891387-0001